Environmental Contamination in Antarctica
A Challenge to Analytical Chemistry

Acknowledgements

The editors gratefully acknowledge the authors of the various contributions to this book and all those who strongly encouraged and supported this project.

Very sincere thanks are also due to Mr. Massimo Delle Femmine and Ms. Clarissa Ferreri for their patience in typing and compiling the many drafts of this book.

Use of the cover image is by kind permission of the Programma Nazionale di Ricerche in Antartide (PNRA) – Italian Antarctic Research Programme.

Environmental Contamination in Antarctica

A Challenge to Analytical Chemistry

Edited by

Sergio Caroli
Istituto Superiore di Sanità
Rome, Italy

Paolo Cescon
CSCTA-CNR, University "Ca' Foscari" of Venice, Italy

David W. H. Walton
British Antarctic Survey,
Cambridge, UK

2001

Elsevier
Amsterdam – London – New York – Oxford – Paris – Shannon – Tokyo

ELSEVIER SCIENCE Ltd
The Boulevard, Langford Lane
Kidlington, Oxford OX5 1GB, UK

First edition 2001

Library of Congress Cataloging in Publication Data
A catalog record from the Library of Congress has been applied for.

British Library Cataloguing in Publication Data
Environmental contamination in Antarctica : a challenge to
analytical chemistry
1. Environmental chemistry – Antarctica 2. Pollution –
Antarctica
I. Caroli, Sergio II. Cescon, Paolo III. Walton, D. W. H.
(David Winston Harris), 1945–
577.1′4′09989

ISBN: 0080431992

∞ The paper used in this publication meets the requirements of ANSI/NISO Z39.48-1992 (Permanence of Paper).
Printed in The Netherlands.

Dedication

In memory of Professor Felice Ippolito who enthusiastically promoted research in Antarctica.

Contents

Contributors

Maria Luisa Abelmoschi
University of Genoa, Department of Chemistry and Industrial Chemistry, Section of Environmental and Analytical Chemistry, Via Dodecaneso 31, 16146 Genoa, Italy

Franca Baffi
University of Genoa, Department of Chemistry and Industrial Chemistry, Section of Environmental and Analytical Chemistry, Via Dodecaneso 31, 16146 Genoa, Italy

Carlo Barbante
Ca' Foscari University of Venice, Centre of Chemistry and Environmental Technology – CNR, Dorsoduro 2137, 30123 Venice, Italy

Claude F. Boutron
CNRS, Laboratory of Glaciology and Geophysics of the Environment, 64, rue Molière,University Campus, P.O. Box 96, 38402 Saint Martin d'Hères cedex, France; and Section for Formation and Research in Mechanics and Physics, Joseph Fourier University of Grenoble, University Campus, P.O. Box 68, 38041 Grenoble, France

Stefano Caimi
Istituto Superiore di Sanità, Viale Regina Elena 299, 00161 Rome, Italy

Jean-Pierre Candelone
Department of Applied Physics, Curtin University of Technology, G.P.O. Box U1987, Perth, 6845, Australia

Gabriele Capodaglio
Ca' Foscari University of Venice, Centre of Chemistry and Environmental Technology – CNR, Dorsoduro 2137, 30123 Venice, Italy

Adriana Y. Cantillo
National Oceanic & Atmospheric Administration (NOAA), 1305 East West Hwy., Silver Spring, MD 20910 USA

Serglo Caroli
Istituto Superiore di Sanità, Viale Regina Elena 299, 00161 Rome, Italy

Alessio Ceccarini
University of Pisa, Department of Chemistry and Industrial Chemistry, Via Risorgimento 35, 56126 Pisa, Italy

Paolo Cescon
Ca' Foscari University of Venice, Centre of Chemistry and Environmental Technology – CNR, Dorsoduro 2137, 30123 Venice, Italy

Michael A. Champ
Texas A&M University, 4601 North Fairfax Drive, Suite 1130, Arlington, VA 22042, USA

Roberto Frache
University of Genoa, Department of Chemistry and Industrial Chemistry, Section of Environmental and Analytical Chemistry, Via Dodecaneso 31, 16146 Genoa, Italy

Roger Fuoco
University of Pisa, Department of Chemistry and Industrial Chemistry, Via Risorgimento 35, 56126 Pisa, Italy

Pietro Giuliani
ENEA, Via Anguillarese 301, 00060 Santa Maria di Galeria, Roma, Italy

Karl-Heinz Grobecker
European Commission, Joint Research Centre, Institute for Reference Materials and Measurements (IRMM), Retieseweg B-2440 Geel, Belgium

Klaus Gustav Heumann
Johannes Gutenberg University of Mainz, Institute of Inorganic Chemistry and Analytical Chemistry, Becherweg 24, D-55099 Mainz, Germany

Sung Min Hong
Polar Science Laboratory, Korean Ocean Research and Development Institute, Ansan, P.O. Box 29, Seoul, 425-600, Korea

Carmela Ianni
University of Genoa, Department of Chemistry and Industrial Chemistry, Section of Environmental and Analytical Chemistry, Via Dodecaneso 31, 16146 Genoa, Italy

Gerard N. Kramer
European Commission, Joint Research Centre, Institute for Reference Materials and Measurements (IRMM), Retieseweg B-2440 Geel, Belgium

Milo Kuneshka
ENEA, Via Anguillarese 301, 00060 Santa Maria di Galeria, Roma, Italy

Gunnar G. Lauenstein
National Oceanic & Atmospheric Administration (NOAA), 1306 East West Hwy., Silver Spring, MD 20910 USA

Emanuele Magi
University of Genoa, Department of Chemistry and Industrial Chemistry, Section of Environmental and Analytical Chemistry, Via Dodecaneso 31, 16146 Genoa, Italy

Jorge Marcovecchio
Instituto Argentino de Oceanografía, Av. Alem 54, 8000 - Bahía Blanca, Argentina

Jean Pauwels
European Commission, Joint Research Centre, Institute for Reference Materials and Measurements (IRMM), Retieseweg B-2440 Geel, Belgium

Laura Pineschi
University of Parma, Istituto di Diritto Internazionale, Via dell'Università 12, 43100 Parma, Italy

Kevin J. R. Rosman
Curtin University of Technology, Department of Applied Physics, G.P.O. Box U1987, Perth, 6845, Australia

Giuseppe Scarponi
University of Ancona, Institute of Marine Sciences, Via Brecce Bianche, 60131 Ancona, Italy

Oreste Senofonte
Istituto Superiore di Sanità, Viale Regina Elena 299, 00161 Roma, Italy

Francesco Soggia
University of Genoa, Department of Chemistry and Industrial Chemistry, Section of Environmental and Analytical Chemistry, Via Dodecaneso 31, 16146 Genoa, Italy

Patricia Smichowski
Comisión Nacional de Energía Atómica, Unidad de Actividad Química, Av. Libertador 8250, 1429 - Buenos Aires, Argentina

Luana Testa
ENEA, Via Anguillarese 301, 00060 Santa Maria di Galeria, Roma, Italy

Clara Turetta
Ca' Foscari University of Venice, Centre of Chemistry and Environmental Technology - CNR, Dorsoduro 2137, 30123 Venice, Italy

Katja Van de Velde
CNRS, Laboratory of Glaciology and Geophysics of the Environment, 54, rue Molière, University Campus, P.O. Box 96, 38402 Saint Martin d'Hères cedex, France

Cristian Vodopivez
Instituto Antártico Argentino, Cerrito 1248, 1010 - Buenos Aires, Argentina

David W. H. Walton
British Antarctic Survey, High Cross, Madingley Road, Cambridge, CB3 0ET, United Kingdom

Preface

The English seaman James Cook, credited with discovering Antarctica in 1772, wrote three years later that in his view no one would ever gain anything of value from such an inhospitable and primeval land. From a modern perspective, it is clear that he was very wrong, but scientific understanding at the time did not engender any other outlook. What Cook did not discern was that Antarctica would attract a great deal of attention owing to its biological resources and their potential exploitation. This exploitation, primarily of the seals, ran its course in the 19th century and it was not until almost the 20th century that genuine scientific interest, as well as national political aspirations, became the driving forces behind the impressive pace at which exploration of the ice continent progressed in the early twentieth century. For many decades in the 1800s hunters harvested seals heavily and systematically; after 1900 they turned their attention to whales. More recently, the catches of fish and of krill have grown, endangering the stocks of some species. On the other hand, reserves of natural gas, oil and coal and important metals such as chromium, cobalt, gold, iron, nickel, platinum and uranium, may be present according to some (but not all) geologists. The ratification of the Antarctic Treaty in 1961 as a result of the goodwill generated by the scientists during the international geophysical year has substantially slowed down the potential for exploitation of Antarctica. This has prompted international cooperation to a degree previously unknown and unthinkable. In spite of this unprecedented agreement and the attendant ban on mining and military uses of the continent, the priority attached to scientific investigations and the proliferation of research stations with the ensuing enhancement in experimental activities, along with the ongoing global pollution of the planet, are progressively affecting the pristine antarctic environment.

The systems now in place, especially the protocol for the protection of the antarctic environment, should ensure that Antarctica will not become the next wasteland. This multi-authored book rather ambitiously surveys the causes and extent of environmental contamination in Antarctica, and looks critically at future prospects. It highlights the key role that modern techniques of analytical chemistry play in achieving reliable empirical data in this field and their impact on shaping legal provisions. Chapter 1 sets forth the basic criteria which should be adhered to when Antarctic materials are sampled and analyzed, while the design and implementation of monitoring protocols and the management of experimental data are dealt with in Chapter 2. In turn, the problems and significance of the determination of trace elements in polar snow and ice are thoroughly discussed in Chapters 3 and 4, with particular emphasis on the use of such data for better understanding of worldwide pollution phenomena and paleoclimatic events. Chapter 5 illustrates the various facets of the quantification of trace elements in the water of the Southern Ocean. Chapter 6 focuses on trace elements, although more specifically from the standpoint of their role in sound environmental management both in general ecosystem terms and in the more local vicinity of research stations. The geochemical cycles of trace elements in sea water are highlighted in Chapter 7, where biomethylation phenomena are examined in particular. The analytical approach followed to quantify trace elements in particulate matter and sediments is the target of Chapter 8. The presence of polychlorobiphenyls in antarctic media and biota is exhaustively debated in Chapter 9, especially as they can be considered clear indicators of global anthropogenic pollution. The four final chapters offer the reader a systematic and detailed strategy for assuring the overall quality of experimental data. Chapters 10 and 11 stress the importance of planning and producing certified reference materials in antarctic matrices for all analytes of interest from an environmental viewpoint. The characteristics and goals of the Antarctic Environmental Specimen Bank are outlined in Chapter 12, while the fundamental support of quality assurance schemes in polar monitoring and research is treated in Chapter 13. Finally, Chapters 14 and 15 take into consideration the legal framework which governs the protection and preservation of the Antarctic environment as prescribed in particular by the Madrid Protocol to the Antarctic Treaty.

The overview of scientific and regulatory aspects set forth in this book, on the one hand, demonstrates how intimately research and legal provisions are interwoven and benefit from each other; on the other hand, it sheds further light on the complexity of environmental contamination in Antarctica and calls for more proactive and resolute action. With this in mind, it is hoped that this work can stimulate further pondering of such priority issues.

Sergio Caroli
Paolo Cescon
David W. H. Walton

Environmental Contamination in Antarctica
A Challenge to Analytical Chemistry
S. Caroli, P. Cescon and D.W.H. Walton, editors

Chapter 1

Environmental chemistry in Antarctica: the quest for accuracy

Sergio Caroli

And now there came both mist and snow,
And it grew wondrous cold:
And ice, mast-high, came floating by,
As green as emerald.

S. T. Coleridge
Rime of the Ancient Mariner

1. Introduction

The twentieth century has witnessed a dramatic increase in the exploration of the Antarctic continent by many countries, partly as a consequence of genuine scientific interest, but also prompted, to a significant extent, by the alluring perspectives of exploiting the natural – and so far intact – resources of this land.

The International Geophysical Year 1957–58 played a crucial role in this context as it led to the establishment of the Antarctic Treaty regime which unequivocally recognized the supremacy of scientific research over political and territorial claims (1). The Antarctic Treaty put much emphasis on the need for international scientific cooperation substantially promoting the peaceful advancement of man's knowledge of this unique continent. The Protocol on Environmental Protection to the Antarctic Treaty reaffirmed, updated and consolidated these concepts and attached the highest priority to the preservation of the pristine conditions of Antarctica and recognized the vital importance of this part of the globe in monitoring environmental phenomena at the planetary scale (2). The environmental monitoring of this remote area of the world, especially when coupled with innovative research, brings about a number of benefits in terms of early prediction of the eventual impact that human activities may have (3). By general recognition, the unrivalled achievements of Antarctic science span vast fields of experimental investigation (e.g., global climate change, stratospheric ozone depletion, anthropogenic pollution, reconstruction of past climate variations). All this was also prompted by the creation in 1958 of the Scientific Committee on Antarctic Research (SCAR) which has acted as the international forum of scientific coordination since.

The nations involved in Antarctic research are currently twenty-six with a global investment of financial resources of approximately US$ 500 million per year (4). With this scenario in mind, it goes without saying that the production of reliable, comparable and defendable experimental data, independently of the scientific discipline considered, is vital. Only when high quality information is available, can valid assumptions be made and realistic models be developed, thus providing assessors with tools that can effectively lead to sound measures to protect health and the environment. The past decade has been marked by an increasing awareness of experimentalists and decision makers alike on quality assurance and its various facets. Quality assurance guidelines, schemes and criteria have proliferated at an astounding pace, their complexity sometimes even causing frustration and discouragement to those who were intended to benefit from their implementation (5).

A successful approach to environmental monitoring and protection is necessarily based on the combined expertise from several disciplines such as biology, geology, oceanography and analytical chemistry. Coordinated and harmonized monitoring programmes in Antarctica are vital to gain reliable insights into temporal changes in the environmental levels of pollutants (6). Chemical measurements, in particular, play a pivotal role in this context, hence lack of accuracy in chemical analysis may eventually result in entirely wrong assessments, with the ensuing disastrous consequences and potentially high social costs.

Waste of time and precious resources, duplication of effort and scientific and legal controversies can be minimized by carefully planning all the steps of the analytical process, from proper sampling and sample storage to the necessary laboratory pretreatment and trustworthy performance of measurements, self-consistent evaluation of the experimental data and their exploitation for subsequent action.

Such considerations are even more stringent because of the often elusive (yet absolutely meaningful) concentrations at which most manmade chemicals are detected in Antarctic matrices. Even the faintest trace of a given contaminant in air, water, soil or biota, in fact, would be probative of an ongoing pollution process that in all likelihood originates from the northern, more industrialized hemisphere, and can have a deleterious impact on the southern moiety of the planet. Nor should it be overlooked that the proliferation of scientific bases, as is shown in Figure 1.1, further enhances the risk of locally spoiling various areas. The gradual degradation of the continent can and must be stopped (7). Current trends in Antarctic legislation pave the way to even more advanced and effective conservation measures that will help make of Antarctica a world park where only scientific research should be allowed to proceed unhindered. Environmental chemistry – with its arsenal of analytical strategies and methodologies – if correctly used, can substantially contribute to preserve the *terra australis incognita* as a (or supposedly so) clean room where a variety of investigations can be undertaken that would otherwise be unthinkable in other overcrowded and polluted regions of the globe.

2. Reliability of experimental information

2.1. Basic aspects

The obtainment of sound environmental data is a complex operation made up of distinct (but all equally important) key steps. Basically, these can be identified as sampling, storage, pretreatment, analysis and data processing (8). In spite of the practically endless combinations of analytes and matrices that can be encountered and the variety of specific analytical problems each of such combinations can give rise to, some general criteria can still be boiled down with specific reference to the characteristics of the Antarctic setting. Other chapters in this book will illustrate in full detail the planning, conduct and outcome of large-scale studies fully incorporating such basic rules. Here it suffices to highlight the principal conditions to be respected for experimental findings to be dependable, meaningful and comparable. Nor should it be overlooked that an essential piece of the puzzle is the involvement of the analyst in the overall process from the very onset, i.e., from the identification of the aims of the study to the preparation of the final report. In this way it will be possible not only to decide beforehand what measures should be undertaken to assure the desired level of quality of the study, but also to achieve this goal with the minimum investment of time and effort still compatible with the preset quality parameters, thus saving precious resources. From this standpoint, due account should be given to the fact that an inherent characteristic of natural environments is their temporal and spatial variability, which combines with sampling and analytical variability to affect experimental data. The better such aspects are understood and duly accounted for, the sounder and more effective the information gained.

2.2. Sampling

By general acknowledgement, environmental analytical chemistry is more liable to significant errors during the field operations and the preanalytical steps than in the actual conduct of determinations. It is a rather common saying in the analytical world that no analyses can be better than the samples themselves. That nowadays this is a commonplace concept does not make it less true. Therefore, validation of sampling procedures is as desirable as for analytical methods. To be fit for purpose, a sample must be representative of the system under investigation, which in turn may well be of a composite nature, as is almost always the case for environmental and biological matrices. This is in practice very difficult to achieve for bulky systems characterized by variability in both time and space. Nevertheless, there is still room in this context for a certain flexibility. In fact, the first (and mandatory) aspect the experimentalist should take into account is what the samples are intended for, i.e., what explicit and implicit demands the experimental data should satisfy. The overall design of the sampling process is thus sketched, while other factors help make a proper balance among different, sometimes contrasting, requirements (minimization of sampling time and costs, available analytical techniques, etc.). Therefore, a sampling strategy must be developed that defines the sites samples will be taken from, the total number of samples to be collected (also dictated by the degree of heterogeneity of the material investigated), the frequency of sampling, the sampling devices and the way samples should be stored and shipped to the laboratory.

There are several possibilities to design a sampling campaign, which can basically entail fortuitous, selected, transect/gradient, grid, random and stratified

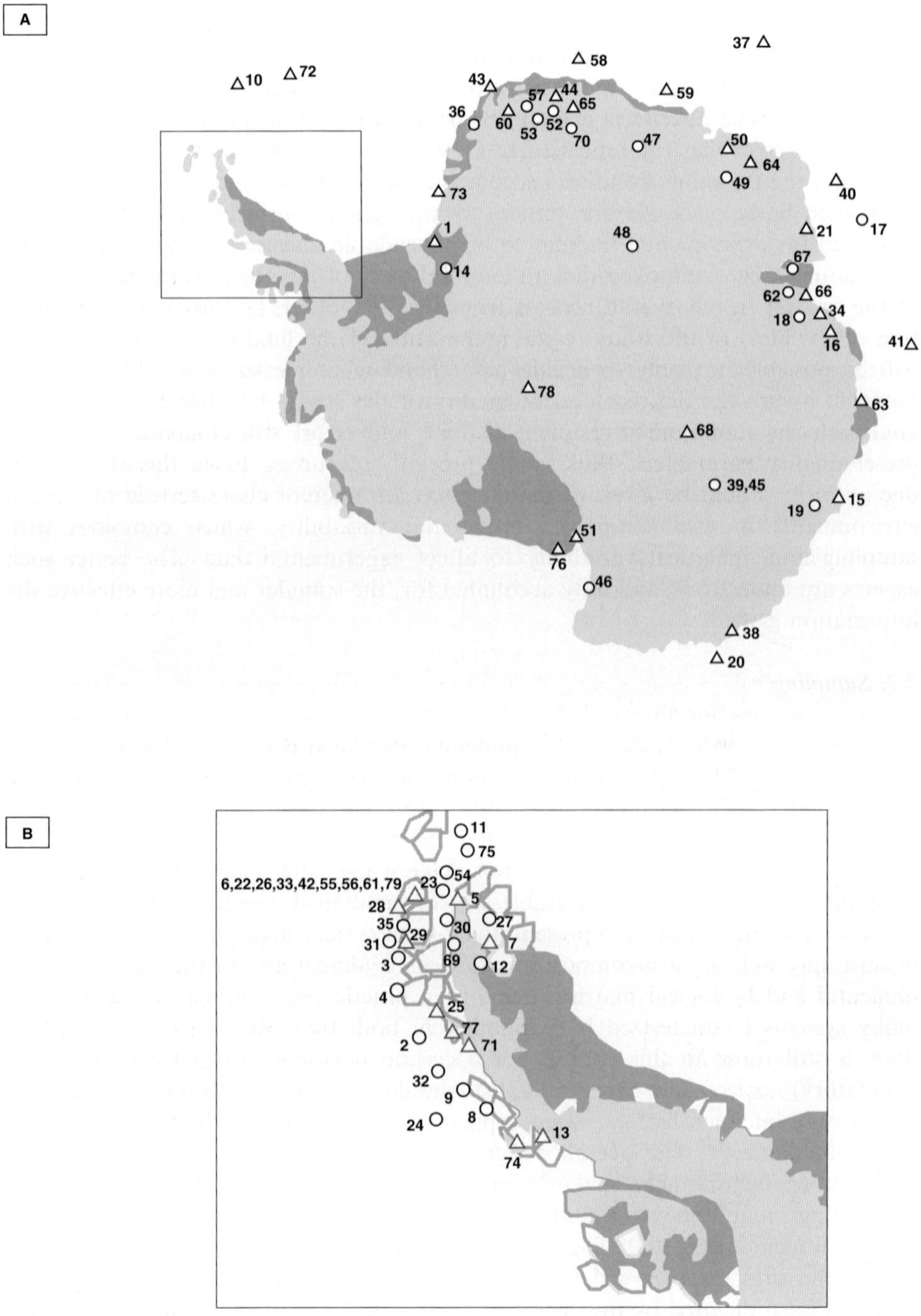

Figure 1.1. Location of scientific bases in the Antarctic continent.

random sampling (9). To these modes, entirely imposed by external circumstances, another one should be added which is of a totally different nature in that environmental bioindicators and artificial habitats can be intentionally exposed to outdoor conditions and withdrawn for analysis after a pre-established period of time. The sample should at least: *i)* represent the properties, in terms of matrix composition and physical state, of the system to be investigated. If these properties change in space and time, the number of samples taken should be adequate to faithfully describe this behavior; *ii)* be compatible, after appropriate pretreatment, with the analytical techniques available; *iii)* keep the original information content intact throughout the subsequent phases of the analytical process; *iv)* imply costs and time demand as low as possible without compromising the quality of the desired data (10). Albeit in principle the systems to be investigated can be either homogeneous or heterogeneous, the former case (e.g., well-mixed gases and liquids) very rarely occurs when dealing with environmental and biological materials. Heterogeneity is therefore the rule. Various approaches have been developed to date to estimate the number of samples necessary not to exceed a given level of sampling uncertainty (see, e.g., 11–13). All of them have pros and cons, but are definitely of assistance once they are tailored to the sampling problem at hand.

1a, Triangles stand for all-year-through bases, circles for summer-only (a detailed view of the Antarctic Peninsula is given in the inset). 1a, Antarctic Mainland and nearby islands: 1, Belgrano II (Argentina); 10, Orcadas (Argentina); 14, Sobral (Argentina); 15, Casey (Australia); 16, Davis (Australia); 17, Heard Is (Australia); 18, Law Base (Australia); 19, Law Dome (Australia); 20, Macquarie Is (Australia); 21, Mawson (Australia); 34, Zhongshan (China); 36, Aboa (Finland); 37, Alfred-Faure (France); 38, Dumont D'Urville (France); 39, Dome C (France); 40, Martin-de-Viviès (France); 41, Port-aux-Francais (France); 43, Neumayer (Germany); 44, Maitri (India); 45, Dome C (Italy); 46, Terra Nova Bay (Italy); 47, Asuka (Japan); 48, Dome Fuji (Japan); 49, Miznho (Japan); 50, Syowa (Japan); 51, Scott (New Zealand); 52, Tor (Norway); 53, Troll (Norway); 57, E-Base (Republic of South Africa); 58, Gough (Republic of South Africa); 59, Marion (Republic of South Africa); 60, Sanae IV (Republic of South Africa); 62, Druzhnaya 4 (Russia); 63, Mirny (Russia); 64, Molo (Russia); 65, Novo (Russia); 66, Progress (Russia); 67, Soyuz (Russia); 68, Vostok (Russia); 70, Wasa (Sweden); 72 Bird Is (United Kingdom); 73, Halley (United Kingdom); 76, McMurdo (United States); 78, South Pole (United States).

1b, Antarctic Peninsula: 2, Brown (Argentina); 3, Cámara (Argentina); 4, Decepción (Argentina); 5, Esperanza (Argentina); 6, Jubany (Argentina); 7, Marambio (Argentina); 8, Matienzo (Argentina); 9, Melchior (Argentina); 11, Petrel (Argentina); 12, Primavera (Argentina); 13, San Martin (Argentina); 22, Ferraz (Brazil); 23, Ochridiski (Bulgaria); 24, Carvajal (Chile); 25, Escudero (Chile); 26, Frei (Chile); 27, Gabriel Gonzalez Vidiez (Chile); 28, O'Higgins (Chile); 29, Prat (Chile); 30, Ripamonti (Chile); 31, Risopatron (Chile); 32, Yelcho (Chile); 33, Great Wall (China); 35, Vicente (Equador); 42, Dallmann (Germany); 54, Macchu Picchu (Peru); 55, Arctowski (Poland); 56, King Sejong (Republic of Korea); 61, Bellingshausen (Russia); 69, Juan Carlos I (Spain); 71, Vernadsky (Ukraine); 74, Rothera (United Kingdom); 75, Signy (United Kingdom); 77, Palmer (United States); 79, Artigas (Uruguay).

This rather complex issue greatly benefits from the proper use of statistical tools. A simple example of such estimates is the one developed by Ingamells who demonstrated that the equation $WR^2 = K_s$ can be advantageously applied to many real instances (here W stands for the weight of the sample, R for the relative standard deviation of the sample composition and K_s for the sampling constant, i.e., the weight of sample necessary not to exceed the sampling uncertainty limit of 1% at a confidence level of 68%). The sampling constant can be ascertained experimentally with a series of samples of different weight W. The costs associated with these sampling schemes can be also estimated by means of mathematical models, such as those proposed by Leemans and by Janse and Kateman (14, 15). Gy, in turn, conceived a model to give a rational basis to sample representativeness (16). This author defines the relative sampling error as the difference between the value of a quantity measured in the batch of collected samples and the true value of the sample. The concept is expressed through the equation $r^2(SE) = m^2(SE) + s^2(SE)$, with $r^2(SE)$ being the mean square, m(SE) the expected value and $s^2(SE)$ the variance. For the sampling to be representative, $r^2(SE)$ must be both accurate and reproducible, i.e., it must not exceed the pre-established value of $r_0^2 = m_0^2 + s_0^2$.

All these facets are made even more difficult by the remoteness of Antarctica, its inherently hostile nature and the obviously high costs associated with the obtainment of samples therefrom, not to mention that there are several other factors (such as legal constraints and technical impediments) which render Antarctic samples unique in the vast majority of cases. The scheme in Figure 1.2 summarizes the main categories of samples that can be collected for environmental studies in Antarctica.

It is mandatory that all steps of the sampling process be carried out in such a way that the loss of analytes from the materials collected or the contamination of the latter with the analytes of interest be minimized (17). Depending on analyte nature and host matrix, sampling tools and containers must be selected and used so as to preserve sample integrity. It is of outmost importance that sampling and storage equipment be cleaned and decontaminated, especially as regards the analytes that will be quantified. The overall cleaning procedure is rather time-consuming and painstaking, all the more so for containers and devices intended for Antarctic matrices such as snow and water, but there are really no alternatives (18). Common sense prescribes that metallic instruments should not be used when trace elements are sought (because of the possible substantial amounts of the same elements that may leach from contact components) and no plastic devices be employed for traces of organics (because of the risk of absorption on internal walls and the leaching of plasticizers). Solids can be sampled using scoops, shovels, pipes, spiers, augers, gravity and box corers, grabs, probes and diggers, while for liquids, a variety of glass and Teflon®-coated bottles are available (e.g., the bucket, Van Dorn, Niskin, Ruttner and Go-Flo types).

Last, but not least, all phases and details of sampling and sample treatment in the field must be faithfully and exhaustively recorded along with details of storage and transport to the laboratory. Description of the appearance of the samples, weather conditions, temperature, devices and materials used and unexpected

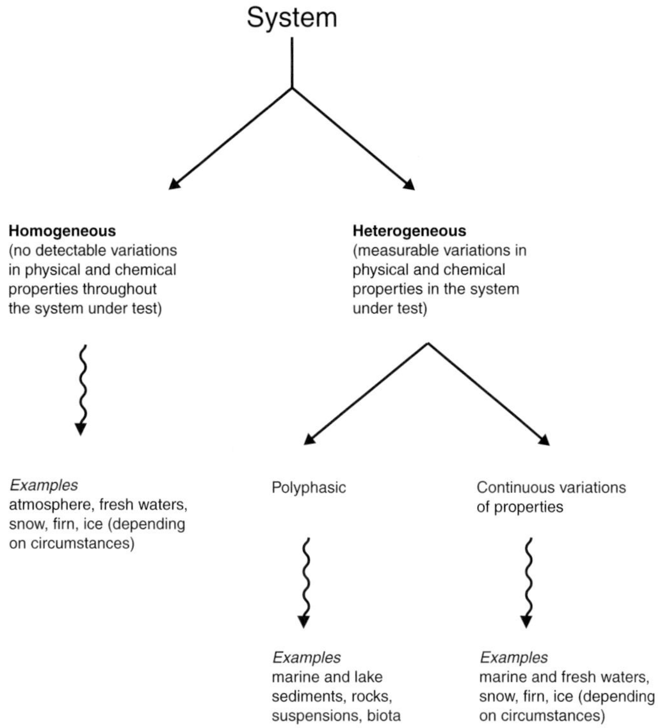

Figure 1.2. Types of systems of interest for Antarctic environmental research and monitoring.

circumstances occurring at any stage are just a few pieces of information among many others that should never be ignored in a quality-control inspired project. Finally, it is strongly advisable that, whenever possible, representative aliquots of samples be set apart for as long as necessary, should the need of any subsequent control arise (see below).

2.3. Sample storage

Once samples have been taken, they should be kept in containers that are specifically intended for the final purpose of the analysis. It is definitely convenient that, whenever possible, the items be collected directly into the vessels which will be used for transport and storage. The containers must be chemically and physically inert, e.g., glass, pyrex, quartz, Teflon®, low- and high-density polyethylene, polypropylene, polycarbonate, polyvinyl chloride and stainless steel, in order that no variations whatsoever are induced in the state and composition of the samples through mutual exchange of trace substances, adsorption on the inner walls or into rubber stoppers and sealing rings, and degradation by light or evaporation. They must be also sturdy enough to avoid any mechanical damage that may eventually lead to leakage of the contents. General conditions of storage, such as temperature, humidity and exposure to light, are crucial to the effective preservation of the samples.

Microbial growth is sometimes overlooked, although it may endanger the integrity of the samples, especially so in the case of biological materials or materials with a high content of water. Sterilization by UV- or γ-irradiation is in such cases a viable solution. Sterilization by heat or by the addition of preservatives (e.g., formaldehyde) and bactericides (e.g., mercuric chloride, thymol, toluene) is to be discouraged because they may trigger adverse effects on the original characteristics of the samples or interfere with the analytical method.

Refrigeration (as a rule at +4°C), freezing (around –20°C) and deep freezing (down to –80°C or lower) can be safely resorted to for several types of samples, such as biological materials. For other materials (e.g., snow, ice, firn) it may become mandatory to keep the samples at the proper temperature until delivery to the analyst. This is, in fact, a major aspect that must be carefully planned beforehand, e.g., in the case of snow, ice and firn, and that can pose serious practical problems when shipping the collected materials to the laboratory, often thousands of miles away. Under such circumstances, very strict precautions must be taken to warrant that by no means will specimens change their physical state with the ensuing changes in the distribution of major, minor and trace components. Moreover, the chain of custody of samples should undergo no interruption and a coded system should be used to univocally identify the samples.

Delivery to the laboratory should be done in the shortest possible time to avoid prolonged contact of samples with the storage container. In the case of Antarctic materials this is seldom feasible because of the distance and the ensuing logistic difficulties, unless samples are analyzed directly in the base premises (this may pose other problems with respect to the instrumentation and facilities available on the spot).

Closely related to this issue, but with an entirely different goal, is the long-term storage of samples peculiar to environmental specimen banks (19). These require the availability of sophisticated facilities to keep the materials selected for future studies under optimal physical and chemical conditions for their preservation. Not only retrospective control of measurements previously done on aliquots of the same materials can thus be envisaged whenever the need for such checks arises, but also other aspects can be investigated at a later stage which today are still ignored or simply cannot be explored with the present analytical methodologies. The management of Antarctic environmental specimen banks requires by definition the adoption of the strictest quality criteria at all possible levels; by the same token, they offer an additional valuable tool to the experimentalist to verify the validity of data obtained in the past. This subject matter is dealt with exhaustively in Chapter 12 of this book.

2.4. Sample handling and pretreatment

After consignment to laboratory, the materials collected for the analysis go through another crucial step before they can actually be presented for the technique selected for the quantification of the analytes of interest, i.e., subsampling and matrix treatment (change of physical state, removal of concomitants or matrix destruction). Only on rare occasions is the total amount of a sample consumed in just one analytical cycle and no manipulation is required.

As regards subsampling, the aliquots chosen for the analysis must still be representative of the entire mass. Homogenization of the original sample may become necessary, unless specific components of the sample are required, such as a given grain size fraction of a sediment or a particular organ or tissue of a living species. For soil and sediment, drying, sieving and grinding may be necessary, although this can be risky for some analytes.

These aspects are not peculiar to the Antarctic context; rather, they are common to all samples independent of their origin, and follow the same basic criteria from a practical standpoint. The only additional consideration that must be made is that materials coming from such a faraway continent have an inherent added value because of the high costs involved in their obtainment and the extreme difficulty (when not impossibility) of their replacement if, for some reason, the original samples are lost, inadvertently spoiled or analyzed without fully respecting sound conditions of quality control and assurance. This obviously calls for special care and precautions in sample processing with the objective of avoiding as far as possible loss, contamination or degradation.

Handling of samples and their physical and chemical treatment prior to analysis so as to entirely preserve their informative contents are today a well explored and adequately mastered province. Hence, a wealth of experience and knowledge is made available, to which the interested reader is referred (see, e.g., 9, 20–22). In brief, if the matrix has to be decomposed to eliminate interfering concomitants and solubilize analytes so as to make them compatible with a given technique (this is often the case with trace elements), then digestion is compulsory. Digestion may take place in closed, pressurized vessels (microwave irradiation ovens, bombs), with the advantage of minimizing losses, amount of reagents and risk of contamination, or in the open (wet ashing, dry ashing, fusion), where the above mentioned phenomena may become substantial. Inertness of vessel materials is another issue of concern, with a wide choice primarily among Teflon®, glass, quartz and glassy carbon. The final decision depends on which reagents (strong acids in the first place) are going to be used for the digestion process and on the physical conditions adopted in terms of temperature and pressure. In particular, it should not be overlooked that Teflon® has many advantageous properties, but it is also prone to the formation of small microscopic cracks on the inner surface with prolonged use, which may become a source of adsorption and release of analytes, thus seriously impacting on the reliability of results. Extraction of the substances under test from matrix components, clean-up, separation by chromatography or derivatization are largely applied when dealing with organic substances. Here the sources of error are basically related to the efficiency of the treatment, especially as regards recovery and specificity. Moreover, preconcentration may become necessary when the naturally incurred levels of a substance are lower than those accessible to the analytical technique available. All these steps are rather critical and can easily lead to results affected by inaccuracy and poor precision.

It is certainly not out of place to summarize here, as set forth in Table 1.1, how those firmly established concepts apply to the specificity of Antarctic samples and

Table 1.1. Pretreatment options for the analysis of Antarctic materials

Host matrix	Analyte type	Preparation	Analytical technique	Comments
Soils, sediments (marine, fresh water)	Trace elements *(in toto)*	Grinding, sieving digestion (e.g., acid-assisted microwave irradiation, high pressure mineralization)	Atomic spectrometry (absorption, emission, fluorescence), mass spectrometry, polarography, voltammetry, neutron activation, X-ray methods	Loss by volatilization and container wall absorption are possible. Physical, chemical and spectral interferences may occur
	Trace elements (chemical species)	Acid extraction (Tessier-based approaches)		
	Organic substances (e.g., polychlorobiphenyls, pesticides, polycyclic aromatic hydrocarbons)	Solvent extraction and clean-up. Derivatization	Separation: gas chromatography, liquid chromatography, electrophoresis. Detection: electron capture, nitrogen phosphorus, flame ionization, flame photometry, fluorescence, diode array, UV, mass spectrometry	Recovery problems are possible
Marine water	Trace elements *(in toto)*	Filtration. Saline matrix removal and analyte preconcentration (e.g., ion-exchange chromatography) or plain dilution with high purity water	Atomic spectrometry (absorption, emission, fluorescence), mass spectrometry, polarography, voltammetry	Filtration and pH adjustment are often performed. Suspended particulate matter can be analyzed separately
	Trace elements (chemical species)	Filtration. Saline matrix removal and analyte preconcentration (e.g., ion-exchange chromatography) or plain dilution with high purity water. Chromatographic separation		Dilution with high purity water is only feasible when the analytical technique has an adequate detection power. Physical, chemical and spectral interferences may occur
	Organic substances (e.g., alkanes, polychlorobiphenyls, polycyclic aromatic hydrocarbons, phthalates)	Solvent extraction and clean-up. Derivatization	Separation: gas chromatography, liquid chromatography, electrophoresis. Detection: electron capture, nitrogen phosphorus, flame ionization, flame photometry, fluorescence, diode array, UV, mass spectrometry	Recovery problems are possible

Table 1.1. (continued)

Host matrix	Analyte type	Preparation	Analytical technique	Comments
Fresh water	Trace elements *(in toto)*	Filtration. Preconcentration (ion-exchange chromatography) or direct analysis	Atomic spectrometry (absorption, emission, fluorescence), mass spectrometry, polarography, voltammetry	Direct analysis can be performed if the analytical technique has an adequate detection power. Filtration and pH adjustment are often performed. Suspended particulate matter can be analyzed separately. Physical, chemical and spectral interferences may occur
	Trace elements (chemical species)	Filtration. Chromatographic separation		
	Organic substances (e.g., alkanes, polychlorobiphenyls, polycyclic aromatic hydrocarbons, phthalates)	Solvent extraction. Derivatization	Separation: gas chromatography, liquid chromatography, electrophoresis. Detection: electron capture, nitrogen phosphorus, flame ionization, flame photometry, fluorescence, diode array, UV, mass spectrometry	Recovery problems are possible
Snow, ice, firn	Trace elements *(in toto)*	Thawing. Preconcentration (ion-exchange chromatography) or direct analysis	Atomic spectrometry (absorption, emission, fluorescence), mass spectrometry, polarography, voltammetry	Direct analysis can be performed if the analytical technique has an adequate detection power. Physical, chemical and spectral interferences may occur
	Trace elements (chemical species)	Chromatographic separation		
	Organic substances (e.g., polycyclic aromatic hydrocarbons, polychlorobiphenyls, phthalates)	Solvent extraction	Separation: gas chromatography, liquid chromatography, electrophoresis Detection: electron capture, nitrogen phosphorus, flame ionization, flame photometry, fluorescence, diode array, UV, mass spectrometry	Recovery problems are possible

Table 1.1. (continued)

Host matrix	Analyte type	Preparation	Analytical technique	Comments
Biota (e.g., mammals, fish, mussels, krill, lichens, mosses)	Trace elements *(in toto)*	Dissection of organs and tissues. Mincing. Digestion (microwave irradiation, high-pressure mineralization, wet digestion)	Atomic spectrometry (absorption, emission, fluorescence), mass spectrometry, polarography, voltammetry	Loss by volatilization and container wall absorption are possible. Physical, chemical and spectral interferences may occur
	Trace elements (chemical species)	Dissection of organs and tissues. Mincing. Extraction		
	Organic substances (e.g., pesticides, polycyclic aromatic hydrocarbons, polychlorobiphenyls, polychlorodibenzo-*p*-dioxins, polychlorodibenzofurans)	Solvent extraction and clean-up. Derivatization	Separation: gas chromatography, liquid chromatography, electrophoresis. Detection: electron capture, nitrogen phosphorus, flame ionization, flame photometry, fluorescence, diode array, UV, mass spectrometry	Recovery problems are possible
Gases and aerosols (atmospheric, occluded in ice, clathrates in water bodies)	Volatile elements. Inorganic and organic substances	Collection on filters, separation from the matrix or other procedures	UV- and IR-spectrophotometry, gas chromatography	Enrichment may pose problems

to the generally much lower concentrations of pollutants contained therein. Nor should it be forgotten that the type of pretreatment is always dictated by the purpose of the final determinations and the analytical methodology. In the case of trace elements, for instance, it is ever more frequently required that the chemical species under which the elements may occur in a given matrix be identified and quantified (23). This challenging task sets new and more severe constraints on pretreatment operations which might easily result in the variation of the pre-existing equilibrium among them, thus rendering erroneous the interpretation of experimental data in terms of ecotoxicity and bioavailability. Another serious limitation to the validation of speciation methods is the present paucity of Certified Reference Materials (CRMs) with certified amounts of chemical forms for given elements. More coordinated and planned action on the side of major manufacturers would certainly result in the availability of a higher number of CRMs of this type in the years to come.

2.5. Sample analysis

The wide variety of instrumental analytical techniques today at the disposal of the practitioner permit even the most challenging determinations in Antarctic samples to be adequately faced. Retrospective considerations lead to the conclusion that Antarctic sciences have significantly benefited from the advancement over the past two decades by environmental analytical chemistry as an autonomous discipline which, among others, allowed much light to be shed on the role Antarctica plays in shaping global change phenomena. Again, as already touched upon in preceding sections, the focus here is on the proper exploitation of existing methods and instrumentation in order that accurate and reproducible experimental measurements can actually be carried out. This requires not only that a given technique is fully mastered and fit for purpose when applied to certain analytes, but also that a total quality system is in place and that the entire laboratory and its management are fully aware of its importance. From this point of view, it is mandatory that standard operative procedures detailing all facets of the laboratory activities be set up and available to the staff, and that clean bench and clean room facilities be installed to efficiently control chemical contamination of samples.

Whatever the analytical method and the determinand may be, the greatest care should be devoted to the proper selection and use of internal standards, careful preparation of blanks and adequate calibration to avoid serious mistakes. Today the Antarctic investigator has access to a multitude of analytical techniques, the scope, detection power and robustness of which were simply unthinkable only two decades ago. For chemical elements they encompass Atomic Absorption Spectrometry (AAS) [with Flame (F) and Electrothermal Atomization (ETA) and Hydride or Cold Vapor (HG or CV) generation], Atomic Emission Spectrometry (AES) [with Inductively Coupled Plasma (ICP), Spark (S), Flame (F) and Glow Discharge/Hollow Cathode (HC/GD) emission sources], Atomic Fluorescence Spectrometry (AFS) [with HC/GD, Electrodeless Discharge (ED) and Laser Excitation (LE) sources and with the possibility of resorting to the important Isotope

Dilution (ID) mode], Mass Spectrometry (MS) [with S, HC/GD, ICP and Thermal Ionization (TI) sources], X-Ray Fluorescence Spectrometry (X-RFS), Particle-Induced X-Ray Emission (PIXE) Spectrometry, Neutron Activation Analysis (NAA), Anodic Stripping Voltammetry (ASV), polarographic methods in general and ion-selective electrodes. In the case of separation of organic substances, on the other hand, one can choose among Gas Chromatography (GC), High Performance Liquid Chromatography (HPLC), Thin Layer Chromatography (TLC), Supercritical Fluid Chromatography (SFC) and Gel Permeation Chromatography (GPC) with several detection systems [Electron Capture (EC), Nitrogen Phosphorus (NP), Flame Ionization (FI), Thermal Conductivity (TC), Flame Photometry (FP), UV, Fluorescence (F) and diode array], Capillary Zone Electrophoresis (CZE) and MS. The analytical potential of all these approaches can sometimes be further enhanced by expediently combining them to achieve on-line separation and quantification of groups of analytes with minimal manipulation (the so-called hyphenated techniques).

It is out of the scope of this chapter even to attempt to summarize the applicability field and the relevant pros and cons of the most popular and powerful analytical techniques as their theoretical and practical aspects are thoroughly reported in many recent books (see, e.g., 24–26). How they can be bent to the specific need of Antarctic environmental research is, in turn, exhaustively illustrated on a case-by-case basis in the chapters which follow. As a general warning, the analyst is well advised not to push an instrumental method beyond its intrinsic limits, in terms of both limits of detection, optimal working range and applicability to specific groups of substances, because otherwise the overall uncertainty associated with the experimental data will increase dramatically.

2.6. *Quality control and quality assurance*

The past two decades have testified to a growing awareness by the experimentalist of the inescapable need for incorporation of quality, in the broadest meaning of the term, into the laboratory management whatever the activities performed might be (routine, research or any combination thereof). No laboratory can survive in the long run if the results it produces are not defendable and traceable back to internationally accepted criteria and standards. The proliferation of quality systems since the late 1970s provides good evidence of what was nothing less than a true revolution in the mentality and working conditions of laboratory personnel and decision makers alike. The models developed so far are designed to satisfy different needs. Among them, of particular importance to an environmental analytical laboratory, are the Standard EN 45001, the ISO/IEC Guide 25 and the OECD Principles of Good Laboratory Practice (27–30). Such models have many aspects in common, although they differ in that they address a variety of issues only partially overlapping each other. On the other hand, they also mutually reinforce one another and greatly assist the laboratory in developing a quality system suited to the specific demands it has to meet. More in detail, through the EN 45001 criteria and the ISO/IEC Guide 25 the laboratory is instructed on how the quality of its analytical activities can be assured, e.g., as regards the overall design of premises, the setting-up and maintenance of technical facilities and the development of guidelines for basic operations (such as handling of samples, filing of documents, application of methods and preparation of reports). The same quality elements of the two preceding models are featured by the OECD principles, which in turn put much more emphasis on the nomination, for each single study, of a study director with full responsibility for its overall planning, conduct and reporting, on the management of archives and on the existence of an independent quality assurance programme. These models were initially conceived to be followed separately so as to provide evidence of the laboratories' technical competence at performing their tasks in specific sectors. At a later stage of their development they evolved towards total quality systems with much broader applicability and reciprocal compatibility. All this goes obviously to the advantage of the laboratory and greatly facilitates the decision-making process as to which model best fits its quality demands. This harmonization process is still in progress, but it will eventually smooth off current difficulties in choosing the scheme to implement, while also discriminating among models less important formally speaking.

Only when a laboratory has implemented a valid quality system is it reasonable to assume that the data generated in that laboratory are acceptable and fit for purpose. But this, in turn, implies that measurements be traceable, i.e., that an unbroken chain of calibrations can be set up to link the actual measurement process to the relevant fundamental units, so as to unequivocally demonstrate that no unexpected factors have impaired the final results.

Chemical analyses violate this basic rule in most cases, as almost without exception samples undergo physical transformation (extraction, digestion, calcination and the like) prior to instrumental determination. This unavoidable circumstance can however be alleviated by resorting to matrix CRMs which allow the analyst to verify the correctness of the experimental results obtained on unknown samples by means of a given procedure through the degree of agreement achieved between the certified figures for given properties in the CRMs and those measured for the same properties in the same analytical run. This, in itself, is not enough to guarantee quality; it is in fact always a conservative approach in that not even an excellent agreement between measured and experimental values in a CRM can give certainty that the data obtained for unknown samples are trustworthy. Nonetheless, the level of confidence increases with the increasing similarity between the CRMs used and the materials under test as regards matrix composition, presence of possible interferents and concentrations of analytes. The use of more CRMs with such characteristics and the obtainment of data in good accordance with the expected ones further improves the probability of success by minimizing the risk that the method adopted is affected by unidentified sources of error.

The uncertainty u_{CRM} which can be attached to a certified value in a CRM is given by the combination of all uncertainty sources relevant to the user (31, 32). This is given by the expression $u_{CRM} = \sqrt{u_{char}^2 + u_{bb}^2 + u_{lts}^2 + u_{sts}^2}$, where u_{char} is the uncertainty of batch characterization, u_{bb} the between-bottle variation and u_{lts} and u_{sts} the uncertainties related to long-term and short-term storage, respectively.

Regretfully, the number and type of matrix CRMs are largely insufficient to cover all possible real life cases. Antarctic research and monitoring suffer even more than other sectors from the paucity of ad hoc CRMs, although there is extremely keen need for them (see Chapters 10 and 11 in this book). Table 1.2 reports the manufacturers and suppliers of CRMs for environmental purposes. Furthermore, it is worth stressing that CRMs are especially intended to assess

Table 1.2. Categories of Matrix Certified Reference Materials for environmental applications and their manufacturers

Category	Manufacturer and country of headquarters
Ashes, dusts, soils, algae, flours	Institute of Radioecology and Applied Nuclear Techniques (IRANT), Slovak Republic
Soils, sludges, sediments, plants, mussels, serum, urine, hair, plankton, cod, krill, rain water, fresh water, ground water, sea water, ashes, fish oil, waste, cow milk	Institute for Reference Materials and Measurements (IRMM) and Bureau Communitaire de Référence (BCR), European Commission, Belgium
Hair, soils, sediments, sea plants, cockle, milk, whey, water, fish, lichens, clover, cabbage, grass, bone, blood, marble, cellulose, Greenland ice sheet precipitation water, air, water, lake sediment, standard light Antarctic precipitation, rye flour, cotton cellulose, hay powder, Vienna mean ocean water	International Atomic Energy Agency (IAEA), Austria
Blood, serum, plasma, urine, bone, tissues, hair, waters, soils, sediments, sludges, ashes, plants, dairy products, cereals, meat, fish, vegetation	Laboratory of the Government Chemistry (LGC), UK
Vehicle exhaust, flour, fish, hair, algae	National Institute for Environmental Studies (NIES), Japan
Dusts, soils, serum, urine, hair, used auto catalysts, soils, sediments, sludges	National Institute of Standards and Technology (NIST), USA
sea water, estuarine water, river water, sediment, fish, lobster	National Research Council (NRC), Canada
Bovine meat, branches, leaves, cabbage, cattle, fly ash, hair, mussel, rice, sediment, serum, simulated rainwater, soil, urine, water	National Research Centre for Certified Reference Materials (NRC-CRM), China
Blood, serum, urine	Recipe Chemicals + Instruments GmbH, Germany
Rainwater, sediment, lake water, river water	National Water Research Institute (NWRI), Canada
Sediments, soils	South African Bureau of Standards (SABS), South Africa

accuracy, while non-certified reference materials are useful to obtain reproducible data (see definitions in Table 1.3).

The latter are hence extremely useful to monitor the performance of an analytical method with time by setting up control charts which allow for the statistical control of measurements (33). Reference materials are necessarily homogeneous and stable. If analyzed at regular intervals, quick and clear information can be gained on any tendency for the analytical process to go out of control when the

Table 1.3. Definitions of terms relevant to quality implementation.

Accuracy	Closeness of agreement between the result of a measurement and a true value of the measurand (34).
Certified Reference Material	Reference Material, accompanied by a certificate, one or more of whose property values are certified by a procedure which establishes its traceability to an accurate realisation of the unit in which the property values are expressed, and for which each certified value is accompanied by an uncertainty at a stated level of confidence (34).
Internal Quality Control	Set of procedures undertaken by laboratory staff for the continuous monitoring of operations and the results of measurements in order to decide whether results are reliable enough to be released. Internal Quality Control primarily monitors the batchwise accuracy of results on quality control materials and precision on independent replicate analysis of test materials (35).
Precision	Closeness of agreement between independent test results obtained under prescribed conditions (36).
Quality	The totality of features and characteristics of a product or service that bear on its ability to satisfy stated or implied needs (37).
Quality Assurance	All those planned and systematic actions necessary to provide adequate confidence that a product or service will satisfy given requirements for quality (37).
Reference Material	Material or substance, one of whose property values is sufficiently homogeneous and well established to be used for the calibration of an apparatus, the assessment of a measurement method, or for assigning values to materials (34).
Traceability	Property of the result of a measurement or the value of a standard whereby it can be related to stated references, usually national or international standards, through an unbroken chain of comparisons all having stated uncertainties (34).
Trueness	Closeness of agreement between the average value obtained from a large series of test results and an accepted reference value (38).
Uncertainty	A parameter associated with the result of a measurement, that characterizes the dispersion of the values that would reasonably be attributed to the measurand (30).

experimental value of the concentration of a given measurand exceeds the pre-established warning and action limits.

Another effective tool to assess the performance of a laboratory and to check the validity of the analytical methods in use is the participation in inter-comparison exercises (39). Obviously, these go hand in hand with the proper use of CRMs and offer another possibility of identifying and eliminating unsuspected sources of error through the exchange of views with other laboratories. In this way the overall proficiency of a laboratory can be not only unequivocally assessed, but also significantly improved, primarily with respect to the general management (training and motivation of the staff included), all facets of sample pretreatment and the actual determination step with the attendant pitfalls in calibration, blank evaluation, signal interferences and instrumental constraints. Quite understandably, the goals of inter-laboratory studies must be clearly defined beforehand so that no ambiguities will hamper the expected benefits to participants in the trials. Yet, crucial as they are for guaranteeing the quality of the work done by a laboratory, CRMs and inter-comparisons are no replacement for good sense and alertness: all stages of the analytical process must be constantly governed, inspired and interpreted through them in order that they can lead to valid conclusions and hence to sound decisions.

From a general viewpoint, the credibility and comparability of environmental analytical data can be assessed through two major parameters, i.e., trueness and precision, which combine into accuracy and express the closeness between an experimental measurement with its uncertainty and the true (or supposedly so) value of a given quantity (40). Such concepts were developed long ago, but are still being refined to accommodate the progress made so far in this sector. Definitions of these and other quantities relevant to this field, for which consensus has been achieved by the scientific community, are listed in Table 1.3.

In conclusion, in light of what has been discussed in this and the preceding sections, it is of primary importance that all sources of variability be taken into consideration – as they can significantly affect the final result – and that the analytical methods in use at the laboratory be fully validated. This implies the existence, availability to all relevant staff and regular updating, of the laboratory quality manual and standard operative procedures, as well as the clear allocation of individual responsibilities. Finally, as regards the premises, clean room facilities of class 100 or better, along with laminar flow clean benches, are nowadays a must for the vast majority of environmental research and monitoring analyses, let alone the highly sophisticated antarctic measurements (18). The protection of the laboratory from severe conditions of heat, dust, humidity, steam, noise, electromagnetic fields and the like is of no minor importance. Rooms, ceilings and walls should be constructed so as to minimize the passage of people, samples and chemicals and exposure to extreme physical conditions; access to the laboratory premises should be restricted only to staff; computers and computerized equipment should also be managed according to the quality assurance principles. In a few words, good housekeeping is essential for the economy and efficiency of laboratory work.

3. A survey of selected recent applications

3.1. Preliminary comments

The push forward undergone by scientific activities in the sixth continent in the past decade, along with the increasing awareness that global change phenomena can be much better observed and monitored from Antarctica than from any other spot on the planet, fostered an impressive number of environmental investigations. The added value of such studies is that they have been progressively incorporating the quality criteria described in the preceding section, thus reaching in general a substantial level of credibility and comparability, with obvious benefits for the self-consistent, harmonized and interactive advancement of knowledge of this part of the world. A selection of these investigations of particular significance under this profile is offered hereafter.

3.2. Sediments and soils

Marine sediments can well reflect the presence of pollution sources in aquatic systems. In the frame of the Italian national programme for research in Antarctica, a project was launched to ascertain the concentration of several major, minor and trace elements (Al, Be, Bi, Cd, Cr, Fe, Mn, Ni, Pb, Sn and Zn plus total carbon) in sediments along the coast of the Ross Sea (Terra Nova Bay and Wood Bay) (41). Samples were collected by means of a stainless steel grab, placed in chemically decontaminated polyethylene containers, immediately frozen at –20°C and kept at this temperature until analysis. All manipulations prior to determination were carried out under contamination-controlled conditions. Sediment was allowed to thaw and all visible exogenous material was removed by hand (small marine organisms and shell fragments in the first place). Sample digestion was achieved by acid-assisted microwave irradiation, and analyses of metals were performed by AAS (either F or ETA), whereas total carbon was measured by an element analyzer. Accuracy of data was checked by including in the analytical cycles CRMs from the National Research Council (NRC), Canada (MESS–1 and PACS–1, Marine Sediments) and from the Istituto Superiore di Sanità (ISS), Italy (MURST-ISS-A1, Antarctic Marine Sediment). Results showed that the clay fraction in the tested sediment is very low. This implies that the potential pollutants cannot be bound by clay minerals, but must rather adhere to the surface of sand and silt grains. It could be concluded, however, that no particular concern is raised by the measured element concentrations as these are thoroughly comparable to those of deep-ocean sediments.

The team of Ahn et al. used a PVC corer to sample habitat sediment relevant to the clam *Laternula elliptica* from the Collins Harbour (Maxwell Bay, King George Island) at a depth of 30 cm (42). Samples were extracted with high-purity HNO_3 after oven-drying. Cadmium, Co, Cr, Fe and Zn were measured by F–AAS and ETA–AAS in an attempt to clarify the mechanism of bioaccumulation of the metals in the organism.

Coastal sediments were sampled in the Ross Sea by a stainless steel grab at a water depth ranging from 0.5 to 540 m and analyzed for their content in Cd, Co, Cr, Cu, Fe, Mn, Ni, Pb and Zn (43). Both total concentrations of these metals after complete dissolution of the sediment in a Teflon® bomb with HF and *aqua regia* and concentrations in three selectively extracted fractions (exchangeable/carbonate, Mn–Fe oxide and organic/sulphide) were quantified with ICP–AES. In the first case, accuracy of measurements was checked by two CRMs, namely the MAG–1 sediment (US Geological Survey) and the BCSS–1 sediment (NRC, Canada). No significant differences in concentration were noted for these elements in sediment samples taken from the same area two years earlier.

Antarctic marine sediments from Terra Nova Bay were analyzed with ICP-AES and X-RFS after extraction with HNO_3 in a microwave oven (44). Accuracy was tested by using the CRM 320 River Sediment, Bureau Communautaire de Référence (BCR), European Commission (EC).

The projects undertaken by Italy in the late 1990s to produce CRMs for environmental research in Antarctica have also prompted similar actions to achieve full traceability to the SI of the certified properties. This was the case of the CRM MURST ISS A1 Antarctic Sediment prepared by ISS and the Institute for Reference Materials and Measurements (IRMM) of the Joint Research Centre (JRC), EC, for which an approach was proposed to obtain traceability for the Cu concentration (45). To this end, after microwave digestion of sediment and separation of the analyte by Ion–Exchange Chromatography (IEC), ID quadrupole ICP–MS was employed. As traceability also implies the concept of "stated uncertainties" (see definitions in Table 1.3), the complete uncertainty budget was estimated. Thanks to this and to the fact that ID is a primary method of measurement, a clear mathematical relationship could be established between Cu concentration and isotope ratios.

Sediments from nine locations (Peter Ioy Island, Powell Island, Antarctic Peninsula, King George Island, Marsh Martin, Bellingshauen, Jubany, Curville Island and Horse Shoe Island) were analyzed for their content in Ba, Cd, Co, Cr, Cu, Fe, Mn, Mo, Ni, P, Pb, Sr, Ti, V and Zn) (46). Samples were air dried, pulverized, sieved and homogenized prior to wet digestion with *aqua regia* and $HClO_4$. High purity chemicals were used throughout and glassware was decontaminated with HNO_3. Accuracy was checked and maintained by means of estuarine and river sediment CRMs, National Institute of Standards and Technology (NIST), USA. Results demonstrate the impact of nearby anthropogenic activities on sediment quality. In particular, at the Marsh Martin location, Ni and V could be traced back to a leakage in fuel barrels.

Metal speciation was performed in sediments and near-shore soils from the Carezza lake (47). Aluminium, Cd, Co, Cr, Cu, Fe, Mn, Pb and Zn were investigated along with humic and fulvic acids because of their complexing ability. Both total and selective solubilization of metals were accomplished. A variety of techniques (ETA–AAS, ICP–AES, thermogravimetry, Fourier–transform IR spectrometry, NMR spectrometry and elemental analysis) were employed. Although the elements under test behave in different ways, they appear to be mostly bound to organic matter and also present as sulphides. The only exception is given by Mn, for which MnO_2 is predominant. Algae were inferred to exert a competitive action with humic substances towards metals because of the presence of carboxylic groups in their chemical structure. This study is clearly inspired by sound awareness of the need for quality control and assurance, but it is unfortunate that no explicit mention of the measures adopted is made.

A complete quality assurance programme including daily calibration checks, surrogates, internal standards, matrix spikes and CRMs, accompanied the analysis of sediments from Arthur Harbor (48). All stages of collection, processing and storage of sediments (sampled by diver or a Smith-McIntyre grab) were governed by clean procedures. High-resolution GC with FI detection and GC-MS were the techniques of choice for the determination of aliphatic and aromatic hydrocarbons, respectively. Diesel fuel spillage and leakage from ships and boats were recognized as the major sources of hydrocarbon contamination in subtidal samples, with concentrations (ng g^{-1}, dry weight) of 29–980 and 267–5227 for polycyclic aromatic hydrocarbons and alkanes, respectively.

The combustion source hypothesis was tested to ascertain the behaviour of hydrocarbons in sediments from a remote sub-antarctic island (King Edward Cove) (49). This site was an important seal hunting ground for decades early in the nineteenth century and then hosted a whale processing factory which closed in 1965. These activities led to significant contamination of the bay with fossil fuel and organic material. Marine sediment cores collected at a depth of 18 m clearly reflect the cessation of industrial operations, while further research is necessary for a better understanding of the fate of petroleum components in cold benthic ecosystems.

Marine and lake sediments and soil samples in the area of Terra Nova Bay and Wood Bay were taken both by a box corer system and manually (50). These materials were placed in polyethylene containers chemically cleaned beforehand and stored at –20°C. GC–MS was used to determine individual congeners and the greatest care was devoted to minimize incorrect peak assignment by selecting a statistically significant time window. Lake sediments showed relatively higher concentrations of these compounds probably because they are trapped in the ice from the atmospheric particulate and subsequently transported to the lake during the deglacial season.

3.3. Marine and fresh waters

As illustrated in detail by Wolff, contaminants in Antarctic snow and ice have their ultimate destination in sea because of the melting process (51). The presence, origin (either local or from other continents) and fate of heavy metals, HNO_3, H_2SO_4 and other substances from fuel combustion and waste burning were reviewed by this author.

Some leaching tests were undertaken to ascertain the extent of trace metals contamination by sampling devices when collecting surface/deep sea water and snow/ice cores (52). Go-flow bottles, polyethylene tanks corers (especially knives and head) as well as all plastic items and Teflon® components of the Differential Pulse (DP) ASV instrumentation used for the analytical determinations had to be preliminarily cleaned according to very stringent procedures. All this resulted in a significant abatement of the contamination by Cd and Pb.

Up to 50 ng l^{-1} of bromoform were measured in water samples from the coasts of the Antarctic Peninsula along with lesser amounts of other bromomethanes (53). The interpretation of the mechanism of exchange of these compounds between sea and atmosphere requires many more data of high quality than presently available.

Polycyclic aromatic hydrocarbons, *n*-alkanes, alcohols, aldehydes, ketones, fatty acid esters and phthalates were quantified in sea water samples taken at various depths under the pack ice in Terra Nova Bay and in the Ross Sea (Wood Bay) (54). Meticulous procedures were followed to clean the go-flow bottles used for sampling and the stainless steel containers intended for storage and transportation at –20°C. The concentrations of these compounds in sea water and in the corresponding pack ice point to wide-range ongoing contamination possibly due to local anthropic activity in the Ross Sea area. However, an important confounding factor may be ascribed to the fact that the go-flow bottles are made of PVC with a Teflon® coating which entails the potential release of organics to the samples.

The extent of hydrocarbon contamination at Factory Cove and Borge Bay (Signy Island) was investigated by quantifying *n*-alkanes and polycyclic aromatic hydrocarbons in sea water samples and in the relevant sediment (55). The former were taken by means of glass stoppered bottles. Results are probative of contamination inputs from a nearby research station, the extent of which decreases generally within a few hundred metres. Total *n*-alkane levels, in fact, passed from 7.6 to 2.6 μg l^{-1} at a distance of 500 m, while polycyclic aromatic hydrocarbons were in the range of 110–216 ng l^{-1} independently of distance. Concentrations in sediments showed a similar pattern, thus indicating that a large fraction of the hydrocarbons were deposited from the water column.

An accidental spill of 1000 l of diesel fuel from Faraday (now Vernadsky) Research Station, Galindez Island, caused contamination of the marine environment by *n*-alkanes and polycyclic aromatic hydrocarbons (56). sea water and limpets (*Nacella concinna*) were monitored from the very first moment up to one year after the event. A well codified procedure was followed to thaw, filter through a 0.45 μm cellulose nitrate membrane and extract with hexane the sea water samples. The limpet tissues were excised after thawing, homogenized, digested in methanolic potassium hydroxide and extracted with hexane. Capillary GC and HPLC were used to analyze *n*-alkanes and polycyclic aromatic hydrocarbons, respectively. Individual components in each group were quantified against proper analytical standards. Although the use of fuels can pose a threat to the Antarctic environment, the results of this study showed that this was not the case at Faraday. After an immediate toxic effect in the intertidal zone, evaporation, solution and dispersion rapidly minimized the spill consequences.

Polychlorobiphenyls (PCBs) were analyzed by GC-MS in the Terra Nova – Gerlach Inlet area to ascertain their coastal depth profile before and after pack ice melting (50). Go-flow Teflon®-lined bottles or a Teflon® pumping system were used to sample water, which was then stored in 20–l stainless steel containers at –20°C. The total concentration of these substances was found to increase by ~70% after pack ice melting, with mean values (pg l^{-1}) passing from 170 (before) to 285 (after).

3.4. Snow, ice and firn

The importance of evaluating the concentrations of trace and ultratrace elements in Antarctic snow and ice cannot be exaggerated as this information is essential to reconstruct their natural levels in the pristine atmosphere and to assess the effect of anthropic activities in more recent times. Such challenging determinations do demand the availability of sophisticated and mature analytical techniques supported by equally well-mastered preanalytical procedures. A significant contribution in this field was made by Townsend and Edwards who developed and optimized a method based on high resolution ICP–MS to quantify Al, Bi, Co, Fe, Mn, Pb, Se and V in samples of ice cores from a high precipitation site at Law Dome (East Antarctica), of snow from a low precipitation high plateau at the Lambert Glacier basin, Princess Elisabeth Land (East Antarctica), and of snow from the iced surface of Ross Sea, where precipitation characteristics are unknown (57). Samples were handled and stored with the strictest precautions, resorting to high–purity reagents and carefully cleaned labware of low-density polyethylene or Teflon®. Sample preparation was performed in class-100 clean room facilities and all instrumental components coming in direct contact with the specimens were carefully decontaminated beforehand. The outer layers of ice cores were preliminarily removed because they were heavily contaminated by the drilling process. The instrumental technique adopted is capable of determining the elements listed above at concentration in the range of pg g^{-1} to fg g^{-1} with minimal manipulation, so that the risk of contaminating the samples in the laboratory is greatly reduced.

The determination of ultra-low levels of Hg in snow and ice from polar region, requires that extremely rigorous precautions be taken to avoid any contamination of samples from laboratory equipment and atmosphere. Ferrari et al. showed how this goal can be achieved by working in a non-laminar flow class 10,000 clean laboratory equipped with clean benches (58). HEPA filters were employed, as usual for such facilities. On the other hand, only 5% of Hg is present in the atmosphere as particulate matter, while 95% of the metal is in the gaseous form. To remove this, 700 m^3 h^{-1} air flow was flushed into the clean laboratory through active charcoal filters so that the air velocity on the clean bench was 0.5 m s^{-1} and the Hg concentration fell down to 2.2 ± 0.6 pg l^{-1}. Clean garments and ultrapure reagents (HNO_3, $SnCl_2$ and water) were imperative throughout the analysis. Teflon® containers were preferred for the storage of chemicals. This complex approach allowed concentrations of Hg as low as 2.5-45 pg g^{-1} to be determined with high confidence by CV–AAS.

A review dealing with the difficulties inherent in field sampling and laboratory analysis of ancient ice and recent snow summarizes the state-of-the-art in this context (59). The major problems are caused by the procedures adopted for the collection of ice cores, i.e., thermal or electrochemical deep drilling, because of the high level of contamination of the outer surface of the cores, all the more serious when a wall-retaining fluid is used to fill the drilling hole. Decontamination is a painful and time-consuming step which requires chiselling of successive veneers of the core from the outside inward under ultraclean conditions. The selection of a suitable, chemically inert material (e.g., polyethylene or Teflon®) and the necessarily sophisticated cleaning methods are further key aspects to be duly taken into account. The message conveyed by these environmental matrices is of extraordinary importance for estimating past climatic events. Many recent and reliable studies point to the occurrence of relatively high concentrations of Cd, Pb and Zn in Dome C and Vostok deep ice cores dating back to the coldest glacial period some 20,000 years BP. This pattern can be ascribed to high fluxes of soil and rock dust into the atmosphere favoured by the increased aridity, stronger winds and lower sea level. Over the past two centuries, anthropic sources have become significant (e.g., Pb is about one order of magnitude more abundant in Antarctic snow than it was several thousand years ago), although better control of atmospheric emissions of heavy metals has now led to a non negligible reduction in their present levels in Antarctica. In full compliance with such rigorous criteria, shallow snow samples were collected in the Hercules Névé, Victoria Land (East Antarctica) by means of a stainless steel auger, sealed in double polyethylene bags and shipped frozen to the laboratory (60). Clean room garments, masks, polyethylene gloves and boot covers were worn all the time during sampling to minimize contamination.

Determination of Cd, Pb and U were performed by magnetic sector high resolution ICP-MS. The lack of CRMs in a snow matrix did not allow for a direct check of the measurement accuracy, but reasonable confidence in the reliability of data was reached through the use of other analytical techniques, i.e., ETA–AAS and DP–ASV. Mean concentrations were found to be (pg g^{-1}) 0.39 for Cd, 5.0 for Pb and 0.04 for U.

Variations in Pb concentration in Dome C ice cores during the Wisconsin/Holocene transition (27,000 to 4000 years BP) were determined with the dependable approach developed by Boutron's team (61). The uncertainty budget for these measurements as caused by contamination events was thought to be ~10% for high Wisconsin and ~50% for the very low Holocene Pb concentrations, respectively. Before anthropogenic contributions started, it is estimated that Pb was primarily due to volcanic activity (Holocene period) and soil dust (late Wisconsin and Holocene periods).

Ultraclean ID-TI-MS was employed to establish Pb concentrations in Vostok deep ice cores spanning the period 155,000 to 26,000 BP (62). Mechanical decontamination of the ice cores and appropriate analytical techniques afforded measurements with an uncertainty of ~5% and ~20% for outer and inner layers of the ice sections, respectively. Values ranging from 2–40 pg g^{-1} were obtained, with higher concentrations during the ice age (Illinois) and the last glacial maximum, the major source of the metal being soil dust.

Laser-excited atomic fluorescence spectrometry allowed for the determination of Pb in Vostok deep ice cores with a precision of 20% (63, 64). Values of 2–40 pg g^{-1} were measured for ages spanning the 155,000–26,000 years BP. The technique permitted sample volumes as low as 20 μl to be dealt with without any preliminary treatment, thus greatly facilitating contamination control.

The determination of the isotopic composition of pg amounts of Pb in snow and ice (useful to reconstruct the pathway of this metal from anthropogenic sources) requires the development of ultraclean procedures for the collection, storage and processing of samples. Surface snow was taken at various locations, while ice cores were drilled on Dome C and at Vostok (East Antarctica) (65). The former were collected by simply pressing acid-cleaned conventional polyethylene bottles vertically into the snow; the latter were obtained by thermal drilling (in the case of the Vostok samples the drilling hole was also filled with kerosene as the retaining fluid). Chiselling was adopted in both instances to eliminate the contaminated outer layers. This and further processing was conducted in the so-called Femtolab, an over-pressured laboratory supplied with HEPA filtered air. Teflon® beakers, carefully precleaned with high-purity reagents were used throughout. Accurate control of blanks was thus achieved, this being crucial for the obtainment of reliable isotopic profiles with the requested precision. ID measurements were carried out by high resolution magnetic sector MS with thermal ion source. The isotopic reference material SRM 981 (NIST) was employed for checking measurement accuracy. As low as 2 pg g^{-1} of Pb could be quantified. At the level of a few tens of pg of Pb a precision of ±02 could be arrived at for the isotopic ratios.

The non-sea spray effects on the levels of Ca, Cl, K, Mg and Na as well as sulfate in snow samples from Terra Nova Bay were ascertained (66). Sampling mode and analytical approach (based on IC) fully complied with criteria previously set up. For Ca and K it was concluded that a nearly uniform background aerosol characterizes the area investigated, which adds to the contribution of marine spray. The oxidation of DMS, in turn, is the dominant source of non-sea spray sulfate.

A chemical baseline for snow at Palmer Station was preliminarily assessed (67). Samples were collected on a sheet of clean polyethylene. Standard EPA protocols were strictly adhered to and clean polyethylene gloves were used throughout. Ionic composition in Br^-, inorganic and organic carbon, Ca^{2+}, Cl^-, F^-, K^+, Mg^{2+}, Na^+, NH_4^+, NO_3^- and SO_4^{2-} was ascertained by means of IC and ETA–AAS. No particular wind direction appears to influence pH and ion concentration gradients. Concurrent aerosol and snowfall along with meteorological data would be necessary to substantiate preliminary assumptions on the chemical baseline.

An overview of the problems related to sampling of snow and ice for quantifying heavy metals was presented by Wolff and Peel (68). Stress was put on the need for careful planning of activities so as not to flood the scientific community with questionable data. The major sources of local pollution were recognized to be manned stations (but only within a few km) and aircraft fuel. The different strategies set up to analyze Cd and Pb by DP–ASV in snow directly at the research station in Antarctica (Terra Nova Bay) or back to the laboratory in Italy were described by Scarponi et al. (69). The former instance led to a faster procedure requiring no blank correction, while the latter case had a definite advantage in that higher sensitivity and repeatability could be obtained. Both procedures were fully compliant with sound quality requisites throughout the analytical chain. Further information on this issue and on the precautions adopted to preserve the integrity of snow samples can be retrieved in a later paper from the same group (70).

In a review focused on the findings of his team over the years, Heumann stressed the role of Antarctica as the natural archive of past events on the planet and highlighted the fact that the very sampling process may be a significant source of contamination of the materials collected due to their extremely low contents in anthropogenic pollutants (53). At such levels even the most conventional analytical steps may be prone to pitfalls. As an example of the higher probability of errors under such conditions, the case of depth profiling of NO^{3-} in firn core samples was reported. ID–MS and IC were used for quantification, but data sets differed by a factor of two as a consequence of faulty calibration. The same team found concentrations of Cd and Tl in surface snow from the Ronne-Filchner ice shelf well below the pg g^{-1} level, whereas those of Co and Pb were in the low pg g^{-1} range. Chloride and several I species (I^-, I_2, IO^- and IO_3^-) were also measured in snow samples from the Ekström ice shelf, with high enrichment factors for the total of I species over Cl^-. As the primary source of these two halogens is the polar sea, a mechanism for their selective transportation to the main inland was worked out. A comprehensive account of ongoing research in this and related fields is given elsewhere in this book (see Chapter 7).

3.5. Biota

Ahn et al. ascertained reference values for Cd, Co, Cr, Cu, Fe, Mn, Ni, Pb and Zn in the digestive glands, gonad, gills, kidney and muscle of the Antarctic clam *Laternula elliptica* by F–AAS and ETA–AAS (42). After careful selection of the site and obtainment of its hydrographic characteristics, samples were hand collected from Collins Harbour (Maxwell Bay, King George Island) by SCUBA divers at a depth of 25–30 m, where the clam occurred at a density of *ca.* 136 individuals m^{-2}. Samples were immediately frozen at –20°C and kept at this temperature until analysis. The clams were thawed at room temperature and dissected into the organs and tissues mentioned above by means of ceramic scissors. Freeze-dry, grinding and high-purity HNO_3-assisted digestion followed in screw-capped Teflon® jars. The Dogfish Muscle CRM (NRC) was resorted to for checking measurement accuracy.

More than twenty years ago the concentrations of As, Cd, Cu, Hg, Ni and Pb were already determined in whole krill, krill muscle tissue, krill products and fillets of the Antarctic fish *Notothenia rossi marmorata*, *Dissostichus eleginoides* and *Notothenia gibberifrons* by adopting an effective quality control scheme based on clean laboratory conditions, quartz and plastic knives and pincers and Bovine Liver and Orchard Leaves CRMs (National Bureau of Standards, now NIST) (71). Samples were taken from the Antarctic Scotia Sea and, after appropriate digestion, were analyzed by several techniques (ETA–AAS, HG–AAS, CV–AAS and DP–ASV), to further increase validity of results. The findings of this study clearly indicated that the elements investigated pose no risk for human consumption of krill.

Liver portions (tip and mid-ventral blubber samples) of seals from the pack-ice of Queen Maud Land were analyzed for their contents in trace elements (Cd, Co, Cr, Cu, Fe, Hg, Mn, Ni, Pb, Zn) and chlorinated hydrocarbon residues (DDE, DDT, dieldrin, PCBs, TDE) (72). Samples were stored at –20°C pending analysis. The metals were quantified by AAS after digestion, as appropriate. The organic compounds were first extracted with hexane, cleaned-up and then assayed by gas-liquid chromatography with EC detection. Concentrations in the high μg g^{-1} level were found for Cd, Fe and Zn, whereas those of the chlorinated compounds were much lower (ng g^{-1} range). Possible correlations with diet were also envisaged.

The scallop *Adamussium colbecki* was sampled from areas adjacent to Terra Nova Bay and Swoya research stations as well as from the Explorers Cove site which is hydrographically isolated from the influence of activities at the McMurdo Base (73). The study aimed at developing a circumpolar baseline for this species as regards trace elements in relation to the different habitat conditions. Determinations of Cd, Cu, Fe, Mn, Ni, Pb and Zn were performed on all tissues combined. The concentrations found were generally lower than those for temperate and tropical scallop species.

The levels of Cd, Cu, Pb, V and Zn in this sentinel organism from the Terra Nova Bay, Ross Sea, were in turn investigated by Minganti et al. (74). After collection the samples were placed in polyethylene bags, frozen at –25°C and shipped to the laboratory where the scallops were individually characterized in terms of shell length, weight of soft parts and sex. Soft tissues were excised, freeze-dried and homogenized. Digestion was performed in a microwave oven with concentrated HNO_3 and determinations were done by ETA–AAS and ICP–AES. The quality control programme included blank analysis to check contamination phenomena and use of the TORT–2 Lobster Hepatopancreas Homogenate CRM (NRC). The values found were (in μg g^{-1}, dry weight) 28.5 ± 6.9, 6.5 ± 4.5, 0.64 ± 0.77, 1.0 ± 0.3 and 88 ± 14 for the five elements in the given order, with significant variations when compared with data of previous campaigns.

Chlordane compounds were detected in the blubbers of Weddel seals caught near to the Syowa Station (75). Sampling, dissection and pretreatment were all conducted so as to prevent any possible contamination from chlorinated hydrocarbons, e.g., by washing the electric devices, knives, polyethylene bags and other materials with high purity acetone. Quantification was performed by GC–MS and values of 12–62 ng g^{-1} were ascertained. Airborne transport can probably account for the presence of these substances in Antarctica.

To investigate the distribution of PCBs, *p,p′* -DDT, *p,p′* -DDE and lindane in the aquatic food web, zooplankton and fish were sampled at Home Beach and South Bay at Ross Island (76). Capillary GC with EC detection was employed. An inverse correlation of these pollutants with fat content was ascertained.

Krill and Emperor penguin feathers from the Dakshin Gangotri region were investigated by GC for their contents in several PCB congeners and organochlorine compounds after hexane extraction (77). Organochlorine contaminants were in the concentration (pg g^{-1}) range of 31.1–166.2 for krill and 30.8–113.6 for feathers. Recovery was estimated through the use of reference samples of copepod homogenate and mussel homogenate supplied by the International Atomic Energy Agency (IAEA), Monaco. Further studies on seasonal changes were deemed to be of the utmost importance to better understand environmental pollution in the area of concern.

The necessity of a stringent exclusion of any possible contamination when analyzing PCBs below the 10 ng g^{-1} level was also stressed by Ballschmiter and coworkers (78). These contaminants were quantified in eggs of penguins and albatrosses from subantarctic areas (Falkland Islands) by GC with EC detection.

Homogenates of seal blubber from animals of different age and sex were analyzed by MS for quantifying their concentrations in polychlorodibenzo-*p*-dioxins (PCDDs) and polychlorodibenzofurans (PCDFs) (79). The area investigated covered the South Shetland Islands. Accuracy of measurements was tentatively estimated to range from 10% to 35% at concentrations of 100 pg g^{-1} and 1 pg g^{-1}, respectively. A similar study on blubber samples of fur seals from the South Georgia region was reported by Oehme and associates (80). High resolution MS was employed and rigorous quality assurance measures were applied. All congeners in all samples were characterized by recovery rates of 50–115%, hence in compliance with international protocols for the determination of PCDDs and PCDFs. The levels of these contaminants were around a few pg g^{-1}.

3.6. Atmosphere and aerosols

The role of atmospheric transportation of trace substances from the northern to the southern hemisphere was exhaustively dealt with by Heumann (53). This allowed a geochemical cycle for nitrates in Antarctica to be established. Also in the case of I species a geochemical cycle was proposed.

PIXE analysis of 89 samples of the fine component of the aerosol collected at Campo Icaro (Terra Nova Bay) was used to quantify Ca, Cl, Cu, Fe, K, Mn, Ni, S, Si, Ti and Zn (81). The authors pointed out that contamination from the nearby base cannot be excluded a priori; this, however, can be easily estimated by means of elemental tracers such as V. The same research group extended the observations of this type in a later study on the aerosol coarse fraction covering also Al, Br, Co and Cr (82). Elemental concentrations were given, with Co being at the lowest and K at the highest level (0.053 ± 0.030 and 90 ± 100 ng m^{-3}, respectively).

Plutonium and ^{90}Sr were quantified in mosses and lichens by Testa and associates (83). Validity of data was supported by the adoption of a quality control system and through participation in international intercomparison trials.

Air samples were taken at Ross Island, Cape Evans, and processed by filtering through polyurethane foam plugs connected in series (76). After extraction with an acetone-hexane mixture, determinations of 24 PCB congeners, *p,p'* -DDT, *p,p'* -DDE and lindane were accomplished by capillary GC with EC detection. Geochemical means (in pg m^{-3}) were 15.2, 2.0, 1.0 and 25.8 for the above compounds in the given order. Lindane concentrations were found to correlate well with seasonal variations (higher levels in summer and autumn), whereas the other chlorinated hydrocarbons showed no dependence on seasons.

The challenges inherent in the analysis of gas bubbles trapped in the ice were reviewed by J. Chappellaz (84). The gas, in fact, should be extracted without losses or contamination and the minute amounts available require sophisticated analytical approaches. On the other hand, unique information can be gained in this way on the composition of the atmosphere as far back as 100,000 years ago. The extraction techniques employed are basically dry extraction, melt extraction and sublimation. Past changes in greenhouse gases can be reliably documented so that global biogeochemical cycles can be better understood.

4. Conclusions

The remoteness and pristine nature of the Antarctic region offers to the international scientific community a unique opportunity to investigate global pollution and climate change as no other place on the earth does. Environmental research and monitoring, if fully integrated, can provide investigators and assessors with invaluable inputs for a straightforward understanding of cause-effect relationships. Rather obviously, many issues relevant to environmental studies are still open to debate and wait for basic clarification. In this regard, questionable experimental information can only cripple the validity of assumptions on the way the Antarctic environment may be affected by anthropic activities and eventually spoil the practical importance of Antarctic investigations. Even worse, this can lead to wrong decisions the impact of which on the preservation of the original characteristics of Antarctica and the correct interpretation of wide-scale phenomena is beyond imagination. The measures, precautions and strategies to avoid such consequences are all inspired by long consolidated quality criteria, the adoption of which should be well within the reach of every experimentalist. Nor does it make sense to argue that quality systems are expensive, as the economic losses ensuing from faulty data and misplaced actions can be higher by many orders of magnitude. Further progress in this field can and must be encouraged, in particular as regards three general aspects, namely: *i)* implementation by the laboratories involved in Antarctic research and monitoring of updated, total-quality schemes tailored to the specific needs of their activities; *ii)* participation of the Antarctic laboratories in proficiency tests and collaborative trials through which sources of procedural errors can be identified and eliminated; *iii)* planning and production of CRMs especially designed for use by the Antarctic scientist through progressive involvement of the major manufacturers. Since none of these three objectives can be easily achieved, fast and full commitment by all parties is essential. This is the only way to attain true international harmonization of environmental research and monitoring in Antarctica, enhance mutual confidence in the output of such activities and, at length, sensibly exploit precious human and financial resources to the benefit of a better understanding and effective protection of the Antarctic environment.

References

1. *Antarctic Treaty*, Washington, 1 December 1959.
2. *Protocol on Environmental Protection to the Antarctic Treaty, Final Report of the XIth Antarctic Treaty Special Consultative Meeting*, Madrid, 3–4 October 1991.
3. M. A. Champ, D. A. Flemer, D. H. Landers, Ch. Ribic, T. DeLaca, The roles of monitoring and research in polar environments. A perspective, *Mar. Poll. Bull.*, 25 (1992), 220–226.
4. G. Weller, C. R. Bently, D. H. Elliot, L. J. Lanzerotti, P. J. Webber, Laboratory Antarctica: research contributions to global problems, *Science*, 238 (1987), 1361–1368.
5. Ph. Quevauviller, E. A. Maier, B. Griepink (Eds.), *Quality Assurance for Environmental Analysis*, Elsevier Science B.V. 1995, xx + 649 pp.
6. R. W. Risebrough, Chemical change in Antarctica – Significance? A perspective, *Mar. Poll. Bull.*, 25 (1992), 227–230.

7. M. Tobias, The next wasteland. Can the spoiling of Antarctica be stopped?, *Ann. N. Y. Acad. Sci.*, 1989, March-April, 18–24.
8. G. Kateman, L. Buydens, *Quality Control in Analytical Chemistry*, John Wiley & Sons, Inc., New York, 1993, xvii + 317 pp.
9. Ph. Quevauviller (Ed.), *Quality Assurance in Environmental Monitoring. Sampling and Sample Pretreatment*, VCH Verlagsgesellschaft mbH, Weinheim, 1995, xv + 306 pp.
10. B. Kratochvil, D. Wallace, J. K. Taylor, *Anal. Chem.*, 56 (1984), 114R.
11. C. O. Ingamells, Derivation of the sampling constant equation, *Talanta*, 23 (1976), 263–264.
12. P. M. Gy, *Sampling of Particulate Materials: Theory and Practice*, Elsevier Science Publishers, Amsterdam, 1982.
13. G. Brands, Theorie der Probenahme II. Probenahme aus geseigerten Stoffen, *Fresenius' Z. Anal. Chem.*, 314 (1983), 646–651.
14. F. A. Leemans, Selection of an optimum analytical technique for process control, *Anal. Chem.*, 43 (1971), 36A–49A.
15. T. A. H. M. Janse, G. Kateman, Enhancement of the performance of analytical laboratories by a digital simulation approach, *Anal. Chim. Acta*, 159 (1984), 181–198.
16. P. M. Gy, *Heterogeneity, Sampling-Homogenization*, Elsevier Science Publishers, Amsterdam, 1991.
17. N. T. Crosby, I. Patel, *General Principles of Good Sampling Practice*, Royal Society of Chemistry, Cambridge, 1995, xii + 68 pp.
18. C. F. Boutron, A clean laboratory for ultralow concentration heavy metal analysis, *Fresenius J. Anal. Chem.*, 337 (1990), 482–491.
19. F. Soggia, S. Dalla Riva, M. L. Abelmoschi, R. Frache, Antarctic environmental specimen bank: a tool for chemical monitoring, *Ann. Chim (Rome)*, 90 (2000), 129–135.
20. M. Stoeppler, *Probenahme und Aufschluss. Basis der Spurenanalitik*, Springer-Verlag, Heidelberg, 1994.
21. C. Watson (Ed.), *Official and Standardized Methods of Analysis*, Royal Society of Chemistry, Cambridge, 3rd Edition, 1994, 778 pp.
22. C. Vandercasteele, C. B. Block, *Modern Methods for Trace Element Determination*, John Wiley & Sons, Inc., New York, 1993, vi + 330 pp.
23. S. Caroli (Ed.), *Element Speciation in Bioinorganic Chemistry*, John Wiley & Sons, Inc., New York, 1996, xxvii + 474 pp.
24. G. W. Ewing, *Instrumental Methods of Chemical Analysis*, McGraw-Hill International, New York, 1987.
25. F. W. Fifield, D. Kealy, *Principles and Practice of Analytical Chemistry*, International Textbook Company, London, 1983.
26. D. A. Skoog, *Principles of Instrumental Analysis*, Saunders College Publishing, New York, 1985.
27. CEN, *General Criteria for the Operation of Testing Laboratories* (European Standard 45001), CEN/CENELEC, Brussels, 1989.
28. ISO/IEC, *General Requirements for the Competence of Calibration and Testing Laboratories*, ISO/IEC Guide 25, ISO, Geneva, 1990.
29. *OECD Series on Good Laboratory Practice and Compliance Monitoring*, Monographs 1–10, OECD, Paris, 1998.
30. S. L. R. Ellison, M. Rosslein, A. Williams (Eds.), *Quantifying Uncertainty in Analytical Measurement*, Eurachem/Citac Guide, 2000.
31. ISO, *Guide to the Expression of Uncertainty in Measurement*, ISBN 92–67–10188–9, Geneva, 1995.
32. J. Pauwels, A. Lamberty, H. Schimmel, The determination of the uncertainty of reference materials certified by laboratory intercomparison, *Accred. Qual. Assur.*, 3 (1998), 180–184.
33. T. H. Hartley, *Computerized Quality Control: Programs for the Analytical Laboratory*, Ellis Horwood, Chichester. 1990, 99 pp.
34. *International Vocabulary for Basic and General Terms in Metrology* (2nd Edition), ISO, Geneva, 1993.
35. *The Harmonised Guidelines for Internal Quality Control in Analytical Chemistry Laboratories*, *Pure Appl. Chem.*, 65 (1993), 2123.
36. ISO, *Terms and Definitions Used in Connection with Reference Materials*, ISO Guide 30, ISO, Geneva, 1992.

37. *Quality Assurance and Quality Management – Vocabulary*, ISO 8402, ISO, Geneva, 1994.
38. *Statistics, Vocabulary and Symbols – Part 1: Probability and General Statistical Terms*, ISO 353-1, ISO, Geneva, 1993.
39. W. D. Pocklington, *Guidelines for the Development of Standard Methods by Collaborative Study* (V Edition), Laboratory of the Government Chemist, Teddington, ISBN 0 94892606 6, 1990, ix + 258 pp.
40. ISO, Accuracy (trueness and precision) of measurement methods and results, *ISO/DIS 5+25*, Geneva, 1991, Part 1.
41. R. Giordano, G. Lombardi, L. Ciaralli, E. Beccaloni, A. Sepe, M. Ciprotti, S. Costantini, Major and trace elements in sediments from Terra Nova Bay – Antartica, *Sci. Total Environ.*, 227 (1999), 29–40.
42. I.-Y. Ahn, S. H. Lee, K. T. Kim, J. H. Shim, D.-Y. Kim, Baseline heavy metal concentrations in the Antarctic clam *Laternula elliptica* in Maxwell Bay, King George Island, Antarctica, *Mar. Poll. Bull.*, 32 (1996), 592–598.
43. B. Cosma, F. Soggia, M. L. Abelmoschi, R. Frache, Determination of trace metals in Antarctic sediments from Terra Nova Bay, Ross Sea, *Int. J. Environ. Anal. Chem.*, 55 (1994), 121–128.
44. T. Gasparics, I. Csató, Gy. Záray, Analysis of Antarctic marine sediment by inductively coupled plasma atomic emission and total reflection X-ray fluorescence spectrometry, *Microchem. J.*, 55 (1997), 56–63.
45. I. Papadakis, P. D. P. Taylor, P. de Bièvre, Establishing an SI-traceable copper concentration in the candidate reference materials MURST ISS A1 Antarctic Sediment using isotope dilution applied as a primary method of measurement, *J. Anal. At. Spectrom.*, 12 (1997), 791–796.
46. I. A. Alam, M. Sadiq, Metal concentrations in Antarctic sediment samples collected during the Transantarctic 1990 Expedition, *Mar. Poll. Bull.*, 26 (1993), 523–527.
47. L. Campanella, T. Ferri, B. M. Petronio, A. Pupella, M. Soldani, B. Cosma, Organic matter and metals in lake sediments at Terra Nova Bay (Antartica), *Ann. Chim. (Rome)*, 81 (1991), 417–437.
48. M. C. Kennicutt II, T. J. McDonald, G. J. Denoux, S. J. McDonald, Hydrocarbon contamination on the Antarctic Peninsula. I. Arthur Harbor – Subtidal sediments, *Mar. Poll. Bull.*, 24 (1992), 499–506.
49. H. M. Platt, P. R. Mackie, Analysis of aliphatic and aromatic hydrocarbons in Antarctic marine sediment layers, *Nature*, 280 (1979), 576–578.
50. R. Fuoco, M. P. Colombini, C. Abete, S. Carignani, Polychlorobiphenyls in sediment, soil and sea water samples from Antarctica, *Int. J. Environ. Anal. Chem.*, 61 (1995), 309–318.
51. E. Wolff, The influence of global and local atmospheric pollution on the chemistry of Antarctic snow and ice, *Mar. Poll. Bull.*, 25 (1992), 274–280.
52. G. Capodaglio, C. Barbante, C. Buretta, G. Scarponi, P. Cescon, Analytical quality control: sampling procedures to detect trace metals in environmental matrices, *Mikrochim. Acta*, 123 (1996), 129–136.
53. K. G. Heumann, Determination of inorganic and organic traces in the clean room compartment of Antarctica, *Anal. Chim. Acta*, 283 (1993), 230–245.
54. P. G. Desideri, L. Lepri, L. Cecchini, D. Santianni, F. Masi, M. Bao, Organic compounds in Antarctic sea water and pack ice, *Int. J. Environ. Anal. Chem.*, 61 (1995), 319–330.
55. G. C. Cripps, The extent of hydrocarbon contaminants in the marine environment from a research station in the Antarctic, *Mar. Poll. Bull.*, 25 (1992), 288–292.
56. G. C. Cripps, J. Shears, The fate in the marine environment of a minor diesel fuel spill from an Antarctic research station, *Environ. Monit. Assess.*, 46 (1997), 221–232.
57. A. T. Townsend, R. Edwards, Ultratrace analysis of Antarctic snow and ice samples using high resolution inductively coupled plasma mass spectrometry, *J. Anal. At. Spectrom.*, 13 (1998), 463–468.
58. Ch. P. Ferrari, A. L. Moreau, C. F. Boutron, Clean conditions for the determination of ultra-low levels of mercury in ice and snow samples, *Fresenius J. Anal. Chem.*, 366 (2000), 433–437.
59. C. F. Boutron, J. P. Candelone, S. Hong, The changing occurrence of natural and man-derived heavy metals in Antarctic and Greenland ancient ice and recent snow, *Int. J. Environ. Anal. Chem.*, 55 (1994), 203–209.
60. C. Barbante, T. Belloni, G. Mezzadri, P. Cescon, G. Scarponi, Ch. Morel, S. Jay, K. van de Velde, Ch. P. Ferrari, C. F. Boutron, Direct determination of heavy metals at picogram per gram levels in Greenland and Antarctic snow by double focusing inductively coupled plasma mass spectrometry, *J. Anal. At. Spectrom.*, 12 (1997), 925–931.

61. C. F. Boutron, C. C. Patterson, Lead concentration changes in Antarctic ice during the Wisconsin/ Holocene transition, *Nature*, 323 (1986), 222–225.
62. C. F. Boutron, C. C. Patterson, V. N. Petrov, N. I. Barkov, Preliminary data on changes of lead concentrations in Antarctic ice from 155,000 to 26,000 years BP, *Atm. Environ.*, 21 (1987), 1197–1202.
63. C. F. Boutron, M. A. Bolshov, V. G. Koloshnikov, C. C. Patterson, N. I. Barkov, Direct determination of lead in Vostok Antarctic ancient ice by laser excited atomic fluorescence spectrometry, *Atm. Environ.*, 24A (1990), 1797–1800.
64. M. A. Bolshov, C. F. Boutron, Determination of heavy metals in polar snow and ice by laser-excited atomic fluorescence spectrometry, *Analusis Magazine*, 22 (1994), M44–M46.
65. W. Chisholm, K. J. R. Rosman, C. F. Boutron, J. P. Candelone, S. Hong, Determination of lead isotopic ratios in Greenland and Antarctic snow and ice at picogram per gram concentrations, *Anal. Chim. Acta*, 311 (1995), 141–151.
66. G. Piccardi, F. Casella, R. Udisti, Non-sea-salt contribution of some chemical species to the snow composition at Terra Nova Bay (Antartica), *Int. J. Environ. Anal. Chem.*, 63 (1996), 207–223.
67. T. P. De Felice, Chemical composition of fresh snowfalls at Palmer Station, Antarctica, *Atm. Environ.*, 33 (1999), 155–161.
68. E. Wolff, D. A. Peel, Assessing global and local pollution for heavy metals in Antarctica, *Analusis Magazine*, 22 (1994), M41–M43.
69. G. Scarponi, C. Barbante, P. Cescon, Differential pulse anodic stripping voltammetry for ultratrace determination of cadmium and lead in Antarctic snow, *Analusis Magazine*, 22 (1994), M47–M50.
70. G. Scarponi, C. Barbante, C. Buretta, A. Gambaro, P. Cescon, Chemical contamination of Antarctic snow: the case of lead, *Microchem. J.*, 55 (1997), 24–32.
71. M. Stoeppler, K. Brandt, Comparative studies on trace metal levels in marine biota. II. Trace metals in krill, krill products and fish from the Antarctic Scotia Sea, *Z. Lebensm. Unters. Forsch.*, 169 (1979), 95–98.
72. T. P. McClurg, Trace metals and chlorinated hydrocarbons in Ross seals from Antarctica, *Mar. Poll. Bull.*, 15 (1984), 384–389.
73. P. A. Berkman, M. Nigro, Extending the mussel watch programme to the Southern Ocean, *Mar. Poll. Bull.*, 24 (1992), 322–323.
74. V. Minganti, R. Capelli, R. De Pellegrini, The concentrations of Pb, Cd, Cu, Zn and V in *Adamussium colbecki* from Terra Nova Bay (Antartica), *Int. J. Environ. Anal. Chem.*, 71 (1998), 257–263.
75. M. Kawano, T. Inoue, H. Hidaka, R. Tatsukawa, Chlordane compounds residues in Weddell seals (*Leptonychotes weddell*) from the Antarctic, *Chemosphere*, 13 (1984), 95–100.
76. P. Larsson, C. Järnmark, A. Södergren, PCBs and chlorinated pesticides in the atmosphere and acquatic organisms of Ross Island, Antarctica, *Mar. Poll. Bull.*, 25 (1992), 281–287.
77. R. Sen Gupta, A. Sarkar, T. W. Kureishey, PCBs and organochlorine pesticides in krill, birds and water from Antarctica, *Deep-Sea Res.*, 43 (1996), 119–126.
78. K. Ballschmiter, Ch. Scholz, H. Buchert, M. Zell, K. Figge, K. Polzhofer, H. Hoerschelmann, Studies of the global baseline pollution, *Fresenius' Z. Anal. Chem.*, 309 (1981), 1–7.
79. A. Bignert, M. Olsson, P.–A. Bergqvist, S. Bergek, C. Rappe, C. de Wit, B. Jansson, Polychlorinated dibenzo-*p*-dioxins (PCDD) and dibenzofurans (PCDF) in seal blubber, *Chemosphere*, 19 (1989), 551–556.
80. M. Oehme, M. Schlabach, I. Boyd, Polychlorinated dibenzo-*p*-dioxins, dibenzofurans and coplanar biphenyls in Antarctic fur seal blubber, *Ambio*, 24 (1995), 41–46.
81. P. Mittner, D. Ceccato, S. Del Maschio, R. Cini, U. Giostra, Multielemental characterization of aerosol at Terra Nova Bay. Preliminary results of the fine component during the 1990–1991 austral summer, *Ann. Chim. (Rome)*, 81 (1991), 605–613.
82. P. Mittner, D. Ceccato, S. Del Maschio, A preliminary characterization of the elemental composition of the aerosol corse fraction at Terra Nova Bay (Antarctica) during the 1990–1991 austral summer, *Int. J. Environ. Anal. Chem.*, 55 (1993), 319–329.
83. C. Testa, D. Desideri, M. A. Meli, C. Roselli, New radiochemical procedures for environmental actinide measurements and data quality control, *J. Radioanal. Nucl. Chem.*, 194 (1995), 141–149.
84. J. Chappellaz, Polar ice bubbles as recorders of past greenhouse gas concentrations, *Analusis Magazine*, 22 (1994), M25–M28.

Environmental Contamination in Antarctica
A Challenge to Analytical Chemistry
S. Caroli, P. Cescon and D.W.H. Walton, editors

Chapter 2

A scientific framework for environmental monitoring in Antarctica

David W. H. Walton, Giuseppe Scarponi, Paolo Cescon

1. Introduction

Environmental monitoring is a fundamental element of basic research, environmental management and conservation. The organized and systematic measurement of selected variables provides for the establishment of baseline data and the identification of both natural and human-induced change in the environment. Monitoring data are important in the development of models of environmental processes, which in turn facilitate progress towards a predictive capability to detect environmental impact or change. The collection and evaluation of monitoring data is essential for the detection of human perturbation within the natural variability of ecosystem processes. Since all environmental monitoring must be based on testable hypotheses it can also contribute to advancement in both basic and applied research.

Environmental monitoring can be done on global, regional or local scales, but the same principles of scientific method should be applied in each context. The basic objective is to detect and measure changes in the environment by collecting time series of data for defined purposes and observing trends in the selected variables (1). Since the purposes of collecting data can be defined by both basic and applied research questions, the general objectives of environmental monitoring are to:

i. further basic understanding of the structure, range of variability in, and interactions within and between natural systems;
ii. obtain baseline information on the environment in order to detect, measure and monitor future environmental changes;
iii. verify predictions concerning the effects of natural phenomena or human activities on variables such as atmospheric processes, ice dynamics, biogeochemical cycling, and ecosystem productivity;
iv. detect possible unforeseen effects of human activities on selected variables;
v. evaluate the effectiveness of existing conservation measures, regulatory mechanisms, and procedures for operating and managing facilities;
vi. assess the consequences of natural and anthropogenic environmental change

on conservation activities, regulatory mechanisms and procedures for operating and managing facilities;

vii. establish whether the environment is in a healthy state.

Antarctica is a continent requiring special care in its management and conservation in order not to destroy the unique features that make it scientifically interesting. Human activities everywhere in the world are, almost by definition, polluting. Antarctica is the cleanest place left on Earth and almost certainly the best place to establish baselines against which to measure global pollution trends. It therefore follows that rigorous monitoring of localised human pollution in Antarctica is needed to ensure that the baseline signal can be adequately distinguished. This chapter attempts to outline the general principles for Antarctic monitoring with reference both to the legal framework created by the Antarctic Treaty and best practice developed elsewhere in the world.

2. Antarctic treaty requirements

The Protocol on Environmental Protection to the Antarctic Treaty (2) calls, under Article 3.2.d and 3.2.e, for regular and effective monitoring to allow assessment of the impacts of ongoing activities. This should include the verification of predicted impacts as well as facilitating early detection of any unforeseen effects of activities, carried on both within and outside the Antarctic Treaty area, on the Antarctic environment and dependent and associated ecosystems. This same issue was addressed in the Antarctic Treaty Consultative Meeting (ATCM) Recommendation XV-5 (3). The Recommendation specifies monitoring programmes relevant to activities such as: (i) waste disposal; (ii) contamination by oil or other hazardous or toxic substances; (iii) construction and operation of stations, field camps, and related ship, aircraft, and other logistic support facilities; (iv) conduct of science programmes; (v) recreational activities; (vi) those affecting the purpose of designated protected areas.

The Recommendation called for a Meeting of Experts to consider and provide advice on these matters to the ATCM. The terms of reference were:

- To Consider monitoring for the following purposes:
  to obtain a regular and verifiable record of activities and environmental data necessary to:
  - assess and quantify impacts of activities, including impacts predicted in the course of environmental impact assessments;
  - provide early warning of negative impacts;
  - identify preventative or remedial measures needed to reduce or eliminate adverse impacts;
  - plan similar activities in the future.
- Topics to be considered by a group of experts:
  - identification of the nature and possible significance of adverse impacts on the values of Antarctica as set forth in Article 3 of the Protocol on Environmental Protection to the Antarctic Treaty (2) which might require monitoring;

- identification of activities and environmental and other data required to detect and monitor possible impacts and to distinguish these impacts from natural variability;
- identification of methodologies and technologies available for monitoring (especially inexpensive and automated systems);
- identification of steps needed to create national and co-operative data systems which would provide for collection, quality control, archiving, evaluation, exchange and retrieval of environmental data;
- identification of existing relevant data sets, including baseline data repositories and programmes which generate these data.

It should be noted that these terms of reference give general guidance to the topics to be considered by the expert group. It would be important for any meeting of experts to refine the scope and increase the clarity of the issues involved to reach a common understanding of types of monitoring programmes to be undertaken and methods to be used.

The Treaty Group of Experts met in June 1992 in Buenos Aires and provided a report to ATCM XVII in November 1992 (4). Their Report contained nine recommendations. The first eight were concerned with the selection of representative facilities for monitoring, development of an international data management system and an Antarctic Data Directory, establishment of national scientific advisory boards, standards to minimise the impacts of fossil fuel combustion, formats for long-term monitoring programmes and baseline surveillance, and ensuring co-ordination of research and monitoring activities. The final recommendation proposed a meeting of technical experts to consider the design of monitoring programmes and protocols, standardisation and quality assurance, applicable technologies and data management procedures. The Scientific Committe on Antarctic Research (SCAR) and the Council of Managers of National Antarctic Programmes (COMNAP) agreed to organise this and report back to the ATCM.

3. Development of international protocols for Antarctic monitoring

Any environmental monitoring should be scientifically defensible, practicable and cost-effective. Monitoring encompasses a wide spectrum of activities, ranging from basic research monitoring to applied monitoring intended to respond to specific environmental decision making or management needs. There is also a distinction between monitoring of global changes and the observed changes in Antarctic environments on one hand and monitoring of local effects caused by the human activities and presence in Antarctica on the other. These two categories of monitoring are also recognised in ATCM Recommendation XV-5 (3). The depletion of the ozone layer is mentioned as an example of global environmental change whilst the potential local impacts on the Antarctic environment are indicated as those engendered by logistic activities.

The first type of monitoring, i.e., basic research monitoring, is a normal part of many ongoing scientific programmes in Antarctica and as such is given considerable attention and support. Although often complex and expensive these activities are well within the objectives of the national Antarctic programmes and established international scientific co-operation. Such basic scientific monitoring provides a source of globally important information. For practical purposes, however, it is often necessary to do monitoring of a more direct and applied nature. This second type of monitoring (i.e., applied environmental impact monitoring) does not have the same scientific tradition in Antarctica. Applied monitoring, driven by practical needs and not the advancement of scientific understanding, is a new field for most Antarctic operators. The most critical step in developing a successful and cost effective monitoring programme in this applied field is the clear definition of information needs and monitoring objectives. The ultimate aim of monitoring is to provide information and not data; so, the emphasis in designing such a project must be equally on both data collection and its analysis.

Since this activity falls outside the scientific career and funding system, it is important to recognise the full organisational and resource implications of any applied monitoring programmes. To implement such programmes effectively the focus of the programme should be well defined, the methods and techniques carefully developed, and the cost effectiveness safeguarded. The full costs of a major monitoring activity, which often will be of a long-term nature, may be beyond the reach of any single Antarctic operator. Hence, a major emphasis should be placed on international co-operation.

The Antarctic Treaty has proved an outstanding model of international co-operation. Article 6 of the Madrid Protocol on Environmental Protection (5) calls for co-operation of the parties in the specific area of the planning and conduct of environmental activities. The scope, magnitude, and expense of the environmental tasks have to be balanced against the support available from the Antarctic operators for important and globally-relevant scientific research and in accordance with the priority accorded to their research as underscored in the Madrid Protocol (Article 3.3). International co-operation in environmental monitoring is, therefore, imperative.

Countries that have undertaken environmental studies in the Antarctic can contribute data already gathered. Work previously done should be reviewed and shared by all countries so that past experiences with the difficulties of working in this continent can be used in planning future work. Such an approach should lead to the development of a set of agreed protocols for measurement of particular variables. The Antarctic Environmental Officers Network (AEON) has now focused on such a protocol manual with the assistance of SCAR and has provided an initial set of monitoring variables related specifically to station impacts.

Operators who have common needs in environmental monitoring and who plan to use the same protocols for techniques, experimental designs, analysis procedures, etc., could co-operate by using the directory of monitoring data sets to examine coherently the common requirements for environmental monitoring in and around their stations. Once a plan is agreed to, countries could further agree to share equipment and expert personnel in a cost-effective manner. Co-operation should also include assessment of the substantial investment in resources that has already been made. This is particularly true in the area of computers and networks where environmental data-base systems and data analysis systems need to be better developed.

International co-operation in the area of environmental monitoring should be aimed at reducing unnecessary duplication of effort by National Programmes and the improvement of planning to reduce initial impacts of all Antarctic operations. Indeed, monitoring geographic areas of concern after accidents (with the potential to damage the Antarctic environment) is fertile ground for international co-operation.

SCAR and COMNAP are both already contributing to the environmental monitoring effort using existing mechanisms. SCAR has actively been studying man's impact on the Antarctic environment for some time. There is a definite need to co-ordinate and make effective the logistic support for such programmes. In a period of diminishing resources operators need to be as cost effective as possible by sharing both data and expertise.

4. Current Antarctic environmental data resources

4.1. Data directory

At present there are no international Antarctic environmental data bases *per se*, although many organisations and individuals within the SCAR membership hold relevant data. In addition, environmental data bases are known to be in various stages of development in Argentina, Chile, Italy, New Zealand, United Kingdom and the United States, and possibly in other countries.

An initial survey of data holdings and activities by AEON yielded an extensive list which was presented at ATCM XXII in Christchurch (6). This will be made available on the COMNAP Web site (http//:www.comnap.aq) and it is intended to update it as further information comes to hand.

The complexity of environmental matrices and the problems due to the spatial-temporal evolution of pollutants and their involvement in biogeochemical cycles calls for the utmost accuracy in data collection, data analysis and environmental control. The first and fundamental requisite to be satisfied in order to give definitive answers to existing environmental problems is the capacity to produce absolutely reliable data, particularly where trace toxic chemical substances are concerned. It is imperative that measured concentrations correspond strictly to the truth. This reminder might appear superfluous, but unfortunately the technical-scientific difficulties involved in the analytical process are often underestimated, as the scientific literature has already amply demonstrated (see for instance refs. 7 through 13).

SCAR experience of pooling marine environmental data from many contributors in the Biological Investigation of Marine Antarctic Species and Stocks (BIOMASS) database (14) showed that one of the major problems was that of data validation to ensure comparability of data. Existing environmental data are unlikely to be comparable between data sets and the value of these data will need to be assessed before embarking upon the collection of new environmental data. It will probably be necessary to establish a new baseline from which to work. Many existing data may need to be excluded from comparisons although recalculation of some data may be possible.

4.2. Development of international protocols for monitoring

Monitoring could be required to be undertaken at a variety of scales, both spatial and temporal, and for a very wide range of variables. Indeed, a significant amount of routine environmental monitoring is already undertaken by several operators in the Antarctic, but it is clear that a lack of standard protocols makes it difficult to compare data from different research groups.

Monitoring is expensive and it is in the interests of all Antarctic countries to be able to maximize the value of data collected. In addition, it will only be possible to gain a good estimate of baselines and change at a regional level when data can be pooled. The establishment of an international framework for monitoring activities is likely to provide both scientific and managerial benefits.

SCAR has proposed that countries engaging in monitoring activities will need to consider how to implement the following:

- obtain international agreement on a data collection and analysis protocol with laboratories that are interested in the specific variables;
- organize and co-ordinate any inter-laboratory calibration scheme and provide international standards as required;
- agree to curate and archive any data obtained under such a protocol;
- make these data available to the scientific community and to the Committee on Environmental Protection (CEP);
- provide an initial summary of the data and a primary evaluation of its environmental consequences to the CEP.

The range of environmental variables that could be monitored is vast. Yet, with limited resources and objectives only limited monitoring is likely to be possible. Some systematic classification is therefore likely to prove useful in deciding on which variables might be most appropriate. For these purposes the variables will be treated as three major fields, i.e., chemicals, biological variables and physical variables.

Within each field it is advantageous to divide the range of possible variables into groups. Although it is difficult to provide any grouping that will meet with universal agreement, one practical way is to base the groups for chemical compounds on assay methods (Table 2.1) and for biological variables on processes (Table 2.2).

Table 2.1 is in no way meant to be either exhaustive or exemplary. It remains difficult at this stage to provide any prioritized list of compounds within each group. The choice of compounds for assay must be based on the scientific reasons for undertaking any particular monitoring exercise. Despite this it seems likely that a number of indicator compounds (for instance those resulting from inadequate incineration of some types of plastics) can be identified as of general concern. Whilst it is recognised that for some compounds either cost or technology

Table 2.1. Chemical groups for monitoring

Chemical groups
Organic compounds
Volatile organic compounds
C_1-C_8 hydrocarbons
Halocarbons
Aromatics (benzene, toluene, xylenes)
Photochemical reaction products
Semi-volatile and particle-related compounds
Polycylcic hydrocarbons
Plasticizers such as phthalates
Silicones
Organochlorines
Chlorinated aromatics such as chlorobenzenes, PCBs, dioxins
Chlorinated pesticides such as DDT, hexachlorocyclohexanes, toxaphenes, chlordanes
Inorganic compounds
Gaseous compounds
Ozone
Nitrogen oxides
Sulphur compounds
Carbon dioxide and monoxide
Ammonia and related compounds
Cations and anions
Alkaline/alkaline-earth cations
Anions (nitrate, nitrite, sulphate, etc.)
Trace elements
Heavy metals
Platinum group elements
Other trace elements
Isotopes
Radio-isotopes

may at present limit the choice, developments in assay techniques may change this at any time. In addition it must be recognized that the continued development of new chemicals elsewhere in the world will eventually result in new pollutants in Antarctica.

Selection of the environmental matrices to be analysed should take into consideration both the general objective of the monitoring and also the opportunity to study the distribution within and the transport between the most important environmental components with respect to the observed chemical substances. Thus a suggested list of important matrices is as follows: sea water (with particular reference to the surface microlayer), marine particulate and sediments, marine ice, marine organisms (especially krill), aerosols, superficial snow and soil and

Table 2.2. Biological indices for monitoring

Diversity	Community structure Species richness
Distribution	Geographical extent Temporal distribution
Development	Population structure Reproductive success
Physiology	Photosynthesis Respiration
Population	Numbers Genetic diversity

lacustrine matrices when present in the area. Of course, even here, a prioritized list of matrices, together with frequency of sampling, must be established in order to optimise the analytical effort and not to overload the participating laboratories unnecessarily. For instance, co-ordinated studies on sea water matrix, marine surface microlayer, marine aerosol and surface snow appear advisable in order to obtain relevant information on matter transport processes and pollutant exchange between the atmosphere and land, particularly in snow deposition (see, e.g., 15–24).

Chemical analytical techniques are always changing as technology produces better equipment. The later papers in this volume amply demonstrate that point. Not infrequently a new technique produces data that are significantly different from previous data. To allow for a change from one technique to another without losing the value of the older data it is suggested that there should be a period when both techniques overlap and that specimen banks should be maintained from which individual historical specimens could be re-analysed by the new technique. This is an issue that the Italian National Antarctic Programme has raised for international discussion and progressed it internationally by developing an Antarctic Environmental Specimen Bank (AESB) (25) and two standard reference materials of marine sediment and krill (26–29) to provide both for quality assurance of data and the basis of a historical archive for future analyses.

The diversity of terrestrial, freshwater and marine organisms in the Antarctic and Southern Ocean is quite considerable and existing knowledge of the biology of most of them is limited. The SCAR/COMNAP monitoring workshops (30) spent considerable efforts trying to identify indicator organisms that would easily provide an accurate indication of the health of an ecosystem, although one might reasonably expect the top trophic level to display the effects of pollution lower down in the food chain. No consensus was possible although there were some indications of where in the food webs one might look for such species. The question remains to be answered: how would the effect be manifested in different species and how could one compare different ecosystems? Numerical data are essential in order that statistical analyses can be used to determine the significance of any changes. Thus all the variables suggested in Table 2.2 are capable of numerical expression.

Table 2.3. Physical variables for monitoring

Wind	Dust Snow drift and accumulation Smoke, fume or aerosol dispersion
Water	Drainage patterns Clarity of freshwaters Snow melt or sublimation Snow albedo
Earth	Soil compaction Permafrost depth and extent Erosion Destruction of geomorphic features
Anthropogenic	Noise Electromagnetic radiation

The selection of physical parameters for monitoring is dependent on the expected form of impacts. Table 2.3 provides one such form of classification of possible major impacts.

The variables selected for the AEON handbook for monitoring the impacts of scientific stations are given in Table 2.4 which also suggests where they are to be measured (31).

5. Design of monitoring programmes

5.1. Sequence of steps in programme design

It is proposed that a sequence of steps be carried out in agreeing the structure of a monitoring programme. A summary of the steps to be considered is shown in Table 2.5.

5.2. Testable hypotheses

All monitoring programmes need to be based on testable scientific hypotheses. Without this rigorous framework it is neither clear what purpose monitoring could have nor how the data could be evaluated. Unless the right questions are asked, it will be impossible to get the right answers. Define the hypothesis in the negative so that it can be refuted. For example: "movement of heavy metals in dry soil is less than one metre per year", or "output of macerated sewage from a station of 40 people into well mixed coastal water causes no change in biodiversity from nutrient enrichment".

5.3. Selection of variables and data collection methods

Monitoring is not the measurement of many variables at many sites in the hope that subsequent analysis will reveal a cause and effect relationship. It is the

Table 2.4. Selected indicators of environmental impacts from human activities at Antarctic Stations. Included in the COMNAP Handbook on Environmental Monitoring (see http.www.comnap.aq/comnap/comnap.nsf/P/PDF/5/)

	Indicator	*Measured in*					
		Waste water (sewage/ grey water)	Fresh or sea water	Soil	Marine sediments	Snow	Other
1.	Suspended solids (total and volatile)	×	×				
2.	BOD	×	×				
3.	COD	×	×				
4.	DO	×	×				
5.	pH	×	×				
6.	Conductivity	×	×				
7.	Nutrients (nitrate, phosphate)	×	×				
8.	Temperature	×	×				
9.	Coliform bacteria	×	×				
10.	Grain size			×	×		
11.	TOC			×	×		
12.	TIC			×	×		
13.	Trace metals (Cd, Cu, Hg, Pb, Zn)			×	×	×	
14.	TPH			×	×	×	
15.	PAH			×	×		
16.	Particulates					×	
17.	Phytoplankton		×				
18.	Waste water Production/emission (quantity/time)	×					
19.	Fuel consumption (quantity/type/time)						×
20.	Waste incineration (quantity/type/time)						×
21.	Hydrocarbons spills (record of spills, amount/type/ location and monitoring of spill area)						×
22.	Station area (monitoring of coverage/use, e.g., photomonitoring)						×

Abbreviations
BOD: Biological Oxygen Demand
COD: Chemical Oxygen Demand
DO: Dissolved Oxygen
PAH: Polyaromatic Hydrocarbons
TIC: Total Inorganic Carbon
TOC: Total Organic Carbon
TPH: Total Petroleum Hydrocarbons

Table 2.5. Flow diagram for designing an environmental monitoring programme for local, regional or global application

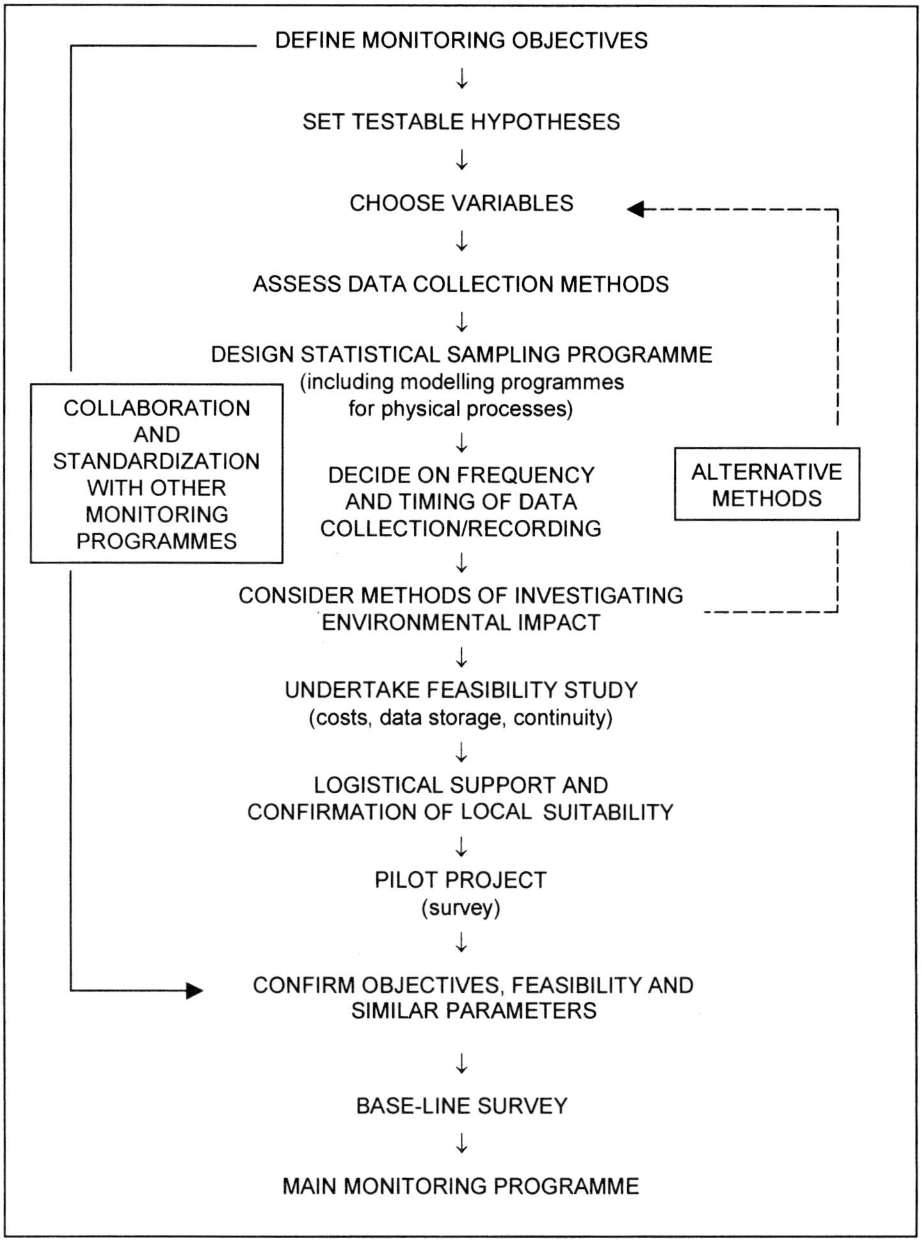

planned measurement of key variables at specific sites and their evaluation in a predetermined fashion to test a particular hypothesis. It is therefore essential to give as much consideration to the selection of the key variables to be measured as it is to the framing of the original hypothesis.

Selecting variables is difficult and yet is crucial to the success of the monitoring exercise. One way is to define "valued ecosystem components" and construct a matrix of interactions between causes and effects. These interactions are then scored for magnitude and relative importance. Those which show the greatest effects are then considered as monitoring variables. This system is used effectively at present for environmental impact assessment in North America and Europe. At present it can be criticised on the grounds of subjective assessment of importance and magnitude, but there are no methods in which environmental quality determinations such as these can be made more independent.

In general the following properties are a good guide when selecting variables for a monitoring programme: show changes considerably in excess of the limits of detection, be directly relatable to a testable hypothesis, be measurable on samples that can be transported without deterioration, be able to sustain the monitoring activity without permanent damage, give information of use to management. It is also advantageous if sampling can be undertaken without dedicated and highly trained personnel, and analyses performed without advanced laboratories and skilled technicians.

5.4. Design of sampling

The sampling programme must be designed with due regard to statistical methods and must recognise the extreme seasonality of the Antarctic environment. It must aim to distinguish between natural background levels (not necessarily measured baseline values) and levels anthropogenically induced by the activity under investigation. The design of the sampling programme must ensure that the number of samples collected does not exceed the available analytical capacity nor exacerbate damage to the environment. The programme should aim for economy of effort by using observational, collection and analytical techniques which can measure more than one useful parameter.

As far as chemical variables are concerned, especially in case of trace determinations, all the steps of sample collection, treatment in the field and storage must be carefully considered in order to maintain sample integrity before analysis. In this respect the selection of appropriate equipment and non-contaminant materials, as well as the application of efficient cleaning procedures, are of paramount importance. For the most critical applications clean chemistry laboratories (equipped with Class 100 laminar flow cabins) must be available in the field or on board.

5.5. Measurement techniques

In general all measurement techniques should be as simple as possible, so that they may be widely applied, and cost-effective in collecting data. Where there is already an agreed international protocol for a particular variable this should be used. Wherever possible non-destructive techniques should be used, especially for biological sampling. If unavoidable destructive techniques are required the over-riding consideration should be the intrinsic value of the sample being measured, i.e., its rarity, the effect of its removal from the ecosystem, etc. Continuous sampling is only necessary when a fine resolution time series is essential for interpretation.

For physical and chemical measurements it is essential that measurements are referenced to standards accepted by all the laboratories undertaking a particular type of measurement. In particular, the use of chemical analytical procedures validated through their application to certified reference materials (CRM) (26) is highly recommended. It is to be noted that a number of CRMs prepared with Antarctic matrices are already available, i.e., marine sediment and krill, or in preparation (26–29). These and other CRMs should be also used routinely by the participating laboratories to assure a periodical assessment of accuracy and repeatability of measurements. These laboratories should also undertake regular intercalibration studies. It is possible that in the first application of these exercises systematic errors will be found, but better results are expected in subsequent rounds together with a general improvement in the performance of laboratories and data quality.

In many countries undertaking the most technically demanding analyses, it is expected that the laboratories will be registered under quality assurance schemes. Accuracy and repeatability should be covered by quality assurance requirements, but resolution and detection limits are critically dependent on technique and instrumentation. Details on the general subjects of quality assurance/quality control, quantification of uncertainty in analytical measurements, accreditation of laboratories, and on the general concepts and strategy for ensuring that analytical chemical measurements are comparable – in one word "traceability" – can be found elsewhere in the chemical literature (26,32–36).

With regard to the problems of blanks and detection limits, it is suggested that both blank and measurement values should be routinely reported and that acceptable data would normally be considered as those more than ten times the value of the blank or the detection limit.

For some types of monitoring, indicator species may prove valuable. When searching for suitable species (especially to assess regional or global trends) it is suggested that the marine ecosystem with its high species diversity may offer greater opportunities than the terrestrial ecosystem, e.g., CCAMLR Ecosystem Monitoring Programme. Not all biological monitoring may be best done at the whole organism level. The use of cellular and organ systems from Antarctic species may provide very useful experimental approaches to toxicity assessments.

5.6. Management of data

This includes the archiving and curation of data and, as importantly, the validation of data based on quality assurance guidelines to ensure comparability between data sets collected by different countries. Publication of data should be encouraged and data should be freely exchanged. At present there is no established route by which this might occur since, unlike the CCAMLR Secretariat, the CEP has no support at all. However, with the establishment of Web sites for both the Treaty and the CEP, new possibilities are opening up for posting data in a Web archive. An initiative from the CEP seems to be called for here.

5.7. Review of the programme

Each monitoring programme should have a review schedule incorporated into the programme design. The first review should be undertaken prior to the start of the programme and consider whether the review schedule proposed is appropriate to the type of monitoring to be undertaken. SCAR could be invited to provide an independent report to the CEP on the monitoring programme, if required. It is presumed that the CEP will evaluate all such reports and make recommendation to ATCMs.

6. Application of monitoring data to environmental management

6.1. Environmental Impact Assessment

The Protocol on Environmental Protection to the Antarctic Treaty requires operators to conduct environmental impact assessment of all field activities before authorising them. The three levels of impact are currently defined as: less than minor or transitory, minor or transitory, more than minor or transitory. The first category requires no action, the second requires an Initial Environmental Evaluation (IEE) and the third category a Comprehensive Environmental Evaluation (CEE). Despite several years of discussion at Treaty meetings, in SCAR and COMNAP and in workshops it has not yet proved possible to agree how to define minor or transitory, thus the usage of the terms differs significantly between parties. What is clear from a scientific view point (if not from a legal one) is that for biological systems the significance of the impacts is determined by timing, place, species and community considerations which are case specific. Thus, one might conclude that the quest for an absolute legal definition or a prescriptive list is a waste of time and a better approach will be to build a body of accepted practice in applying the terms. This possibility was raised at the Lima ATCM in 1999 (37), but has yet to receive unanimous approval.

Both an IEE and a CEE require detailed predictions about the likely impacts on the local and general environment and these will usually be linked to a list of monitoring procedures to be undertaken to validate the accuracy of the predictions. This form of impact assessment monitoring is mandatory under the Treaty legislation, but is limited to a particular period, normally specified in advance. It relies on the scientific understanding of environmental processes gathered by more general scientific monitoring and, if well focused, should contribute to improved modelling and predictive capabilities in the future.

6.2. Establishing baselines

There is a strong scientific interest in the value of establishing global baseline data for overall levels of various classes of pollutants by analysis of Antarctic material. In addition, for Antarctic operators needing to provide monitoring for Environmental Impact Assessment, there is the need to determine the baseline inputs

Table 2.6. Selected compounds representative of typical emission sources and for which baselines should be established

Compounds	Emission sources
Benzo (a) pyrene	Combustion sources such as power generators, vehicle engines, waste incineration
Benzene (alkylated benzenes)	Fuel spills, combustion sources
Dioctylphthalate	Building materials (PVC), packaging materials
Freon F 12	Foam insulation, refrigerators, degreasing of components
Pentachlorophenol/hexachlorobenzene	Vehicle engines, waste combustion, indicator that other compound classes are formed (PCBs, dioxins), treated timber

against which change will be assessed. Clearly there is a very wide range of variables which could be measured and to be practical the field must be narrowed (see Table 2.6).

In order to relate impacts and effects over time a suggested way forward would be to treat baseline variables as several classes within each of which indicator variables would be chosen. For example, let us take the case of establishing a field station with its own generator for summer use by 10 people in a previously pristine area. In each major class detailed baseline (and subsequent) measurements would be made of some feature of a key indicator variable.

An example could be as follows:

flora	detailed maps of the spatial extent of plant communities
fauna	local abundances of a dominant bird species
microbiology	presence of human commensals
organic pollution	polycyclic compounds typical of diesel exhaust plasticizers typical of building materials
inorganic pollution	heavy metals in soil, snow or water nitrate and sulphate in local water
dust	dust content in local snow

In choosing the indicator variables and the exact form of the monitoring data it is essential for chemical and physical measurements to utilise existing knowledge of similar areas elsewhere in the Antarctic and of biology of the same species elsewhere. This may well make it clear that one particular approach to monitoring will be more successful than others.

It is important to recognise that there is a fundamental difference between determining chemical/physical baselines and biological ones. In general, there exists within the ice, soil or sediments a historical record of chemical changes from which it may be possible to assess the natural temporal variability. Rarely is this possible for biology. It is already well established that annual variation in population size, breeding success and growth rates can be very considerable for some, if not all, species. The spot measurements collected for baseline determination may be a poor basis on which to assess change, and a hypothetical baseline may need to be established taking the subsequent monitoring into account.

6.3. Ecotoxicology

Baseline data on the concentration and distribution of a variety of manmade compounds are slowly being acquired for the Antarctic. This activity will increase and will expand to include determination of sources and transport routes. However, for the conservation and management of Antarctic ecosystems this alone is of limited value. Without any knowledge of the biological effects of these compounds it will not be possible to assess critical doses, nor instigate management changes to mitigate exposure. Data on the source, accumulation, uptake and persistence of a compound must be matched by experimental data on its effects on a biological system (does it reduce breeding success, stunt growth, limit winter survival, etc.?).

The Antarctic food chains are unique, both on land and in the sea. It is not therefore possible to directly transfer temperate data and models of toxicity to Antarctica since several of the trophic levels are missing, many organisms live for much longer than in temperate regions and grow much more slowly. In addition it appears that for many compounds little is known of their behaviour (such as chemical speciation, distribution and life time in the environment) at very low temperatures. Research is essential to fill both of these gaps.

A desirable parallel track would be the compilation of a directory of compounds already identified from the Antarctic with details of their known toxic effects. Equally desirable, but, with existing biological knowledge, much more difficult to achieve, is the identification of indicator organisms (for terrestrial, freshwater and marine systems) whose "health" can be taken to represent that of the entire community or ecosystem.

6.4. Research needs

A major objective in relation to understanding environmental impact in the Antarctic is to establish the level of impact (intensity and area) on the environment of scientific stations and their associated activities, field camps, aircraft runways and similar operations. It would be useful to know the actual "footprint" of such facilities of different sizes. To this end, one notes the initiation of research programmes to establish the imprint of facilities in different environments on their surroundings. The objective is to determine the intensity of impact at increasing distances from the centre of the activity, with respect to certain factors (see Table 2.7, adapted from Ref. 38). These impacting factors can then be correlated with their adverse effects on the biota at increasing distances from the source of the impacts, with a view to mapping (with "contours") the intensity of the impacts, until the impact becomes undetectable or negligible. In this way the expected

Table 2.7. Examples of activities and impacts. Based on Abbott and Benninghoff, 1990 (38)

Examples of logistical and scientific activities that should be subject to monitoring efforts	Antarctic features of scientific value which are of special concern	Areas of special importance
Construction Stations Airstrips Harbours Roads	Particularly vulnerable species and communities (e.g., grass, moss and lichen communities and associated invertebrates)	Existing scientific stations and field camps Designated Sites of Special Scientific Interest (SSSIs)
Routine Operations Stations Field camps Ships Helicopters Fixed-wing aircraft Trucks All-terrain vehicles Power generators Heating Water desalinisation Waste disposal	Pollution-free air, water and ice Marine living resources Phytoplankton Antarctic krill and other zooplankton Finfish Squid Penguins and flying birds Marine mammals Coastal benthos Snow algae	Designated Specially Protected Areas (SPAs) Designated historic sites and monuments Bird and seal breeding sites Coastal ice-free areas Continental shelf (fish/krill breeding and fishing areas)
Scientific activities Collection of birds, seals and other biota Experimental harvesting/perturbation Offshore drilling Ice coring Seismic surveys Collection of meteorites/rock samples, fossils, etc. Use of chemicals, acids, radioactive isotopes, etc.	Meteorites Ventifacts Fossils Fresh water sources (snow/ice in the vicinity of stations) Ice-free coasts with sheltered anchorages	Victoria Land dry valleys Ice-free mountain areas Sheltered bays (possible anchorages/station sites) Fresh water lakes, ponds and subsurface waters Glaciers Ice shelves
Commercial activities Fishing Tourism	Mineral occurrences Mountains and other areas with particular recreational and aesthetic value	Sites visited by tourists
Accidents Ship/aircraft/vehicles Fuel lacks Fires Introduction of alien species	Ecosystem integrity	

footprint of such varied activities could be established, thus facilitating the design of monitoring programmes (e.g., number and location of sampling stations).

At its meeting in Switzerland in 1998 the SCAR Group of Specialists on Environmental Affairs and Conservation considered current research needed to support future monitoring tasks (31). At the level of the organism they concluded that emphasis should be on improving the understanding of enzyme activities for biochemical monitoring, respiration and heart rate for physiological monitoring and a range of movements, posture changes and responses for behavioural monitoring. For birds and seals it would be unwise to generalise population responses from the limited data available; a wider range of species need to be considered. In the pollutants field cheap screening methods are needed, and new sensors are urgently required to allow near continuous measurements.

6.5. Opportunistic monitoring

Despite the best attempts to prevent them, accidents continue to happen. On some occasions a disaster can be utilised to provide an unplanned opportunity for monitoring impacts. The most recent major environmental disaster was the sinking of the M/V *Bahia Paraiso* close to Palmer Station. The monitoring and damage assessment undertaken by the USA and Argentina have provided very detailed data on the effects of hydrocarbons on sea birds, the persistence of oil residues in sediments, the rate of recovery of inter-tidal communities, etc. All such opportunities should be exploited to the full.

In addition there are at present a considerable number of historical and abandoned work sites in the Antarctic. They offer an exceptional opportunity to investigate the persistence of compounds over long periods (up to 80 years), the natural rates of recovery after disturbance, etc. Such information is difficult to obtain and the opportunities offered by these datable work sites should be exploited before operators remove them as part of their waste management operations.

7. Monitoring and management

Scientific monitoring can fulfil a variety of objectives, but the value of the data collected is determined by their use. Basic research requires their interpretation to increase our understanding of environmental and ecosystem processes. Applied research should evaluate them in terms of management decision making. Poorly organised monitoring, with inadequate objectives and protocols, will be a constant drain on resources with no realisable value at the end of the exercise. Monitoring is by its very nature long-term and requires a funding commitment for longer time periods than the typical research project.

Valuable initiatives are possible at present which will enhance our opportunities of managing the Antarctic in a sensitive and responsible way into the next century. The establishment of a framework for monitoring pollution and its impacts, an increase in the agreed list of banned compounds, the compilation of a data directory of environmental data, the creation of a publication series for monitoring data, the development of ecotoxicological research, are all new proposals of considerable significance.

If basic research incorporates monitored changes in a better understanding of processes, then applied research, resulting from monitoring significant or deleterious changes induced by human activities, provides the information from which management decisions must flow. It is in these ways that monitoring underpins research, environmental management and conservation and it is on these criteria that it will be judged to be of value or not. All this information will provide invaluable input into any state of the Antarctic environment report that is prepared in the future.

8. Conclusions

The need for scientific monitoring in the Antarctic has been clearly identified and its value to both basic and applied research is recognised. A data directory of existing Antarctic environmental data has already been established and is being added to (see latest web version on http.www.comnap.org).

To maximise the value of Antarctic monitoring data, agreement among national operators on adoption of internationally recognised protocols for data and sample collection and analysis is required. In this respect the publication of the first volume of an environmental monitoring handbook is welcome. Measurement techniques must be internationally agreed, with any chemical analyses firmly related to interlaboratory standards and verifiable detection limits.

Monitoring programmes should be designed around testable scientific hypotheses, utilising existing knowledge and models of system processes, wherever possible. They should be designed to help managers discern when environmental impact or change is due to natural environmental variability and when such impact or change is due to human activity in Antarctica. The selection of variables to be monitored should be based on assessment of the importance and magnitude of the expected impact on and change in environmental quality attributes.

All monitoring programmes should be subject to regular independent scientific review and have termination dates. Countries would be expected to make data publicly available and provide a primary evaluation of the environmental significance of any changes or trends. The monitoring studies will initially define baselines against which the effects of continuing activities, such as scientific stations, must be assessed. In this respect it will be useful to establish the pollution "fooprint" of existing stations. In addition, they will contribute to improved environmental decision making by management, and to a basic understanding of the Antarctic environment. Organised and reliable monitoring data will be required for any future state of the Antarctic environment report.

The monitoring requirements for the environmental impact assessment process can be organised under, and contribute data to, the system outlined here.

Monitoring programmes are expensive, require long-term commitment to provide adequate data sets and should be funded with funds distinct from those allocated for basic scientific research.

Acknowledgements

We are grateful to many scientists in SCAR for discussion of these ideas and especially to those who attended the two workshops on environmental monitoring.

References

1. R. E. Munn, Global Environmental Monitoring Systems (GEMS): Action Plan for Phase 1. Toronto, Scientific Committee of Problems of the Environment, 1973.
2. ATCM, Protocol on Environmental Protection to the Antarctic Treaty. XI Antarctic Treaty Special Consultative Meeting, Madrid, Spain, 1991.
3. ATCM, Human Impact on the Antarctic Environment: Environmental Monitoring in Antarctica. Recommendation 5, XV ATCM, Paris, France, 1989.
4. ATCM, Report of the First Meeting of Experts on Environmental Monitoring in Antarctica. Information paper 9, XVII ATCM, Venice, Italy, 1992.
5. ATCM, Protocol on Environmental Protection to the Antarctic Treaty. SCAR Bulletin no. 110, pp.1–20, 1993.
6. ATCM, Summmary of Environmental Monitoring Activities in Antarctica. Information paper 54, XXII ATCM, Tromsø, Norway, 1998.
7. C. C. Patterson, D. Settle, The Reduction of Order of Magnitude Errors in Lead Analyses of Biological Materials and Natural Waters by Evaluating and Controlling the Extent and Sources of Industrial Lead Contamination Introduced During Sample Collecting, Handling, and Analysis. In *Accuracy in Trace Analysis: Sampling, Sample Handing and Analysis*, P. La Fleur (Ed.), NBS Spec. Publ. (U.S.) 422 (1976), 321–351.
8. M. Zief, *Contamination Control in Trace Element Analysis*, Wiley, New York, 1976.
9. B. K. Schaule, C. C. Patterson, Lead Concentrations in the Northeast Pacific: Evidence for Global Anthropogenic Perturbations. *Earth Planet. Sci. Lett.*, 54 (1981), 97–116.
10. K. M. Bruland, Trace Elements in Sea-water. In *Chemical Oceanography*, J. P. Riley, R. Chester, (Eds.), Vol. 8, Chapter 45, Academic Press, London, 1983, pp. 157–220.
11. C. F. Boutron, A Clean Laboratory for Ultralow Concentration Heavy Metal Analysis. *Fresenius J. Anal. Chem.*, 337 (1990), 482–491.
12. G. Capodaglio, G. Toscano, P. Cescon, G. Scarponi, H. Muntau, Collaborative Sampling Error Assessment of Trace Metal Determination in sea water. *Ann. Chim.*, 84 (1994), 329–345.
13. G. Capodaglio, C. Barbante, C. Turetta, G. Scarponi, P. Cescon, Analytical Quality Control: Sampling Procedures to Detect Trace Metals in Environmental Matrices, *Mikrochim. Acta*, 123 (1996), 129–136.
14. M. R. Thorley, P. N. Trathan, The History of the BIOMASS Data Centre and Lessons Learned During its Lifetime. In *Southern Ocean Ecology – The BIOMASS Perspective*. S. Z. El-Sayed (Ed.), Cambridge University Press, Cambridge, 1993, pp. 313–322.
15. R. Cini, G. Loglio, Adsorption and Pollutants Transport by Marine Aerosol. *Mar. Pollut. Bull.*, 34 (1997), 501–504.
16. R. Cini N. Degli Innocenti, G. Loglio, G. Orlandi, A. M. Stortini, U. Tesei, Spectrofluorimetric Evidence of the Transport of Marine Organic Matter in Antarctic Snow via Air-Sea Interation. *Int. J. Environ. Anal. Chem.*, 55 (1994), 285–295.
17. R. Cini, N. Degli Innocenti, G. Loglio, C. Oppo, G. Orlandi, A. M. Stortini, U. Tesei, R. Udisti, Air-Sea Exchange: Sea Salt and Organic Microcomponents in Antarctic Snow. *Int. J. Environ. Anal. Chem.*, 63 (1995), 15–27.
18. R. Cini, R. Udisti, G. Piccardi, G. Loglio, N. Degli Innocenti, A. M. Stortini, B. Pampaloni, A Simple Model for K and Ca Enrichment Interpretation in Antarctic Snow. *Int. J. Environ. Anal. Chem.*, 71 (1998), 265–287.
19. F. Casella, R. Udisti, G. Piccardi, The Oceanic Source Contribution to the Snow Composition, as Function of Elevation, at Two Coastal Stations in the Terra Nova Bay Area (Antarctica). *Terra Antartica Reports*, 1 (1997), 77–80.

20. G. Scarponi, G. Capodaglio, C. Turetta, C. Barbant, G. Toscano, P. Cescon, Evolution of Cd and Pb Content in Antarctic Coastal sea water During the Austral Summer. *Int. J. Environ. Anal. Chem.*, 66 (1997), 23–49.
21. G. Capodaglio, C. Turetta, G. Toscano, A. Gambaro, G. Scarponi, P. Cescon. Cadmium, Lead and Copper Complexation in Antarctic Coastal sea water. Evolution During the Austral summer. *Int. J. Environ. Anal. Chem.*, 71 (1998), 195–226.
22. G. Scarponi, C. Barbante, C. Turetta, A. Gambaro, P. Cescon, Chemical Contamination of Antarctic Snow: the Case of Lead. *Microchemical J.*, 55 (1997), 24–32.
23. C. Barbante, C. Turetta, G. Capodaglio, G. Scarponi, Recent Decrease in the Lead Concentration of the Antarctic Snow. *Int. J. Environ. Anal. Chem.*, 68 (1997), 457–477.
24. C. Barbante, C. Turetta, A. Gambaro, G. Capodaglio, G. Scarponi, Sources and Origins of Aerosols Reaching Antarctica as Revealed by Lead Concentration Profiles in Shallow Snow. *Ann. Glaciol.*, 27 (1998), 674–678.
25. F. Soggia, C. Ianni, E. Magi, R. Frache, Antarctic Environmental Specimen Bank. In *Environmental Contamination in Antarctica: A Challenge to Analytical Chemistry*, S. Caroli, P. Cescon, D. W. H. Walton (Eds.), Elsevier, Amsterdam (this Book, Chapter 12, p. 305).
26. S. Caroli, Environmental Chemistry in Antarctica: the Quest for Accuracy. In *Environmental Contamination in Antarctica: A Challenge to Analytical Chemistry*, S. Caroli, P. Cescon, D. W. H. Walton (Eds.), Elsevier, Amsterdam (this Book, Chapter 11).
27. S. Caroli, O. Senofonte, S. Caimi, P. Kárpáti, Comparative Study of Marine Sediment from Antarctica by Low-pressure Discharge Atomic Emission Spectrometry and Inductively Coupled Plasma-based Spectrometry. *J. Anal. Atom. Spectrom.*, 11 (1996), 773–777.
28. S. Caroli, O. Senofonte, S. Caimi, P. Robouch, J. Pauwels, G. N. Kramer, Certified Reference Materials for Research in Antarctica: The Case of Marine Sediment. *Microchem. J.*, 59 (1998), 136–143.
29. S. Caroli, O. Senofonte, S. Caimi, P. Pucci, J. Pauwels, G. N. Kramer, A Pilot Study for the Preparation of a New Reference Material Based on Antarctic Krill. *Fresenius J. Anal. Chem.*, 360 (1998), 410–414.
30. M. C. Kennicutt, II, J. A. Sayers, D. W. H. Walton, G. Wratt, Monitoring of Environmental Impacts from Science and Operations in Antarctica. A Report for SCAR and COMNAP, 1996.
31. SCAR, Tenth Report of the Group of Specialists on Environmental Affairs and Conservation. SCAR Report no. 17, 1999 (full text of AEON Handbook available on http.www.comnap.aq).
32. M. Valcarcel, A. Rios, Traceability in Analytical Chemistry. *Analyst*, 120 (1995), 2291–2297.
33. B. King, Traceability of Chemical Analysis. *Analyst*, 122 (1997), 197–204.
34. B. King, Traceability of Trace Analysis. *Ann. Chim.*, 87 (1997), 199–210.
35. W. Wegscheider, H. J. Zeiler, R. Heindl, J. Mosser, Quantifying Uncertainty in Sampling and Analytical Measurements. *Ann. Chim. (Rome)*, 87 (1997), 273–283.
36. B. Nijenhuis, The Challenge of Quality Assurance and Accreditation. *Ann. Chim.*, 87 (1997), 233–240.
37. ATCM, Final Report of XXIII Antarctic Treaty Consultative Meeting. Resolution 1: Guidelines for Environmental Impact Assessment in Antarctica. Lima, Peru, 1999.
38. S. B. Abbott, W. S. Benninghoff, Orientation of Environmental Change Studies to the Conservation of Antarctic Ecosystems. In *Antarctic Ecosystems: Ecological Change and Conservation.* K. R. Kerry, G. Hempel (Eds.), Springer, Berlin, 1990, pp. 394–403.

Environmental Contamination in Antarctica
A Challenge to Analytical Chemistry
S. Caroli, P. Cescon and D.W.H. Walton, editors

Chapter 3

Trace element determination in polar snow and ice. An overview of the analytical process and application in environmental and paleoclimatic studies

Carlo Barbante, Clara Turetta, Gabriele Capodaglio, Paolo Cescon, Sungmin Hong, Jean-Pierre Candelone, Katja Van de Velde, Claude F. Boutron

1. Introduction

Over the past few decades, there has been a continued international effort to decipher the unique atmospheric archives stored in the snow and ice layers which have been accumulating in the Greenland and Antarctic ice caps over time (1). Amongst the most interesting species to be investigated in these archives are heavy metals. Many of them are indeed toxic and have been dispersed into the atmosphere of our planet by human activities since antiquity (2–4). Studying these frozen archives for heavy metals has the potential to enable the past and recent history of the pollution of the atmosphere by these metals to be documented on a planetary scale from ancient times to the present. In addition, it should provide a better understanding of the relative importance of their different natural sources together with an insight into past changes in the role of these natural sources during the most recent climatic cycles (5–7).

Deciphering these archives has unfortunately proved to be a formidable challenge. This is especially due to the extremely low levels of contaminants in polar snow and ice, which are, by far, the purest natural water found on the Earth's surface. Most of the elements are present in polar snow and ice at extremely low concentrations, sometimes below the pg/g level. Although several elements have been analysed in Greenland and Antarctic snow and ice, still there is much to do to completely understand the biogeochemical cycles of such substances during past climatic eras and their behaviour during most recent times in which human beings have exerted a strong influence on their global presence in the environment. As an example of this, Pb concentration in Antarctic ice dating from several millennia ago is typically about 0.5 pg/g, i.e., less than 10^{-12} µg (6). It means that 1000 metric tonnes of ice contains only 500 µg of Pb. In Table 3.1 a comparison of concentrations found in Greenland and Antarctic snow and ice is reported to better illustrate this concept.

The crucial problem in the challenging task of trace element determination is

Table 3.1. Typical trace element concentrations in Greenland and Antarctic ice and snow

Element	Measured concentration (pg/g)			
	Greenland ancient ice	Greenland recent snow	Antarctica ancient ice	Antarctica recent snow
Ag		0.08–5.0 [a]		
Ba			1–30 [b]	
Bi	0.028 [c]	0.03–0.9 [c]	0.001 [b]	0.001–0.017 [b]
Cd	0.3 [d]	0.3–3 [d]	0.1–3 [e]	0.1 [f]
Co		0.65–17.2 [a]		
Cu	0.8–6.4 [g]	0.78–13.6 [a]	1.0 [h]	3.5 [f]
Hg		<0.05–2 [i]	0.02–2.1 [j]	0.13–0.50 [k]
Mo		0.08–7.0 [a]		
Pb	0.5 [l]	10–150 [m]	0.3–35 [n]	2–12 [o]
Pd	0.01 [p]	0.38 [p]		0.57 [p]
Pt	0.01 [p]	0.10 [p]		0.67 [p]
Rh	0.0007 [p]	0.04 [p]		0.04 [p]
Sb		0.21–4.3 [a]		
U		0.21–15.4 [a]		0.04 [a]
Zn	25 [m]	30–150 [m]	5–100 [q]	1.5 [f]

a) From ref. (62). b) From ref. (53). c) From ref. (43).
d) From ref. (35). e) From ref. (40). f) From ref. (72).
g) From ref. (3). h) From ref. (44). i) From ref. (99).
j) From ref. (5). k) From ref. (58). l) From ref. (2).
m) From ref. (34). n) From ref. (6, 64). o) From ref. (27–29).
p) From ref. (63). q) From ref. (24).

the control of contamination; it is mandatory to make sure that no artifact contamination affects the samples at each individual step of the analytical procedure from field sampling to laboratory analysis. If this condition is not fulfilled, the data might be erroneous by several orders of magnitude and then lead to junk science (8). It especially implies that the work is conducted by highly experienced researchers who have fully understood that ultra-trace analysis is by no means a routine business, and that it is much better for science to obtain a small number of reliable data than a large number of misleading "garbage" data. These researchers should, of course, not work inside a conventional laboratory, since such a contaminated environment will inevitably destroy the integrity of the samples, but they will have to build special clean laboratories flushed with filtered air (9, 10). Also, they should follow very strict criteria for choosing labware which will be directly in contact with the samples, and thoroughly cleaning it (8, 10). Last but not least, it will be of the utmost importance that they are able to evaluate the blanks unequivocally for the overall analytical procedure; it means that they must be able to determine quantitatively the exact amount of each investigated chemical element which is added to the sample at each individual step of the procedure. Such blank determinations are very difficult and will often take a significant portion of the working time of the investigators.

The push forward in this field of research came from the landmark work of Clair Patterson and his co-workers at the California Institute of Technology. They were the first to successfully address the problem of obtaining reliable data for Pb in Greenland and Antarctic snow and ice in a famous paper published in the late 1960s (11). At that time, other investigators proved to be unable to emulate this exceptional pioneering work which provided the first clear evidence of massive pollution of the atmosphere of our planet by Pb.

Later, a close interaction between leading scientists in the field and Clair Patterson led to the setting up of a large network of collaborative effort between research groups in Australia, Belgium, France, Italy, Russia and the United States, which allowed for extensive progress to be achieved. Nowadays, these international efforts have resulted in a wealth of fascinating data for a wide range of heavy metals both in Greenland and Antarctica, which were published in top scientific journals. Such cornucopia could never have been explored without the key influence of Clair Patterson, whose memory will remain alive in the minds of many scientists for the years to come.

This chapter will present an overview of the main aspects of the investigation of heavy metals in polar snow and ice, especially as regards sample collection and preparation and laboratory analysis. Some examples of the results obtained will also be given.

2. Ice and snow archives

The Antarctic ice cap represents the largest water reservoir on the earth: if totally melted, this huge ice volume (about 3×10^7 km^3) would raise the world ocean level by several tens of meters. The mean annual temperatures range from about –10°C on the coast to below –55°C in the central part of the continent. Antarctica is one of the driest places on Earth, with annual precipitation rates amounting to only a few centimeters of water equivalent per year.

On the other side of the world the Greenland ice cap is much smaller, with an ice volume of about 3×10^6 km^3. Here the temperatures and the mean precipitation rates are higher than in Antarctica (about –30°C in the central areas and from about 1 m of water equivalent per year in the southern parts to about 10 cm in the north-eastern areas, respectively).

These two peculiar areas are both geographically located in strategic positions (see Figure 3.1). Antarctica, on the far end of the southern hemisphere, centered on the South Pole, senses the effects of the preponderance of oceanic surfaces, small population and limited industrialisation. From a meteorological point of view this continent is affected by the surrounding zone of cyclonic storms and high precipitation, but also by surface winds flowing radially outwards. This geographical insulation and its peculiar characteristics preserve Antarctica from heavy pollution.

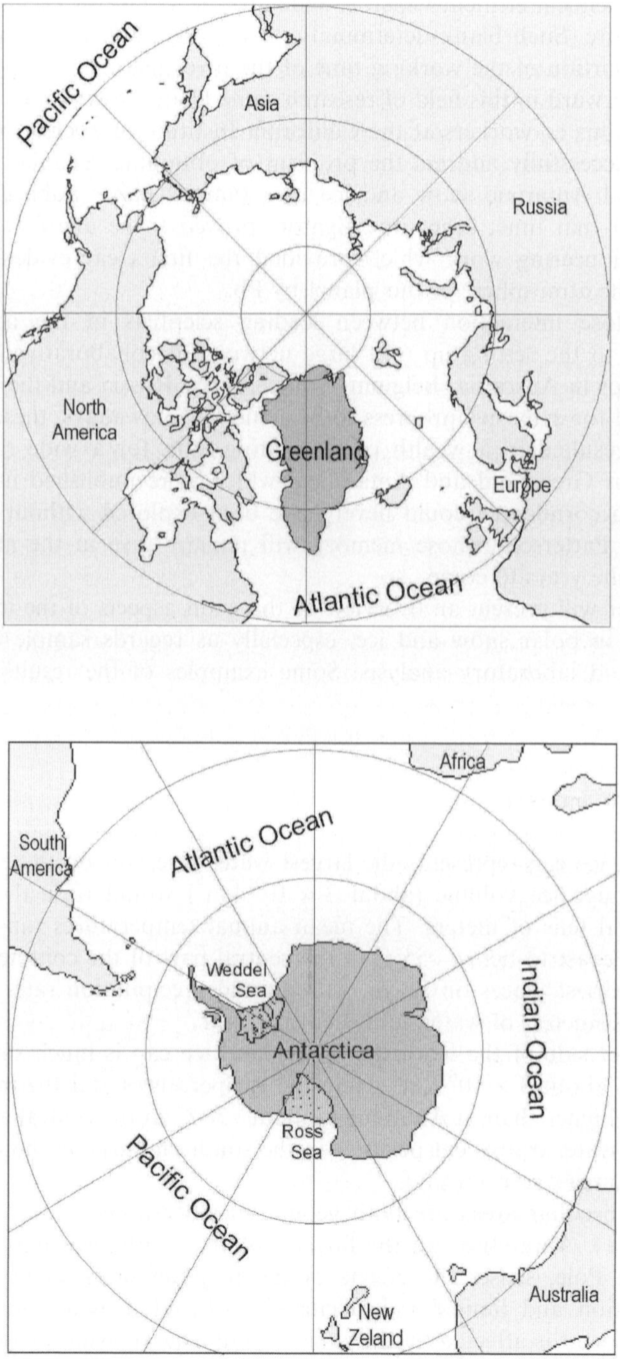

Figure 3.1. Maps showing geographical locations for (a) Greenland (north hemisphere) and (b) Antarctica (south hemisphere).

Greenland, conversely, is characterised by the fact that it is in the northern hemisphere, which contains the largest part of the landmass of the whole Earth (about 67%) and most of the population of our planet. This location, rather close to the industrialised areas of north America, western Europe and northern Russia, does not prevent pollution from easily reaching the ice cap. These opposite peculiarities in terms of geographical location and atmospheric regimes underline why scientists are so interested in conducting parallel research programmes on both ice caps.

Atmospheric precipitation, in the form of snow and ice, has accumulated continuously for thousands of centuries on polar ice caps. The snow density at the surface is about 0.3 g cm^{-3} and it increases with depth; below the so-called close-off depth the firn (snow with density higher than 0.4 g cm^{-3}) turns into ice (density 0.83 g cm^{-3}). At this depth (between 95–115 m in the central east Antarctica) air in the pores of the firn gets trapped in ice as bubbles preserving fascinating information on the gaseous composition of past atmospheres (12–16).

Besides atmospheric gases trapped in the air bubbles, these ice layers also contain a lot of impurities. These compounds are initially present in the local atmosphere as aerosols or reactive gases trapped during snowfall. In-cloud and below-cloud processes together with post-snow fall dry deposition are the mechanisms by which the impurities are scavenged from the atmosphere and collected into the snow (17–22). Chemical impurities, e.g., heavy metals, can originate from natural sources such as rock and soil, volcanoes, biomass, oceans, or cosmic dust and, for most recent layers, from human activities involving processes of fuel burning, mining and transportation.

Once emitted into the atmosphere these substances can travel for thousands of kilometers from point sources, being partially scavenged along the transport pathways and finally deposited on the polar ice sheets.

3. Sampling and sample treatments

Depending on the scientific goals, different sampling techniques can be chosen to have access to these valuable archives. Surface snow sampling is usually carried out in order to determine the spatial distribution of trace constituents, while shallow sampling, involving the excavation of snow pits, is preferable whenever the seasonal behaviour of different chemical species has to be explained. Finally, deep firn and ice cores can be drilled in order to reconstruct past changes (up to several millennia ago) in the composition of the ancient atmosphere of the Earth (23). Moreover, other aspects such as the remoteness of sampling sites and the extreme climatic conditions of the ice caps must be carefully considered as severe constraints from the logistical and financial point of view. For such reasons it is preferable to join forces in the framework of international programmes for such highly demanding sampling activities. One can cite, e.g., the European Project for Ice Coring in Antarctica (EPICA), the International Trans-Antarctic Scientific Expedition (ITASE) and the Greenland Ice Coring Project (GRIP).

3.1. Shallow sampling

Surface or near-surface sampling for heavy metal analysis can be carried out by plunging ultra-clean containers (bottles) directly into the snow, downwind of the operators and away from local contamination sources (such as scientific stations, tractors or helicopters). The sampling bottles are then capped, sealed in double polyethylene (PE) bags and stored frozen until the analysis in the laboratory. During operations researchers must wear special clean-room disposable clothing and PE gloves over their warm garments to prevent contamination from themselves. For heavy metal measurements the material of containers (wide-mouth bottles) must be made of plastic such as low density polyethylene (LDPE) or fluoroethylenepropylene (FEP) (see below) (24–27).

Deep hand-dug pits are used to collect clean samples down to a depth of 10–15 m. Using appropriate clean-room clothing and ultra-clean shovels and tools (28), operators may dig pits and collect various sized blocks (28, 29) or several samples by pushing ultra-clean plastic cylinders horizontally into the walls, from the surface to the bottom of the pit (27, 30).

3.2. Deep sampling

From about ten meters down to the close-off depth it is necessary to use electromechanical drilling systems (31) to collect samples. The outside surface of cores collected in this way can often be contaminated by heavy metals because of the contact between the drill body and the snow or ice core itself. This fact can be minimized by using special devices, such as plastic coated drills, and by carefully manipulating samples. In any case, it is necessary to proceed to a decontamination of the cores because the outer layer can be heavily contaminated (see below); this operation is usually feasible because the high density of the firn or ice prevents transfer of contaminating metals to the inner part of the sample section.

Beyond the close-off depth large drills, either electromechanical (32) or thermal (33), are used. Once more, due to the technical operations, the ice cores suffer contamination problems. In fact, to counteract the enormous pressure encountered at great depth and prevent closure of the hole, it is necessary to fill it with special fluids (usually kerosene mixed with freon or *n*-butyl acetate), so it is unavoidable that some fluid gets into the core. The contaminated outside layer, once roughly cleaned, can be used to measure oxygen or hydrogen isotopes, but for trace element it is necessary to decontaminate the ice core with the procedure described below.

3.3. Preparation of the sample

As mentioned at the beginning of this chapter, most of the chemical compounds in polar snow and ice are present at extremely low concentrations (below the pg/g level). It took many years to realize that it was necessary to develop new analytical methodologies to study these very peculiar samples. Surface or shallow sampling can usually be well controlled, as we have seen in the previous sections, provided that sampling items are carefully cleaned and operators strictly follow the clean sampling instructions.

3.3.1. Decontamination of snow and ice cores

For deeper samples, collected in the form of snow and ice cores, contamination is more or less always present. In this respect, decontamination of the snow and ice cores is of paramount importance in order to give reliable results. Decontamination consists of eliminating the contaminated outside concentric layers and recovering the presumably uncontaminated inner core. This really important operation has to be carried out in strict accordance to clean room protocols (2, 6, 27, 29, 34, 35). Usually, the ice core to be decontaminated is fixed horizontally on a LDPE speed lathe under a high purity air laminar flux, in a cold room at –15°C (6, 34). Beginning from the outside and moving towards the center of the core, successive veneers of ice are chiseled by a series of ultra-clean plastic or stainless steel knives depending on the hardness of the snow or ice; the material obtained is collected in ultra-clean LDPE bottles and then analyzed to quantify the trace element content. Usually, three-four outer layers and one inner ice core can be obtained.

It is of paramount importance to check whether the heavy metal concentration in the inner core reflects the original value in the ice. This can be done by investigating changes in the concentrations of heavy metals from outside toward the center of each core section. Only if a clear concentration plateau is obtained in the central part of the core it can be assumed that the mean plateau value represents the original one in the ice (6, 24, 36). As an example, an ice core section collected in Dome C (east Antarctica) in the framework of the EPICA program has recently been decontaminated and the samples obtained analyzed in order to determine the chemical concentration of several elements. Typical outside–inside concentration profiles are shown in Figures 3.2 and 3.3. In most cases, good concentration plateaus are obtained, as reported in Figure 3.2 for Mn in the 196.35–196.9 m section and for V in the 169.4–169.95 m section. In a few cases, however, metal concentrations decrease continuously from the outside to the innermost part of the ice core, as reported in Figure 3.3 for Fe and Pb in the 196.35–196.9 m. In this case the concentration obtained in the inner core must be considered as the upper limit of the original concentration in the ice or snow.

3.3.2 Ultra-clean laboratories

Another aspect requiring careful consideration is the contamination contribution from the analysis apparatus, i.e., not only the labware, but also the laboratory itself. The working environment where samples are prepared must be at least a Class 100 clean room as classified by Federal Standard 209 (37). In these environments two different areas should be available: the first for sample handling, standard solution preparation and rinsing of the plastic items; the second for ultrapure reagent production (e.g., HNO_3 by subboiling evaporation) and the displacement of clean acid baths used during the material cleaning. For safety reasons, this

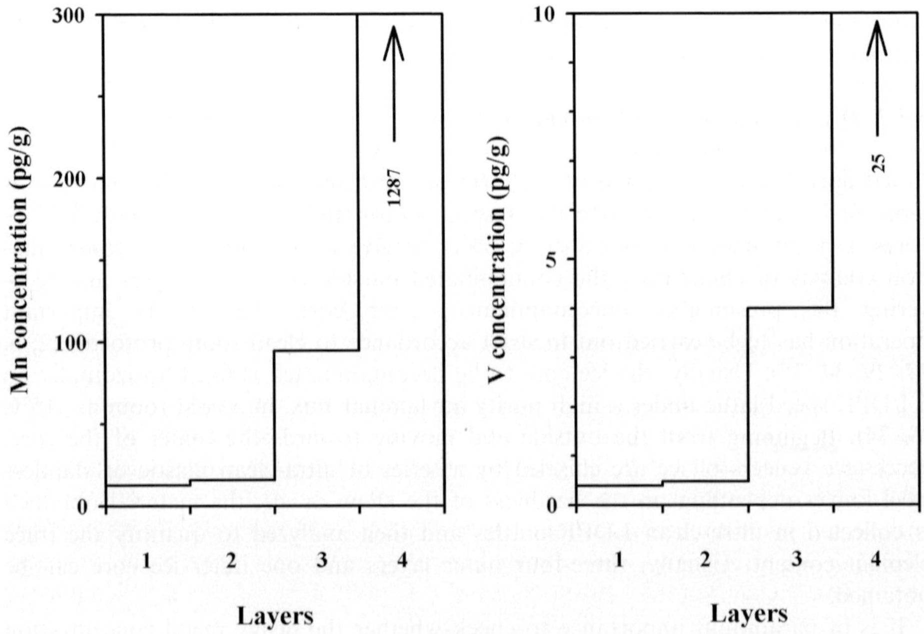

Figure 3.2. Changes in Mn and V concentrations from the outside (layer 4) to the inside (layer 1) in two sections (depths, 196.35–196.9 m and 169.4–169.95 m) of the EPICA core, electromechanically drilled at Dome-C, east Antarctica.

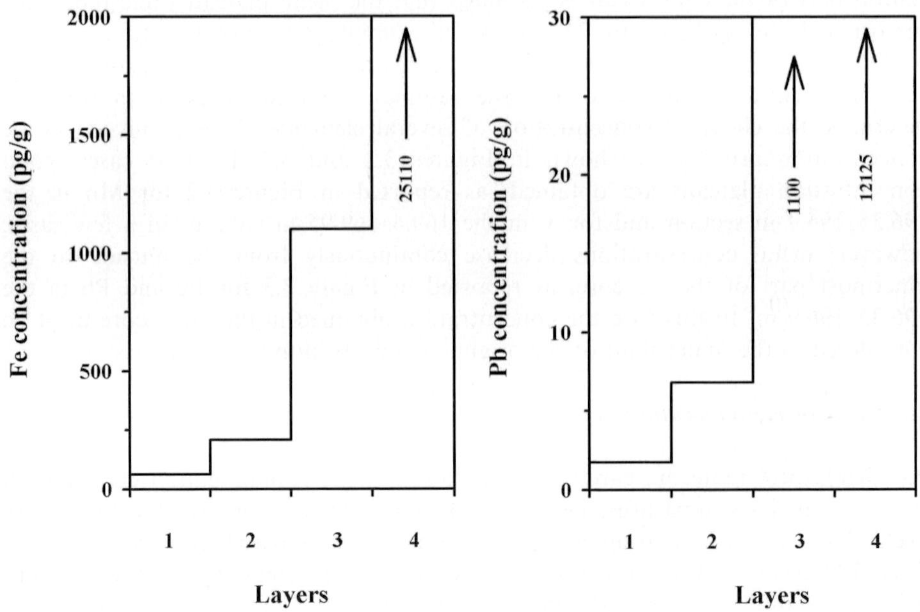

Figure 3.3. Changes in Fe and Pb concentrations from the outside (layer 4) to the inside (layer 1) in one section (depths: 196.35–96.9) of the EPICA core, electromechanically drilled at Dome-C, east Antarctica.

part should be equipped with a device for the extraction of exhaust fumes. Clean laboratories must have an entrance-room acting also as a dressing-room in which, before entering the working area, it is necessary to put on special decontaminated garments including PE gloves and special covers for shoes. The air inside the clean room, built with polyvinilchloride (PVC) or other plastic walls, is pressurized, preventing outside contaminated air admission. Furthermore the pressurized air must be filtered through high efficiency particulate filters, able to remove 99.999% of particles greater than 0.5 μm (8–10). Under such conditions the transfer of airborne particulate to the samples is kept to a minimum.

3.3.3. Ultra-pure reagents and chemicals

The availability of ultra-pure reagents is an important prerequisite for reliable analysis of ultra-low heavy metal concentrations (9). First of all it is important to have ultra-pure water for bottle cleaning and for the preparation of standard solutions. This water can be produced by coupling a reverse osmosis system Milli-RO, with a bed of Milli-Q ion-exchange resins (both from Millipore, Bedford, MA, USA), or by passing tap water through a set of activated charcoal and a mixed bed of ion-exchange resins (from Maxy, La Gardé, France). Table 3.2 reports a

Table 3.2. Trace element concentration of ultra-pure Milli-Q and Maxy waters

	Element concentration (pg/g)				
	Milli-Q		Maxy		Detection limit [c] (in pg/g)
	ICP-MS [a]	ETA-AAS [b]	ICP-MS [a]	Other techniques	
Ag	0.07 (0.01)		0.06 (0.01)		0.03
Bi	0.03 (0.04)		0.02 (0.01)		0.03
Cd	0.8 (0.2)	<0.01	0.6 (0.1)	<0.05 [f]	0.3
Co	1.55 (0.04) [d]		1.47 (0.03) [d]		0.09
Cu	0.9 (0.3)	0.3	0.6 (0.2)	<0.1 [e]	0.6
Mo	0.32 (0.09)		0.64 (0.07)		0.21
Pb	1.2 (0.2)	0.1	1.04 (0.03)	0.27 [g]; 0.28 [g]	0.09
Pd	0.20 (0.02)		0.24 (0.03)		0.09
Pt	0.05 (0.01)		0.043 (0.003)		0.009
Sb	0.15 (0.05)		0.15 (0.02)		0.06
U	0.12 (0.03)		0.06 (0.03)		0.09
Zn	2.1 (0.3)	0.3	1.8 (0.3)	0.3 [e]	0.9

a) From ref. (62).
b) From ref. (44).
c) From Maxy water data measured by ICP-MS, according to the 3s criterion (100).
d) In brackets, standards deviations, n = 5.
e) From ref. (10).
f) From ref. (40).
g) From ref. (48).

typical concentration of heavy metals in ultra-pure water as detected by several analytical techniques.

Ultra-pure HNO_3 (70%) is extensively used both for cleaning field sampling equipment and labware and during the analytical procedures themselves. In the past, HNO_3 produced by the National Institute of Standards and Technology (NIST, Gaithersburg, MD, USA) with the non-boiling purification system was used because of the very low heavy metal concentrations (38). Since this kind of reagent is no longer available from NIST, a double subboiling distillation unit is currently being tested in the laboratories of the University of Venice. Preliminary results show that the concentration of heavy metals in HNO_3 is comparable with that produced by NIST.

Other chemicals used throughout the analysis of heavy metals in polar snow and ice by different analytical techniques have to fulfil strict specifications for ultra-clean procedures as described in specific papers (6, 39–45).

LDPE containers are used to collect shallow snow samples or to store them after melting. LDPE represents a compromise between chemical resistance during strong cleaning treatments, low content of impurities and a low price (see Table 3.3). In any case, before use, all the plastic bottles and items must be extensively cleaned following tested procedures (10, 27).

Briefly, items are cleaned as follows: rough rinse with tap water to remove dust; remove grease with chloroform and rinse with ultra-pure water; immerse in a first acid bath (Merck "Suprapur" HNO_3 in ultra-pure water, 1 + 3v/v, 50°C, 2 weeks) and rinse with ultra-pure water; immerse in a second acid bath (NIST HNO_3 diluted in ultra-pure water, 1 + 1000v/v, 50°C, 2 weeks) and rinse with ultra-pure water; immerse in a third acid bath (NIST HNO_3 diluted in ultra-pure water, 1 + 1000v/v, 50°C, 2 weeks); finally, bottles are rinsed several times with ultra-pure

Table 3.3. Summary of the average element content of 12 plastic materials. Adapted from ref. (101)

Material	Number of elements	Total concentration (μg/g)	Major constituents
ETFE	32	1007	Cl, Pb, Si
FEP	25	241	Ca, K, Mg
HDPE	22	654	Ca, Si, Zn
LDPE	18	23	Ca, Cl, K
PC	10	85	Al, Br, Cl
PMP	14	178	Ca, Mg, Zn
PP	21	519	Ca, Cl, Mg
PS	8(8 nd *)	4	Al, Na, Ti
PSF	16(12 nd)	17	Ca, Fe, Na
PVC-rigid	7(11 nd)	2541	Ca, Mg, Sn
PVC-tubing	9	280	Fe, Sb, Zn
TFE	24	19	Ca, Cu, Fe, Pb

* nd = Not detected

water, filled with a diluted ultra-pure HNO_3 fresh solution (1 + 1000) and stored inside double polyethylene acid clean bags, while other tools remain in the last bath until use. Before use, LDPE bottles are rinsed several times and conditioned for at least 4 hours with ultra-pure water.

4. Analytical techniques

The ideal analytical technique to be used in the challenging task of reconstructing past changes and recent variations in the concentration of trace substances in polar snow and ice should present several important features. Of these, extremely low detection limits, multi-element capability, low sample consumption and the possibility to avoid, as far as possible, any preconcentration step which could be a source of contamination are the most appreciated. Nevertheless, there is currently no technique with all the special features listed above; several instrumental methods have been used in the past for trace element determination in polar snow and ice (see Table 3.4).

Laser Excited Atomic Fluorescence Spectrometry (LEAFS) (40, 41, 46–48), Thermal Ionisation Mass Spectrometry (TIMS) (6, 42, 43, 45, 49–53), Electrothermal Atomization Atomic Absorption Spectrometry (ETA–AAS) (24, 28, 54, 55), Differential Pulse Anodic Stripping Voltammetry (DPASV) (27, 49, 56, 57),

Table 3.4. Analytical techniques used for the determination of trace elements in polar snow and ice

Technique	Trace element	Relevant References
Atomic Fluorescence Spectrometry (AFS)	Hg	(5, 58)
Differential Pulse Anodic Stripping Voltammetry (DPASV)	Cd, Pb	(27, 49, 56, 57)
Electrothermal Atomization Atomic Absorption Spectrometry (ETA-AAS)	Al, Ca, Cd, Cu, Fe, K, Mg, Mn, Na, Pb, Zn,	(24, 28, 54, 55)
Inductively Coupled Plasma Mass Spectrometry (ICP-MS)	Ag, Bi, Cd, Co, Cu, Mo, Pb, Pd, Pt, Sb, U, Zn	(59–62)
Instrumental Neutron Activation Analysis (INAA)	Al, As, Na, REE, Sc, Se	(102, 103)
Laser Excited Atomic Fluorescence Spectrometry (LEAFS)	Bi, Cd, Pb	(40, 41, 46–48)
Thermal Ionisation Mass Spectrometry (TIMS)	Bi, Pb, Pb isotopes	(6, 42, 43, 45, 49–52)

Atomic Fluorescence Spectrometry (AFS) (5, 58), quadrupole and double-focusing magnetic sector ICP–MS (59–63) have all been used in the past for the determination of trace element in polar snow and ice. Of those techniques only LEAFS, double-focusing magnetic sector ICP–MS and DPASV have proved capable of direct determination on the analytes in the samples at the required levels.

The other techniques used are less sensitive and could require preconcentration by extraction by the chloroform-dithizone method (6, 45, 64), non-boiling evaporation (39) or by adsorption on W wire loops (44, 65). Preconcentration and extraction methods require great care to be taken in order to avoid contamination and are often time-consuming, and require a fairly large volume of samples.

However, it must be kept clear in mind that direct instrumental detection methods for trace substances are physically relative methods which require calibration, during which systematic errors, caused for instance by spectral and non-spectral interferences, may occur. Relative methods are in fact matrix-dependent and would require the analysis of Certified Reference Materials (CRMs) in order to guarantee the good quality of the analytical data. Unfortunately, CRMs are not available for polar snow and ice and hence the only way to assure the quality of the data is, whenever possible, to make careful intercomparisons of the techniques able to measure the same analytes with different approaches.

In the following sections the analytical techniques will be briefly described, underlining their potentialities in trace element determination in polar snow and ice. Some key examples of the results obtained will also be presented.

4.1. TIMS

TIMS has been used in the past few decades for the analysis of heavy metals in polar snow and ice of Antarctica and Greenland (11, 49–53, 66, 67). It is in fact an absolute technique, whose results can give not only the total concentration of the metal with unrivalled accuracy and precision, but also quantify the isotopic composition of the analyte. This last feature constitutes a precious tool for probing the sources of emission of heavy metals (e.g., Pb) to the global atmosphere throughout the centuries.

The first paper to deal with the use of TIMS in trace element determination in polar snow and ice was the pioneering work of Murozumi et al. (11). It described the use of Isotope Dilution Mass Spectrometry (IDMS) for the determination of Pb in Antarctic snow samples. Considering the extremely low concentration of Pb that was anticipated, they used about 20 kg of melted snow sample to minimize the analytical blank and obtain the amount of Pb needed for the quantification. In his excellent study Murozumi et al. (11) first showed the Pb pollution of polar regions by comparing the Pb concentration in layers of Greenland snow and ice dating back to 800 BC with modern snow layers. Deep strata were found to contain a 200-fold lower Pb concentration than contemporary snow. This study started a sort of revolution in the approach to trace element determination in environmental matrices, particularly as regards the aspect of sampling and sample handling and treatments. The paper was so innovative for that time that several attempts were made without success to confirm the results (68–70). Confirmation arrived only twelve years later in a landmark paper by Ng and Patterson (45) who measured Greenland ice collected in Camp Century using only 300 g of ice. The innovative approach to trace element determination in ice described in that paper provided a useful references for all the subsequent measurements of heavy metals in polar snow and ice (6, 55, 64, 71).

Recently, thanks to the ultra-clean procedures adopted and the high sensitivity obtained, successful attempts have been made to measure Pb concentration and its isotopic composition in just a few millilitres of snow and ice samples (42, 51–53, 66). In this case, samples of snow or ice, after decontamination (if necessary), were melted and transferred to LDPE bottles under strict conditions of contamination control (42). The volume of the aliquots depends on the expected concentration in the sample, but it must be chosen in order to have at least about 10 pg of Pb. This means that for old Holocene ice, in which the Pb concentration is about 0.5 pg/g (6), at least 20 g of sample must be provided. Melted samples are poured into a preconditioned 15 ml PFA beaker, spiked with 20 pg of ^{205}Pb, acidified with 10 µl of ultra-pure concentrated HNO_3 and with 3 µl of Si gel-added H_3PO_4. The use of ^{205}Pb-enriched tracer (half-life 15×10^6 year), which is not naturally present in the lead mass spectrum, has the great advantage of allowing the simultaneous determination by IDMS of both concentration and isotopic composition. After evaporation, a small droplet is finally transferred to an acid-cleaned Re filament of the mass spectrometer together with few µl of, Al solution to increase the ionization efficiency of the instrument. About 10 samples can be measured in the same session together with blanks and reference standards. Isotopic composition and Pb concentration are finally determined simultaneously in each samples by mass spectrometry, measuring the ion currents reaching the detectors after different isotopes have been dispersed by the magnetic field of the instrument. A systematic study of blank contribution has also allowed for the determination of the amount of Pb added to the samples during the single steps of the entire analytical procedure (HNO_3, H_3PO_4, silica-gel, beakers); procedural blanks in the order of 2 pg have been obtained, allowing absolute detection limits of just a few pg to be achieved.

Lead isotopes vary widely in their abundance in nature and can be used to identify the source of Pb in past polar snows. Lead isotopic abundance patterns have been used by Rosman et al. (51) to identify and quantify the contribution to the contamination of the Arctic from Eurasia and United States during the 1970s and 1980s. Automobile emissions have been the major source of atmospheric Pb during the twentieth century, since the metal has been widely used as anti-knocking additive in gasoline. During the mid-1960s the United States used an increasing amount of more radiogenic Pb from the Mississippi Valley region as an additive in their gasolines, while the contribution from Eurasia and Canada remained at lower values, with an isotopic ratio ^{206}Pb/^{207}Pb of 1.14 (see Figure 3.4). The variation imposed in the States was reflected in a change in the isotopic fingerprint of Pb emitted into atmosphere and hence in the isotopic composition of anthropogenic Pb as revealed in Greenland snow. With a simple linear mixing model it was also possible to estimate that the United States emissions dominated during the early- to mid-1970s, while, as a consequence of the switch to unleaded

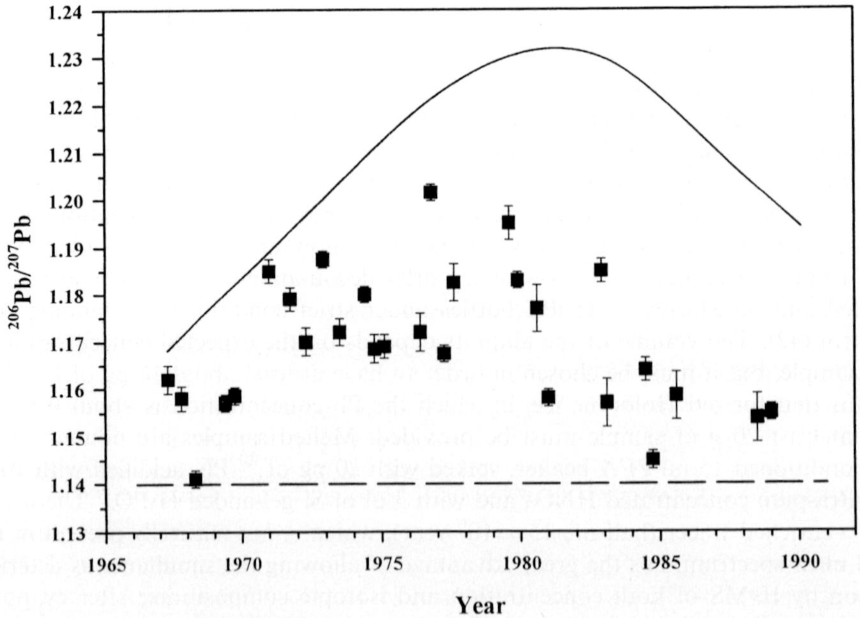

Figure 3.4. Measured $^{206}Pb/^{207}Pb$ isotope abundance ratio in Summit (central Greenland) snow during the past three decades. Trends in USA (——) and Eurasian/Canadian (- - - - -) emissions are also shown. Adapted from ref. (51, 66).

gasoline the contribution of United States had almost disappeared in the late 1980s. The contribution of Eurasian and Canadian sources remained fairly constant during the same period, becoming worthy of note during 1980s.

Studies carried out in Antarctic snow and ice (52, 53) on the isotopic composition of Pb have proved unequivocally that recent snow layers, and hence the present Antarctic troposphere are contaminated by anthropogenic Pb. Isotopic fingerprinting also proved two different sources: a natural component associated with aeolian dust and an anthropogenic component. The possible origin of this Pb-rich aerosol is hypothesised to be South America.

Elements other than Pb have been measured by TIMS in polar snow (43, 49, 50, 53), showing the high potential of the technique. Bismuth determination (43) is characterized an accuracy comparable to that of LEAFS and takes advantage of the simultaneous ionisation of Bi and Pb during the Pb isotopic abundance measurements. Determination of Ba was also carried out on some Antarctic samples (53) showing its possible use as a reference element in monitoring the contribution of terrestrial dust level.

4.2. ETA–AAS

A considerable advance in trace element determination in polar snow and ice was made during the 1980s by the use of ETA–AAS. The limitation of this technique (i.e., only one element at time can be determined) is counterbalanced by its high detection power, which for many elements provides sufficiently accurate determination in the pg range (10, 39, 65).

Although the detection power of ETA–AAS could also be improved by about one order of magnitude through electronic signal addition processing, there are a series of different possibilities based on on-line and off-line chemical and physical preconcentration methods which can greatly improve the detection power of the entire methodology in the analysis of polar snow and ice (39, 44, 45, 65).

Chemical preconcentration of the samples (e.g., extraction into chloroform and dithizone) was used in the past by several authors in order to determine the very low concentration of Pb in ancient Antarctic ice (6, 45, 64). Although this method gave reliable results, it needed a huge amount of sample (500–1000 g) and a long period of sample treatment with organic solvents. Another method is based on the preconcentration onto W wire loops that can be placed directly in a graphite furnace of instrument AAS for the atomisation of adsorbed metals for the analysis (44, 65). To minimise the contamination problems from the air of the laboratory the whole preconcentration procedure was carried out inside a vertical laminar flow clean bench. Detection limits of 0.01, 0.47, 0.22 and 0.24 pg/g were obtained for Cd, Cu, Pb and Zn respectively (44). These extremely low detection limits have recently enabled the concentrations of heavy metals in Antarctic ancient ice and recent snow to be determined (28, 72).

Non-boiling evaporation (39) on ultra-clean FEP Teflon® beakers has also been extensively used in the past for the preconcentration of heavy metals in Greenland and Antarctic samples (2, 3, 55). In this case, 30–100 g of melted snow sample are allowed to evaporate into a drop of 50–100 μl, at a temperature of 80–90°C inside a clean bench. Then the sample is treated with 1 ml of 0.1% ultra-pure HNO_3 and 50 μl are transferred into the graphite furnace for the analysis, following tested operating parameters (39). Coupling a chemical or a physical preconcentration method with the detection power of ETA–AAS has provided a wealth of fascinating information in the field of heavy metals determination in polar snow and ice.

The history of Cu production was revealed by analysing GRIP ice core samples taken at Summit, Greenland (3). The results obtained show that the ice in central Greenland has retained a trace of the history of Cu production, in particular in ancient times, testifying that the fall-out from anthropogenic Cu on the Greenland ice cap as a whole was ten times greater over a period starting 2500 years ago and ending at the industrial revolution than over the period from the industrial revolution to the present (3). These studies opened the door to a quantitative approach to paleometallurgy, which played an important part in the development of human cultures during ancient times.

Particular attention has also been given to the reconstruction of recent Pb emissions into the atmosphere, as detected in Greenland and Antarctic snow and ice. Owing to its well known toxicity and wide dispersion in the environment, caused in turn by the massive use as a component in a variety of products such as batteries, pigments, rolled and extruded items, cables sheaths and above all gasoline additives in the form of highly volatile alkyl-Pb compounds, Pb is classified as one of the most toxic metals dispersed in the environment. This metal is considered by far one of the older hemispheric-scale anthropogenic polluting substances emitted into the atmosphere. Greenland ice dating from 7760 to 470 years BP contains the first well documented traces of anthropogenic Pb released to the atmosphere by Hellenic and Roman civilizations (2). These statements were also confirmed by comparing the Pb isotopic pattern in the same samples with those of ores from ancient Roman mines in southern Spain (4). Concentrations found in Greenland ice dating from Graeco-Roman times are about four times higher than those detected during the early Holocene (0.55 pg/g) (2). After a decrease, due to the decline and fall of the Roman Empire, the Pb concentration in Greenland ice increased continuously during the centuries, reaching a value of about 10 pg/g between the eighteenth and nineteenth centuries (11, 35, 73).

In more recent years, starting from the Industrial Revolution up to the mid-1960s, Pb concentration in Greenland ice and snow has increased by about 20 fold even if at different rates of increase during the years (see Figure 3.5) (11, 35, 55, 73). Up to the 1940s coal and wood combustion and non-ferrous metal production accounted for most of the anthropogenic Pb emissions into the atmosphere (74). From the 1940s onwards the widespread use of alkyl-Pb additives in gasoline become predominant in the whole budget of Pb emission to the atmosphere. This origin has been confirmed by recent studies on organo-Pb compounds in Greenland snow (see below) (75–77). As a consequence of the massive diffusion and use

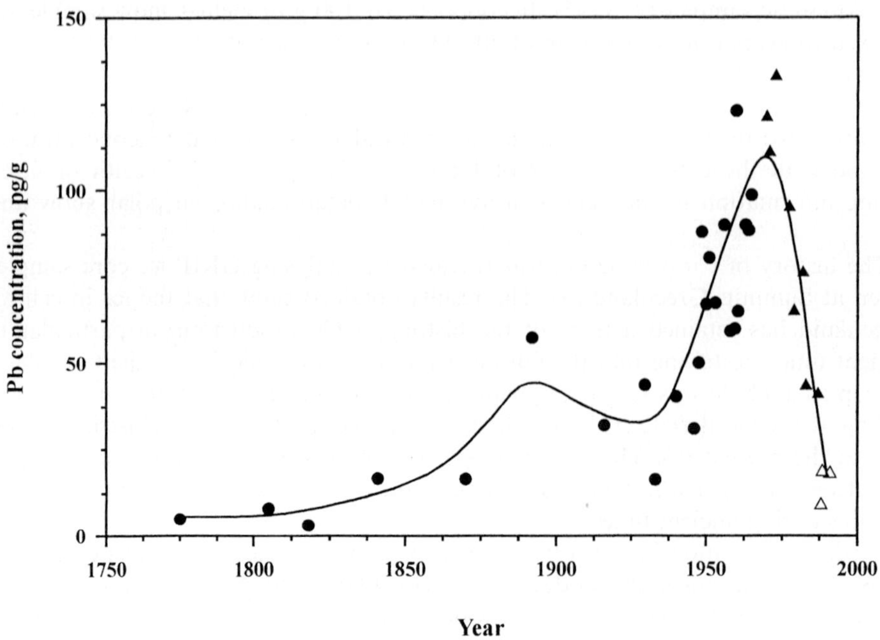

Figure 3.5. Lead concentration changes in central Greenland (Summit) snow and ice from the 1770s to present. Adapted from ref. (35, 55).

of automotive gasoline in industrialised countries, the concentration of Pb detected in Greenland snow increased to values of about 200 pg/g in two decades (11, 35, 73). Then policy initiatives, started in the early 1970s in the United States and Canada and later in European countries, taken in order to stop this worrying trend, were reflected in a marked decrease by a factor of 7.5 of Pb concentration in Greenland snow (35, 55). The automobile source of most of the anthropogenic Pb found in Greenland snow, together with the determination of the relative contributions of USA and Europe, has also been confirmed using the different isotopic signatures of European and USA Pb used as gasoline additive (see Figure 3.4) (51, 66).

In Antarctica the Pb concentration in recent snow is about eight to ten times lower than in Greenland (25–29, 71, 78–81). The remoteness of the continent, the distance from highly industrialised countries, the difficulty of transport processes across both the equator and the Antarctic convergence zone, restrict the deposition of Pb-enriched aerosol particles (82, 83). As a consequence of the pollution revealed in the northern hemisphere, some countries of the southern hemisphere also began to use unleaded or low-Pb fuels, although about ten years later. The effect of these initiatives on the Pb content of the Antarctic ice cap has been recently convincingly shown by Wolff and Suttie, who determined Cd, Cu, Pb and Zn by ETA–AAS in several snow samples dating from 1920s to 1980s (28, 72). The Pb deep concentration profile as reported in Figure 3.6 shows that the concentration during the 1920s was about 2.5 pg/g (about five times higher than Antarctic

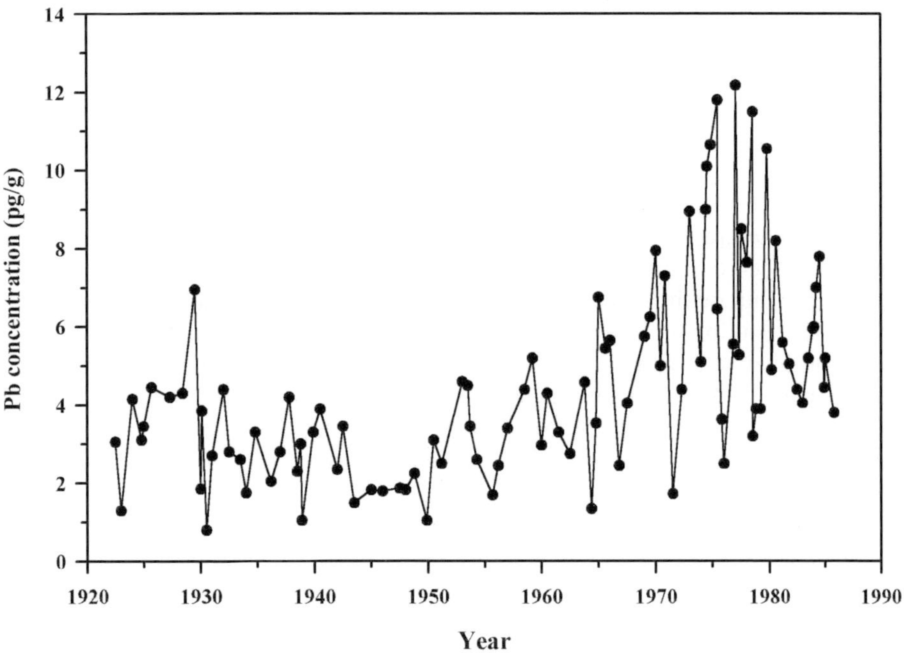

Figure 3.6. Lead concentration changes in Coats Land (Antarctica) snow since 1920s. Adapted from ref. (28).

Holocene ice, i.e., 0.5 pg/g (6), highlighting how at the beginning of this century the remote atmosphere of the Antarctic continent was already contaminated by this toxic metal. In the following years, after a slight decrease during the 1930s, probably due to the economic depression, the Pb concentration in Antarctic snow showed a continuous increase up to a value of about 6 pg/g during 1970s, reflecting the huge increase of antiknocking Pb additives used in the gasoline. Afterwards the banning of Pb additives in the gasoline of southern hemisphere countries resulted in a reduction of the Pb content in the snow of Antarctica. This trend was also recently confirmed by an independent study covering the years 1965–1991 (see below) (27).

4.3. LEAFS

The ultra-sensitive technique of LEAFS was developed and exploited for the analysis of snow and ice to be carried out at the Institute of Spectroscopy of the Russian Academy of Science (ISAN) (46). This technique allows the direct determination of toxic metals in polar snow and ice to be carried out at and below the pg/g level in sample volumes of less than 100 µl. This is an important issue considering that very often the volume of sample obtained after the decontamination procedures amounts to just a few tens of milliliters in the best case and that this must be subdivided into different aliquots for a complete chemical characterisation.

The instrument for LEAFS comprises the following basic units: a Tunable Dye Laser (TDL), an electrothermal atomizer and a detection system. The TDL is pumped by an XeCl excimer laser and can be tuned to the wavelength corresponding to the excitation interval of Pb and other heavy metals. The electrothermal atomizer consists of graphite electrodes fixed to a water-cooled metal holder. Samples of 20–50 µl are injected into the cup and then evaporated and atomised following different programmes depending on the element to be analysed (40, 41, 47, 48). The free atoms of the analyte are excited in the analytical zone above the graphite cup by the resonance radiation of TDL. The fluorescence radiation is then collected at an angle of 90° in the detection system constituted by a telescopic unit with an entrance slit, a monochromator and a photomultiplier tube. The resulting current, proportional to the analyte concentration in the sample, is finally measured. Considering that the contamination problem can be critical with instruments of such sensitivity, the analytical parts of LEAFS (atomizer, detector, sample preparation table) are held under a filtered air flow. The operator works inside this clean area wearing clean-room clothing, gloves and mask.

Unlike IDMS, LEAFS is not an absolute technique. The instrument must be calibrated using synthetic standards with concentrations similar to those expected in the real samples. The preparation of these standard solutions (from 0.1 to hundreds of pg/g) poses a formidable challenge to analytical chemists due to the extremely low concentrations involved and the subsequent risk of contamination of samples. Standard preparation is carried out by subsequent dilutions of synthetic standard stock solutions in ultra-clean PE bottles. The detection limits obtained with this technique for Bi, Cd and Pb (0.05, 0.07 and 0.18 pg/g respectively) (40, 41, 47, 48) are among the best ever obtained for these three elements.

Although it cannot be considered a routine technique, the extraordinary detection limits coupled with the exceptionally low sample volume, make the LEAFS a very powerful tool for the direct determination of heavy metals, such as Bi, Cd and Pb and more recently of Ag, Au and Hg, in ultra-pure snow and ice of polar regions. Cadmium and Pb have been determined in several snow and ice samples and the results obtained were compared with those obtained with different techniques in order to check the accuracy of the method (7, 40, 41, 48). Lead concentration in Antarctic ice samples covering the period from 34,000 to 3800 years BP were measured by LEAFS (41) and results were compared with those previously obtained by IDMS (6). LEAFS and IDMS data, although measured by completely independent methodologies, were in very good agreement, strongly supporting the accuracy of both sets of results.

Bismuth, considered a potential tracer of volcanic activity, has been extensively studied thanks to the exceptional sensitivity of LEAFS. So far, little attention has been paid to this metal whose atmospheric cycle is likely to be still undisturbed by man. This element can be of extraordinary importance, since it can represent untransformed tracers of specific natural sources and of atmospheric transport pathways. Although data on the occurrence of Bi in the environment are still very scarce, it is indeed likely that this metal is an excellent tracer of volcanic emission into atmosphere, since concomitant emissions from other natural sources are of little importance (84–86). Investigations of the occurrence of this metal in polar archives could produce very valuable time series of volcanic activity in both hemispheres and relevant data on the transport patterns of volcanic aerosols in the atmosphere (87).

4.4. ICP–MS

Over the past decade Inductively Coupled Plasma (ICP) sources, in particular coupled with Mass Spectrometry (MS) instruments, have shown an immense potential for multielement analysis in environmental samples (88). These capabilities have been obtained thanks to the combination of the great ionization energy of a plasma source with the high sensitivity and selectivity of the mass spectrometric detector. Since polar snow and ice are considered as the purest material on the earth surface, these environmental matrices constitute the ideal samples for ICP–MS since potential interferences formed in the plasma are kept at a minimum level.

Quadrupole ICP–MS has been used for the determination of trace elements in snowfall in the remote Scottish Highlands at the ng/g level after preconcentration of samples by non-boiling evaporation (59, 60). Trace elements have been also determined in Arctic snow (Ellesmere Island) coupling a quadrupole ICP–MS with an Electrothermal Vaporization (ETV) system (61). In spite of the considerable improvement in terms of sensitivity achieved through the use of the ETV sample introduction system, the detection limits of quadrupole ICP–MS are still inadequate for the direct determination of trace elements in the snow and ice of remote polar regions.

A considerable improvement has been obtained thanks to the use of double focusing magnetic sector ICP–MS (89) which, thanks to the very low background signal and the high ion transmission, offers an extreme sensitivity at and below the pg/g level. Furthermore, the possibility of working at higher resolutions ($m/\Delta m = 3000$ or 7500) than a quadrupole instrument, allows most of the analytes to be separated from interfering species formed in the plasma. Direct determination of a wide range of trace elements in polar snow is hence potentially feasible, provided that considerable care is taken in the collection, handling and treatment of the samples in order to avoid contamination (62). The high purity of the snow and ice matrices is very attractive for direct trace metal determination by double focusing magnetic sector ICP-MS, although spectral interferences should be carefully considered. This requires that an accurate study be made on possible species forming in the plasma (63).

Melted snow and ice samples are nebulised at atmospheric pressure and transported into the plasma by an Ar flow. Ions formed in the high temperature plasma are extracted by a special interface with two concentric cones which allow ions to be accelerated and separated by a mass spectrometric analyser and finally detected by a secondary electron multiplier.

This technique has been used with success for the direct, simultaneous determination of Ag, Bi, Cd, Co, Cu, Mo, Pb, Pd, Pt, Sb, U and Zn at the low and sub pg/g level in Greenland and Antarctic snow (62). Good detection limits (as reported in Table 3.2) for most of the analysed elements and a repeatability of the measurements ranging between 8 and 25% were reported.

In order to give an overview of the great capability of the technique some data, relative to samples collected in Greenland snow, are reported in Table 3.5. The different seasonal mechanism of atmospheric transport from remote continents to the Greenland ice sheet accounts for the wide dispersion of the concentration

Table 3.5. Summary statistics for concentration of trace elements in 68 Greenland snow samples covering a 4 year-time period (Winter 1990 to Spring 1995). Adapted from ref. (62)

	Element concentration (pg/g)	
	Median (minimum–maximum)	Mean (SD)
Ag	0.38 (0.08 – 5.0)	0.60 (0.75)
Bi	1.7 (0.29 – 21.2)	2.5 (3.2)
Co	5.3 (0.65 – 17.2)	5.8 (3.6)
Cu	3.7 (0.78 – 13.6)	4.6 (3.0)
Mo	1.6 (0.08 – 7.0)	1.6 (1.21)
Pd	0.9 (0.05 – 4.2)	1.1 (0.81)
Pt	0.55 (0.08 – 1.5)	0.61 (0.34)
Sb	0.7 (0.21 – 4.3)	0.9 (0.60)
U	0.8 (0.21 – 15.4)	1.8 (2.7)
Zn	30 (2.0 – 207)	47 (40)

values reported. As an example of this seasonal influence, Cu and Zn deep concentration profiles are reported in Figure 3.7, together with the estimated period of deposition of the snow. For comparison purposes, results for Cu, after preconcentration of the samples (39), and Zn obtained by ETA–AAS on the same samples are also reported in the plots. Concentration profiles for Cu and Zn show spring and summer maxima, which can be attributed to the transport and scavenging of polluted air masses coming from lower latitude countries during these periods of the year.

Comparing the two analytical techniques used, it has been ascertained that Cu values measured by double focusing ICP–MS are systematically lower than those measured by ETA–AAS: the slope of the plot of double focusing magnetic sector ICP–MS *vs.* that of ETA–AAS is 0.53 ± 0.08 (95% confidence interval), with a correlation factor of 0.860. The cause of this systematic discrepancy for Cu is not well understood at the moment, but is probably due to the poor storage of standard solutions. As regards Zn, the agreement between the two independent techniques used is good: the slope of the plot of ICP–MS *vs.* that of ETA–AAS is 1.16 ± 0.13 (95% confidence interval), with a correlation factor of 0.916.

In recent times double focusing magnetic sector ICP–MS has been applied to the determination of Pd, Pt and Rh in polar and Alpine snow and ice (63). The interest in these elements, widely known as Pt-Group Elements (PGEs), has increased in the recent years owing to their use as active compounds in catalytic converters for cars. From the determination of Pd, Pt and Rh in dated Greenland ancient ice and recent snow, it has been shown that this pollution can now be detected even in the free troposphere of the remote Arctic indicating that there is a large scale pollution of the whole northern hemisphere. Preliminary investigations of Pd, Pt and Rh in Antarctic snow and ice (90) have shown concentration values slightly above the natural background levels, highlighting a possible pollution of the remote polar areas of the southern hemisphere by PGEs; however, further studies are currently underway to provide more evidence for this assumption.

From an analytical point of view the determination of PGEs is hampered by their extremely low concentrations (below the pg/g) and also by spectral interferences which strongly affect measurements (91). Thus the selection of the isotopes to be used for the analysis requires a critical investigation on possible interferents. In this respect, ^{106}Pd, ^{195}Pt and ^{103}Rh were selected as a good compromise between natural abundance and freedom from interfering compounds formed under plasma conditions. The interference study carried out has shown that for polar samples the contribution of interfering species is negligible in most of the samples (63). Anyway, by monitoring the elements forming the interferences, accurate correction equations can be set up.

From the analysis of Pd, Pt and Rh in old Greenland ice samples and recent snow, it has been shown that the concentrations of PGEs in present day snow are about 100 times above natural levels, testifying to a large scale pollution of the troposphere of the northern hemisphere for these elements (91). As shown in Figure 3.8, by combining the Greenland time series with available data on world demand for Pt, it appears that a large fraction of the observed increase for this metal originates from the abrasion and deterioration of ageing automobile catalytic

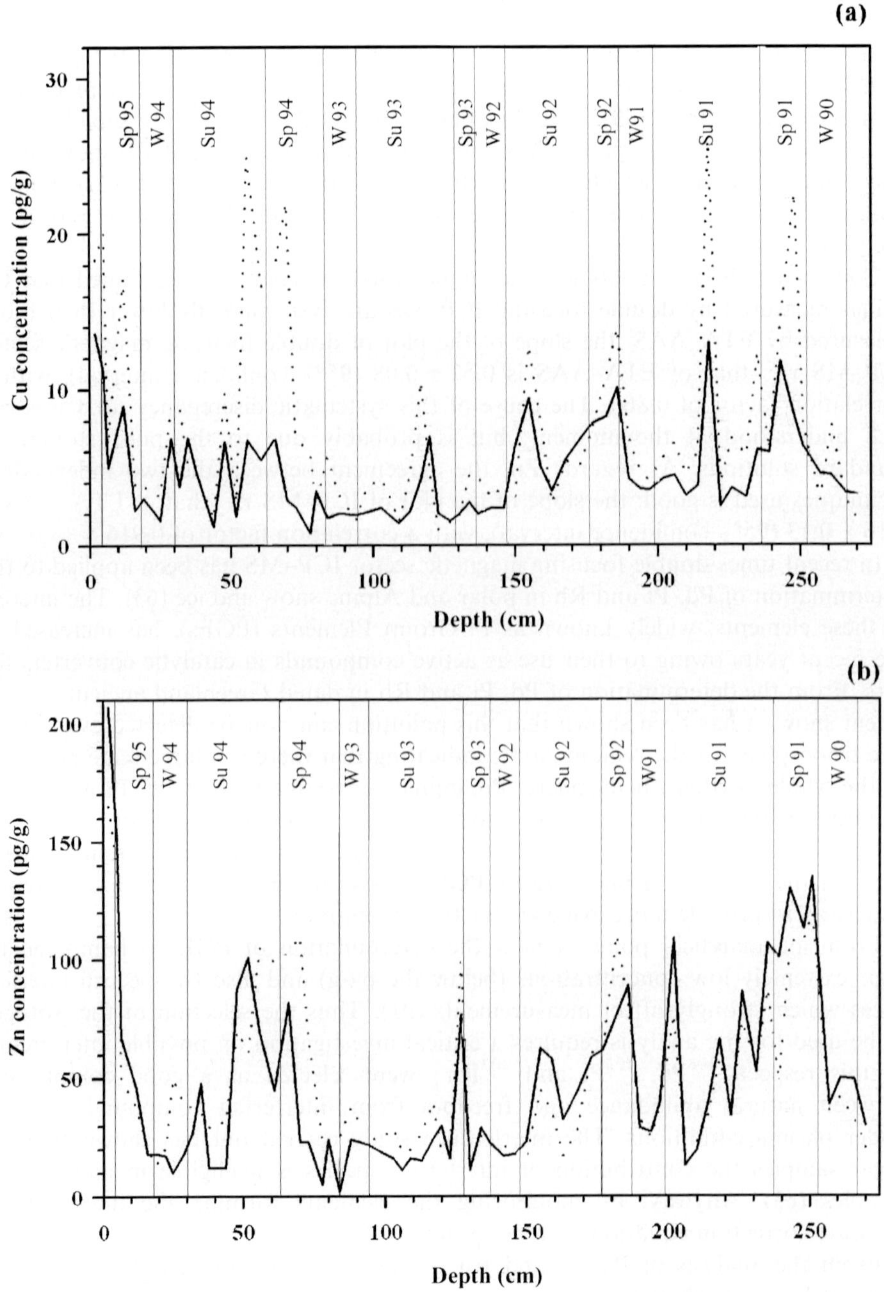

Figure 3.7. Cu (a) and Zn (b) concentration profiles at Summit (Central Greenland) obtained by double focusing magnetic sector ICP-MS (——). ETA-AAS results (- - - -) are showed for comparison. Datation is also reported. W = winter; Sp = spring; Su = summer. Adapted from ref. (62).

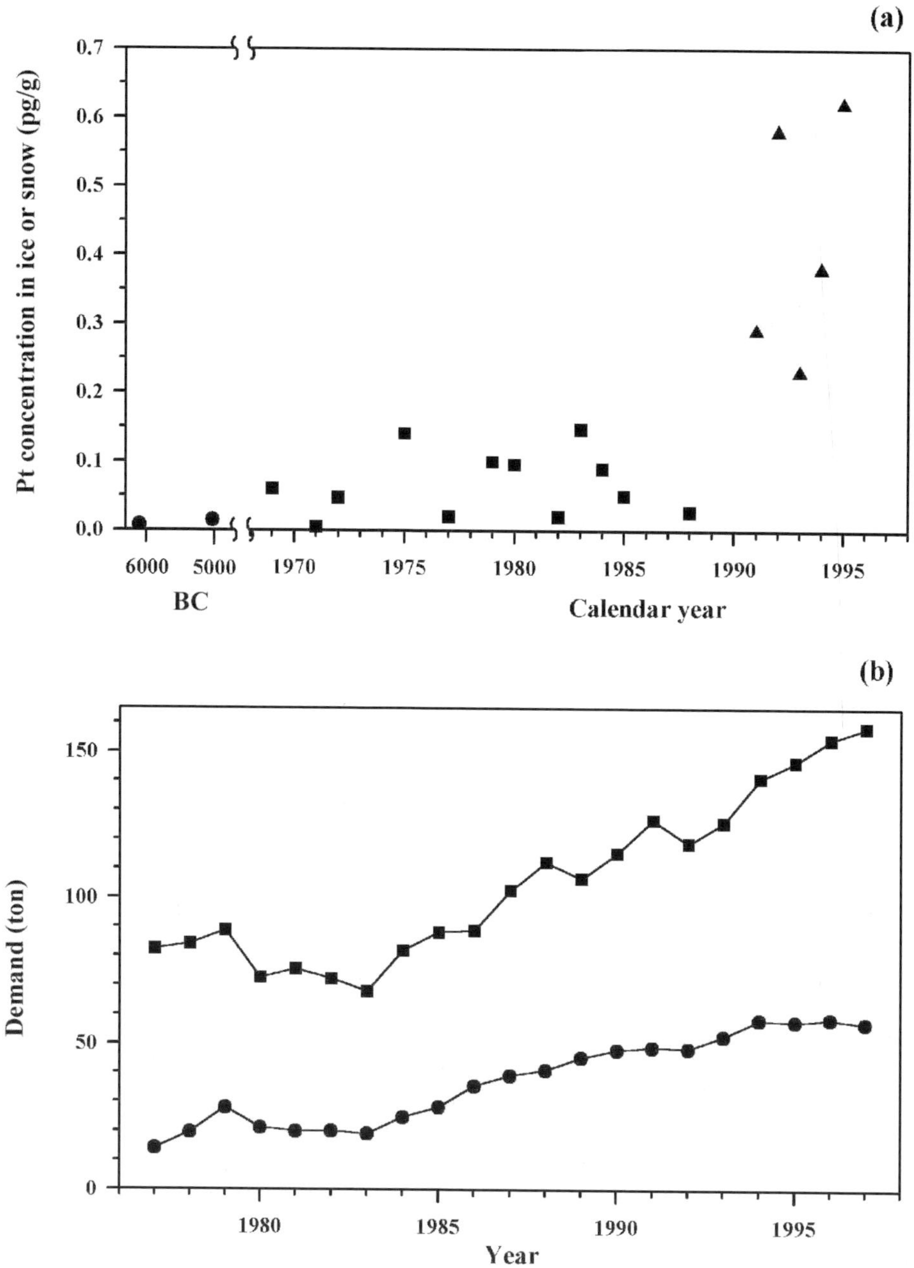

Figure 3.8. (a) Changes in Pt concentrations as function of ice and snow age in central Greenland (Summit); (●) ancient ice, (■) snow from the shallow core; (▲) snow pit samples. (b) Data on changes in world demand for Pt; (■) total demand; (●) demand for automobile catalytic converters. Adapted from ref. (27).

converters. The fact that this pollution can be seen even in the free troposphere of the remote Arctic indicates that there is a man-induced large scale pollution of the whole northern hemisphere.

4.5. DPASV

DPASV (92) is largely appreciated for its high sensitivity and for the possibility of working in remote areas such as polar ice caps in order to check samples and sampling procedures directly in the field and to obtain reliable results on the occurrence of heavy metals in polar snow and ice (49, 56, 93, 94). This electroanalytical technique is widely used for trace metal analysis in sea water (80); the high dissolved salt content makes this matrix the ideal medium in which to perform trace metal analysis by DPASV. Because of its high sensitivity, at and below the pg/g level for Cd and Pb, it is also suitable for analysis of trace metals in polar snow (49, 56, 93, 94). Nevertheless, in spite of much effort to develop this methodology, it is still restricted to a minority of laboratories owing to the operator skills required, the low throughput of samples and the high sample volume needed for each determination.

The technique is based on controlled potential electrolytic deposition of analytes on a Thin Mercury Film Electrode (TMFE) prepared beforehand on the surface of a Rotating Glassy Carbon Disk Electrode (RGCDE). This procedure acts as a sort of preconcentration of analytes which form an amalgam on the surface of the RGCDE. Successively, applying a potential scanning in the differential pulse mode, the analytes are redissolved in the solution while the diffusion current involved in the process is recorded (57). The concentration of metals in the sample is then back calculated by the height of the peak current developed during the stripping phase.

Analyses can be carried out in clean chemistry laboratories both on site, in order to check sampling procedures immediately after collection and after a long storage period (27, 49, 56, 57). Successful results obtained from experiments performed on site confirmed that this technique is the only one which can be used in remote areas, allowing checks on samples/sampling procedures and measurements directly immediately in the field.

This methodology has been used for the preliminary assessment of the Cd, Cu, Pb and Zn content in Antarctic snow samples (93). Comparing these results with those recently obtained on the same elements and considered reliable (27–29, 72) it seems that contamination problems had a serious effect on these determinations. Nevertheless, this pioneering study opened the door to in-field measurements, enabling the researchers to check the quality of their data in the field. In this respect, DPASV has also been used in the Arctic, on board a research vessel, with the aim of determining seasonal variations of Cd, Cu, Pb and Ni (56) in meteoric precipitation, showing once more the high adaptability of the technique to extreme field conditions. Comparison of the results obtained in the field for Cd and Pb with those measured in the laboratory has been made by Völkening and Heumann (49) and Barbante et al. (62) using IDMS and a double focusing magnetic sector ICP–MS for laboratory studies, respectively.

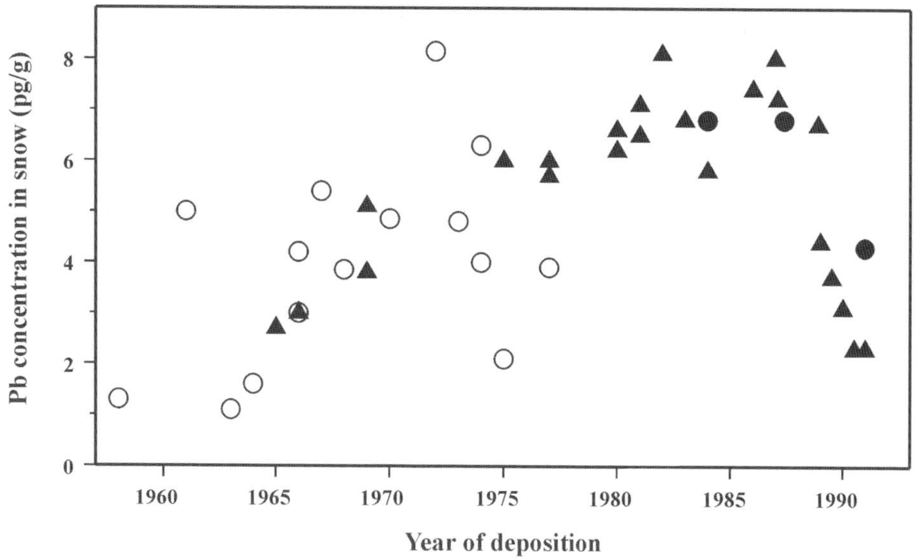

Figure 3.9. Temporal trend of Pb concentration in Antarctic snow from Victoria Land for the period 1965–1991. (○) Stake D55, (●) McCarthy Ridge, (▲) Styx Glacier plateau. Adapted from refs. (29, 91).

More recently, DPASV has been used for the simultaneous determination of Cd and Pb in recent east Antarctic snow (Victoria Land) (27, 57). The deep concentration profile for Pb, extended from the mid 1960s to the beginning of the 1990s (see Figure 3.9) shows a very interesting pattern. After a continuous increase up to the mid 1980s, from 2.7 pg/g to a maximum of 8.1 pg/g, a net decrease is observed down to about 3 pg/g in 1991. The present results confirm and extend previous initial evidence (28, 30, 57) concerning the reverse trend occurring in the Pb concentration of Antarctic snow over the last two decades. Moreover, considering also all the available data from the literature (28–30, 57) it is worth noting that the recent historical trend of Pb concentration in Antarctic snow is quite similar, even if in a lower range of concentrations, to that already found in the Arctic (Central Greenland) (55, 95) (Figure 3.5). By comparing separately the temporal trend of Pb concentration reported in the literature for Coats Land (Atlantic sector of east Antarctica) (28) and that obtained for Victoria Land (Pacific sector), respectively, with the gasoline Pb consumption data in South America and in Oceania it was also possible to identify the geographical origin of the Pb-rich aerosol reaching the Antarctic continent (27, 81).

4.6. Coupled techniques for the analysis of organometal compounds

Coupled techniques for the analysis of polar snow and ice were used a few years ago in order to achieve good evidence of the influence of human activities and in particular of the use of gasoline additives, in the Pb pollution of the Arctic region

(75–77). Organo-Pb compounds have in fact been added to gasolines since the 1920s mainly in the form of apolar methylated and ethylated tetraalkyl-Pb in order to act as antiknocking compounds in modern engines. During combustion there is an incomplete breakdown of the molecules and hence ionic trialkyl-Pb and dialkyl-Pb species are formed and emitted into the environment.

Many analytical problems were faced by researchers involved in this field, mainly because of the limited amount of sample available (usually only 50–100 g of the central part of the decontaminated snow core) and the very low concentrations encountered. In fact, considering that the Pb concentration level in Greenland snow dating back to the last century is in the range 1–200 pg/g (55) and considering that the fraction of total organo-Pb compounds is about 0.1–0.5% (96) the concentrations to be detected are in the low fg/g level.

To detect organometal compounds in environmental samples, it is necessary to resort to hyphenated analytical techniques which couple the separation power of Gas Chromatography (GC) or Liquid Chromatography (LC) with an element-sensitive detector, such as an AAS or an AES instrument (97). In particular, in the case of extremely low concentrations in snow samples the detection power of traditional atomic spectrometric techniques is usually insufficient to detect and quantify the organometal level in polar samples. For this reason the use of Microwave-Induced Plasma Atomic Emission Spectrometry (MIP–AES), with experimental detection limits at the 0.01 pg/g level, would be preferable (75). However, a specific sample preparation procedure involving preconcentration and derivatisation of the analytes is needed (75). Sample preparation consists first of complexing ionic organo-Pb compounds with diethyldithiocarbamate (DDTC), yielding at this stage a preconcentration of about 200 times; complexes are then extracted with hexane and derivatised by propylation using a Grignard reagent. Finally, after the decomposition of the excess of the derivatising reagent, the extract is further concentrated by about 50 times using an on-line solvent venting technique and analysed by means of a GC–MIP–AES coupled technique (75).

Results obtained on 18 fresh snow samples collected in Greenland from January to August 1987 have shown a seasonal behaviour in the concentrations of ionic ethyl-Pb species; lower concentrations were found in summer and much higher (by a factor of 5) in winter and early spring. Seasonal differences were also found to be stronger for methyl-Pb species, which proved to be totally absent during the summer months. These patterns reflect the extension of the polar front into mid-latitudinal populated areas (such as Europe) in winter and spring (98), which can strongly contribute to the transport of polluting substances into the Greenland ice sheet (76).

On a longer time scale (see Figure 3.10) the total organo-Pb deep profile covering the period 1920–1989 shows a continuous increasing trend from the 1920s up to the early 1970s. Afterwards, a rapid increase (by a factor of 5) from the early 1970s to the early 1980s was registered. In the more recent snow layers a well marked decrease was finally noted testifying to a decrease in the atmospheric contamination from organo-Pb compounds of the northern hemisphere (77). This behaviour has been correlated with the data on the distribution of the organo-Pb compounds in gasoline, the consumption of leaded gasoline in Europe and USA

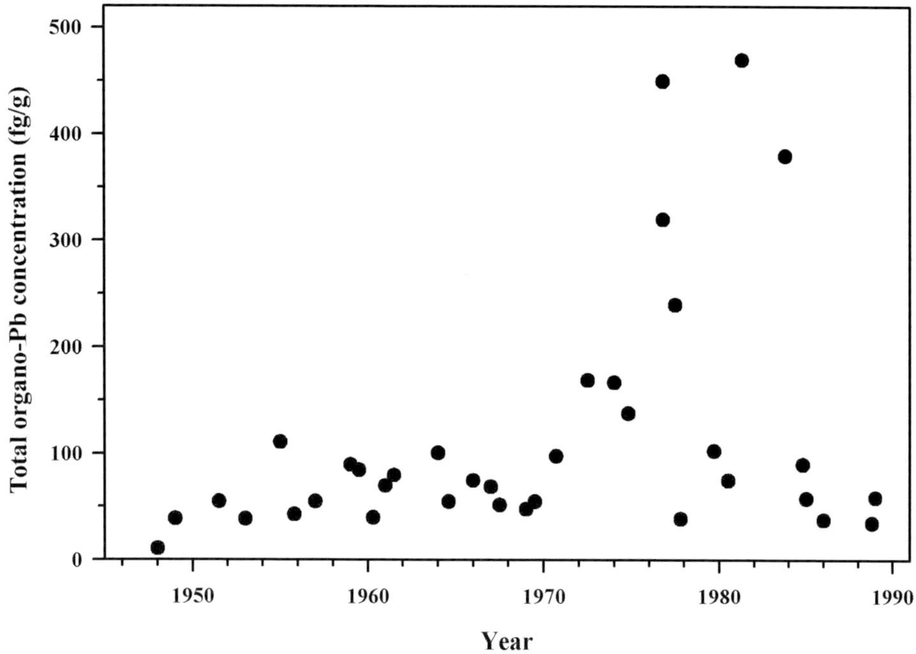

Figure 3.10. Changes in total organo-Pb compounds in central Greenland (Summit) snow from the 1930s. Adapted from ref. (77).

and with the isotopic composition of total Pb in selected samples (77), confirming how policy initiatives taken in order to lower the content of Pb in gasolines were promptly reflected in an improvement of the quality of air as testified once more in polar snow and ice.

5. Conclusions

The study of trace element distribution in polar regions provides outstanding information on the history of global scale atmospheric pollution caused by man. Lead concentration profiles in Greenland and Antarctic snow and ice certainly contribute to a better knowledge of the worldwide dispersion of toxic substances emitted by several sources at mid latitudes and scavenged to the ground up to the polar ice caps. Together with Pb, other heavy metals, such as Cd, Cu and Zn and more recently the PGEs have revealed their presence in the snow of these pristine environments, with concentrations above the natural background levels, pushing researchers to new objectives in the field of analytical chemistry in order to develop novel approaches for the determination of these analytes. In particular, the study of the occurrence of PGEs in polar snows constitutes an absolute novelty in terms of the effects of the use of gasoline of new formulation on the environment.

The very few laboratories in the world that accepted this challenge have used sophisticated sampling procedures in the field and powerful analytical techniques such as double focusing magnetic sector ICP–MS, LEAFS and TIMS, to produce a wealth of reliable data useful in the difficult task of deciphering snow and ice archives.

Acknowledgements

This study was carried out in the framework of projects on environmental contamination of the Italian National Programme of Researches in Antarctica (PNRA) and financially supported by ENEA through cooperation agreements with the University of Venice. This is a *European Project for Ice Coring in Antarctica* (EPICA) publication No. 25. This work is a contribution to the joint *European Science Foundation* (ESF/EC) scientific programme, funded by the European Commission under the Environment and Climate Programme (1994–1998) contract ENV4-CT95-0074 and by national contributions from Belgium, Denmark, France, Germany, Italy, The Netherlands, Norway, Sweden, Switzerland and United Kingdom. The authors gratefully acknowledge useful discussions with Giulio Cozzi.

References

1. C. F. Boutron, Historical reconstruction of the earth's past atmospheric environment from Greenland and Antarctic snow and ice cores, *Environ. Rev.*, 3 (1995), 1–28.
2. S. Hong, J. P. Candelone, C. C. Patterson, C. F. Boutron, Greenland ice evidence of hemispheric scale pollution for lead two millennia ago by Greek and Roman civilizations, *Science*, 265 (1994), 1841–1843.
3. S. Hong, J. P. Candelone, C. C. Patterson, C. F. Boutron. History of ancient copper smelting pollution during Roman and Medieval times recorded in Greenland ice, *Science*, 272 (1996), 246–249.
4. K. J. R. Rosman, W. Chisholm, S. Hong, J. P. Candelone, C. F. Boutron, Lead from Carthaginian and Roman Spanish mines isotopically identified in Greenland ice dated from 600 BC to 300 AD, *Environ. Sci. Technol.*, 31 (1997), 3413–3416.
5. G. M. Vandal, W. F. Fitzgerald, C. F. Boutron, J. P. Candelone, Variations in mercury deposition to Antarctica over the past 34,000 years, *Nature*, 362 (1993), 621–623.
6. C. F. Boutron, C. C. Patterson, Lead concentration changes in Antarctic ice during the Wisconsin/Holocene transition, *Nature*, 323 (1986), 222–225.
7. C. F. Boutron, S. N. Rudniev, M. A. Bolshov, V. G. Koloshnikov, Patterson, C. C., Barkov, N. I., Changes in cadmium concentrations in Antarctic ice and snow during the past 155,000 years, *Earth Planet. Sci. Lett.*, 117 (1993), 431–444.
8. C. C. Patterson, D. M. Settle, in *Accuracy in Trace Analysis*; P. L. Fleur (Ed.); Natl. Bur. Stand. Spec. Publ., 1976; Vol. 422, 321–351.
9. J. R. Moody, NBS clean laboratories for trace element analysis, *Anal. Chem.*, 54 (1982), 1358A–1376A.
10. C. F. Boutron, A clean laboratory for ultralow concentration heavy metal analysis, *Fresenius J. Anal. Chem.*, 337 (1990), 482–491.
11. M. Murozumi, T. J. Chow, C. C. Patterson, Chemical concentrations of pollutant lead aerosols, terrestrial dusts and sea salts in Greenland and Antarctic snow strata, *Geochim. Cosmochim. Acta*, 33 (1969), 1247–1294.

12. J. M. Barnola, P. Pimienta, D. Raynaud, Y. S. Korotkevich, CO_2-climate relationship as deduced from the Vostok ice core: A reexamination based on new measurements and on a reevaluation of the air dating, *Tellus*, 43B (1991), 83–90.
13. J. Schwander, J. M. Barnola, C. Andrié, M. Leuenberger, A. Ludin, D. Raynaud, B. Stauffer, The age of the air in the firn and the ice at Summit, Greenland, *J. Geophys. Res.*, 98 (1993), 2831–2838.
14. M. Legrand, J. Jouzel, D. Raynaud, in *Topics in Atmospheric and Interstellar Physics and Chemistry*; C. F. Boutron (Ed.); Les Editions de Physique: Paris, France, 1994, 453–477.
15. T. Blunier, J. A. Chappellaz, J. Schwander, A. Dallenbach, B. Stauffer, T. Stocker, D. Raynaud, J. Jouzel, H. B. Clausen, C. U. Hammer, S. J. Johnsen, Asynchrony of Antarctic and Greenland climate change during the last glacial period, *Nature*, 394 (1998), 739–743.
16. B. Stauffer, T. Blunier, A. Dallenbach, A. Indermuhle, J. Schwander, T. Stocker, J. Tschumi, J. A. Chappellaz, D. Raynaud, C. U. Hammer, H. B. Clausen, Atmospheric CO_2 concentration and millennial-scale climate change during the last glacial period, *Nature*, 392 (1998), 59–62.
17. C. E. Junge, International Association of Hydrological Sciences, 1977.
18. L. A. Barrie, Atmospheric particles: their physical and chemical characteristics, and deposition processes relevant to the chemical composition of glaciers, *Ann. Glaciol.*, 7 (1985), 100–108.
19. L. A. Barrie, in *Seasonal Snowpacks. NATO-ASI Series*; T. D. Davies, M. Tranter, H. G. Jones (Eds.); Springer-Verlag: Berlin, 1991; Vol. G28, 1–20.
20. C. I. Davidson, in *The Environmental Record in Glaciers and Ice Sheets*; H. Oeschger, C. C. Langway, J. (Eds.); Wiley: New York, 1989, 29–51.
21. J. W. C. White, in *The Environmental Record in Glaciers and Ice Sheets*; by H. Oeschger and C. C. Langway, J. (Ed.); Wiley: New York, 1989, 85–89.
22. E. W. Wolff, Signals of atmospheric pollution in polar snow and ice, *Antarctic Science*, 2 (1990), 189–205.
23. J. R. Petit, I. Basile, A. Leruyuet, D. Reynaud, C. Lorius, J. Jouzel, M. Stievenard, V. Y. Lipenkov, N. I. Barkov, B. B. Kudryashov, M. Davis, E. Saltzman, V. I. Kotlyakov, Four climate cycles in Vostok ice core, *Nature*, 387 (1997), 359–360.
24. C. F. Boutron, C. C. Patterson, N. I. Barkov, The occurrence of zinc in Antarctic ancient ice and recent snow, Earth Planet. Sci. Lett., 101 (1990), 248–259.
25. E. D. Suttie, E. W. Wolff, Seasonal input of heavy metals to Antarctic snow, *Tellus*, 44B (1992), 351–357.
26. E. D. Suttie, E. W. Wolff, The local deposition of heavy metal emissions from point sources in Antarctica, *Atmos. Environ.*, 27A (1993), 1833–1841.
27. C. Barbante, C. Turetta, G. Capodaglio, G. Scarponi, Recent decrease in the lead content of Antarctic snow, *Int. J. Environ. Anal. Chem.*, 68 (1997), 457–477.
28. E. W. Wolff, E. D. Suttie, Antarctic snow record of southern hemisphere lead pollution, *Geophys. Res. Lett.*, 21 (1994), 781–784.
29. U. Görlach, C. F. Boutron, Variations in heavy metals concentrations in Antarctic snows from 1940 to 1980, *J. Atmos. Chem.*, 14 (1992), 205–222.
30. G. Scarponi, C. Barbante, C. Turetta, Recent variations in the lead content of Antarctic snow, *Terra Antartica Reports*, 1 (1997), 103–106.
31. F. Gillet, D. Donnou, C. Girard, A. Manouvrier, C. Rado, G. Ricou; G. Holdsworth, K. C. Kuivinen, J. H. Rand (Eds.); U.S. Army Cold Regions Res. Eng. Lab. Spec. Res. Rept, 1984, 73–80.
32. W. Dansgaard, S. J. Johnsen, H. B. Clausen, D. Dahl-Jensen, N. S. Gundestrup, C. U. Hammer, C. S. Hvidberg, J. P. Steffensen, A. E. Sveinbjörnsdottir, J. Jouzel, G. Bond, Evidence for general instability of past climate from a 250–kyr ice core record, *Nature*, 364 (1993), 218–220.
33. J. Jouzel, N. I. Barkov, J. M. Barnola, M. Bender, J. Chappellaz, C. Genthon, V. M. Kotlyakov, V. Lipenkov, C. Lorius, J. R. Petit, D. Raynaud, G. Raisbeck, C. Ritz, T. Sowers, M. Stievenard, F. Yiou, P. Yiou, Extending the Vostok ice-core record of palaeoclimate to the penultimate glacial period, *Nature*, 364 (1993), 407–412.
34. J. P. Candelone, S. Hong, C. F. Boutron, An improved method for decontaminating polar snow and ice cores for heavy metals analysis, *Anal. Chim. Acta*, 299 (1994), 9–16.
35. J. P. Candelone, S. Hong, C. Pellone, C. F. Boutron, Post industrial revolution changes in large scale atmospheric pollution of the Northern Hemisphere by heavy metals as documented in central Greenland snow and ice, *J. Geophys. Res.*, 100 (1995), 16,605–16,616.

36. C. F. Boutron, F. M. Ducroz, U. Gorlach, J. L. Jaffrezo, C. I. Davidson, M. A. Bolshov, Variations in heavy metals concentrations in fresh Greenland snow from January to August 1989, *Atmos. Environ.*, 27A (1993), 2773–2779.
37. Federal Standard 209, U.S., 1988.
38. P. J. Paulsen, E. S. Beary, D. S. Bushee, J. R. Moody, Inductively coupled plasma mass spectrometric analysis of ultrapure acids, *Anal. Chem.*, 60 (1988), 971–975.
39. U. Görlach, C. F. Boutron, Preconcentration of lead, cadmium, copper and zinc in water at the pg g^{-1} level by non-boiling evaporation, *Anal. Chim. Acta*, 236 (1990), 391–398.
40. M. A. Bolshov, C. F. Boutron, F. M. Ducroz, U. Görlach, O. M. Kompanetz, S. M. Rudniev, B. Hutch, Direct ultratrace determination of cadmium in Antarctic and Greenland snow and ice by laser atomic fluorescence spectrometry, *Anal. Chim. Acta*, 251 (1991), 169–175.
41. M. A. Bolshov, C. F. Boutron, A. V. Zybin, Determination of lead in Antarctic ice at the picogram-per-gram level by Laser Atomic Fluorescence Spectroscopy, *Anal. Chem.*, 61 (1989), 1758–1762.
42. W. Chisholm, K. J. R. Rosman, C. F. Boutron, J. P. Candelone, S. Hong, Determination of lead isotopic ratios in Greenland and Antarctic snow and ice with picogram per gram concentrations, *Anal. Chim. Acta*, 311 (1995), 141–151.
43. W. Chisholm, K. J. R. Rosman, J. P. Candelone, C. F. Boutron, M. A. Bolshov, Measurement of bismuth at pg g^{-1} concentrations in snow and ice samples by thermal ionisation mass spectrometry of lead isotopic ratios in Greenland and Antarctic snow and ice with picogram per gram concentrations, *Anal. Chim. Acta*, 347 (1997), 351–358.
44. E. D. Suttie, E. W. Wolff, Preconcentration method for electrothermal atomic absorption spectrometric analysis of heavy metals in Antarctic snow at sub ng kg^{-1} levels, *Anal. Chim. Acta*, 258 (1992), 229–236.
45. A. Ng, C. C. Patterson, Natural concentrations of lead in ancient Arctic and Antarctic ice, *Geochim. Cosmochim. Acta*, 45 (1981), 2109–2121.
46. V. M. Apatin, B. V. Arkhangelskii, M. A. Bolshov, V. V. Ermolov, V. G. Koloshnikov, O. M. Kompanetz, N. L. Kuznetsov, E. L. Mikhailov, V. S. Shishkovskii, C. F. Boutron, Automated laser excited atomic fluorescence spectrometer for determination of trace concentrations of elements, *Spectrochim. Acta*, 44B (1989), 253–262.
47. M. A. Bolshov, S. N. Rudniev, J. P. Candelone, C. F. Boutron, S. Hong, Ultratrace determination of Bi in Greenland snow by laser atomic fluorescence spectrometry, *Spectrochim. Acta*, 49B (1994), 1445–1452.
48. M. A. Bolshov, V. G. Koloshnikov, S. N. Rudniev, C. F. Boutron, U. Görlach, C. C. Patterson, Detection of trace amounts of toxic metals in environmental samples by laser-excited atomic fluorescence spectrometry, *J. Anal. At. Spectrom.*, 7 (1992), 99–104.
49. J. Völkening, K. G. Heumann, Determination of heavy metals at the pg/g level in Antarctic snow with DPASV and IDMS, *Fresenius J. Anal. Chem.*, 331 (1988), 174–181.
50. K. G. Heumann, Determination of inorganic and organic traces in the clean room compartment of Antarctica, *Anal. Chim. Acta*, 283 (1993), 230–245.
51. K. J. R. Rosman, W. Chisholm, C. F. Boutron, J. P. Candelone, U. Görlach, Isotopic evidence for the sources of lead in Greenland snows since the late 1960s, *Nature*, 362 (1993), 333–335.
52. K. J. R. Rosman, W. Chisholm, C. F. Boutron, J. P. Candelone, C. C. Patterson, Anthropogenic lead isotopes in Antarctica, *Geophys. Res. Lett.*, 21 (1994), 2669–2672.
53. K. J. R. Rosman, W. Chisholm, C. F. Boutron, S. Hong, R. Edwards, V. Morgan, P. Sedwick, Lead isotopes and selected metals in ice from Law Dome, Antarctica, *Ann. Glaciol.*, 27 (1997), 349–354.
54. F. Batifol, C. Boutron, M. de Angelis, Changes in copper, zinc and cadmium concentrations in Antarctic ice during the past 40,000 years, *Nature*, 337 (1989), 544–546.
55. C. F. Boutron, U. Görlach, J. P. Candelone, M. A. Bolshov, R. J. Delmas, Decrease in anthropogenic lead, cadmium and zinc in Greenland snows since the late 1960s, *Nature*, 353 (1991), 153–156.
56. L. Mart, Seasonal variations of Cd, Pb, Cu and Ni levels in snow from the eastern Arctic Ocean, *Tellus*, 35B (1983), 131–141.
57. G. Scarponi, C. Barbante, P. Cescon, Differential pulse anodic stripping voltammetry

for ultratrace determination of cadmium and lead in Antarctic snow, *Analusis*, 22 (1994), M47–M50.

58. G. M. Vandal, W. F. Fitzgerald, C. F. Boutron, J. P. Candelone, in *Ice Core Studies of Global Biogeochemical Cycles*; R. J. Delmas, C. Hammer, D. Kley, P. Mayewski (Eds.); Kluwer: Dordrecht, 1995, 401–415.
59. T. D. Jickells, T. D. Davies, M. Tranter, S. Landsberger, K. Jarvis, P. W. Abrahams, Trace elements in snow samples from the Scottish Highlands: sources and dissolved/particulate distributions, *Atmos. Environ.*, 26A (1992), 393–401.
60. T. D. Davies, M. Tranter, T. D. Jickells, P. W. Abrahams, S. Landsberger, K. Jarvis, C. E. Pierce, Heavily–contaminated snowfalls in the remote Scottish Highlands: a consequence of regional scale mixing and transport, *Atmos. Environ.*, 26A (1992), 95–112.
61. R. E. Sturgeon, S. N. Willie, J. Zheng, A. Kudo, D. C. Grégoire, Determination of ultratrace levels of heavy metals in Arctic snow by electrothermal vaporization Inductively Coupled Plasma Mass Spectrometry, *J. Anal. At. Spectrom.*, 8 (1993), 1053–1058.
62. C. Barbante, T. Bellomi, G. Mezzadri, P. Cescon, G. Scarponi, C. Morel, S. Jay, K. Van de Velde, C. Ferrari, C. F. Boutron, Direct determination of heavy metals at picogram per gram level in Greenland and Antarctic snow by Double Focusing Inductively Coupled Plasma Mass Spectrometry, *J. Anal. At. Spectrom.*, 12 (1997), 925–931.
63. C. Barbante, G. Cozzi, G. Capodaglio, K. Van de Velde, C. Ferrari, A. Veysseyre, C. F. Boutron, G. Scarponi, P. Cescon, Determination of Rh, Pd and Pt in Polar and Alpine snow and ice by double focusing ICP–MS with microconcentric nebulization, *Anal. Chem.*, 71 (1999), 4125–4133.
64. C. F. Boutron, C. C. Patterson, V. N. Petrov, N. I. Barkov, Preliminary data on changes of lead concentrations in Antarctic ice from 155,000 to 26,000 years BP, *Atmos. Environ.*, 21 (1987), 1197–1202.
65. E. W. Wolff, M. P. Landy, E. D. Suttie, *Anal. Chem.*, 53 (1981), 1566–1569.
66. K. J. R. Rosman, W. Chisholm, C. F. Boutron, J. P. Candelone, S. Hong, Isotopic evidence to account for changes in the concentration of lead in Greenland snow between 1960 and 1988, *Geochim. Cosmochim. Acta*, 58 (1994), 3265–3269.
67. K. J. R. Rosman, W., Chisholm, C. F. Boutron, J. P. Candelone, J. L. Jaffrezo, C. I. Davidson, Seasonal variations in the origin of lead in snow at Dye 3, Greenland, *Earth Planet. Sci. Lett.*, 160 (1998), 383–389.
68. C. C. Patterson, *Science*, 183 (1974), 553–554.
69. E. W. Wolff, D. A. Peel, The record of global pollution in polar snow and ice, *Nature*, 313 (1985), 535–540.
70. C. F. Boutron, in *Toxic metals in the atmosphere*; J. O. Nriagu, C. I. Davidson (Eds.); Wiley: New York, 1986, 467–505.
71. C. F. Boutron, C. C. Patterson, Relative levels of natural and anthropogenic lead in recent Antarctic snow, *J. Geophys. Res.*, 92 (1987), 8454–8464.
72. E. W. Wolff, E. D. Suttie, D. A. Peel, Antarctic snow record of cadmium, copper and zinc content during the twentieth century., *Atmos. Environ.*, 33 (1999), 1535–1541.
73. J.-P. Candelone, S. Hong, C. F. Boutron, in *Heavy Metals in the Environment*; R. J. Allan, J. O. Nriagu (Eds.); Page Bros.: Norwich, U.K., 1995, 575–577.
74. J. O. Nriagu, Global inventory of natural and anthropogenic emissions of trace metals to the atmosphere, *Nature*, 279 (1979), 409–411.
75. R. Lobinski, C. F. Boutron, J. P. Candelone, S. Hong, J. Szpunar-Lobinska, F. C. Adams, Speciation analysis of organolead compounds in Greenland snow at the femtogram-per-gram level by capillary gas chromatography/atomic emission spectrometry, *Anal. Chem.*, 65 (1993), 2510–2515.
76. R. Lobinski, C. F. Boutron, J. P. Candelone, S. Hong, J. Szpunar-Lobinska, F. C. Adams, Northern Hemispheric organic lead emissions in fresh Greenland snow, *Environ. Sci. Technol.*, 28 (1994), 1459–1466.
77. R. Lobinski, C. F. Boutron, J. P. Candelone, S. Hong, J. Szpunar-Lobinska, F. C. Adams, Present century snow core record of organolead pollution in Greenland, *Environ. Sci. Technol.*, 28 (1994), 1467–1471.
78. C. F. Boutron, C. C. Patterson, The occurrence of lead in Antarctic recent snow, firn deposited over the last two centuries and prehistoric ice, *Geochim. Cosmochim. Acta*, 47 (1983), 1355–1368.

79. G. Scarponi, C. Barbante, C. Turetta, A. Gambaro, P. Cescon, Chemical contamination of Antarctic snow: the case of lead, *Microchem. J.*, 55 (1997), 24–32.
80. G. Scarponi, G. Capodaglio, C. Barbante, P. Cescon, The anodic stripping voltammetric titration procedure for study of trace metal complexation in sea water, in *Element Speciation in Bioinorganic Chemistry*; S. Caroli (Ed.); Wiley: New York, 1996, 363–418.
81. C. Barbante, C. Turetta, A. Gambaro, G. Capodaglio, G. Scarponi, Sources and origins of aerosols reaching Antarctica as revealed by concentration profiles in shallow snow, *Ann. Glaciol.*, 27 (1997), .
82. G. E. Shaw, Antarctic aerosols: a review, *Rev. Geophys.*, 26 (1988), 89–112.
83. G. E. Shaw, in *The Environmental Record in Glaciers and Ice Sheets*; H. Oeschger, C. C. Langway, J. (Eds.); Wiley: Chichester, 1989, 13–27.
84. C. C. Patterson, D. M. Settle, *Geochim. Cosmochim. Acta*, 51 (1987), 675.
85. G. Lambert, M. F. Le Cloarec, M. Pennisi, *Geochim. Cosmochim. Acta*, 52 (1988), 39.
86. M. F. Le Cloarec, P. Allard, B. Ardouin, W. F. Giggenbach, D. S. Sheppard, *Earth Planet. Sci. Lett.*, 108 (1992), 19.
87. S. Hong, J. P. Candelone, C. F. Boutron, Deposition of atmospheric heavy metals to the Greenland ice sheet from the 1783–1784 volcanic eruption of Laki, Iceland, *Earth Planet. Sci. Lett.*, 144 (1996), 605–610.
88. A. Montaser, *Inductively coupled plasma mass spectrometry*, 1998, Wiley-VCH, New York, pp 964.
89. L. Moens, N. Jakubowski, Double-Focusing Mass spectrometers in ICP-MS, *Anal. Chem.*, (1998), 251A–256A.
90. C. Barbante, G. Cozzi, G. Capodaglio, A. Gambaro, G. Scarponi, P. Cescon, 6th National Conference on Biogeochemical Cycles of Contaminants, Venice, 27–28 April 1998, Book of Abstracts, 35–40.
91. M. Krachler, A. Alimonti, F. Petrucci, K. J. Irgolić, F. Forastiere, S. Caroli, Analytical problems in the determination of platinum-group metals in urin by quadrupole and magnetic sector field inductively coupled plasma mass spectrometry, *Anal. Chim. Acta*, 363 (1998), 1–10.
92. J. Wang, Stripping analysis, principles instrumentation and applications, 1985, Wiley, VCH, Deerfield Beach, Florida, pp 160.
93. M. L. Landy, An evaluation of differential pulse anodic stripping voltammetry at a rotating glassy carbon electrode for the determination of cadmium, copper lead and zinc in Antarctic snow samples, *Anal. Chim. Acta*, 121 (1980), 39–49.
94. L. Mart, H. W. Nürberg, P. Valenta, *Fresenius' Z. Anal. Chem.*, 300 (1980), 350–362.
95. S. Hong, J. P. Candelone, C. F. Boutron, Greenland ice history of the pollution of the atmosphere of the northern hemisphere for lead during the last three millennia, *Analusis*, 22 (1994), M38–M40.
96. R. J. A. Van Cleuvenbergen, F. C. Adams, *Environ. Sci. Technol.*, 26 (1992), 1354–1360.
97. R. Lobinski, W. M. R. Dirkx, J. Szpunar-Lobinska, F. C. Adams, Speciation analysis of organolead compounds by gas chromatography with atomic spectrometric detection, *Anal. Chim. Acta.*, 286 (1994), 381–390.
98. L. A. Barrie, Arctic air pollution: an overview of current knowledge, *Atmos. Environ.*, 20 (1986), 643–663.
99. C. F. Boutron, G. V. Vandal, W. F. Fitzgerald, C. P. Ferrari, A forty year record of mercury in central Greenland snow, *Geophys. Res. Lett.*, 25 (1998), 3315–3318.
100. G. L. Long, J. D. Winefordner, Limit of detection, a closer look at the IUPAC definition, *Anal. Chem.*, 55 (1983), 712A–724A.
101. J. R. Moody, R. Lindstrom, *Anal. Chem.*, 49 (1977), 2264–2267.
102. C. F. Boutron, M. Leclerc, N. Risler, Atmospheric trace elements in Antarctic prehistoric ice collected at a coastal ablation area, *Atmos. Environ.*, 18 (1984), 1947–1953.
103. M. De Angelis, N. I. Barkov, V. N. Petrov, Aerosol concentration over the last climatic cycle (160 kyr) from an Antarctic ice core, *Nature*, 325 (1987), 318–321.

Environmental Contamination in Antarctica
A Challenge to Analytical Chemistry
S. Caroli, P. Cescon and D.W.H. Walton, editors

Chapter 4

Natural isotopic variations in lead in polar snow and ice as indicators of source regions

Kevin J. R. Rosman

1. Introduction

Our understanding of the extent to which the polar regions have been contaminated with heavy metals has increased enormously during the past three decades (1), but our knowledge of the origin of these pollutants is still rather limited. Mass spectrometry is a technique which can utilise natural isotopic tracers for this purpose and is now proving to be effective in polar regions.

Mass spectrometry is commonly used to measure both the mass and abundances of isotopes and molecules (2, 3), but here we will focus on the measurement of the abundances of isotopes.

The measurement of isotopic abundances began early this century following the discovery of neon isotopes by J.J. Thompson in 1912. F.W. Aston developed the mass spectrometer into a quantitative instrument for measuring isotopic abundances and by 1935 the isotopic composition of most elements was known. The first International Table of Stable Isotopes was drawn up in 1936, while the latest table of Isotopic Compositions of the Elements appeared recently (4). Lead is an element for which there was early evidence of natural variations in its isotopic composition (5); these were ultimately used to measure the age of the Earth (6). Natural variations have been reported in 43 other elements although many relate to exceptional samples. There are 18 elements in which variations are not uncommon, although most of these elements have relatively light atoms with atomic numbers less than 16 (4).

Early applications of mass spectrometry were in the fields of nuclear physics and geochronology. Identification by Nier et al. (7) of ^{235}U as the fissionable component of natural U led to the construction of isotope separators which were ultimately used to produce enriched isotopes of most elements following World War II (1939–45). The ready availability of enriched isotopes was particularly important for the development of isotope geology where they are needed for geochronology and isotope geochemistry (8). These enriched materials were used by Patterson during the 1950s to accurately measure Pb and U in terrestrial rocks and meteorites and resulted in the first accurate age for the Earth. Isotopic techniques developed during this period eventually led to the first reliable measurements of Pb in Greenland ice by Murozumi et al. (9).

The pioneering measurements performed in Patterson's CIT laboratory by Murozumi et al. (9) showed that the concentration of Pb in Greenland ice increased from <1 pg/g ice at 800 BC to >200 pg/g by 1965. The early concentration was consistent with estimates based on dust levels (35 μg/kg) which predicted ~0.4 pg/g although additional Pb from volcanic emanations was expected. In comparison, the dust concentration measured in Antarctica was ~2 μg/kg suggesting Pb at ~0.03 pg/g ice, but this was below the detection limits at the time. It was, to a large extent, through an understanding the sources of contamination, provided by the use of mass spectrometry, that this incredible feat of measurement was accomplished.

Since the 1960s there have been significant improvements in the techniques used to sample and to process snow and ice samples, making it possible to analyse much smaller samples than ever before with an even higher degree of reliability. Perhaps the most significant advance was the mechanical decontamination procedure developed in Patterson's CIT laboratory (10) and later refined in Boutron's LGGE laboratory (11). This approach has proved to be successful even for measuring Pb in Greenland and Antarctic ice cores where concentrations are at sub-pg g^{-1} levels (12).

2. Isotopic composition of lead

Lead has four stable isotopes: ^{204}Pb, ^{206}Pb, ^{207}Pb and ^{208}Pb, with abundances typically 1.4, 24.1, 22.1 and 52.4, percent respectively (4). Lead displays large natural variations in isotopic composition depending on its origin, because the isotopes ^{206}Pb, ^{207}Pb and ^{208}Pb are the stable end products of three different radioactive decay series starting with ^{238}U (half-life 4.5×10^9 years), ^{235}U (half-life 7.1×10^8 years) and ^{232}Th (half-life 1.4×10^{10} years), respectively. While the composition of Pb in crustal rocks containing U and Th steadily changes over geological time, Pb in ores remains constant because at the time of ore formation the Pb is separated from U and Th because of their different geochemical properties (8). Consequently, young ores and crustal rocks will generally have a ^{206}Pb/^{207}Pb ratio ~1.20 or greater while older Pb ores have a significantly lower value. The isotopic composition of some common ores are shown in Table 4.1. The ^{206}Pb/^{207}Pb ratio shows a range of 35% in these ores and yet the modern mass spectrometer can measure this ratio with an accuracy of better than 0.05%.

Lead is present at high concentration in emissions from oil combustion, iron and steel production, primary non-ferrous metal production and waste incineration (13). Often complications of identifying sources by their isotopic composition arise when the emission are mixtures of Pb from different primary origins with different isotopic compositions. The importance of Pb as an isotopic tracer was demonstrated by Chow and Johnstone (14) more than three decades ago when they showed that petrol was the major source of pollution in the Los Angeles Basin. It has also been used since, but not extensively. Maring et al. (15) used it to trace polluted air masses through western Europe, while Sturges and Barrie (16) followed pollution across the Canada-USA border. Hopper et al. (17) successfully

Table 4.1. The ratios of isotopic compositions of some selected ores used for industrial purposes

Source of ore	$^{206}Pb/^{207}Pb$	$^{208}Pb/^{207}Pb$	$^{204}Pb/^{206}Pb$	reference
Recently mined				
Broken Hill, Australia	1.0399	2.3165	0.06248	[18]
Mt Isa, Australia				
British Columbia, Canada	1.064	2.3349	0.05995	[19]
New Brunswick, Canada	1.160	2.4319	0.05441	[19]
Idaho, USA	1.052	2.3249	0.06078	[19]
Missouri, USA	1.385	2.5927	0.04591	[19]
Taxco & Durango, Mexico	1.195	2.4917	0.05311	[19]
Shuikoushan, China	1.1779	2.4709	0.05422	[20]
Cerro de Pasco, Peru	1.200	2.4917	0.05302	[19]
Rammeisberg, Germany	1.1678	2.4444	0.05481	[21]
Altay, Kazakhstan, Russia	1.131	2.465	0.05868	[19]
Ancient workings				
Rio Tinto, Spain	1.1637	2.4482	0.05494	[22]
Mazarron region, Spain	1.1955	2.4865	0.05341	[22]
Laurion, Greece	1.2027	2.4771	0.05296	[23]
Mendips & Bristol, Britain	1.1803	2.4567	0.05415	[24]

characterised sources of European emission. Lead isotopes in lake sediments and peat bogs have also been used to identify historical sources of pollution (e.g., 25, 26). In all of these studies μg-sized samples were available for analysis ensuring that accurate and precise measurements could be taken.

Until recently the sensitivity of isotopic measurements has not been sufficient to measure Pb isotopic ratios in Greenland and Antarctic snow and ice. The techniques developed for these measurements and the results obtained will be the main subject of this chapter.

3. Mass spectrometry

3.1. Isotope dilution mass spectrometry

Murozumi et al. (9) determined the concentration of Pb in Greenland and Antarctic snow and ice by Isotope Dilution Mass Spectrometry (IDMS) (27). They added pure ^{208}Pb tracer to their samples before chemically processing them. Both the mass of tracer and the water sample were measured. After thoroughly mixing the tracer with the sample Pb was isolated in a chemically pure form and the isotopic ratios were measured in a mass spectrometer. Simple comparison of the ^{208}Pb abundance with the abundance of other isotopes in the mass spectrum allowed the amount of Pb in the sample to be determined. The technique is not limited to Pb, but is applicable to any element with two or more stable or long-lived radio-isotopes. Also any isotope of the element can serve as the tracer. The sensitivity of the technique primarily depends upon the ability of the mass spectrometer ion source to produce a measurable beam of ions from the sample. Then, the greatest sensitivity will be achieved when the tracer isotope has the lowest natural abundance.

IDMS is a definitive analytical method not commonly used by chemists, except in Standards Laboratories (e.g., National Institute of Standards and Technology, Institute of Reference Materials and Measurements) where concentrations must be certified. The Earth Sciences also use IDMS when high accuracy is required, for instance where element ratios are needed to calculate the age of rocks.

The low concentrations of Pb found in Greenland and Antarctic snow and ice makes reliable concentration and isotopic composition measurements difficult to determine. Contamination with anthropogenic Pb during sample collection or drilling must be minimised, then extreme precautions must be taken to access a contamination-free sample (12, 28). Sensitive analytical methods which can analyse pg quantities of Pb are also required. A number of different methods meet this requirement; however, discussion in this chapter will be limited to Thermal Ionisation Mass Spectrometry (TIMS) because this is the only technique, to date, to be successfully used to measure isotope abundances in polar ice. IDMS is an integral part of the technique used to measure the isotopic composition of the samples.

TIMS offers very high sensitivity allowing even sub-pg/g Pb concentrations to be accurately measured (29). Concentrations as low as 0.03 pg/g have been reported in ice from Law Dome in Antarctica (30).

3.2. Isotope ratios measurements

3.2.1. Instrumentation

The modern TIMS instrument consists of an ion-source chamber, flight-tube, sector magnet, and ion-collector chamber. Depending upon the manufacturer and age of the instrument as many as 20 samples can be mounted together in the ion-source chamber and analysed sequentially. The ion-collector chamber may include an array of up to 9 multiple Faraday collectors as well as secondary electron amplification devices allowing ion currents from $\sim 1 \times 10^{-10}$ A to less than $\sim 1 \times 10^{-18}$ A to be measured. Since these instruments are designed to measure isotope abundance ratios with high accuracy and precision, the mass spectrum is well-resolved and the spectral peaks are broad with flat tops.

Samples may be analysed automatically, where the instrument takes full control of all aspects of the measurement, or manually where sensitive operations, such as raising the sample temperature and identifying the sample ion beam are controlled by the analyst. An extensive description of modern isotope ratio mass spectrometry is given by Habfast (31).

3.2.2. Mass spectral analysis

Chisholm et al. (29) describe most of the current procedures used to measure Pb isotopes in polar ice at pg/g concentrations. Because of the limited quantities of

Table 4.2. Measurement of the amount and isotopic composition of Pb in a sample using a ^{205}Pb spike (29)

Isotope	(i)	204	205	206	207	208
Measured spectrum	(i/205)	0.1430	1.0000	2.4668	2.2025	5.2108
^{205}Pb spike	(i/205)	0.0005	1.0000	0.0607	0.0079	0.0186
Sample composition	(meas-spike)	0.1425		2.4061	2.1945	5.1922
		$\frac{^{206}Pb}{^{207}Pb}$	$\frac{^{208}Pb}{^{207}Pb}$	$\frac{^{206}Pb}{^{204}Pb}$		
Sample	(i/j)	1.0964	2.3660	16.8869		

$$\text{Sample Pb(pg)} = \frac{\text{Atoms(sample)}}{\text{Atoms(spike)}} * \frac{\text{Atomicweight(sample)}}{\text{Atomicweight(spike)}} * \text{Added spike(pg)}$$

$$= 9.133 * \frac{207.21}{205.10} * 9.89 = 91.3\text{pg}$$

sample generally available for analysis and the low Pb concentration involved, the use of a ^{205}Pb tracer, which has a half-life of 1.5×10^7 years, has distinct advantages. The preparation of this isotope has been described by Parrish and Krogh (32). Because the ^{205}Pb isotope does not occur naturally, a single measurement with the tracer yields both the isotopic composition and the amount of Pb present. The cost of ~US$ 5000 per µg makes its routine use for polar ice analyses with TIMS feasible because only pg amounts are needed for each analysis. Table 4.2 illustrates how the isotopic composition and amount of Pb present are computed from a mixed spectrum of sample and tracer.

3.2.3. Sensitivity

The sensitivity of the mass spectrometry method depends upon the efficiency of the ionisation process, the size of the analytical blank and the precision of the isotopic ratios. Rosman et al. (30) report sample-processing blanks of 0.10 ± 0.05 pg when measuring Antarctic ice. These blanks are exceptionally low and difficult to maintain.

A major advance in the technology of TIMS ion sources, allowing greater than expected efficiencies to be achieved, occurred in 1969 (33) when "silica gel" ionisation enhancers were introduced. Although this ionisation mechanism is still poorly understood efficiencies of up to ~10% have been reported for Pb (34).

3.2.4. Sources of contamination

Contamination of the samples can occur during sampling, storage, decontamination, chemical processing and during analysis. Sources of Pb include aerosols, drilling fluid, core and sample storage containers, stainless steel from chisels, and chemical reagents including organic solvents, water, acids and silica gel. The transfer of contaminants from storage containers to samples is normally minimised during storage by keeping the samples frozen.

Removal of the outer contaminated parts of an ice core or snow sample is fundamental to the reliable analysis of metals. This has been amply demonstrated on ice cores by Ng and Patterson (10) who showed that outer layers were very contaminated, but could be effectively removed by mechanical means. This was later demonstrated by measurements on Dome C and Vostok ice core samples by Patterson and Boutron (12) and Boutron et al. (35). A similar approach was successfully used to decontaminate blocks of Antarctic snow (28, 36). Candelone et al. (11) refined the decontamination of ice cores by constructing a polyethylene lathe. The procedural blank for the decontamination process with the lathe was determined on an artificial core, and found to be 0.11 pg/g.

Clean laboratories and associated techniques have been described by Patterson and Settle (37), Murthy (38), Moody (39, 40), Loss and Rosman (41), Boutron (42) and Chisholm et al. (29).

Since water is the base for diluting acids, it must be relatively free of impurities. Figure 4.1 shows measurements on water produced from a reverse-osmosis/Milli-Q system. Here three different amounts of water have been evaporated to dryness in a teflon beaker and analysed by IDMS. The gradient of the line fitted through the points yields a Pb concentration of 0.16 pg/g. The intercept of ~1 pg reflects contributions from the mass spectrometer filament material, the beakers used for evaporation, laboratory aerosols, and other reagents added to the water sample prior to evaporation. Other reagents can be analysed in the same way.

The production of ultraclean reagents from commercially available reagents can

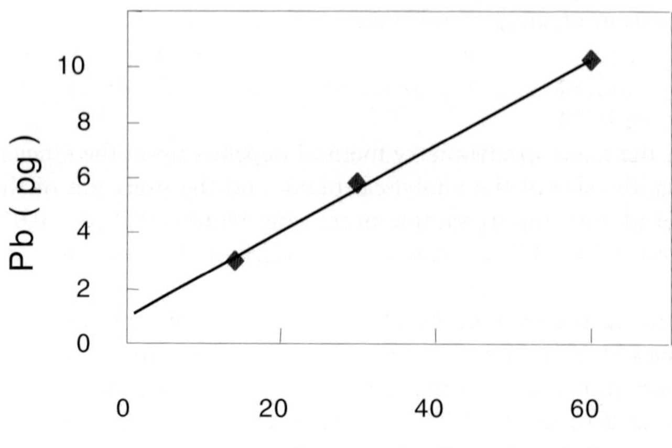

Figure 4.1. Measurement of the Pb concentration of laboratory water produced by a reverse-osmosis/ MilliQ water system. The concentration, given by the gradient of this line, was 0.16 pg/g (27).

be carried out in various ways. Nitric acid is most easily produced by sub-boiling distillation in a quartz still, and HF in a teflon still of similar design or by double-bottle distillation (38). Hydrochloric acid and HBr can be produced by bubbling HCl and HBr gases into high purity water. With these methods reagent Pb concentrations of ~1 pg/g can be achieved.

All labware used for preparing and storing samples must be rigorously clean. The selection of materials and cleaning procedures for labware have been described, for instance, by Settle and Patterson (37) and Boutron (42). For the isotopic analysis of Greenland and Antarctic snow and ice sample preparation involves the evaporation of the sample in a preconditioned beaker, as reported by Chisholm et al. (29). The Pb contributions from extraneous sources can be reduced to ~0.10 ± 0.05 pg (30).

3.2.5. Accuracy

IDMS is an inherently accurate method of measuring Pb concentrations. The technique involves measurements of a mass spectrum that is simple and cannot be mistaken for any other element. The measurements are of atom ratios. The Pb tracer is calibrated using IDMS by mixing the tracer with a solution prepared gravimetrically from high purity Pb metal. Normally two independently prepared primary solutions of the Pb metal are set up as a check, and this redundancy is maintained throughout the calibration process. Although the IDMS technique is capable of accuracies of better than 0.1% these are not generally achieved in measurements of polar ice. Limitations occur with the calibration of the tracer solution where only a small quantity of tracer can be used for this purpose. The concentration of the final working tracer solution is typically ~0.5 ng/g, then the tracer can be dispensed in pg amounts using a micropipette. Calibration of the tracer entails an enormous dilution of the primary standard solution and precautions must be taken to ensure that neither the tracer solution nor the standard have been contaminated. In view of the other problems associated with the measurement of pg amounts of Pb in polar snow and ice ~10–15% is an acceptable accuracy, even using IDMS.

Mass spectrometers are complex instruments and cannot be assumed to measure accurately. Accuracy in the isotopic composition is established through regular measurements of NIST Standard Reference Materials which are certified for accuracy (43). The SRM 981 Pb standard is often used for this purpose. Each batch of samples measured must therefore include the isotopic standard. This is most important for measurements of Pb in polar ice where a secondary electron amplification devices such as a Daly collector, electron multiplier or channeltron is needed to measure small ion currents ($<1 \times 10^{-13}$A). Significant measurement-nonlinearities in devices at this level have been reported (44). The accuracies available with a Daly collector system operating in current mode is indicated by Chisholm et al. (29). They report a bias, obtained by comparing the measured $^{206}Pb/^{207}Pb$ ratio with the certified value (NIST SRM 981) of 0.24 ± 0.06% per amu with an enhancement in the lighter isotopes. After correcting the other ratios by this amount there is agreement for the $^{206}Pb/^{204}Pb$ ratio, but the $^{208}Pb/^{207}Pb$

ratio measured is ~0.1% lower. Taking account of an 0.04% uncertainty in the certified ratio and a measurement precision, which is typically larger than 0.1%, this difference is generally not significant.

3.2.6. Isotopic systematics

When two samples of Pb with different isotopic compositions are mixed together the results will lie on a straight line joining the points when presented on a three-isotope plot of say $^{206}Pb/^{207}Pb$ *vs.* $^{208}Pb/^{207}Pb$. This is a useful plot when trying to identify the isotopic composition of sources. Also the position of the sample along the mixing line for two components reflects the atomic proportion of each component in the mixture (see the Appendix).

4. Greenland and Antarctica

Measurements of the isotopic composition of Pb in Greenland and Antarctic snow and ice are at present limited to the author's laboratory at the Curtin University of Technology. Most of these data are obtained from Greenland samples. The only other data on heavier elements are for Nd and Sr measured in ice from the Last Glacial Maximum taken from Vostok and Dome C in Antarctica.

4.1. Greenland

4.1.1. Snow

The first measurements of Pb isotopes in Greenland snow were reported in 1993 (45). The samples were taken from a 10.7 m long, 10.5 cm diameter, snow core drilled at Summit, central Greenland, in 1989 (72°35′ N, 37°38′ W, mean annual accumulation rate 21.5 g cm^{-2} year^{-1}). Cores were drilled with a polycarbonate auger to minimise the Pb contamination. The core contained snow deposited between the years 1967 and 1988. The ~3.23 km elevation of the site provided representative samples of free tropospheric aerosols. An expanded data set and a more complete description and interpretation of these data were later reported by Rosman et al. (46). The latter included samples from the upper part of a 70 m snow core including snow deposited between 1960 and 1974. Data on all four Pb isotopes were given for these samples ($^{206}Pb/^{207}Pb$, $^{208}Pb/^{207}Pb$ and $^{206}Pb/^{204}Pb$). Aliquots of these samples were also analysed for heavy metals by Boutron et al. (47) who showed there was a reduction in the Pb concentration in Greenland snow after ~1970, which they attributed mainly to the reduction in the use of alkyl-leaded petrol.

Results of the isotopic analyses are shown in Figure 4.2. The $^{206}Pb/^{207}Pb$ ratio increases from 1960 reaching a peak value between 1975 and 1980, then decreases back to the earlier values. The change was attributed to mixing of USA and Eurasian Pb arriving at Summit. The higher values for the USA were due to the increased use of Mississippi Valley Pb petrol. Isotopic signatures, taken from the literature, were used to apportion the contributions (Figure 4.3), which showed

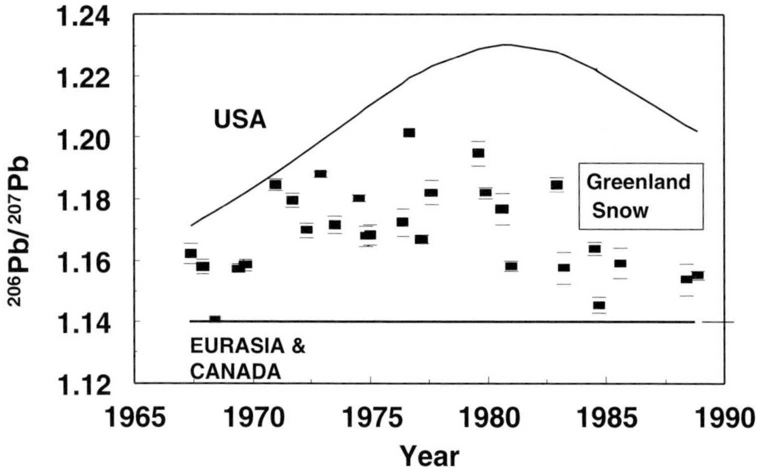

Figure 4.2. The $^{206}Pb/^{207}Pb$ ratio in snow from Summit, central Greenland, deposited over three decades. Published data, represented by solid lines for the USA Eurasia and Canada, show how the isotopic composition of Pb in aerosols and petrol has changed during the same period (45).

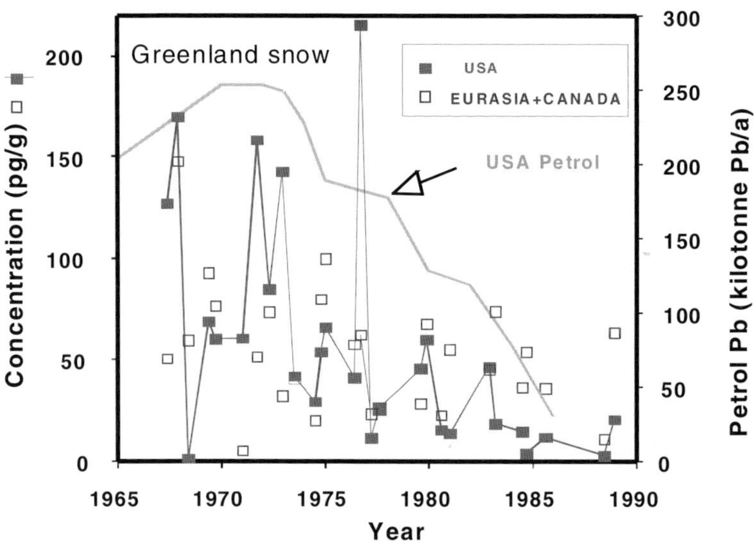

Figure 4.3. The concentration of USA and Eurasian-Canadian Pb in the Summit snow (46). These concentrations were determined using the proportional contribution from each source, calculated from the $^{206}Pb/^{207}Pb$ ratios measured in the Summit snow and the source signatures from the literature, shown in Figure 4.2. The Pb consumption in the USA gasoline over the period is shown for comparison (R. Fiat, personal communication).

that the USA contribution was correlated with Pb consumption in gasoline. However, the literature values are few and rather scattered in time and point to the need for more systematic monitoring of the isotopic composition of emissions from different parts of the world. Irrespective of this difficulty, the isotopic measurements provided indisputable evidence of USA petrol-Pb in the atmosphere particularly during the early 1970s.

4.1.2. Seasonal variations

The measurement of Pb isotopes in Greenland snow at Summit showed a long term trend in the $^{206}Pb/^{207}Pb$ ratio with superimposed short term fluctuations (46). It was suggested that the latter were probably due to seasonal changes in weather patterns transporting aerosols to Summit from different source regions. Further evidence for seasonal effects was provided by measurements of Pb isotopes in individual snowfalls at Dye 3 (48). Dye 3 is located in southern Greenland (65°11′ N, 43°50′ W), at an elevation of 2479 m and approximately 150 km from the seacoast. Fresh snow was sampled from at least eighteen precipitation events between August 13, 1988 and August 6, 1989. The results given in Figure 4.4 show clear evidence of a seasonal trend in the isotopic ratios. The $^{206}Pb/^{207}Pb$ ratio changes from ~1.15 in spring/mid-summer to ~1.20 in winter. There is clear evidence in the snow of Pb originating from North America in the autumn/winter and from Eurasia in spring/mid-summer.

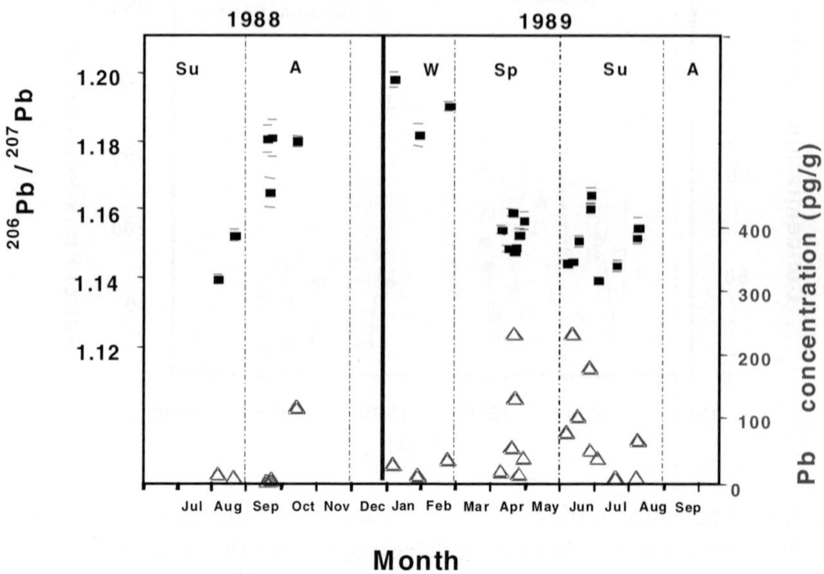

Figure 4.4. The $^{206}Pb/^{207}Pb$ ratio and concentration of Pb in fresh surface snow collected at Dye 3 during 1988 and 1989 showing seasonal changes in isotopic composition that reflect different geographical sources. The four seasons are labeled Su, A, W and Sp (48).

4.1.3. Pollution in ancient ice

The availability of ancient ice from the European Greenland Ice-Core Project (GRIP) and the USA Greenland Ice Sheet Project (GRIP) has allowed ancient Pb pollution to be investigated. The importance of the Greenland ice archive for this purpose was pointed out by Murozumi et al. (9). A peak of Pb production during Greek and Roman times (49) gave a strong indication that evidence of an enhanced Pb concentration in the ice might be found. Hong et al. (50) reported measurements of Pb concentrations in 2 k-years old Greenland ice confirming the existence of this peak. Lead isotope measurements were made on the same samples to identify the source of this pollution (51).

To establish the pre-human background composition, Pb in ice with an age of 7–9 k-years was analysed. Figure 4.5 shows that between 680 BC and 366 BC the $^{206}Pb/^{207}Pb$ ratio decreased significantly providing the first evidence of anthropogenic Pb in the ice. The ratio reached a minimum value between 143 BC and 36 AD, then increased, returning to almost background values by ~200 AD before decreasing once more. The figure shows that the isotopic ratios were mirrored by changes in the Pb concentration. These data form a linear trend on a plot of the $^{206}Pb/^{207}Pb$ *vs.* the $^{208}Pb/^{207}Pb$ ratio in Figure 4.6 suggesting a mixture of at least two sources with different isotopic compositions. An examination of the isotopic composition of Pb from ancient mining regions, constrained by historical evidence of active mining regions, led to the identification of two source regions in southern Spain. These two regions (south-east and south-west), with isotopic compositions

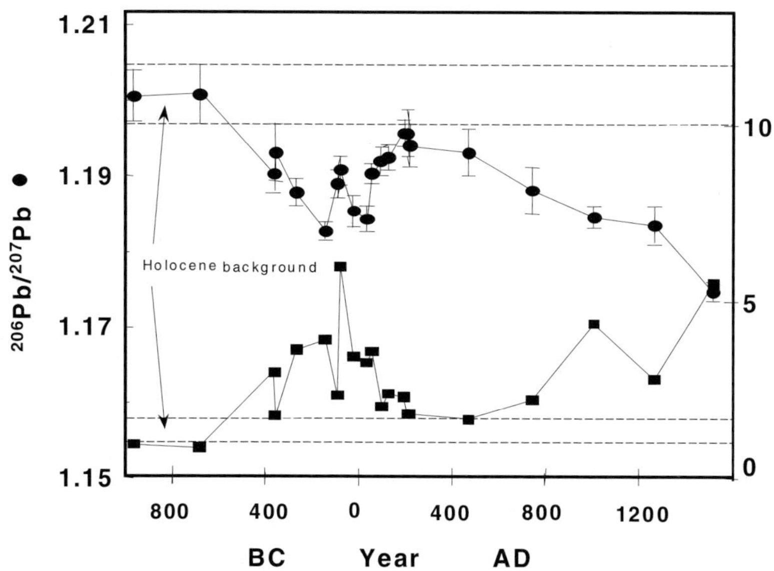

Figure 4.5. The $^{206}Pb/^{207}Pb$ isotopic ratio (?) and concentration (?) of Pb in Greenland ice. Corresponding values for the Holocene background, based on measurements of five ~8 k-years old samples, fall between the dotted lines (51). The dip in isotopic ratio after ~600 BC is definitive evidence of anthropogenic Pb emissions.

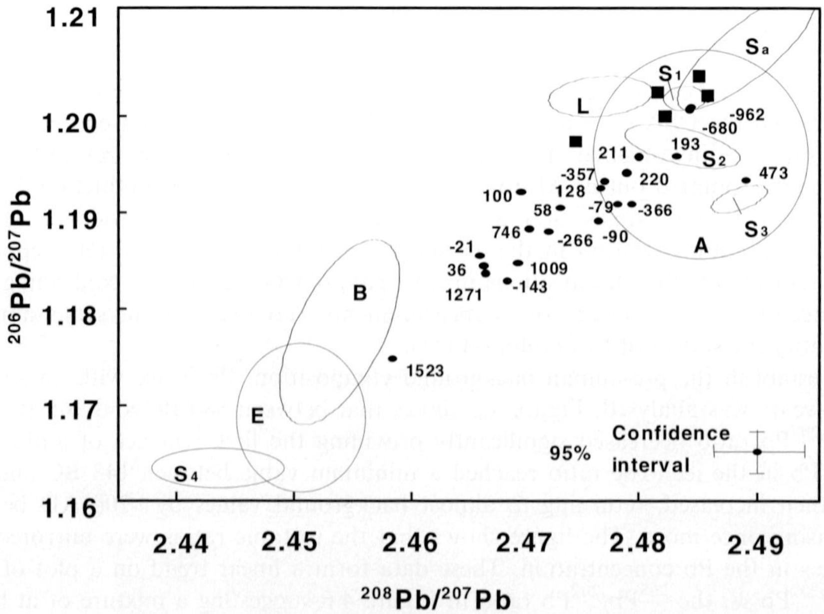

Figure 4.6. Isotopic composition of Pb in Greenland ice between 7313 BC and 1523 AD. Sample dates are shown, with BC dates given a negative prefix. Lead ores are shown as ellipses: A, Aegean; B, British; E, Europe; L, Greek ; S, Spanish: S_1 (Cabo de Garta region); S_2 (Cabo de Garta and Mazarron regions); S_3 (Cartagena region); S_4 (includes Rio Tinto region); Sa, Atlantic ocean sediments near the western Sahara. Five samples used to represent the Holocene background are shown as filled squares (51). The linear trend of measured isotopic ratios indicates a mixture of contributing sources.

near extensions of the linear trend of the data, were extensively mined by the Romans, and emissions from those regions are consistent with the historical evidence. Based on the isotopic ratios it was possible to assign ~70% of the emissions to Rio Tinto (south-west Spain) between 336 BC and 35 AD (51).

4.2. Antarctica

Sample locations where Pb isotopes have been determined are shown in Figure 4.7.

4.2.1. Surface snow

The first measurements of Pb isotopes in Antarctica were made on surface snow by Rosman et al. (52). The samples included snow blocks whose Pb concentrations had already been measured by Boutron and Patterson (28). The Pb isotope measurements provided independent evidence that Antarctica was highly polluted with Pb and suggested likely sources of this pollution.

The snow blocks for this study were collected in 1983–1984 at locations 33 km, 103 km and 433 km inland from the French coastal station at Dumont d'Urville

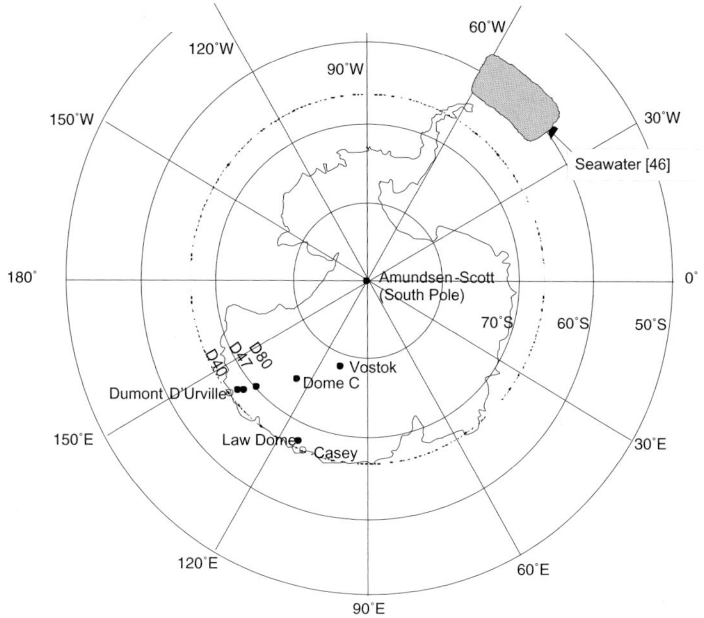

Figure 4.7. The Antarctic region, showing sites (•) where the isotopic composition of Pb has been measured.

and at the South Pole. The isotopic composition of the Holocene background Pb ($^{206}Pb/^{207}Pb$ ~1.252) was determined by analysing 7.5 k-years old ice from an ice-core thermally drilled at Dome C (location map, Figure 4.7). All the surface snow had a lower $^{206}Pb/^{207}Pb$ ratio than the Holocene background indicating that it was contaminated with anthropogenic Pb, which supported the conclusion reached by Boutron and Patterson (28) based on geochemical arguments. However, measurements on snow from two inland sites (Stakes 47 and 80) had different Pb concentrations, but gave similar isotopic ratios, thus suggesting a uniform modern background isotopic composition for East Antarctica ($^{206}Pb/^{207}Pb$ ~1.163). This Pb was also identified (52) as a component of the Pb measured in sea water by Flegal et al. (53).

The snow collected 33 km from Dumont d'Urville was found to contain a $^{206}Pb/^{207}Pb$ ratio significantly lower than the background showing that it was polluted by base emissions, probably from leaded fuel which was purchased in Australia ($^{206}Pb/^{207}Pb$ as low as ~1.07). An alternative mechanism for lowering the ratio in the coastal snow was the transfer of Australian emissions through the atmosphere which was consistent with atmospheric general circulation models (54). Snow at the South Pole was also polluted by emissions from the Amundsen-Scott base, located only 7.3 km away.

Rosman et al. [52] suggested that the modern background Pb ($^{206}Pb/^{207}Pb$ ~1.163) originated from South American cities, since this was consistent with geochemical data on Na, Al and Ca levels (55) and evidence from Sr and Nd isotopes in Dome C ice dated at the Last Glacial Maximum (56). However, no suitable Pb isotopic data from South American sources were available to test this hypothesis.

4.2.2. Snow and ice from Vostok, Dome-C and Law Dome

Vostok A limited number of samples of ice core from Vostok, Dome C and Law Dome have been analysed for Pb isotopes. Chisholm et al. (29) reported a decontamination profile on an ice core from 500 m depth at Vostok, which was shown to be contaminated with Pb from at least two different sources. This core, thermally drilled using kerosene as a wall retaining fluid, was decontaminated at the California Institute of Technology in 1984–1985. This profile is still of interest in itself because it illustrates the additional information provided by an isotopic analysis. Figure 4.8 shows that the Pb concentration decreases sharply from the surface, then much more slowly to the fifth layer. No material from the inner core was available for analysis. The $^{206}Pb/^{207}Pb$ ratio decreases from the surface reaching a minimum value in the second layer, then progressively increases towards the core. These data indicate there were at least two different sources of Pb contamination in the core. It appears that the inner layers were first contaminated with fluid containing Pb with $^{206}Pb/^{207}Pb$ ratio $< \sim 1.14$, after which the surface was contaminated with Pb having a ratio of ~1.16. No plateau of concentration or isotopic ratio was reached, indicating that layer 5 could also be contaminated. Higher isotopic ratios found at Dome C lend support to this hypothesis.

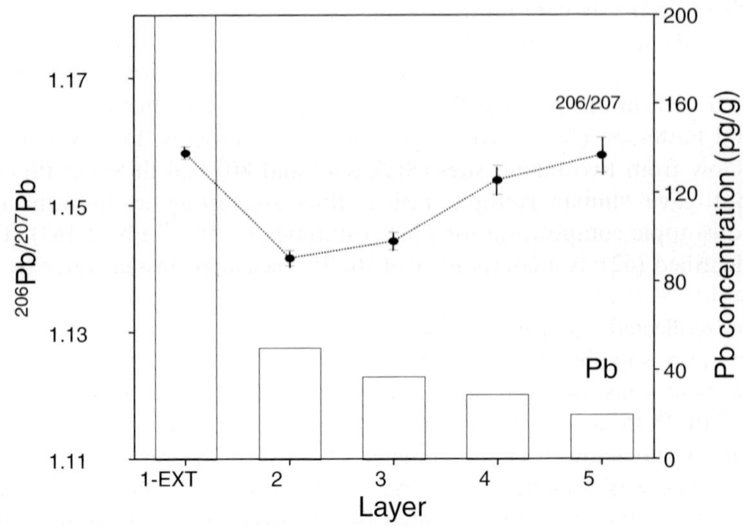

Figure 4.8. Profile of the $^{206}Pb/^{207}Pb$ isotopic ratio (●) and Pb concentration (histograms) across an Antarctic ice core section. This core was taken at Vostok from 500 m depth using a thermal drill in a hole filled with kerosene as a wall retaining fluid. The change in the isotopic ratio suggests two different sources of contamination (29).

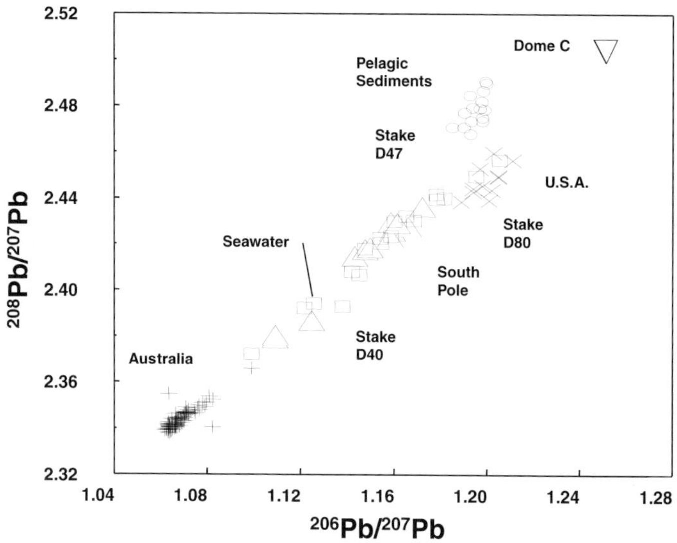

Figure 4.9. The plot [52] shows isotope ratios of Pb in Antarctic snow (△), Dome C dust (▽), Antarctic sea water (□) [53], pelagic sediments adjacent to South America (○) [57], and gasoline and aerosols from Perth, Australia, during 1982-83 (+) [*K. Rosman*, unpublished data] and from the United States during 1971–74 (×) (19, 58).

Dome C Two samples of ice from Dome C have been analysed for Pb isotopes. Both were taken from the same dry, thermally drilled core at depths of 308 m (7.5 k-years BP) (52) and 515 m (14 k-years BP) (29). Both samples displayed good plateaus of concentration and isotopic ratio. The inner cores had Pb concentrations of 2.1 and ~9 pg/g, with $^{206}Pb/^{207}Pb$ ratios of 1.252 and 1.202, respectively. These data suggest that $^{206}Pb/^{207}Pb$ ratios are lower near the time of the Glacial Maximum. The relationship of the isotopic composition of 7.5 k-years ice to the surface snow, pelagic sediments near South America, sea water and some principal sources of anthropogenic Pb is shown in Figure 4.9.

Law Dome A preliminary survey of Pb isotopes in selected ice core samples from Law Dome, East Antarctica, was recently reported by Rosman et al. (30). Samples were taken from two cores, DSS and DE08, drilled at sites with very high ice-accumulation rates of ~60 kg m^{-2} year^{-1} and 1200 kg m^{-2} year^{-1}, respectively. The Pb concentrations encountered in all the samples were <1.5 pg/g, with one measured at 0.03 pg/g, assuming a decontamination blank of 0.1 pg/g (11). Because only ~10 mL sized samples were available for analysis the measurement precision on the isotopic ratios was low for some samples. Contamination was also evident in some samples. The isotopic composition of the contaminating Pb ($^{206}Pb/^{207}Pb$ ~1.05–1.08) found in the outer part of the decontamination profile (Figure 4.10) was typical of the Pb being used in gasoline in Australia at the time. The results of analysing the cores are shown in Figure 4.11. The age of each sample is given by calendar year. Where a plateau of concentration was not

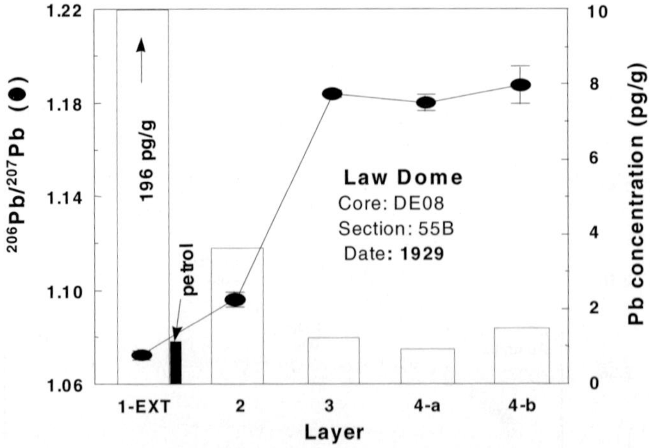

Figure 4.10. The decontamination profile for sample 1929 from the DE08 ice core showing the concentration (histogram) and $^{206}Pb/^{207}Pb$ isotopic ratio (●) for each concentric layer of the core. The range of isotopic ratios in Australian petrol sampled in Perth between 1989 and 1995 is also shown (30).

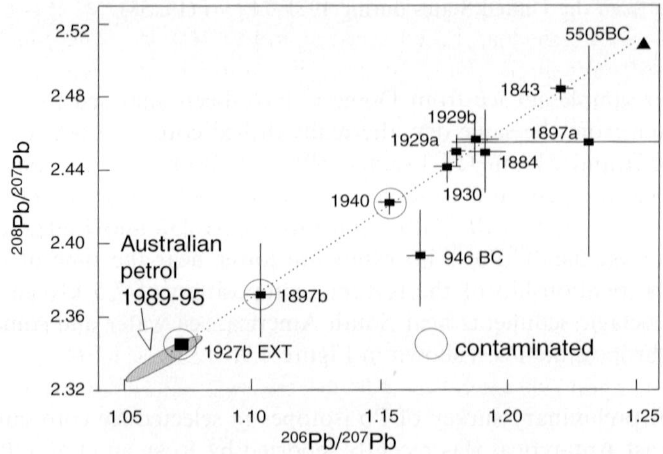

Figure 4.11. A three-isotope plot for ice from the Law Dome (■). The dotted line is drawn through the higher precision measurements. Samples which yielded a poor decontamination plateau are circled. Also included are data for Holocene ice from Dome C (▲) and Australian petrol (shaded). (30)

achieved in the decontamination process the data are circled. These samples lie in a mixing line with Australian petrol at one end and Holocene ice at the other.

The isotopic data are too few to make a sensible comparison of isotopic data between the hemispheres, although the available data suggest that the isotopic ratios are similar.

5. Isotopic tracers of other elements

Measurements of two other heavy elements isotopic tracers have been reported. Grousset et al. (56) identified the source area from which dusts in the Last Glacial Maximum-section (~18 k-years BP) of the Dome C ice core originated by comparing their $^{87}Sr/^{86}Sr$ and $^{143}Nd/^{144}Nd$ isotopic ratios with samples from potential source areas. The tracers involved are radiogenic ^{87}Sr from the decay of ^{87}Rb (half-life 4.9×10^{10} years) and radiogenic ^{143}Nd from the decay of ^{147}Sm (half-life 1.1×10^{11} years). Outer parts of a 5.5 m length of the ice core (depth ~550–590 m) were removed by melting with deionised water, then the remaining ice was melted and evaporated to dryness yielding ~2.5 mg of red dust particles. The dust was digested with a mixture of HF, $HClO_4$ and HNO_3 and Sr and Nd were isolated using standard chemical techniques.

Potential sources in Antarctica, Australia, Africa and South America were investigated, but the isotopic ratios clearly showed that the origin of the dust was South America (Patagonia). Although the Patagonian pampas are much smaller than either the Great Sandy desert (Australia) or the Namibian and Kalahari deserts (southern Africa), Grousset et al. (56) argued that the effective area was much greater during the LGM when sea-level was ~120 m lower and the broad Argentine continental shelf was exposed to weathering. This however was inconsistent with experimental data from another study (59) which indicated that there were only fresh particles and no marine diatoms in Dome C LGM ice. The conclusion was also inconsistent with atmospheric general circulation models which gave preference to an Australian source for the dust reaching East Antarctica.

Recently another five ice core samples from Dome C and Vostok were analysed for Sr and Nd isotopes by Basile et al. (60), who showed that the source area of continental dust deposited in East Antarctica during glacial stages 2 (~18 k-years), 4 (~60 k-years) and 6 (160 k-years) was the same. Moreover, they measured Argentine continental shelf sediments and discounted this as a source of dust found in the ice-cores. They attributed enhanced dust loading during the glacial periods to fluvioglacial erosion and deposition processes on the Patagonian plateau coupled with intensified aeolian deflation and transport. They concluded that the dust has a predominant Patagonian origin with the possible admixture of 10–15% of dust from southern Africa and/or Australia.

6. Conclusion

The use of heavy isotope tracers for investigating sources of pollution reaching the polar regions of the Earth is in its infancy. This is particularly the case for Antarctica where the quantity of pollutants encountered is lower than elsewhere. For this reason this chapter includes a description of the methods used to make isotopic measurements, although it is focused on Pb.

Techniques with very high sensitivity are needed for these measurements, but to a large extent they are already available and only require additional tuning to manage the smaller samples. The greatest measurement challenges are presented by elements such as Pb with ubiquitous anthropogenic sources that can contaminate samples and produce inaccurate results. Ultra-clean environments and procedures are needed in which samples can be safely decontaminated and processed for analysis. A great deal of attention must also be given to the clean collection and storage of samples. Although there is an increasing awareness of the importance of the quality of the samples being analysed, more research is needed for significant progress in isotopic measurements.

Data for only three heavy isotopic tracers (Nd, Pb and Sr) are available at present and these are at an early stage in their development. Deep ice and near surface snow from Greenland has been analysed and the potential of using isotopic tracers to identify the sources of Pb has been demonstrated. The number of measurements on Antarctic snow and ice are few. The Pb concentration is one-two orders of magnitude lower than in Greenland, making the measurements much more difficult. The predominant sources of pollutants, other than Antarctic stations, are still unknown. However, both isotopic and geochemical methods have shown that there is a significant anthropogenic contribution.

Only a few Pb isotope measurements have been made on deep ice cores and these show signatures similar to Greenland. Neodymium and Sr isotopes have been made on LGM-dust from East Antarctica to identify its origin. The results identify a South American (Patagonian) source for this dust, although further investigation is needed to determine the transport mechanisms.

Acknowledgements

My thanks to Claude Boutron for inviting me to participate in the challenge of high precision measurement of almost nothing in polar snow and ice and to the Australian Research Council and Australian Science Advisory Committee for their support in this venture which is improving our understanding of the Earth's present and past environments.

8. Appendix

Mixing isotopic ratios

If χ is the proportion of Pb with different $^{206}Pb/^{207}Pb$ ratios ($6/7_A$ and $6/7_B$) contributing to a binary mixture with components A and B, then

$$6/7_A\,\chi + 6/7_B\,(1-\chi) = 6/7_M.$$

Therefore

$$\chi = (6/7_M - 6/7_B)/(6/7_A - 6/7_B).$$

Also

$$\chi = (8/7_M - 8/7_B)/(8/7_A - 8/7_B).$$

Equating χ, gives

$$6/7_M = [(6/7_A - 6/7_B)/(8/7_A - 8/7_B)]\ 8/7_M + [8/7_A\ 6/7_B - 8/7_B\ 6/7_A].$$

Clearly, this gives a straight line when $6/7_M$ is plotted against $8/7_M$. Note that the gradient of the line is a 6/8 ratio, but it is that of neither A or B, and lies between the two. The proportion of each type of Pb in any mixture is indicated by the position of the mixture on the line joining the composition of each constituent. The complete isotopic composition of each component of the mixture needs to be taken into consideration if the atomic ratio of the two components is required.

The atom ratio for the components, A/B, is then given by

$$A/B = [(1\text{-}6/7_M/6/7_B)/(6/7_M/6/7_A - 1)](Ab6_B/Ab6_A),$$

where $Ab6_A$ and $Ab6_B$ represent atomic proportions of ^{206}Pb in samples A and B, respectively.

References

1. C. F. Boutron, *Environ. Rev.* 3 (1995), 1–28.
2. H. E. Duckworth, R. C. Barber, V. S. Venkatasubramanian, Mass Spectroscopy. 2nd Edition. Cambridge University Press, Cambridge. 1986, pp. 337.
3. S. K. Aggarwal, H. C. Jain., *Introduction to Mass Spectrometry*. Indian Society for Mass Spectrometry, India. 1997, pp. 479.
4. K. J. R. Rosman, P. D. P. Taylor, *Pure & Appl. Chem.* 70 (1998), 217–235.
5. F. W. Aston, *Mass Spectra and Isotopes*. 2nd Edition. London: Edward Arnold, 1942.
6. C. C. Patterson, *Geochim. Cosmochim. Acta* 10 (1956), 230–237.
7. A. O. Nier, *Rev. Sci. Instr.* 1940, 11, 212.
8. G. Faure, In *Principles of Isotope Geology*, New York: Wiley, 1986.
9. M. Murozumi, T. J. Chow, C. C. Patterson, *Geochim. Cosmochim. Acta* 33 (1969), 1247–1294.
10. A. Ng, C. C. Patterson, *Geochim. Cosmochim. Acta* 45 (1981) 2109–2121.
11. J-P. Candelone, S. Hong, C. F. Boutron, *Anal. Chim.* Acta 229 (1994), 9–16.
12. C. F. Boutron, C. C. Patterson, *Nature* 323 (1986), 222–225.
13. J. O. Nriagu, *Nature* 279 (1979), 409–411.
14. T. J. Chow, M. S. Johnstone, *Science* 1965, 147, 502–503.
15. H. Maring, D. M. Settle, P. Buat-Ménard, F. Dulac, C. C. Patterson, *Nature* 330 (1987), 154–156.
16. W. T. Sturges, L. A. Barrie, *Nature* 329 (1987), 144–146.
17. J. F. Hopper, H. B. Ross, W. T. Sturges, L. A. Barrie, *Tellus* 43B (1991), 45–60.
18. J. R. Richards, I. R. Fletcher, J. G. Blockley, *Mineral Deposita* 16 (1981), 7–30.
19. T. J. Chow, C. B. Snyder, J. L. Earl, United Nations FAO and International Atomic Energy Association Symposium, Vienna, Austria. (IAEA-SM-191/4). *Proceedings*, 1975, pp. 95–108.
20. R. H. Brill, K. Yamasaki, I. L. Barnes, K. J. R. Rosman, M. Diaz, *Ars Orientalis*, 11 (1979), 87–109.
21. K. H. Bielocki, G. Tischendorf, *Contrib. Min. Petrol.* 106, (1991), 440–461.
22. Z. Stos-Gale, N. H. Gale, J. Houghton, R. Speakman, *Achaeometry* 37 (1995), 407–415.
23. N. H. Gale, *Thera and the Aegean World II*, (Ed. Doumas C.) London, 1980, 161–196.
24. B. M. Rohl, *Achaeometry* 38 (1996), 165–180.
25. J. R. Graney, A. N. Halliday, G. J. Keeler, J. O. Nriagu, J. A. Robbins, S. A. Norton, *Geochim. Cosmochim. Acta* 59 (1995), 1715–1728.
26. W. Shotyk, D. Weiss, P. G. Appleby, A. K. Cheburkin, R. Frei, M. Gloor, J. D. Kramers, S. Reese, W. O. Van Der Knaap, *Science* 281 (1998), 1635–1640.
27. R. K. Webster, Mass spectrometric isotope dilution analysis. In *Methods in Geochemistry* (Eds. A. A. Smales and L. R. Wager), 1960, pp. 202–246.

28. C. F. Boutron, C. C. J. Patterson, *Geophys. Res.* 92 (1987), 8454–8464.
29. W. Chisholm, K. J. R. Rosman, C. F. Boutron, J-P. Candelone, S. Hong, *Anal. Chim. Acta.* 311 (1995), 141–151.
30. K. J. R. Rosman, W. Chisholm, C. F. Boutron, S. Hong, R. Edwards, V. Morgan, P. Sedwick, *Ann. Glaciology* 27 (1998), 349–354.
31. K. Habfast, In *Modern Isotope Ratio Mass Spectrometry*, (Ed. Platzner, I.T.) Chichester: Wiley, 1997, pp. 11–82.
32. R. R. Parrish, T. E. Krogh, *Chem. Geol.* 66 (1987), 103–110.
33. A. E. Cameron, D. H. Smith, R. L. Walker, *Anal. Chem.* 41 (1969), 525–526.
34. H. Gerstenberger, G. Haase, *Chem. Geol.* 136 (1997), 309–312.
35. C. F. Boutron, C. C. Patterson, V. N. Petrov, N. I. Barkov, *Atmospheric Environment* 21 (1987), 1197–1202.
36. E. D. Suttie, E. W. Wolff, *Tellus* 44B (1992), 351–357.
37. C. C. Patterson, D. M. Settle, In *Proceedings of the 7th Materials Research Symposium*, NBS Special Publication 422. US Government Printing Office, Washington, DC, 1976, pp. 321–351.
38. T. J. Murphy, In *Proceedings of the 7th Materials Research Symposium*, NBS Special Publication 422. US Government Printing Office, Washington, DC 1976, pp. 509–539.
39. J. R. Moody, *Anal. Chem.* 54 (1982), 1358A–1376A.
40. J. R. Moody, *J. Roy. Soc.* W.A. 79 (1996), 29–32.
41. R. D. Loss, K. J. R. Rosman, Curtin University of Technology, School of Physical Sciences. Report No. SPG 464/1987/AP143, 1987, p. 51.
42. C. F. Boutron, *Fresenius J. Anal. Chem.* 337 (1990), 482–491.
43. E. J. Catanzara, T. J. Murphy, W. R. Shields, E. L. Garner, *J. Res. NBS* 72A (1968), 261–267.
44. K. J. R. Rosman, W. Lycke, R. Damen, R. Wertz, F. Hendrickx, P. De Bievre; *Int. J. Mass Spectrom. Ion Proc.* 79 (1987), 61–71.
45. K. J. R. Rosman, W. Chisholm, C. F. Boutron, J-P. Candelone, U. Gorlach, *Nature* 362 (1993), 333–334.
46. K. J. R. Rosman, W. Chisholm, C. F. Boutron, J-P. Candelone, S. Hong, *Geochim. Cosmochim. Acta* 58 (1994), 3265–3269.
47. C. F. Boutron, U. Görlach, J-P. Candelone, M. A. Bolshov, R. J. Delmas, *Nature* 353 (1991), 153–156.
48. K. J. R. Rosman, W. Chisholm, C. F. Boutron, J-P. Candelone, J-L. Jaffrezo, C. I. Davidson, *Earth Planet. Sci. Lett.* 160 (1998), 383–389.
49. C. C. Patterson, *Lead in the Human Environment*. Report prepared by the Committee on Lead in the Human Environment, National Academy of Sciences, Washington, DC. 1980, pp. 265–349.
50. S. Hong, J-P. Candelone, C. C. Patterson, C. F. Boutron, *Science* 265 (1994), 1841–1843.
51. K. J. R. Rosman, W. Chisholm, S. Hong, J-P. Candelone, C. F. Boutron, *Environ. Sci. Technol.* 31 (1997), 3413–3416.
52. K. J. R. Rosman, W. Chisholm, C. F. Boutron, J-P. Candelone, C. C. Patterson, Geophys. *Res. Lett.* 21 (1994), 2669–2672.
53. A. R. Flegal, H. Maring, S. Niemeyer, *Nature*, 365 (1993), 242–244.
54. S. J. J. Joussaume, *Geophys. Res.*, 98 (1993), 2767–2805.
55. R. J. Delmas, J. R. Petit, *Geophys. Res. Lett.* 21 (1994), 879–882.
56. F. E. Grousset, P. E. Biscaye, M. Revel, J-R. Petit, K. Pye, S. Jousel, Joussaume, *J. Earth Planet. Sci. Lett.* 111 (1992), 175–182.
57. T. J. Chow, C. C. Patterson, *Geochim. Cosmochim.* Acta 26 (1962), 263–308.
58. M. B. Rabinowitz, G. W. Wetherill, *Environ. Sci. Technol.* 6 (1972), 705–709.
59. L. H. Burckle, M. Gayley, M. Ram, J-R. Petit, *Geology* 16 (1989), 326–329.
60. I. Basile, F. E. Grousset, M. Revel, J-R. Petit, P. E. Biscaye, N. I. Barkov, *Earth Planet. Sci. Lett.* 111 (1997), 573–589.

Environmental Contamination in Antarctica
A Challenge to Analytical Chemistry
S. Caroli, P. Cescon and D.W.H. Walton, editors

Chapter 5

Trace metals in Antarctic sea water

Gabriele Capodaglio, Carlo Barbante and Paolo Cescon

1. Introduction

Considering the Southern Ocean as the area delimited by the Subtropical Front to the north, it represents over 20% of the world's ocean area. It is characterized by the Antarctic Circumpolar Current (ACC) flowing eastward driven by wind stress acting on the Southern Ocean (1). South of the ACC there are two cyclonic gyres located in the Weddell and the Ross Seas (see Figure 5.1) (2). Significant exchanges of water masses take place between the Southern Ocean and the remaining oceans of the world. North Atlantic Deep Water (NADW) enters the Southern Ocean from the north and mixes with adjacent water masses to generate the Circumpolar Deep Water (CDW) that enters the Weddell and Ross Gyres (3, 4). The circulation within or on the edges of the cyclonic gyres generates two important water masses which affect the characteristics of world oceans. The first consists of water produced by the equatorialward transport of the surface layer across the ACC, that sinks at the ACC front and generates intermediate water which spreads through the world ocean (5). The other is the Antarctic Bottom Water (AABW) the dispersion of which represents a major process that ventilates the deep part of the world ocean (6).

CDW is generated by processes of advecting and mixing of water masses, which have obtained their characteristics in remote areas; in a few locations around Antarctica CDW mixes with local cold waters and leads to the formation of AABW that flows to the north through gaps in the mid-ocean ridges and determines the characteristics of the abyssal waters of the world ocean (7). Therefore, the characteristics of the Southern Ocean are strongly affected by water generated in remote oceanic areas and at the same time the Antarctic sea water, modified by local processes, generates water masses that affect the characteristics of all oceanic bottom waters.

These considerations lead to the conclusion that the characterisation of Antarctic water masses represents one of the most important challenges for oceanographers. Considerable studies are being undertaken to characterize the Southern Ocean from a physical point of view, but only limited studies are being carried out to define the chemical characteristics and trace component distribution in particular is insufficiently examined. Moreover, most studies focus on the Weddell Sea, while other key areas, such as the Ross Sea, are inadequately researched (8).

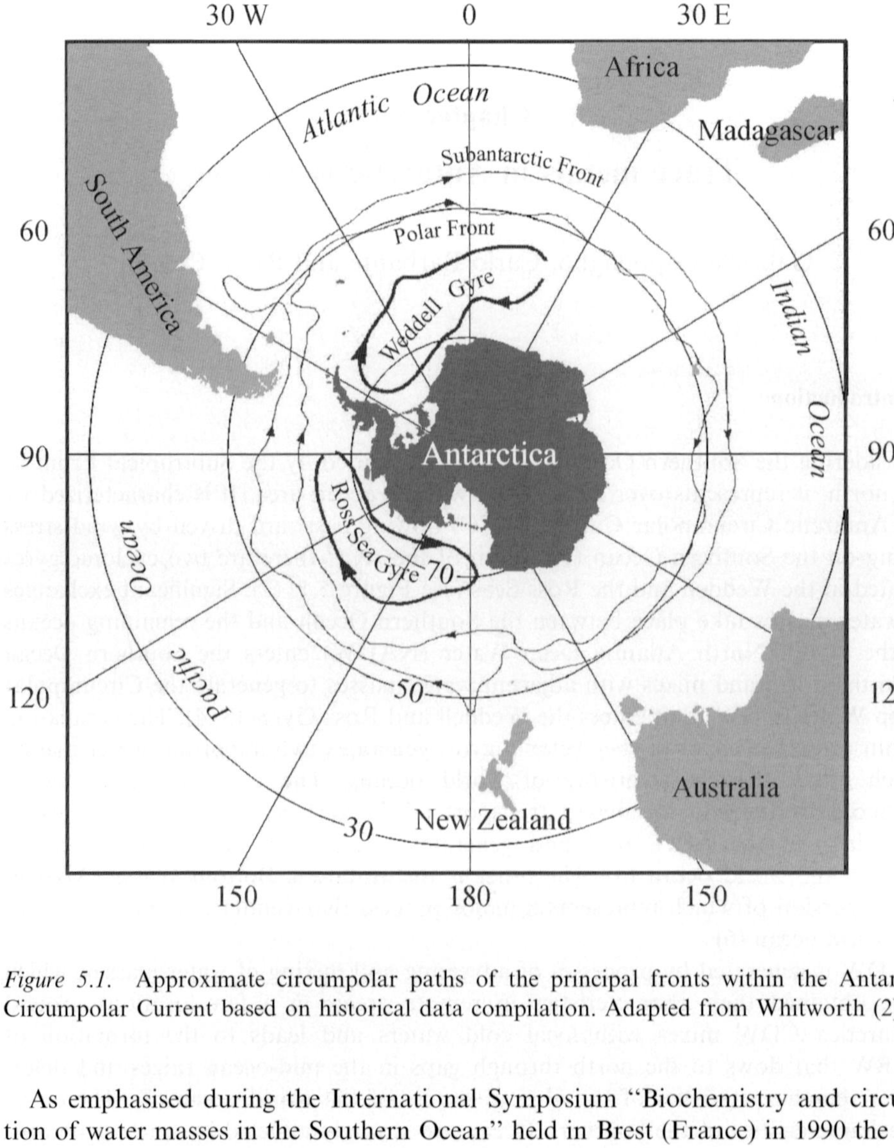

Figure 5.1. Approximate circumpolar paths of the principal fronts within the Antarctic Circumpolar Current based on historical data compilation. Adapted from Whitworth (2).

As emphasised during the International Symposium "Biochemistry and circulation of water masses in the Southern Ocean" held in Brest (France) in 1990 the use of chemical tracers could represent a powerful tool in describing the dynamics of the Southern Ocean (8). During the symposium it was also shown that the interpretative models frequently used to describe the circulation of tracers in other oceanic areas cannot be applied to the Southern Ocean and modified models must be developed.

Trace elements represent the chemical tracers that indicate an interface research area between physics, chemistry and biology. To understand the role of trace elements in biochemical cycles, and their utility as tracers, it is first necessary to establish the primary routes whereby they enter the Southern Ocean; in other words, to define the composition of water masses entering and flowing out of the target area.

An especially important gap in the present knowledge of processes changing the characteristics of water masses within the Southern Ocean centers on the effect of the seasonal cycle, the formation and dissolution of pack ice and its associated biological processes. One aim must be to explain the paradox of the mismatch of dissolved nutrient concentration and the primary production in the pelagic ecosystems (9). One important hypothesis that needs further testing is the possibility of Fe limitation of nutrient uptake (10).

The possibility that trace elements can represent one limiting factor to the biological activity introduces the necessity to identify and quantify their bioavailable chemical forms. Therefore, use of analytical methods with the ability to differentiate individual species, oxidation states and association with inorganic and organic ligands is as important as an accurate determination of the total concentration of these elements.

Improvements in the field of environmental analytical chemistry over the past few decades have provided marine science with knowledge of the distribution patterns and chemical behavior of trace elements in sea water (11, 12). In particular, awareness of the risks of contamination of samples during the entire analytical process (collection, storage, treatment and analysis) has prompted the development of better analytical methodologies with the aim of performing measurements under careful contamination control (13–22). Studies conducted under stringent, non-contaminating conditions have led to the discovery of vertical profiles of trace elements that show patterns which are consistent with known biogeochemical and/or physical processes active in the oceans (12, 13).

Furthermore, with the improvement of instrumental analytical methods it is possible to face the problem of identifying the chemical forms of trace metals and/or of discriminating classes of metal species (individual species, oxidation states, association with inorganic and organic ligands). Methodologies are now available with speciation capability to assess the environmental impact of metals or to study the biogeochemical processes involving metals in sea water (23–27).

Different approaches can be applied to study metal speciation and all have some advantages and limitations. It is of particular interest to consider direct methods for speciation, e.g., procedures to differentiate metal forms on the basis of physical, chemical and biological properties. Some techniques are able to determine chemical forms of elements selectively, i.e., potentiometry with ion-specific electrodes, chromatographic methods and some spectroscopic methods such as Electron Paramagnetic Resonance (EPR) or fluorescence. However, the majority of methodologies allow for the separation or differentiation of fractions on the basis of the particular property of the technique used. In physical techniques the separation is based on the size of fractions, in chemical techniques it is based on the difference of chemical affinity or chemical reactivity and in bioassay techniques it is based on the different interaction of metal species with living organisms. A review introducing the different approaches to the study of metal speciation was recently reported (28).

Electroanalytical methods are frequently applied to the study of metal complexation by organic ligands; in particular, two voltammetric titration methodologies to study trace metal complexation in sea water have been developed. In both cases the procedure involves the titration of ligands present in the sample with the metal under study. Differential Pulse Anodic Stripping Voltammetry (DPASV) can be used to determine the electroactive fraction of metal (labile metal, i.e., ionic plus the inorganic complexes of metal) during the titration (29–35). The other procedure provides for the competitive equilibration of samples with chemically well characterized ligands or with active sites located on the surface of solid matter (adsorption on colloidal matter), after which titration is followed by Adsorptive Cathodic Stripping Voltammetry (ACSV) or DPASV (23, 36–39). Whatever the procedure adopted, an independent measurement of the total metal content is required and a suitable treatment of titration data is necessary to determine the fraction of metal uncomplexed with organic ligands, the total ligand concentration and the related conditional stability constant.

Here the discussion focuses on the analytical procedure adopted to determine trace metals concentration in sea water in the dissolved phase. Particular attention will be given to the procedures preceding the analytical measurement (sampling, sample treatment and storage), the analytical determination of total concentration by DPASV and Inductively Coupled Plasma Mass Spectrometry (ICP-MS); the contamination control procedure will also be discussed. The direct DPASV procedure for determining metal complexation in sea water is reported in detail and after a discussion of theoretical aspects an outline of the experimental procedure is presented. Finally, an overview of the distribution in the Southern Ocean of some metals of particular interest is examined and the evaluation of trace metals distribution is carried out also by comparison with results obtained in different geographical areas.

2. Sample processing

2.1 Sampling

To study physical, chemical and biological processes that control the flow of chemicals through the ocean a detailed knowledge of spatial and time scale distribution of the chemicals is required. As indicated by Johnson et al. (40), the ideal procedure for the evaluation of analyte distribution is to make measurements *in situ* by a probe which could be lowered from the surface to the bottom to give a continuous profile of the water column or operate unattended for long periods of time. The application of probes able to carry out *in situ* measurements reduces problems related to the handling of samples (problems deriving from sample transformation or contamination during collection or storage) and permits real time analysis and therefore detailed description of complete ecosystems and interfaces (41, 42).

Only recently have devices been introduced which are capable of precise measurements of some trace elements; their application is limited because the sensitivity of sensors is not adequate to determine trace metals at concentrations present in oceanic areas. A submersible probe for trace metal determination in the water column has been described by Tercier et al. (43) and used to determine Cd, Cu, Pb and Zn concentrations in sea water at natural levels. Shipboard methods were developed to produce large data sets and to avoid storage and transport. Flow systems using voltammetric determination have been introduced for continuous trace metals monitoring; they consist of computer-controlled electrochemical instruments fitted with units for pretreatment of samples and introduction of internal standards (44, 45). Brainina et al. integrated a similar system into a flow-through immersion device (46). Flow injection analysis methods using chemiluminescent and fluorimetric detection have also been developed for a number of metals (47, 48).

Because of problems related to the non-availability of instruments to make *in situ* measurements for trace elements, chemical oceanographers are still dependent on the collection of water samples at the sites and at the depths of interest. However, in oceanographic studies where trace species must be determined, sampling by general purpose devices is normally unsuited. Therefore, samplers tailored to specific requirements are adopted (13, 49).

Procedures for water sampling may be instantaneous or continuous, the choice of procedure being made on the basis of analytical requirements, the aim of the study and the characteristics of the system. Three general approaches can be used, i.e.: 1) pumping water to the surface from the requested depth; 2) sampling by bottles lowered to an appropriate depth and closed by a signal from the surface; 3) adsorbing the elements or compounds of interest on an appropriate material lowered to the desired depth.

Pumping systems have been used less frequently than the discontinuous sampling devices, usually only to sample shallow water or when large volume samples or continuous profiles are required (50, 51). Collection of very large samples at different depths is necessary to detect natural or anthropogenic radionuclide elements and trace organic pollutants. Measurements of salinity (or conductivity), temperature and depth carried out by devices capable of *in situ* determinations (normally reported as CTD probes) and other probes have shown a generally fine structure in the water column; this is missed if samples are collected by bottles at a limited number of depths. Sampling by pumping in conjunction with continuous analysers could permit detailed vertical and horizontal profiling; however, this is linked to the availability of continuous analysers to detect trace elements and, of course, the water depth examined is necessarily limited. Flow Injection Analysis (FIA) methods are being developed to determine some trace metals including Co, Cu, Mn, Ni and Zn (44, 45, 47, 48).

Instantaneous samples are collected by bottles along the water column at different depths. Single or multiple instantaneous samples are collected by bottles bound to a cable, closure of samplers at desired depths being obtained in cascade by messengers running along the cable. An approximately simultaneous collection of multiple samples at different depths is obtained by bottles connected in a rosette sampler and closing the bottles at the desired depths by signals transmitted by a conducting cable. These procedures allow a general characterization of the water column by physical and chemical measurements to be carried out at the same time as the sample collection by a multiparametric probe.

The principal problem in determining trace elements is the contamination of samples by the devices used to gather water and by the ship used to reach the sampling area. The samples collected by the rosette system are therefore normally used to determine the majority of chemical and physical-chemical parameters (nutrients, salinity, particulate), but are not useful for determining trace elements. In this case two different approaches are used in collecting sea water samples from surface/shallow depth and deep water, respectively.

In the collection of surface water it is preferable that researchers move away from the ship by small non-metallic boats and gather samples using plastic pumps for temporary storage in a high capacity plastic containers (20–50 l). The procedure consists in reaching the sampling area by dinghy, stopping the engine and rowing up-wind while sampling uncontaminated water from the prow. Before collecting the sample, the polyethylene tank (cleaned as reported below for sample containers) is conditioned two or three times with sea water.

Deep water samples (more than 20 m from the surface) can be collected directly from the ship, using a sampling bottle immersed to the desired depth by a non-metallic hydrowire, normally a Kevlar cable because of its mechanical characteristic of low extendibility.

A few studies have been carried out to compare different sampling devices (52). The results emphasized that to minimise sample contamination the surface/volume ratio for the sampler must be as low as possible. So samplers larger than 8 l are therefore recommended and considerable care must be taken during cleaning and conditioning. Bottles of the close-open-close type, internally coated in Teflon® and with pressurization capability, are recommended for collecting samples to determine trace elements. These bottles enter the sea in the closed position, open themselves automatically at a fixed depth of about 10 m and close at the desired depth by the "messenger" system. This procedure avoids contamination of the internal part of the sampler by the surface microlayer which is particularly rich in trace metals.

Subsurface water (0.5–1 m depth) was collected while moving away from the ship in a rubber dinghy, rowing up-wind and using a submersible Teflon® pump for temporary storage in a high capacity polyethylene container (50 l) (53). To test the sampling procedure a few experiments were carried out on Antarctic sea water in which samples were also collected, for reference, by direct immersion of storage bottles in sea water (avoiding pumping and filtration). The samples were analyzed for total Cd and Pb concentrations (53).

Deep water samples were collected directly from the ship, using a 20–30 l sampling bottle (Teflon®-coated and with pressurization capability, Go-Flo, General Oceanics, USA) which was immersed by means of a non-metallic (Kevlar) hydrowire. To sink the bottle a plastic covered ballast was used, attached to the wire at least 20 m below the bottle.

The Go-Flo devices are first repeatedly rinsed by ultrapure water to remove dust particles; the cleaning procedures continue in the clean laboratory, with operators wearing polyethylene gloves and clean garments. The procedure adopted consists of repeated washings with 1:100 ultrapure water diluted ultrapure HCl (Suprapur grade, Merck, Darmstadt, Germany). At the end of each washing the bottle is filled with the same solution and left standing for a few days before repeating the treatment. After three or four treatments the sampling bottle is rinsed abundantly and repeatedly with ultrapure water. Bottles are stored dried and sealed in multiple clean polyethylene bags to protect them from dust particles during storage and transport. Before use the samplers are conditioned by filling them several times with sea water collected offshore.

As regards the submersible Teflon pumps, the parts in contact with the sample are cleaned in a similar way to that used for the storage containers (see below). The pumping system is conditioned with sea water by pumping and discarding at least 10 l before collecting the sample. Tests carried out to verify samplers leaching before use are reported below in the contamination control section.

2.2. Sample treatment

If analysis is not performed immediately after sampling or the analytical methodology has particular requirements, a convenient treatment and storage procedure must be applied so that the sample maintains its characteristics. The treatment will depend on the analysis to be carried out and the information to be obtained from the sample. An efficient treatment and storage procedure avoiding sample contamination or alteration is important both when measurements of total metal concentration and speciation studies are carried out. In the latter case control must be more accurate; indeed, contamination can affect both the total metal concentrations and the distribution between different species. Control must therefore be extended to all elements with a potential effect on analyte speciation. In general, for trace metals analysis in sea water, particulate matter is removed from the water sample by filtration or centrifugation. Reagents may then be added to the sample and the water sample can be stored in an appropriate container to minimise change in the sample composition.

A long time contact of the dissolved fraction with particulate matter can produce changes in the distribution of chemical forms of heavy metals in solution. Any change in the equilibrium conditions after collection can promote or remove dissolved metals (54) or desorption of adsorbed metals operated by particulate. Biological activity involves photosynthesis and respiration which will change the carbon dioxide content of the water and its pH. All the equilibrium reactions affected by pH will be altered, e.g., the reactions of precipitation, complexation and redox involving heavy metals.

The aim of sample treatment is frequently the differentiation of water components on the basis of their physical-chemical properties. More separation techniques can be applied to divide trace elements in sea water in fraction (55). The major assumption made by applying separation techniques is that removal of one or more components from a sample does not disturb the solution equilibria, but frequently there is evidence that this is not true: e.g., it was observed that after removal of particles from water there was regrowth of filterable particles. However, operationally defined separation techniques are a useful means of comparing the characteristics of different samples. The problems and the advantages associated with specific separation methods for trace elements speciation has been reviewed (55).

Filtration is the more commonly applied practice for differentiating dissolved from particulate matter, although the distinction is arbitrary. The dissolved fraction will contain filterable matter defined on the basis of pore size, although the so-called dissolved phase will contain all the suspended matter (colloids or particles) not retained by the filters. The dimensional distribution of retained particles can change as a function of the different type of membrane filter used. In particular, the conventional cellulose ester membrane filters did not show an accurate size distribution of particles considering the nominal pore size; for filters of 0.45 μm pore size it was observed that filters retained more than their nominal rating, while the smallest pore size filters appeared less efficient at retaining particles greater than the nominal rating (56).

The sample pretreatment of Ross Sea water (Antarctica) adopted was the filtration of samples through a conventional filter of 0.45 μm pore size. The treatment was carried out as soon as possible after sampling and inside a clean laboratory that was available on board. A Teflon® apparatus developed for on-line pressure filtration (e.g., Sartorius SM 16540, Göttingen, Germany) and a cellulose nitrate membrane filter (e.g., Sartorius SM 11306, 142 mm diameter) can be used for pressure filtration with maximum pressure at 0.5 bar. Technically the filtration is performed by using pure nitrogen to exert a slight pressure inside the 50 l polyethylene tank used for temporary storage of the subsurface sample, or the Go-Flo bottles used to collect deep samples. The sample is forced to pass through the filtration apparatus and, after discarding a first aliquot of about 3–4 l, the filtrate was collected directly in the storage bottle.

To prevent any contamination risk, all the sample treatments must be performed in clean laboratories (class 100, US Federal Standard 209D) and the materials in contact with the sample must follow a cleaning procedure. In compliance with this, the filters are soaked before use in 1:10 diluted ultrapure grade HCL for few days and rinsed with ultrapure water. The procedure is repeated twice, then the membranes are stored in 1:100 diluted ultrapure acid until use. The filtration apparatus is subjected to a drastic cleaning procedure, similar to that described below for storage containers, the first hot treatment HNO_3 following repeated soaking with HCl (ultrapure grade) solution, before one final rinsing with ultrapure water. Some investigators prefer to collect the samples directly into the storage bottles avoiding, when possible, the filtration step (especially under open-sea conditions or for short-term storage) (15, 16).

2.3. Storage

Prolonged preservation of the integrity of samples with respect to metal speciation is a very difficult task and in principle it would be better to carry out the analyses on board immediately after collection and filtration. One important aspect for analytical chemists practising trace metal determination in sea water is the development of methods that can be used aboard the ship to produce large data sets. However, analytical methodologies so far available to determine trace metals in sea water are normally tedious and require a long time for the analyses, so that measurements on board, when possible, are restricted to relatively few samples mainly to make tests of contamination and assess reference values to compare with results obtained later in the laboratory ashore. As a consequence, storage of samples for periods of several months is often essential.

Storage of water samples to detect trace metals is normally carried out in plastic containers; fluorinated plastic materials i.e., Teflon®, fluorinated ethylene propylene (FEP), perfluoroalkoxy polymers (PFA) or polyethylene are used because if opportunely treated and conditioned they guarantee contamination-free samples. However, FEP or Teflon® bottles are preferable to those made of polyethylene when speciation studies are carried out. In fact, polyethylene over long periods of storage can release plasticizers (above all phthalates or amines), which behave as ligands and modify the complexation equilibria of the solution. When Hg is to be determined plastic materials must be avoided because they are permeable to gases and vapours; glass or quartz are therefore used.

Before use bottles must follow a cleaning and conditioning procedure to avoid any contamination during the storage. For new bottles the cleaning procedure starts with a first wash with tap water, followed by rinsing and cleaning with acetone. Then bottles are left in a heated (50°C) detergent solution for 10 days to maximize the degreasing stage. After rinsing with ultrapure water, bottles are immersed in a heated (50°C) 10% solution of HNO_3 (analytical reagent grade) where they are kept for two weeks in order to allow heavy metals to be released from the walls of the containers. After further rinsing, bottles are enclosed in polyethylene bags and transferred to the clean laboratory for final treatment. They are filled with a 10-fold diluted solution of ultrapure HCl (Merck Suprapur). After a week this step is repeated for another week. Again, rinsing with ultrapure water is performed and bottles are filled with a 1:100 diluted solution of ultrapure HCl and sealed in double polyethylene bags until sampling. Used bottles are subjected to repeated washings with small aliquots of concentrated HNO_3 (Merck Suprapur). After rinsing with ultrapure water, the final treatment described above is carried out.

Two procedures are normally used to preserve sample integrity as much as possible: storage of the sample frozen at the temperature of –20°C or lower, and acidification of the sample by addition of mineral acids (pH $\leq$ 2). The two procedures are about equivalent if total dissolved metal concentration needs to be determined, while acidification of the sample must be avoided if a speciation study is to be carried out because this procedure would certainly alter the complexation equilibria of metals in sea water. There is some experimental evidence that freezing of samples does not significantly alter the results of the complexation studies carried out on heavy metals according to the methodology discussed here (53).

3. Analytical methodologies

3.1. Voltammetric methodology

DPASV instrumentation can easily be operated on a ship, provided that a clean laboratory is available, to determine dissolved concentration of some metals of environmental interest at the natural levels in oceanic waters. The methodology was used in polar regions under ice-breaking conditions to take measurements on board (57, 58).

DPASV equipped with the Thin Mercury Film Electrode (TMFE) plated onto a Rotating Glassy Carbon Disc Electrode (RGCDE), is one of the most sensitive and powerful techniques at present available for the determination of ultra-trace metals in real time samples (59, 60). Cadmium, Cu, Pb and Zn are the most frequently determined metals at subnanomolar or even picomolar concentration in sea water, without any preconcentration step (17, 61, 62), but other elements can also be determined (63–66). Detection limits lower than 0.5 ng/l are normally reached (17, 60).

As reported above, one fundamental characteristic of DPASV with respect to the speciation problem is the ability to differentiate the total dissolved metals in two fractions on the basis of electrochemical properties of chemical forms in which the metals are present. When the methodology is applied to analyze an untreated (undigested) sample, the measurement is sensitive to the so-called "electroactive" fraction of total dissolved metal concentration, which, under carefully controlled conditions (especially as regards the deposition potential and the rotation speed of the electrode), is practically composed of the ionic and the inorganically complexed fractions of metal. The metal aliquot strongly bound to organic ligands remains undetected (59, 67, 68).

In this section the instrumentation and the experimental procedure to determine the Cd, Cu and Pb total dissolved concentration is described in some detail. Several reviews regarding the general theoretical and experimental aspects of anodic stripping voltammetry and its application in sea water (or natural water) analysis can be found in the literature (15–17, 41, 59, 62, 67, 68). Details of the speciation procedure are reported below.

3.2.1. DPASV instrumentation

The instrumentation used consists of a polarographic analyzer with DPASV capability (Mod. 384B, EG&G PARC, Princeton, USA) equipped with an electrochemical cell developed for ultra-trace metal determination in sea water (17, 32) (Model Rotel 2, EG&G). The electrochemical cell comprises a RGCDE as the working electrode, on which a thin Hg film is deposited, an Ag/AgCl,KCl (saturated) as the reference electrode and a Pt wire as the auxiliary electrode. Both the reference and auxiliary electrodes are inserted inside small FEP tubes, filled with saturated KCl/AgCl solution and fitted with porous Vycor® tips. The cell cup containing the sample and the head supporting the electrodes are in Teflon®. The cell compartment is separated from all the other mechanical and electronic components (the cell controller and the electrode motor) by a Plexiglass box.

To avoid possible contamination from atmospheric particles during sample manipulation and analysis, the equipment is installed in a clean chemistry laboratory, inside a Class 100 laminar flow area, or, during oceanographic cruises, under a laminar flow area in a clean chemistry laboratory container (ISO20) available on board.

3.1.2. Preparation of TMFE

The TMFE is prepared by electrolysis on the surface of the RGCDE just before the beginning of the determination. The glassy carbon surface is polished with wetted alumina powder (0.075 mm grain size, or lower) on a filter paper held against the electrode rotating at 1000 rpm. Afterwards the electrode is rinsed for five minutes with 1:200 diluted ultrapure HCl and then two or three times with ultrapure water. The Hg film is prepared by controlled potential electrolysis of a $Hg(NO_3)_2$ solution. The electrolytic solution consists of 2.5 ± 10^{-2} *M* KCl (ultrapure) and 10^{-4} *M* Hg^{2+} that is purged with N_2 for at least 15 minutes before the start of electrolysis. The salt is prepared by oxidizing hexadistilled Hg with HNO_3. The film deposition is carried out applying a potential of -1.00 V for 20 minutes and rotating the electrode at 4000 rpm. After film deposition, a rest time of 30 s is allowed to pass and then a differential pulse potential scan is carried out from the deposition potential of –1.0 V until a potential of –0.18 V is reached with a scan rate of 10 mV s^{-1}, a pulse height of 50 mV and a pulse frequency of 5 s^{-1}.

If the voltammogram obtained does not show any peak and the base current is sufficiently low (300–400 nA), then it is possible to continue with the analysis of the samples, otherwise the Hg film is destroyed and prepared again. To avoid alteration of the complexation equilibria due to residual ionic Hg and to clean the electrode assembly, the latter is rinsed with a purged aliquot of the sample before transferring the cell cup containing the sample in the measurement position.

3.1.3. Total metal concentration

Total dissolved metal concentration is determined on filtered, acid digested samples. The digestion is normally carried out by addition of 100 μl of HCl (30% Suprapur Merck or 32% Ultrapure NIST) to approximately 50 ml of sea water directly into the same Teflon® vessel subsequently used as the cup of the electrochemical cell (the exact volume of sample is determined at the end of measurement, see below). A pH of about 2 is obtained and digestion is carried out at room temperature for at least 48 h.

If the samples contain large amounts of dissolved organic matter, such as those collected in coastal or highly productive areas, the acid digestion alone may be not sufficient to release the metals completely; in these cases, acidified samples are subjected to UV irradiation by high power Hg lamp (1.2 kW) for at least 3 h.

After the proper mineralization procedure, measurements of Cd, Cu and Pb are carried out separately according to the procedure outlined below.

The aliquot of digested sample placed in the electrochemical vessel is purged for 15–20 minutes with N_2, then it is rapidly transferred and screwed under the cell head of the electrochemical device where the Hg film electrode has already been prepared, tested and rinsed. The electrolytic pre-concentration step during the analytical measurement is carried out by constant potential electrolysis. The deposition potential is set at –0.95 V for Cd and Pb and at –0.85 V for Cu. The deposition time depends on the metal concentration, a time of 20 min normally being sufficient to determine Cd and Pb concentration in samples collected in

Antarctic sea water; sometimes a shorter time is sufficient for Cu. A rotation rate of 4000 rpm for the working electrode is used during the deposition step. This is followed by a quiescent period of 30 s and then by a potential scan between the deposition potential and the final potential, fixed at –0.18 V for Pb and Cd and at –0.15 V for Cu. The scan rate is set at 10 mV s^{-1}, the pulse height at 50 mV and the pulse frequency at 5 s^{-1}. At the end of the scan the potential is held at –0.20 V for 5 min, while the electrode is rotating, to allow the amalgamated metals to be removed completely from the working electrode. To test electrode stability and repeatability of the voltammogram, a second measurement is performed on the sample solution. The quantification is carried out by the multiple standard addition method using three/four metal additions. The volume of the sample is finally measured by a graduated cylinder.

The blank of the acid employed in the digestion step is evaluated by repeating the acid treatment on a 3 ± 10^{-2} *M* KCl solution prepared with ultrapure water and KCl. The voltammetric measurement is carried out before and after the acid treatment and the concentration increment observed gives the blank value to be used for correction. Using the Merck Suprapur HCl the Cd and Pb blanks are found to be 17 and 19 p*M*, respectively (35, 57), while the blank for Cu is under the detection limit (69). Using the NIST ultrapure acid all the blanks are below the detection limit.

3.2. Spectroscopic techniques

The very low concentrations expected in the analysis of trace elements in offshore and coastal Antarctic sea water can be also detected thanks to the high detection power of spectroscopic techniques such as Electrothermal Atomic Absorption Spectrometry (ETA-AAS) and Inductively Coupled Plasma Atomic Emission Spectrometry (ICP-AES) or ICP-MS. However, the saline matrix which constitutes the ideal medium in which to perform electrochemical measurements poses severe problems to the direct analysis of sea water because of possible signal suppression and/or undesired matrix effects.

Preconcentration and separation procedures, by means of off-line and on-line techniques, are possible solutions to these problems (70–75), provided that an ultraclean approach is adopted in order to avoid contamination of the samples from laboratory air, labware and the operator. In all these methodologies the pH of the sea water sample is adjusted to a certain value, depending on the chelating system used; then samples are mixed with the resin carrying the chelating agent (off-line) or injected through a small column packed with the resin (on-line). During this step several trace elements are retained by the chelating system and finally eluted by means of an appropriate reagent into the spectroscopic detector for quantification.

Off-line preconcentration systems are less expensive and easier to handle, but often require very large sample volume (500–1000 ml), in some cases hampering the analytical determination due to the low amount of sample available. Furthermore, off-line methods are more contamination-prone, since they require a great number of subsequent steps of sample treatment. In order to reduce sample consumption and preparation time and to lower the possibility of contamination of the sample, on-line preconcentration systems, based on FIA techniques, have been developed and used for the determination of trace elements in sea water (70, 71, 74). Many papers rely on the analysis of Certified Reference Materials (CRMs) in Open Ocean sea water (e.g., NASS-3 and NASS-4) (76–78), testifying to the high reliability of preconcentration methods with matrix separation in terms of accuracy and precision. Considering that concentrations of open ocean and CRMs are close to those found in pristine environments, these coupled techniques can be used for the analysis of trace elements also in Antarctic sea water.

Recently thanks to the high detection power of double-focusing ICP-MS, new direct methods for the multi-elemental analysis in sea water have been developed (77, 79). These methods are based on the idea of leaving the sample as far as possible untouched, thus reducing the risk of contamination during sample processing. Nevertheless, a dilution of about five to ten times is necessary in order to alleviate severe suppression of the sensitivity by the high salt concentration. This dilution obviously constitutes a critical step considering that the concentration of some trace elements in ultrapure water used for the dilution could be of the same order of magnitude as in Antarctic sea water. This methodology as been used up to now for the determination of 17 trace elements in the NASS-4 CRM. Preliminary results (79) show good agreement with certified values, with the exception of Fe and Zn for which higher values were obtained, as shown in Table 5.1. Although

Table 5.1. Analytical results for the NASS-4 CRM.

Element	Concentration (μg l^{-1})	
	Certified values with confidence interval	Found values with standard deviation (n = 5)
Ba	–	6.22 ± 0.08
Bi	–	0.0003 ± 0.0002
Cd	0.016 ± 0.003	0.026 ± 0.005
Co	0.009 ± 0.001	0.011 ± 0.002
Cr	0.115 ± 0.010	0.14 ± 0.02
Cu	0.228 ± 0.011	0.22 ± 0.02
Fe	0.105 ± 0.016	0.68 ± 0.2
Mn	0.380 ± 0.023	0.38 ± 0.02
Mo	8.84 ± 0.60	8.41 ± 0.1
Pb	0.013 ± 0.005	0.014 ± 0.002
Pd	–	0.006 ± 0.001
Pt	–	0.0025 ± 0.0003
Sb	–	0.21 ± 0.01
Sn	–	1.65 ± 0.8
U	2.68 ± 0.12	2.55 ± 0.15
V	–	1.28 ± 0.08
Zn	0.115 ± 0.018	0.18 ± 0.06

this method takes advantage of the high detection and resolving power of double-focusing ICP-MS and it could be considered preferable in terms of minimum sample handling, considerable care must be adopted in order to keep ultrapure water blanks under strict control.

Spectroscopic methods have been used with success in the past for the analysis of antarctic sea water (71, 80–82) using an on-line preconcentration system coupled with both ETA-AAS and ICP-AES. Preconcentration factors, ranging between 60 and 100 and obtained using an Amberlite XAD-2 resin (71), allowed for the determination of Cd, Cu, Fe, Mn, Ni and Zn in Antarctic sea water. Spectroscopic methods using the preconcentration of analytes were also compared to ACSV in Antarctic sea water samples (80). The results obtained were in good agreement, showing that both methodologies are suitable for heavy metal determination at the ng/l level.

3.3. Contamination control

Trace metals in sea water are present at very low concentrations (normally lower than 10^{-9} *M*); thus, the study of metal distribution and speciation poses problems that have been overcome only in the last few decades. A prerequisite for determination of trace metals is the collection and processing of uncontaminated samples for the element of interest. Improvements in analytical procedures have led to a decrease by some orders of magnitude in the reported concentration of some elements, especially so for ubiquitous and contamination-prone elements, such as Fe or Pb. These changes lead to the concept of oceanographic consistency as a criterion for accepting trace chemical measurements (83).

If speciation studies are carried out, the control must be extended to all the elements and organic compounds that can potentially determine equilibrium displacements. Contamination can be so important as to render any study meaningless. From this standpoint it is of primary importance to keep all the analytical steps under rigorous contamination control, from preparation of the sampling equipment to instrumental analysis. Details about the cleaning procedures for sampling equipment, the storage containers and the pretreatment materials are reported above in the sections describing the sampling procedures, the storage and the sample pretreatment. Here the focus is on the control procedures adopted to assess contamination problems during sampling and storage and due to chemical treatments. The laboratory characteristics necessary to avoid sample contamination are described in the following sections.

3.3.1. Sampling procedure

The first step in studies to detect trace metals in natural waters is the collection of samples; problems of contamination during this phase jeopardize every subsequent step. From this standpoint, testing and optimisation of the sampling procedure is fundamental.

Sampling devices must be verified for contamination before use by controlling the leaching from materials during cleaning procedures. The efficiency of commercial samplers used to collect surface and deep sea water can be checked by comparing the results obtained with those gained by systems especially designed to gather uncontaminated samples. Some intercomparison studies of sampling devices are reported in the literature (22, 52, 84, 85), results showing that when a convenient cleaning treatment of the sampling device is applied, when bottles of volume higher than 8–10 l are used and when samplers present certain characteristics, the problems of sample contamination can be kept under control. Generally closed-open-closed type sampling bottles are preferable to avoid contamination from the surface microlayer.

An intercomparison exercise of sampling devices showed that samples collected by 20–30 l Teflon-coated Go-Flo bottles guarantee low metal release (52). Although some researchers found problems of contamination for Pb and Zn using the same sampling device, the results obtained by this investigation agree with those previously reported (14, 34, 86). Bruland et al. did not detect any contamination for Cd, Cu, Ni and Zn when they used these devices (14). Results of Pb concentration in the Pacific Ocean reported by Capodaglio et al. using the same samplers, were in agreement with those obtained by Flegal et al., the values of which were obtained analyzing samples collected by a device especially designed to gather uncontaminated water to determine Pb, values considered oceanographically consistent (13, 34, 87).

Although the choice of appropriate samplers for collecting samples to determine trace metals is fundamental, adequate cleaning treatment and conditioning of the devices is also important. Along with the cleaning procedure, tests to assess the system efficiency must be carried out, also when especially designed devices are used (52). The procedures adopted to verify samplers used to collect sea water during expeditions in Antarctica have previously been reported (88). Figure 5.2 shows the trend of Cd and Pb concentration in ultrapure water after temporary storage in devices used to collect deep sea water. The samplers were cleaned with diluted HCl solutions after each test until the water contamination could be considered negligible.

Surface water samplers can be tested by comparing results obtained on samples collected by the sampling device with those obtained on water aliquots collected directly in the storage bottles, as shown in Table 5.2. The procedure can be applied when filtration can be avoided (particularly when working in open waters). The operation can be carried out by hand and with the use of long polyethylene gloves, in the case of surface waters, or by a plastic telescope bar with the sampling bottle inserted at its end (16).

3.3.2. Laboratory and chemicals

Airborne particles are one important source of sample contamination. Therefore, the laboratory must enable samples to be treated and measurements to be carried out in a satisfactorily dust-free environment. Improvements in existing laboratories can be obtained by installing a laminar flow hood, but optimal improvement involves the design of clean rooms in which air is pumped through a dust-stop prefilter and then forced by a blower through an absolute filter inside the working

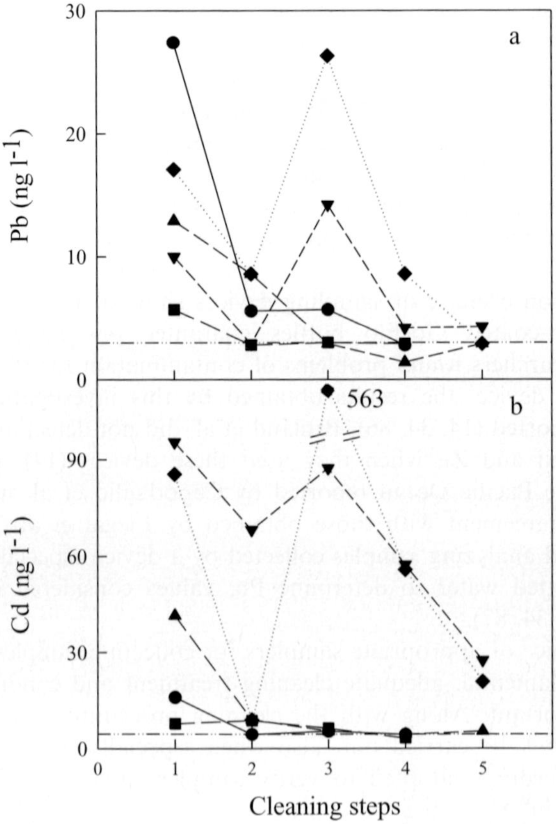

Figure 5.2. Cadmium (a) and Pb (b) concentrations in the ultrapure water used to condition Go-Flo samplers (different symbols refer to different bottles). Concentration was detected in aliquots of 50 ml of water where 300 μl of satured KCl were added(– · · · – stands for the blank). Adapted from Capodaglio et al. (88).

area, allowing researchers to work under a laminar flow pattern in which contamination problems from the atmosphere are greatly reduced (class 100 laboratory). To avoid the release of particles from the laboratory walls, these are coated with plastic or plastic paint; all metallic furniture must be avoided, or, if this is not possible, it must be carefully coated with metal-free plastic paint. The filtered blown air produces a slight positive pressure that stops dust entering the working area. The cleanest area is preceded by a pre-room to avoid direct contact to the external area; normally, the floor of this room is covered with a sticky plastic mat to retain the dust attached to the researchers' and visitors' shoes. To protect samples from problems arising from contamination, all handling of sampling equipment and instrumentation devices must be performed in this kind of environment.

To handle and to treat sea water samples collected during expeditions in Antarctica and other oceanographic campaigns, a mobile clean laboratory was set up in an ISO-20 standard container and installed on board one of the research vessels

Table 5.2. Comparison of sampling methods for surface sea water. Concentration of Cd and Pb in surface sea water of Terra Nova Bay (Antarctica) (53).

Sample no.	Total dissolved metal concentration (n*M*)			
	Cd		Pb	
	Sampling by Teflon® pump[a]	Manual sampling[b]	Sampling by Teflon® pump[a]	Manual sampling[b]
10	0.18	0.20	0.116	0.113
16	–	–	0.029	0.024
17	–	–	0.165[c]	0.038
31	0.08	0.10	0.034	0.034
37	0.08	0.08	0.029	0.029
46	0.24	0.26	0.036	0.031

[a]Filtered sample
[b]Non-filtered sample
[c]Contaminated sample

(Figure 5.3). This laboratory is subdivided into two parts: the first is equipped as a general chemical laboratory, contains the sampling bottles stand and constitutes the pre-room for the second part; this last is the internal clean area equipped with two laminar flow hoods class 100 (US Federal Standard 209D) (89).

Reagents and water employed to clean all items and to perform the analysis must be extremely pure. Only reagents with a very low heavy metal content can be used (90). Ultrapure water is produced in a quartz sub-distillation apparatus or with special ion-exchange devices (15, 91, 92). The blanks of pure water and reagents must be checked periodically in order to estimate the metal content added to the samples during analysis.

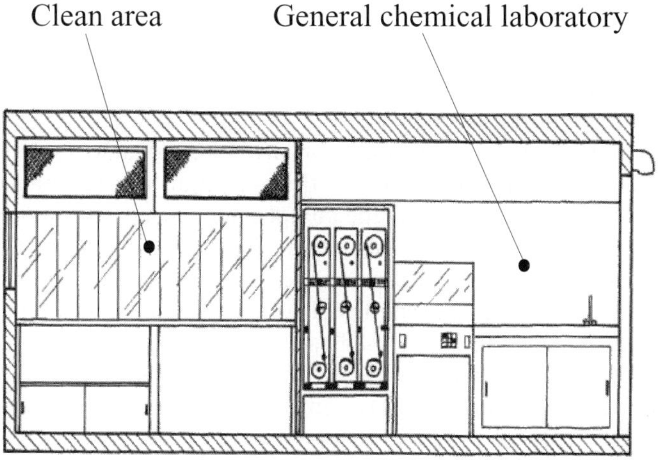

Figure 5.3. Clean chemistry laboratory on board. Reprinted from Capodaglio et al. (52), with permission of Società Chimica Italiana, Rome.

Table 5.3. Mean analytical results for the NASS-4 CRM obtained by DPASV measurements at the beginning of 1998.

Element	Concentration ($\mu g\ l^{-1}$)	
	Certified values with confidence inteval (95%)	Found values with standard deviation (n = 8)
Cd	0.016 ± 0.003	0.016 ± 0.002
Pb	0.013 ± 0.005	0.012 ± 0.003
Cu	0.228 ± 0.011	0.230 ± 0.014
Zn	0.115 ± 0.018	0.110 ± 0.015

Laboratory quality control schemes should be applied to guarantee consistent results of the adopted analytical procedure by testing the overall calibration procedure. The routine analyses of sea water CRMs are carried out in general, at least once per week (93, 94). Tables 5.1 and 5.3 show recent results obtained on the NASS 4 CRM by ICP-MS and DPASV measurements, respectively (93).

Participation in international intercomparison exercises can ensure that a laboratory is working with the required accuracy (22, 85, 94, 95). These exercises can be used to compare results obtained on an unknown sample with the values obtained by many other qualified laboratories that have analyzed the same sample using different instrumental approaches. Intercomparison campaigns are now being organized in which the overall procedure from sampling to chemical analysis, carried out independently by each of the participating laboratories, is being tested (84, 85, 96). Past experience in these intercalibration exercises showed that relatively few laboratories involved in trace metal determination adopt the stringent clean procedures (during the whole analytical procedure, from sampling to analysis) required to obtain uncontaminated samples and accurate results.

4. Speciation procedure

sea water contains a variable amount of dissolved organic matter complexing metals, originated by different processes; the maximum amount of metal that can be complexed is normally known as the complexing capacity. Primarily electroanalytical techniques are used to evaluate this characteristic of aquatic systems. If organic complexes or inorganic forms of the metals are selectively detected by an analytical technique, this can be employed to study the complexation equilibria in solution. Several metals have been used to evaluate the complexing capacity, e.g., Co and Zn, but the most frequently used transition metal is Cu because it forms strong complexes with organic ligands and it is easily determined by voltammetric techniques (30, 97–103).

The application of DPASV to evaluate metal complexation derives from the assumption that metals form non-electroactive complexes with strong ligands present in solution; therefore, when it is used to analyze untreated samples, the measurement does not depend on the total dissolved metal concentration, but only on the fraction reducible during the deposition step (labile or electroactive fraction). With careful selection of the experimental conditions (deposition potential, electrode rotation speed) the labile fraction turns out to consist principally of the ionic and inorganically complexed metal to be examined. The procedure involves the titration of organic ligands by adding aliquots of a standard metal solution to the untreated sample and, after a suitable equilibration period is allowed to pass, the DPASV measurement is carried out to detect the labile metal concentration. The first part of the titration curve shows a low response due to complexation of added metal with free ligand present in the sample. Beyond the end-point, when sufficient titrant excess is present, the metal added is not complexed by organic ligands and gives the normal DPASV response. Consequently, a straight line is finally obtained in the upper region of the curve. To obtain the fraction of metal which is uncomplexed with organic ligands, the total ligand concentration and the related conditional stability constant, an independent measurement of total metal content and a suitable treatment of titration data are required. A schematic graphic representation of the titration curve to evaluate the complexing capacity is shown in Figure 5.4a.

Interpretative models to evaluate the titrimetric method for study of metal complexation in sea water have gradually been refined over the last fifteen years (29, 36, 104).

4.1. Theoretical aspects

The composition and structure of organic matter present in sea water and with complexing properties is complicated. Single chemical species cannot be isolated and studied separately; however, these components can be distinguished in several groups possessing operationally defined similar properties and each group can be treated as a single structure whose properties are the sum of the properties of the individual components, plus the possible synergistic actions between them (105). Applying the fractionation of dissolved metals on the basis of the ASV response (ASV-labile, composed of electroactive species; inert forms, composed of non-electroactive organic complexes) the current signal will derive from all the electroactive species weighted according to their kinetic and thermodynamic properties.

Although the macromolecules constituting the dissolved organic matter often contain a large number of complexing sites, the experimental titration curve is evaluated by simple interpretation models. Semi-empirical models are frequently sufficient for comparative purposes and to describe the environmental properties of the metals studied.

The titration of organic ligands present in the sample by the metal to be studied forms non-electroactive complexes, following the general reaction:

$$mM^{n+} + iL^{p-} \Leftrightarrow (M_mL_1)^{n-p} \quad (1)$$

The equilibrium of the complexation reaction (1) is defined, in terms of the concentration constant, K, by the expression (the charges will be omitted here and in the following):

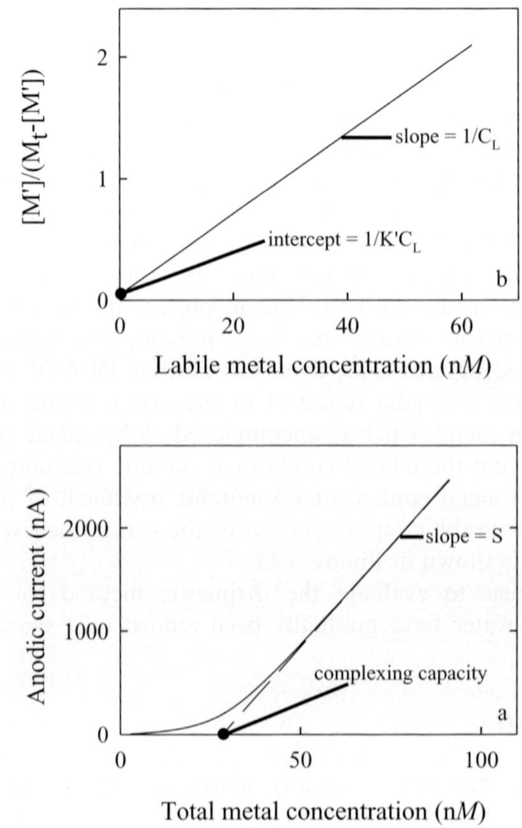

Figure 5.4. Titration curve of one ligand complexing the metal M: (a) with a 1:1 stoichiometric ratio; (b) plot of the transformed titration data.

$$K = \frac{[M_m L_l]}{[M]^m [L]^l} \tag{2}$$

The interpretation models more frequently used in studying metal complexation in sea water by DPASV are discussed here with particular emphasis on the formation of complexes presenting a 1:1 stoichiometric ratio between metal and ligand. In particular, the discussion focuses on the complexation reactions involving one metal and one ligand as well as one metal and more ligands. A more rigorous and complete theory of the titrimetric method considering also the competition between different metals and the general formation of complexes with other than 1:1 stoichiometry has already been reported (104).

Single 1:1 complex formation. Consider the simple 1:1 complexation reaction between the metal, M and the ligand, L, to form the complex ML. If one assumes that M participates in side-reactions with inorganic ligands (or weak organic ligands) X_i (Cl^-, OH^-, CO_3^{2-}, ...) to form complexes MX_i, that L participates in side-reactions with proton H^+ and major cations, M_j (Ca^{++}, Mg^{++}, ...), and that ML participates in side-reactions with inorganic ligands, e.g., with OH^- and Cl^- and proton to form mixed complexes, the equilibria involved can be represented as follows:

$$
\begin{array}{ccccc}
 & MX_i & & H_nL & MH_n\,L \\
+\,X_i & \Updownarrow & +\,nH & \Updownarrow & \Updownarrow\;+\,nH \\
 & M & + & L \Leftrightarrow & ML \\
 & & +\,M_j & \Updownarrow & \Updownarrow\;+\,Y_m \\
 & & & M_jL & MY_mL
\end{array}
\qquad (3)
$$

The stoichiometric stability constant becomes:

$$K = \frac{[ML]}{[M][L]} \qquad (4)$$

where K is related to the thermodynamic stability constant K^* by the following relationship:

$$K = K^* \frac{\gamma_M \gamma_L}{\gamma_{ML}} \qquad (5)$$

γ_i being the activity coefficient of the species i.

The side reactions can considerably change the concentration of the free metal ion and the free organic ligand. To account for their effect on the main equilibrium, the side-reaction coefficients α_M, α_L and α_{ML} as defined by Ringbom et al. need to be introduced into the stability constant (106). The side reaction coefficients are, of course, functions of the equilibrium constants of the side reactions; if [M′] is the conditional concentration of M (all the forms of M not complexed with L), [L′] is the conditional concentration of L (all the forms of L not coordinated by M) and [ML′] is the conditional concentration of ML (the metal complex present in all the forms, even as mixed complexes, then the mathematical relationships to describe α_M, α_L and α_{ML} are:

$$\alpha_M = \frac{[M']}{[M]} = \frac{[M] + \sum_i [MX_i]}{[M]} = 1 + \sum_i [X_i]K_i \qquad (6)$$

$$\alpha_L + \frac{[L']}{[L]} + \frac{[L] = \sum_n [H_nL] + \sum_j [M_jL]}{[L]} = 1 + \sum_n [H]^n \beta_n + \sum_j [M_j]K_{M_j} \qquad (7)$$

$$\alpha_M L = \frac{[ML']}{[ML]} = \frac{[ML] + \sum_n [MH_nL] + \sum_m [MY_mL]}{[ML]} = 1 + \sum_n [H]^n \beta_n + \sum_m [Y_m]K_{y_m} \qquad (8)$$

In terms of side-reaction coefficients, the concentrations of the free metal ion and the free ligand will be defined by $[M] = [M']/\alpha_M$, $[L] = [L']/\alpha_L$ and

$[ML]=[ML']/\alpha_{ML}$ respectively. Considering the complex, ML, not involved in side reactions, i.e., $[ML']=[ML]$, one can define the conditional stability constant *K*' as:

$$K' = \frac{[ML]}{[M'][L']} = \frac{K}{\alpha_m \alpha_L} \tag{9}$$

With careful selection of experimental conditions, the DPASV technique can be used to determine the inorganic forms of metals and weak organic complexes. Considering that the latter does not contribute substantially to the metal complexation because the concentration of ligands and the stability of complexes are low, the electroactive metal concentrations are constituted by ionic forms and inorganic complexes. In consequence of this, the ASV labile fraction can be equated to the conditional concentration [M'].

The labile metal concentration, [M'], is evaluated during titration by the peak current, i_p, obtained after each addition. Sensitivity, S, represents the slope of the titration curve measured at high values of titrant added, where organic ligands have been saturated. The titration curve becomes thus a straight line (Figure 5.4a):

$$[M'] = i_p/S \tag{10}$$

The concentration of complexed metal, [ML], after each addition is calculated by the mass balance as the complement of [M'] to the total concentration after each addition, M_t. The value of M_t is obtained by the total dissolved concentration initially present in the sample, C_M, determined by independent measurement and by the amount of metal added, M_a, after each addition. On the other hand, the complexed metal concentration represents the difference between the analytical concentration of the ligand, C_L, and the ligand not complexed by the metal, [L']:

$$[ML] = M_t - [M'] = C_M + M_a - [M'] \tag{11}$$

$$[ML] = C_L - [L'] \tag{12}$$

The theoretical equation describing the titration curve, [M'] *vs.* M_t, is available by solution of one equation system considering the equations 9,11 and 12. By the expression of the conditional stability constant and the mass balance for the ligand one can obtain:

$$[ML] = \frac{K' C_L [M']}{1 + K'[M']} \tag{13}$$

If one considers the mass balance for the metal and the equation 13 one can obtain:

$$M_l = [M'] + \frac{K' C_L [M']}{1 + K'[M'[} \tag{14}$$

By rearrangement of equation 14 one obtains the theoretical equation decribing the titration curve

$$K'[M']^2 + [1 + K'(C_L - M_t)M'] - M_t = 0 \tag{15}$$

The complexation parameters C_L and K' can be obtained by a non-linear fitting procedure of the experimental data to the equation 15; however, the procedure requires initial values for the parameters to be estimated.

A different approach to evaluate the values of K' and C_L involves rearrangement of equation 14, giving the following expression (29, 36, 107).

$$\frac{[M']}{M_t - [M']} = \frac{[M']}{C_L} + \frac{1}{K'C_L} \tag{16}$$

i.e., the ratio of free to bound metal concentration is linearly related to the free metal concentration. As described above, in each point of the titration curve one can determine the labile metal concentration, [M′] and the complexed metal concentration, [ML], so as to compute $[M']/(M_t - [M'])$ and plot it *vs.* [M′]. In this way a transformed, linear plot of titration data is obtained and by a simple linear regression it is possible to compute both the ligand concentration, as the reciprocal of slope, and the conditional stability constant from the intercept, taking into account the C_L value just obtained from the slope (Figure 5.4b).

Multiple 1:1 complexes between one metal and more ligands. If more ligands (L_1, L_2 ... L_i) compete in complexing the metal M, forming complexes (ML_1, ML_2 ... ML_i) with 1:1 stoichiometry, the equilibrium problem can be described by one system of 2i + 1 equations:

$$[ML_1] = K_1[M'][L_1] \tag{17}$$

$$[ML_2] = K_2[M'][L_2] \tag{18}$$

- - - - - - - - - - - - - - - - - -

- - - - - - - - - - - - - - - - - -

$$[ML_i] = K_i[M'][L_i] \tag{19}$$

$$M_t = C_M + M_a = [M'] + \sum_i [ML_i] \tag{20}$$

$$C_{L1} = [L_1] + [ML_1] \tag{21}$$

$$C_{L2} = [L_2] + [ML_2] \tag{22}$$

- - - - - - - - - - - - - - - - - -

- - - - - - - - - - - - - - - - - -

$$C_{Li} = [L_i] + [ML_i] \tag{23}$$

where K'_i, C_{Li}, $[L_i]$ and $[ML_i]$ are, respectively, the conditional stability constant for the ith complex, the total concentration for the ith ligand, the conditional concentration for the ith ligand and the equilibrium concentration for the ith complex; the other variables have the same meaning as defined above.

Considering the expression of the conditional stability constant and the mass balance for the generic ith ligand one obtains:

$$[ML_i] = \frac{K'C_{L_i}[M']}{1 + K_i'[M']} \tag{24}$$

Substituting the equation (24) in the metal mass balance, equation (20), for all the i ligands, the theoretical equation of the titration curve is obtained (104):

$$M_t = [M'] + \sum_i \frac{K_i' C_{L_i} [M']}{1 + K_i'[M']} \tag{25}$$

The theoretical equation in the transformed form, $[M']/(M_t - [M'])$ *vs.* $[M']$, becomes:

$$\frac{[M']}{M_t - [M']} = 1/\sum_{i=1}^{n} \left(\frac{C_{L_i}}{[M'] + 1/K_i} \right) \tag{26}$$

The transformed plot of titration data is, of course, no longer linear.

Studies of metal speciation in oceanic water show that experimental titration data fit models which consider one class of ligands (33–36, 57, 108) or two classes of ligands (32, 37, 69, 109). Attempts to fit experimental data to models which consider more than two classes of ligands did not improve goodness of fitting.

Equation (26) when two classes of ligands are present becomes:

$$\frac{[M']}{M_t - [M']} = 1/\left[\frac{C_{L_1}}{([M'] + 1/K'_1)} + \frac{C_{L_2}}{([M'] + 1/K'_2)} \right) \tag{27}$$

or also (36, 37):

$$\left(\frac{[M']}{M_t - [M']} \right) = \frac{[M']}{C_{L_1} + C_{L_2}} + \frac{[ML_1]}{K'_1(C_{L_1} + C_{L_2})([ML_1] + [ML_2])} + \frac{[ML_2]}{K'_2(C_{L_1} + C_{L_2})([ML_1] + [ML_2])} \tag{28}$$

In these cases the parameters relative to the complexation (C_{L_1}, C_{L_2}, K'_1 and K'_2) are obtained by fitting the experimental data to equation (27) by the non-linear fitting algorithms (e.g., Marquart-Levenberg).

Initial values for the parameters used in the fitting procedure can be estimated by considering the two limiting situations obtained at low and high metal concentration, respectively. At low metal concentration, if $K'_1 \gg K'_2$, complexes with stronger ligands (L_1) will predominate. The plot of $[M']/(M_t - [M'])$ as a function of $[M']$ approximately follows a straight line, typical for a system with one ligand, and the initial approximated values of L_1 and K'_1 can be calculated from the slope and the intercept obtained by application of the linearization procedure (equation 16) to the first few measurements of the titration (37). At high metal concentration (the final part of the titration), both ligands tend to be completely saturated, then, $[ML_1] = C_{L_1}$ and $[ML_2] = C_{L_2}$. T the plot of $[M']/(M_t - [M'])$ *vs.* $[M']$ tends to a straight line defined by the equation obtained by substituting in the equation (28, 29):

$$\left(\frac{[M']}{M_t - [M']} \right)_{[M'] \to \infty} = \frac{[M']}{C_{L_1} + C_{L_2}} + \frac{C_{L_1}}{K'_1(C_{L_1} + C_{L_2})^2} + \frac{C_{L_2}}{K'_2(C_{L_1} + C_{L_2})^2} \tag{29}$$

whose slope and intercept are $1/(C_{L_1} + C_{L_2})$ and $(C_{L_1}/K'_1 + C_{L_2}/K'_2)/(C_{L_1} + C_{L_2})^2$, respectively (Figure 5.5). By using the L_1 and K'_1, as previously evaluated, and the

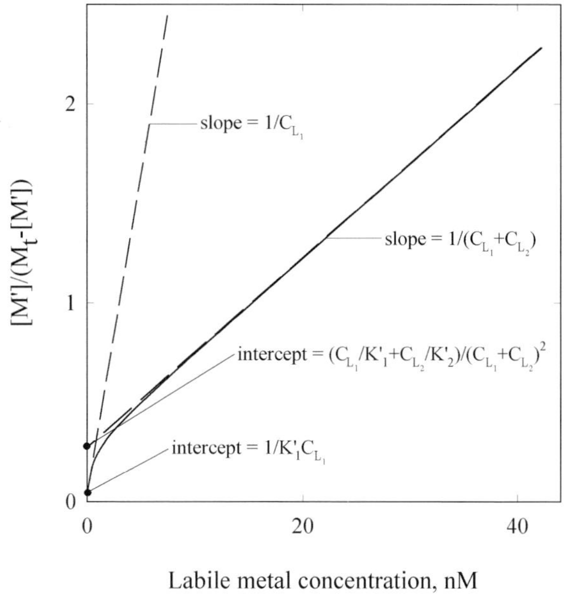

Figure 5.5. Plot of the transformed titration curve of one sample containing two ligands complexing the metal M.

experimental values for the latter slope and intercept, the initial values of L_2 and K'_2 can be obtained.

The models described have been applied to estimating the extent of Cd, Cu, Pb and Zn complexation in oceanic waters (32–35, 38, 57, 69, 108–110).

4.2. Experimental aspects

The titrimetric procedure for evaluating the organic complexation is carried out by numerous standard additions. Typically, 15–20 standard additions are necessary; therefore, the procedure is completely automated by interfacing the electrochemical equipment with a robot (MasterLab System, Model 9000, PerkinElmer, USA) controlled by a personal computer. The PC controls two polarographic analyzers, acquires and stores all the voltammetric data ready for subsequent processing and provides the necessary additions of the metal standard solution by collecting one clean tip, aspirating the correct reagent and injecting the aliquot of reagent to the polarographic cell (Figure 5.6). A typical titration curve obtained for Antarctic coastal sea water titrated with Cu is shown in Figure 5.7.

The procedure for studying metal complexation in sea water includes both the determination of the total content in an aliquot of the sample subjected to acid digestion using the procedure reported above, and the voltammetric titration of organic ligands in an untreated aliquot of the sample with the metal of interest (samples destined for speciation measurements are not subject to acidification or

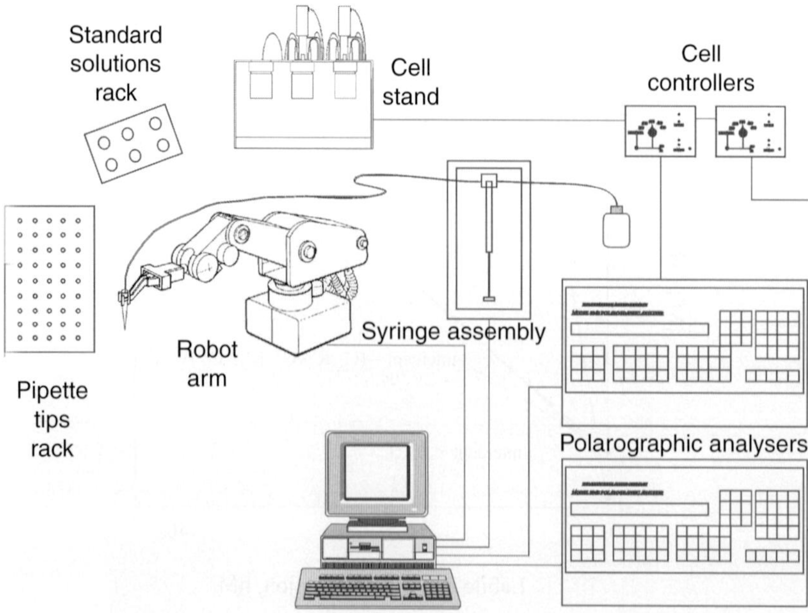

Figure 5.6. Scheme of the automated electrochemical instrumentation for the titrimetric procedure to evaluate the metal complexation in sea water.

other treatment, except filtration, when prepared for storage). The titration to evaluate the complexation of each metal is carried out on separate aliquots of the sample, freshly defrosted and untreated. The voltammetric sequence is mostly the same as that described above in the context of total concentration measurement and only the differences are outlined here. The deposition potential is selected from pseudopolarograms obtained by plotting the anodic stripping current *vs.* the applied potential during the deposition step (see, e.g., Figure 5.8). Values of –0.95, –0.75/–0.80 and –0.80/–0.85 V are selected for Cd, Cu and Pb, respectively, in this way.

To obtain accurate values for the parameters of complexation it is necessary to follow the titration curve with sufficient precision and to reach the linear part of the titration curve with certainty (as required by the theory outlined in Section 4.1). To satisfy these requirements numerous metal standard additions (at least 10, but preferably between 15 and 20) are performed throughout the titration experiment. After each standard addition and before the voltammetric measurement a period of 15 min for Cd and Cu, and 25–30 min for Pb are allowed to pass to reach the chemical equilibrium (typically, one titration requires about 20 h).

When metals are present at very low concentrations (especially Pb), the initial measurement on the unspiked sample is carried out with a higher deposition time (normally 30–40 min), to enhance sensitivity, and it is repeated two or three times to ensure conditioning of the working electrode and repeatability. The currents are then normalized to the deposition time of the rest of the titration, given the linear relationship observed between the two quantities (17, 34, 111).

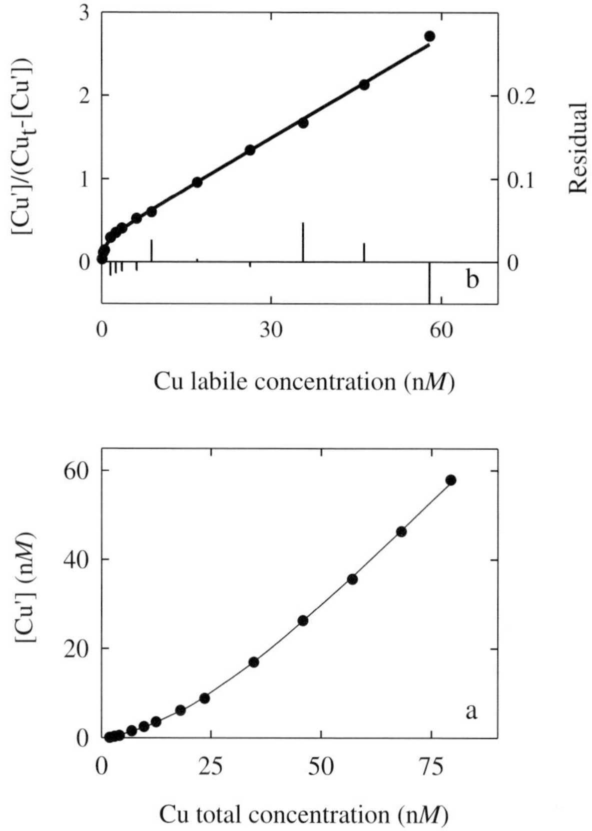

Figure 5.7. Titration curve (a) and transformed plot (b) for one sample collected in the Ross Sea during the 1988–99 expedition. The bar graph in (b) plots the difference between experimental data and fitting curve.

4.3. Limits of the DPASV approach to the metal speciation

The dissolved organic matter present in sea water presumably consists of a complex mixture, many of the macromolecules having complexing capability for metals. The complexation parameters derived from these simplified models applied to the titration data may not therefore be completely correct, from a thermodynamic point of view, with respect to the actual organic ligands present in the sample. However, if the experimental data fit such simple models well, taking into account the above considerations (Section 4.1) about the operational approaches to metal speciation, determination of these parameters allows us to describe the metal complexing properties of natural ligands and their apparent influence on metal speciations.

Application of the ASV technique to studying the complexation of trace metals with organic ligands in natural waters has been the subject of some criticism. This has centered on the possible reduction of organic complexed metals due to the

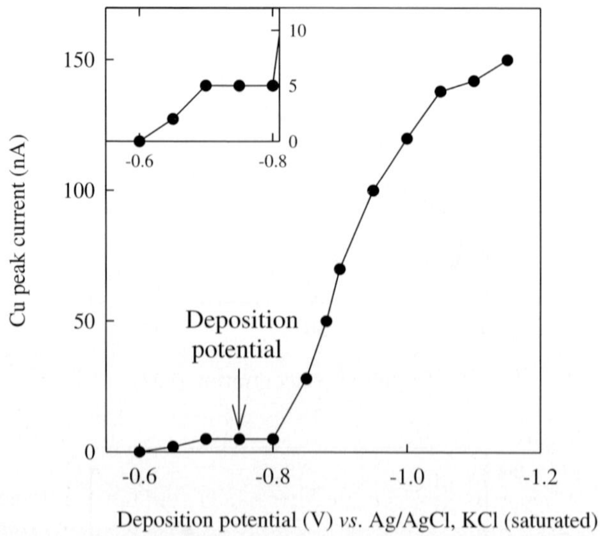

Figure 5.8. Pseudopolarographic experiment performed on a sample collected *ad hoc* in the North Adriatic Sea (3 miles offshore). Reprinted from Capodaglio et al. (69), with permission of Gordon and Breach Publishers, Reading, UK.

dissociation of complexes at the electrodic interface (kinetic contribution to the stripping current due to the dissociation of complexes in the diffusion layer) during the preconcentration step or a direct reduction of metal-organic complexes, thus determining overestimation of the labile fraction (105, 112, 113). The problem of adsorption of organics on the electrode surface, thereby decreasing sensitivity, has also been raised (105).

These criticisms and possible action to eliminate or minimize them were discussed when the methodology was applied to study the complexation of Cd, Cu, Pb and Zn (32–34, 53, 104, 108). The results showed that Cu and Pb complexes were kinetically inert, with k_d values of between 10^{-3} and 10^{-7} s^{-1}, which means that the lifetime of metal complexes, expressed by $1/k_d$, is some orders of magnitude higher than the residence time (1–100 ms) of complexes in the diffusion layer when Rotating Disk Electrodes (RDEs) are used. It can therefore be concluded that the reduction process is not appreciably affected by dissociation reaction inside the diffusion layer. Experiments showed instead that Cd complexes present a kinetic lability when Hanging Mercury Drop Electrode (HMDE) or RDE methods are used at low rotation speed (53). The results emphasized that dissociation from the electrode interface determines an underestimation of the conditional stability constant when low rotation speeds are used. To minimize the risk with respect to this problem the RDE method is normally used at the highest rotation speed.

The direct electrochemical reduction of metal-complexes during the preconcentration step is possible only when a sufficient negative potential is applied to the working electrode. By careful selection of the deposition potential only the most labile complexes are reduced. The potential can be selected through a pseudo-polarographic experiment in order to perform the plating step at potentials negative enough to reduce only inorganic forms of the metal (32, 33, 57, 69).

The use of a TMFE in which adsorption of organics has not been observed in analyzing sea water is recommended.

In short, the accurate selection of experimental conditions can considerably reduce the incidence of these problems on the results of measurements. In particular, the type of electrode, the deposition potential and the hydrodynamic conditions at the electrode surface (which influences the diffusion layer thickness) are the most important.

However, some considerations need to be given to the significance of results of metal speciation using an operational approach. Some studies have shown that the application of different analytical methodologies and procedures determines complexes with different stability as a consequence of different detection windows. The peak depletion due to metal complexation and the detection limit of the technique represent the limits to the determination of metal complexation by the titration procedure with subsequent ASV. In particular, one can measure a peak depletion only if the difference between the two peak heights is greater than the accuracy with which one can detect the peak current (this represents the lower window limit). The upper limit is related to the ability to detect the labile fraction of metal, e.g., [M′] > Limit of Detection (LoD). A detailed discussion of the detection window relative to the titrimetric ASV method has been previously reported (104) and it was concluded that considering the 1% reproducibility with which one can measure the peak current (at low metal concentration in the first part of the titration curve) and a 10^{-11} *M* LoD, one can define a detection window expressed as $K'C_L$ values with a lower limit of 10^{-2} and an upper limit of 10^2. For $K'C_L < 10^{-2}$, the titration curve is practically superimposed on the calibration curve, until the total amount of metal is doubled or even tripled with respect to the ligand quantity. On the contrary, in cases of $K'C_L > 10^2$ very strong complexes are formed. The complexation reaction here is quantitative at each step and no detectable metal is revealed until the ligand is almost completely titrated, after which the added metal remains totally free in solution and its concentration follows a straight line with the slope identical to the curve obtained in absence of ligands. Finally, in cases in which values of the $K'C_L$ product are included in the range $10^{-2} \leq K'C_L \leq 10^2$, titration curves show intermediate shape between the two extremes and the complexation reaction can profitably be studied (33).

For a ligand concentration at the nanomolar level, as frequently detected in oceanic waters, the stability constants which can be explored range between about 10^7 and 10^{11} M^{-1}. In the case of stability constants higher than 10^{11} M^{-1} only the ligand concentration can be evaluated by direct titration (end-point detection), the constant remaining undefined (but $>10^{11}$ M^{-1}). In the case of $K' < 10^7$ M^{-1} no complexation can be observed at all (when reasonable quantities of titrant are added) and neither C_L nor K' are obtainable.

Considering the doubts and criticisms directed at operational speciation procedures because of the potential perturbation of the equilibrium of the system

Table 5.4. DPASV measurements of metal complexation in sea water using EDTA as model ligand

Element	Actual values		Experimental values	
	Ligand (n*M*)	Log *K*′	Ligand (n*M*)	Log *K*′
Cd (108)[a]	10.0	13.8	10.0	>12
Cu (32)	10.0	9.2	10.4±1.6	8.6±0.1
Pb (34)	28.1	8.5	27.3	8.6
Zn (33)	9.5	7.6	10.0	7.9

[a]measurements carried out in 0.1 *N* KCl.

analyzed, it was recommended that two or more independent techniques should be used in parallel. A few researchers evaluated ASV measurements by comparing the results with those of at least one other procedure of speciation (114, 115). In particular, the results obtained by the ASV procedure were compared to the fractionation method based on adsorption of organic complexes on C_{18} column (114). Reasonably comparable concentrations were obtained between the non-labile metal fraction detected by ASV and the fraction sorbed onto the C_{18} column. When Donat and Bruland compared results obtained by DPASV and DPCSV to detect the complexation of Zn in oceanic water, they reported excellent agreement between the values of ligand concentrations obtained by two methods, although some differences were observed in the values of conditional stability constants, probably due to different detection windows for the two techniques (115).

Tests to evaluate the accuracy of the DPASV approach to metal speciation have previously been reported using model ligands to study the complexation of Cd, Cu, Pb and Zn in sea water (32–34, 108). Reported results show that the ligand concentrations and conditional stability constants obtained are in agreement with the theoretical data, as set forth in Table 5.4.

5. Review of literature data

As reported above the study of trace metal distribution and speciation indicates an interface research area among many disciplines. Many trace metals are required as micronutrients, as cofactors in the enzymes and for various metabolic functions in living organisms. Although the hypothesis that metals can represent a limiting factor for phytoplankton growth has so far been regarded as speculative, some laboratory experiments indicate that trace metals (Co, Cu, Fe, Mn, Ni and Zn) can act as a selective force that may regulate the phytoplankton diversity (24). Studies carried out in potentially productive regions showed that Fe can represent the limiting factor for primary production (8). On the other hand, biological activity can be inhibited by an excess of some of these elements (116, 117).

The distribution of trace metals is governed by input and removal processes superimposed upon physical processes. The reactive trace metals, elements presenting a low concentration relative to the crustal abundance, are normally classified in two groups on the basis of oceanic profiles (118). The nutrient type elements (e.g., Cd and Zn), whose distribution is controlled by biological activity and decomposition of organic matter, follow the profiles of major nutrients; they are depleted in surface waters and are regenerated in intermediate and deep waters where processes of mineralization take place. The concentration increases from the relatively young deep waters of the North Atlantic to the older deep waters of North Pacific (12). Scavenged metal distribution is controled by external sources (Al and Pb) and it is characterized by surface concentration maxima corresponding to higher external sources. Their concentration in deep waters is appreciably higher in the younger waters compared with the concentration observed in the older deep waters (12).

The main removal process for oceanic components is via sedimentation and burial; thus, the interaction of dissolved metals with particles in sea water is a major indication of their concentration and distribution in the world's oceans. In open ocean areas the particle cycle is driven by the biological production of particles in the surface layers, which after processes of mineralization and packaging reach the necessary size and density to fall to the ocean bottom. On the basis of this consideration, one can say that in the open ocean area the biogeochemical cycle of trace metals determining their distribution and speciation is frequently dominated by biological processes. In coastal areas or particular geographical zones, other phenomena, e.g., inorganic precipitation, can take place.

In the last few decades many studies have been carried out to evaluate the distribution of dissolved trace metals in sea water (12, 119–122) and more recently studies of metal speciation have been reported (24, 33, 34, 108–110, 123). However, few data sets are available for the Southern Ocean (124–133), and studies of trace metal speciation are limited to a few papers (35, 57, 69, 134).

The majority of the investigations were carried out in the Weddell Sea, the Weddell/Scotia confluence and the Indian sector of the Southern Ocean (125, 127, 129, 132, 135). Very few data are available on trace metal distribution in the Ross Sea area; early measurements on Cu distribution in surface waters between New Zealand and the Ross Sea were reported by Boyle and Edmond and more recently some investigations were carried out by Martin et al., also during the oceanographic campaigns as a part of the Italian National Programme of Researches in Antarctica (PNRA) (35, 57, 69, 124, 130, 131, 133, 134, 136, 137).

Studies carried out in oligotrophic areas of major oceanic gyres showed a marked surface depletion of major nutrients and nutrient type trace metals, but this is not the case in Antarctic waters and some other areas presenting nutrient-rich waters. The high levels of nutrients and the simultaneous low primary production in surface Antarctic waters constitute the so-called "Antarctic Paradox".

The dominant processes controlling metal distribution in the Southern Ocean, in particular the effects of local phenomena on the water composition, such as formation and melting of pack ice and bed rock erosion due to glacier flow, should be clarified. Here an overview is given of the distribution of some trace metals of particular interest.

5.1. Cadmium

Studies carried out in the Weddell Sea showed relatively high Cd concentration ranging between 0.5 and 1 n*M*. The authors emphasized differences of concentration between the different areas examined and they hypothesized that this reflected the age of the water (127). Samples collected in the centre of the Weddell Sea showed little surface depletion and, at the same time, the authors observed that the suspended Cd concentration was considerably higher in surface and subsurface samples compared with the samples collected deeper than a few hundred meters. It can be hypothesized that the high suspended Cd level found in the surface layer was generated by biota which actively accumulate it. On the basis of these observations and the particle composition and distribution Westerlund and Öhman concluded that Cd distribution is linked to easily decomposed carbon particles and not inert silica particles. Studies carried out in the coastal areas of the Ross Sea (Gerlache Inlet and Wood Bay, bays in the Terra Nova area), which are covered by ice until mid January, show a homogeneous vertical distribution of Cd until the beginning of the summer at a level of about 0.6–0.9 n*M* (136, 137). These concentrations are consistent with the deep waters of the Atlantic and Indian sectors of the Southern Ocean, the subantarctic regions and the North Pacific and with the global distribution of dissolved Cd (119, 125, 127, 128, 132, 138, 139). A subsequent marked depletion of concentration in surface waters was observed, reaching levels about 0.1 n*M* at the end of the summer. An analogous observation was made by Frache et al., who studied the metal distribution along the water column in the Wood Bay. They also observed a simultaneous increase of particulate Cd (130).

In their study of the Weddell Sea, Westerlund and Öhman tried to measure speciation by comparing the total dissolved Cd concentration and the recoverable metal concentration using an imminodiacetic resin on samples at natural pH (127). The results showed that, especially for samples collected over the Filchner shelf, the Cd recovered by Chelex resin in the upper 500 m of the water column was appreciably lower than the total concentration. They concluded that this layer probably contains organic matter complexing this element and they related this to the particle composition along the water column.

A more exhaustive study was carried out to consider the evolution of Cd complexation during one austral summer (134). The results showed that the labile fraction was initially higher than 90% of total dissolved concentration; subsequently, at the same time as the decrease in total dissolved concentration, the inorganic Cd fraction was reduced to a minimum of 8% of the total.

Cadmium-complexing ligands were detectable only after the middle of December when the pack ice break-up and the phytoplankton bloom had started, initially in the first surface layers and gradually through the whole water column. The results showed that the metal was complexed by one single class of ligands. The free ionic metal concentration along the water column was calculated by values of α_{Cd}, between 30.8 (value calculated for $T = 0°C$) and 30.5 (value calculated for $T = -2°C$). It ranges from 16 to 32 p*M* for samples collected before 26 December, while for samples collected after that date the concentration ranges from 0.3 p*M*

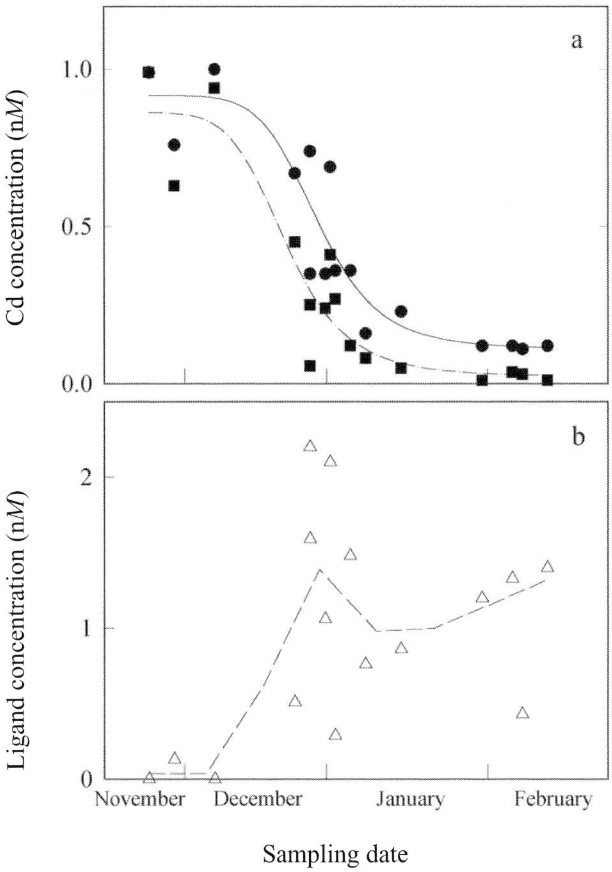

Figure 5.9. Summer variation of Cd concentration and speciation in the Gerlache Inlet during 1988–1999 and 1990–1991 expeditions: (a), ●, total dissolved concentration; ■, labile fraction concentration; (b), Δ, ligand concentration and values obtained by filtering ligand concentration data. Reprinted from Capodaglio et al. (134), with permission of Gordon and Breach Publishers, Reading, UK.

(subsurface sample collected on 30 January) to 26 p*M* (sample collected at a depth of 250 m on 6 January).

The results emphasize that the surface summer depletion of dissolved Cd concentration is associated with a different speciation for this element. Figure 5.9 reports the total surface concentration, the labile fraction and the Cd ligand concentration data for samples collected during the 1988–1989 and 1990–1991 campaigns in the Gerlache Inlet and emphasizes the correspondence between the rapid increase of ligand concentration and the depletion of total dissolved Cd concentration. Indeed, a negative correlation between the total dissolved concentration and the ligand concentration ($r = -0.61$) was observed. The study highlights a correlation between ligand concentration and chlorophyll ($r = 0.87$) determined by *in situ* measurement of fluorescence (134).

The results described above are in agreement with studies of Cd complexation carried out in different oceanic areas. These showed that a large amount of Cd in surface and subsurface layers is organically complexed. Sakamoto-Arnold et al. emphasized that the ASV-labile fraction of Cd within the upper 250 m ranged from 8 to 67%, while the inorganic forms were 99% of the total dissolved Cd at depths greater than 300 m (47). Bruland, studying the organic complexation of Cd in the Northern Pacific gyre by DPASV, detected one class of ligands in the photic layer present at concentration of about ten times lower than the values obtained in the coastal area of the Ross Sea (108). Despite the difference of ligand concentration probably due to the different hydrological characteristics of the two areas, the ligand distribution along the water column showed the same trend. Therefore, on the basis of ligand distribution one can hypothesize that organic matter complexing Cd consists of labile matter which is quickly decomposed along the water column and related to primary production.

The results of measurements carried out on surface samples collected during two oceanographic expeditions at the Terra Nova Bay were analyzed by a multivariate statistical approach. The Principal Components Analysis was used to observe association between variables and it showed an opposition between the Cd concentration (total and labile) and the ligand that complexes it (134). These results show that the metal speciation could affect its distribution. It, in particular, could emphasize the direct involvement of complexation in the transfer of Cd from the dissolved phase to the particulate affecting the total dissolved distribution.

5.2. Lead

The mean total dissolved concentration determined in surface waters shows a high variability as a function of time and position, probably dependent on hydrodynamic or local processes. Measurements carried out at the Terra Nova Bay showed a high variability of Pb concentration in coastal waters (24–114 p*M*), while the surface distribution in the off-shore area showed a much more constant concentration (mean value 28 ± 3 p*M*) (57). Several data sets are available for open sea in the Weddell Sea areas: Westerlund and Ohman reported a mean value of 13 p*M* for surface water, and Flegal et al. reported values for water collected in the Weddell/Scotia Sea ranging between 10 and 103 p*M* (127, 140). The low Pb concentration in surface waters and knowledge of global atmospheric circulation supported the idea that Antarctica is a relatively pristine continent. However, Flegal and co-workers analyzed isotopic composition to reveal a significant anthropogenic contribution to the Pb concentration in sea water and indicated an efficient scavenging process, due to intense primary production, as responsible for this low concentration.

Studies carried out in coastal areas of the Terra Nova Bay show that the Pb concentration changes as a function of time. The evolution of Pb distribution was followed during the austral summer in the Gerlache Inlet and Wood Bay (136, 137). Results showed that the mean total dissolved concentration in the Gerlache Inlet ranges from 90 p*M*, at the beginning of the summer when the larger part of Terra Nova Bay was still covered by pack ice, to 30 p*M* in the superficial 100 m at the end of the summer. An analogous trend was observed in Wood Bay, in which the total dissolved concentration of Pb decreased from a mean value of 34 p*M* in November to 16 p*M* at the end of January. In the latter study a homogeneous distribution of this metal was observed through the water column. Considering that in both cases the December/January period corresponds to the maximum primary production, the trend agrees with the hypothesis of Flegal et al. that Pb depletion is mediated by biological process.

Very few studies of Pb complexation by organic ligands have been carried out in oceanic areas (34, 57, 134, 141). The inorganic Pb fraction (ASV-labile) detected in the Gerlache Inlet did not change during the season, represented 39% of the total mean dissolved amount and its concentration was well correlated to the total dissolved concentration (134).

The studies of Pb complexation in sea water always showed the presence of a single class of ligands. The results of the investigation carried out on surface water collected at Terra Nova Bay reported concentrations between 0.25 and 0.40 n*M* in open sea areas and between 0.47 and 0.91 n*M* for coastal samples (57). The ligand concentrations and levels of inorganic fraction determined in off-shore waters were comparable to those measured in the surface waters of the Eastern North Pacific Ocean (34). The seasonal study carried out in the Gerlache Inlet reported different ligand concentrations as a function of sampling date, the mean concentration ranged from 0.66 ± 0.28 n*M* at the beginning of the summer to 1.2 ± 0.3 n*M* at the end of season. The distribution through the water column was practically homogeneous and showed no clear trend. Although the seasonal increase of ligand concentration points to some relation of this with the evolution of biological activity in the studied area, the presence of organic ligands complexing Pb throughout the season and their homogeneous distribution along the water column seem to suggest a refractory nature for this organic matter with a lifetime longer than the annual cycle.

Using values of α_{Pb} ranging between 18.6 (value calculated for $T = 0°C$) and 17.7 (value calculated for $T = -2°C$) the calculated free ionic Pb concentration ranged from 0.3 to 4.1 p*M*; the minimum values (0.3–0.7 p*M*) were calculated for subsurface samples (10–50 m) collected after 26 December, while the higher values (1.1–4.1 p*M*) were calculated for samples collected earlier or at greater depths. Considering the uncertainty which may affect evaluation of the ionic concentration, one can conclude that during the summer variations are detectable only for surface layers.

De Gregori et al. reported results of Pb complexation in a coastal area of the South Pacific Ocean (141). They observed that the labile fraction ranged between 30 and 50% as a function of distance from the coast, but they did not report any data on concentration of ligands complexing Pb and the level of metal concentration was decidedly higher than in the samples collected at Terra Nova Bay. The results are therefore difficult to compare with those reported above.

5.3. Copper

The distribution of Cu in sea water is intermediate between that of nutrient-type elements and that of scavenged elements; in surface waters of oligotrophic regions it is present at a low concentration (0.5 n*M*) that increases about linearly with the depth (12).

Copper is one of the more frequently studied elements in oceanic waters because it is a biologically essential metal that becomes toxic when it reaches too high a concentration and it forms stable complexes which heavily affect its bioavailability. The Cu concentration in surface waters is strongly dependent on upwelling phenomena. The early data set in the Southern Ocean reported by Boyle and Edmond showed a clear increase in Cu concentration along a transect south of New Zealand across the circumpolar current where upwelling of deep water takes place with values ranging between 0.98 and 3.25 n*M* (124). This yields a high variability of concentration in surface waters, while a regular trend can be observed for deep oceanic waters with a net cumulative increase in the Cu concentration as the deep water traverses from the Atlantic to the North Pacific (12).

Westerlund and Öhman observed two levels of Cu in the Weddell Sea: a lower level (about 2 n*M*) over the Filchner Shelf and in the surface water and a higher level corresponding to the Weddell Deep Water (about 2.6 n*M*) and Antarctic Bottom Water (about 2.8 n*M*). For the surface waters they suggest that concentration is affected by dilution from melting ice. A more complex vertical distribution was shown in the Weddell/Scotia Sea area, with the surface concentration ranging between about 2 n*M* in the Scotia Sea and about 4 n*M* in the Weddell Sea (125, 126). The authors emphasized a covariance between Cu and silicate when deep waters were examined, while no relationship was observed for surface waters indicating that their variability in the first 100 m was due to different processes. On the basis of Cu concentration in suspended matter and the high concentration of SiO_2 in settling particles, the same authors concluded that the sedimentation of Cu is due to inert particles rich in SiO_2. They also supported this idea by the covariance between Cu and SiO_2.

Few data are available for Cu concentration in the Ross Sea: sometimes only surface waters were examined and water collected along vertical profiles was analyzed in only a few cases (69, 124, 130, 131, 134). Results of measurements carried out on samples collected during three different campaigns at Terra Nova Bay showed Cu concentration between 0.9 and 4.8 n*M* as a function of sampling area and time (69). The mean concentration for water collected between January and the beginning of February was lower (1.8 ± 0.5 n*M*) than the mean concentration measured in November, December or later in February (3.1 ± 0.9 n*M*) (134).

The seasonal study carried out in the Gerlache Inlet reported total dissolved Cu concentration ranging from 1.6 to 4.6 n*M* (134). The Cu distribution showed a subsurface depletion in January, which extended through the water column in February. However, the concentration during the season did not show a clear trend like that observed for Cd.

The total dissolved amount of the metal always presented a surface concentration higher than the minimum observed at a depth of 10–25 m (1.6–2.2 n*M*) compared with the surface concentration (2.0–3.6 n*M*); the concentration further increased to values of between 2.2 and 3.7 n*M* near the bottom. The Cu concentration along a vertical profile in the Ross Sea offshore at Cape Adare, an area affected by the coastal Antarctic current, was reported by Abollino et al. (131).

Also in this case results showed a significant minimum at a depth of about 150 m in spite of the fact that the values that they reported along the entire water column were about half the concentrations measured in different oceanic areas. This low concentration measured in deep water seems in contrast to the cumulative Cu increase from the Atlantic to the North Pacific (12).

Analogous surface maxima were observed studying the distribution of Cu along profiles of the Pacific Ocean and Indian Ocean (83, 135). In both cases the authors emphasized the presence of a minimum of concentration at a depth of about 500 m and they explained this as the effect of an important local surface source. In particular, Boyle et al. provided evidence that the surface maxima may be transient features resulting from the advecting of Cu-rich near-shore surface water into the more central regions of the oceans, while Saager et al. hypothesized the contribution of atmospheric particles to the surface concentration (19).

It must be stressed that although the trend observed in the coastal zone of the Ross Sea was not so marked and regular as the results of Saager et al., the effect of local phenomena can be assumed. In fact, glacier transport and the ice pack formation/dissolution cycle can play a fundamental role in the composition of surface coastal sea water. However, more detailed information about the local sources (eolian dust composition and deposition rate, glacier composition and dissolution rate, effect of pack ice dissolution and formation) are necessary to establish the origin of the surface water enrichment.

Studies to evaluate Cu complexation by organic ligands have been carried out in oceanic areas and the results always showed that organic complexation strongly affects its speciation (109, 110, 123). Studies carried out both in the Pacific Ocean and Atlantic Ocean emphasised that Cu is complexed by two classes of organic ligands. One of them is a low concentration strong ligand located at a depth corresponding to the chlorophyll maximum and seems to dominate Cu speciation in the euphotic zone. Coale and Bruland showed that in the northern Pacific Ocean the inorganic Cu fraction varies between 0.1% in the euphotic zone and 30–40% in the deeper water where the stronger ligands were absent (110).

Very few studies of Cu complexation by organic ligands have been carried out in the Southern Ocean. Measurements in the surface water of Terra Nova Bay confirm the presence of two ligands complexing this element (69). The investigation carried out in the Gerlache Inlet to study the evolution of metal speciation during the austral summer showed the vertical distribution of the stronger ligand observed in the Pacific and Atlantic Ocean (134). The stronger ligand (L_{1Cu}) presented a concentration in the nanomolar order of magnitude, reaching a maximum value of 19 n*M* at the end of December. This coincided with the phytoplankton bloom and seemed to follow a vernal stratification. The vertical distribution of the weaker ligand (L_{2Cu}) did not show an evident trend, but in the surface/subsurface waters there was a clear increase in average concentration during the summer. The mean value ranged from 26 ± 3 n*M* until the beginning of December to 60 ± 10 n*M* in February. For the latter class of ligands the temporal trend throughout Terra Nova Bay seems to have been affected by seasonal evolution; in fact, the mean concentration for samples collected in the Bay during 1987–1988, 1988–1989 and 1989–1990 was 30 ± 5 n*M* and did not show any particular trend (69). The vertical distribution and its presence also at the beginning of the summer suggests a refractory nature for these compounds and a lifetime longer than the mixing time of the water column.

The labile metal fraction (principally inorganic forms) was strongly dependent on depth and time. In shallow waters, with the stronger ligand(s) present, it was lower than 1%, while a value always higher than 5% (up to a maximum of 40%) was observed at depths greater than 100 m and early in November and December.

In the same study the free ionic metal concentration (Figure 5.10) was calculated by using values of α_{Cu} between 6.9 (value calculated for T = 0°C) and 6.4 (value calculated for T = –2°C). The values varied in the first 25 m by about four orders of magnitude (from 0.01 to 140 p*M*) as a function of the time, and about the same difference was observed between the surface and the deeper water at the end of January. The lower values (0.01–1 p*M*) were calculated for subsurface samples (0.5–10 m) collected after 26 December, while the higher values (0.04–0.14 n*M*) were calculated for samples collected at greater depths (100–290 m).

Westerlund and Öhman estimated the inorganic Cu in the Weddell Sea by

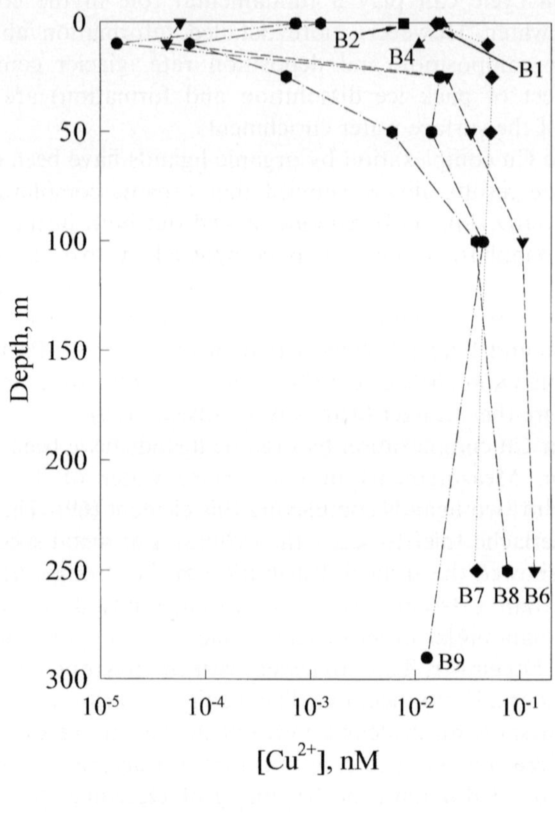

Figure 5.10. Profile evolution of ionic Cu concentration in the Gerlache Inlet during the 1990–1991 campaign. B1, November 24; B2, November 29; B4, December 7; B6, December 26; B7, January 6; B8, January 30; B9, February 11. Reprinted from Capodaglio et al. (134), with permission of Gordon and Breach Publishers, Reading, UK.

recovery of the metal using an imminodiacetic resin at natural pH (127). The results showed that it represented 30% of the total mean dissolved concentration without a clear trend, although some differences were observed when comparing shelf waters and open sea waters.

There is evidence that the ionic Cu represents the bioavailable form of this metal, and a high variation of the ionic metal concentration could determine plankton species succession in the local community (142–144). Brand et al. showed that in water where the $[Cu^{2+}]$ reached values as high as 10^{-10} M only eucaryotic algae maintained their maximum reproduction rate; procaryotic cyanobacteria reduced their reproductive rate when the ionic Cu was higher than 10^{-11} M (116). Di Tullio and Laws speculated that in upwelling waters the high concentration of free Cu can cause a decline in cyanobacteria abundance (145). Studying the seasonal production cycle in the McMurdo Sound, Knox showed that there is a succession of phytoplankton populations during the summer, dominated by diatoms in the early summer until mid-December followed by the *Phaeocystis* bloom in December and by a diatom bloom in January–February (146). Measurements of picoplankton carried out at Terra Nova Bay, very close to the Gerlache Inlet, showed an increase in abundance of microbial populations from January to February (147).

5.4. *Iron*

Iron is one of the essential elements for biological systems with functional roles in oxygen transport and electron transfer systems. It is a ubiquitous element present at nM level in sea water; it presents difficult problems of contamination during sampling and through all the analytical steps. Oceanographically consistent data describing its distribution and concentration in marine environments have therefore been reported only recently (10, 148–152). Very few investigations report Fe distribution in the Southern Ocean (126, 130, 132, 133) following the observation that Fe may represent a limiting element to phytoplankton growth in surface areas containing high levels of major nutrients, but relatively low primary production (10, 149, 153).

The geochemical behaviour of Fe is frequently related to oxygen minima as observed in the Pacific and Indian Oceans (150). Fe(III) is the stable oxidation state in oxygenated sea water and it is relatively insoluble when present in the form of hydrous Fe oxide, while Fe(II) may be the dominant form in anaerobic waters, given its higher solubility. Therefore, the oxygenation of waters strongly affects Fe distribution; its concentration ranges between 0.05 nM in surface waters of oligotrophic areas to 1 μM in deep anoxic waters (154, 155).

There is evidence that in oceanic water Fe as well as Mn are removed via oxidative scavenging by biogenic or organically coated particles (156). On the other hand, Fe and Mn oxyhydroxides on settling biogenic particles are important carriers for other trace elements like Cd, Co, Cu and Zn (157). Regenerative fluxes from reducing sediments can contribute considerably to the dissolved Mn in the overlying waters; however, it seems that rapid oxidation prevents the build-up of gradients for dissolved Fe.

Iron may be supplied to the euphotic zone from advective and diffusive processes within the ocean as well as by atmospheric deposition of particulate matter to the ocean surface. In coastal areas the water composition can be affected by the contribution of rivers and in polar regions the glacier effect, in terms of ice melt and erosion during the ice flow, can be important.

Dissolved and particulate concentrations of Fe have been shown to be quite low in the euphotic zone of the North Pacific oligotrophic and eastern equatorial Pacific waters. The dissolved concentration is normally lower than 0.1 n*M* and the particulate is about 0.2 n*M*. The concentration along the water column shows a nutrient-type vertical profile characterized by surface depletion and increase with depth. Iron concentration reaches typically values >0.5 n*M* at depths below 1000 m (158).

Gordon et al. showed that in the equatorial Pacific Ocean the Fe flux is dominated by upwelling of Fe-rich Equatorial Undercurrent waters in correspondence of which a peak of concentration was observed (154).

In the centre of the north Pacific subtropical gyre, Bruland et al., on the basis of a vertical profile, emphasized a significant aeolian contribution to the dissolved Fe concentration in the surface mixed layer (surface concentration of 0.35 n*M* compared with the minimum of 0.02 n*M* at 70–100 m) (159). Taking into account the study of Hutchins et al. demonstrating that Fe assimilated by plankton in such oceanic waters is recycled on a timescale of days, they concluded that a substantial part of Fe entering through the atmospheric input is recycled and is retained in the oligotrophic mixed layer (160).

The lower euphotic zone (depths of 70–100 m), which is isolated from direct atmospheric inputs, is subjected to intensified processes of Fe scavenging that determine the extremely low concentration of 0.02 n*M*. At depths below 100 m, dissolved Fe exhibits the characteristic nutrient-type distribution observed in other zones of the Pacific Ocean (10, 161). The same authors emphasized that in regions where new production is high and intensified scavenging occurs within the surface mixed layer, the dissolved Fe concentration assumes concentrations similar to those they observed in the central gyre at depths of 70–100 m (159).

Saager et al. reported the vertical distribution of Fe in one area of the Indian Ocean characterized by seasonal upwelling and a broad oxygen minimum zone in intermediate waters (150). The dissolved Fe-profile exhibited a maximum (5.1 n*M*) in the oxygen minimum zone, while lower values were determined both in surface waters (0.3 n*M*) and deep water (around 1 n*M*). They concluded that the distribution of Fe is largely driven by regional sources and sinks and it is characterized by a short residence time. Although its involvement in biological processes is known, its distribution contrasts with that of nutrient-type trace metals. That is the result of the high reactivity of this element and its own redox chemistry.

Martin et al. reported the results of investigations carried out in three upwelling areas of open ocean rich in major nutrients where atmospheric dust-Fe input is known to be low (i.e., the north-east Pacific and the southern Ocean) (161).

The available nitrate is usually considered the factor limiting phytoplankton growth. Some oceanic areas are characterized by high concentration of nitrate, high light levels and low primary production which suggests that some other factors must be responsible for the low phytoplankton growth. On the basis of these considerations, and considering that Fe is required for the synthesis of chlorophyll and nitrate metabolism, Martin et al. hypothesized that Fe is the limiting factor for the phytoplankton activity. When Fe becomes available, diatoms quickly bloom, chlorophyll levels increase and nutrient stocks are rapidly depleted.

They present the results obtained in two extreme stations in the Ross Sea, one near shore with local Fe source and one offshore in deep water. The coastal station was characterized by large standing crops of Particulate Organic Carbon (POC) and low concentrations of major nutrients, which indicates good growth conditions. The same authors reported that shallow and near shore waters as well as ice can be rich sources of Fe (133). Thus, it was not surprising that particulate Fe levels were high and the addition of Fe at the 5 n*M* level had little effect on nitrate uptake and chlorophyll synthesis. The offshore station was 500 km east of Cape Adare and 650 km north of the Ross Ice Shelf, i.e., far from shallow-bottom Fe sources. Nevertheless, in spite of the stable water column and high light levels, chlorophyll and POC concentrations were both relatively low and no evident surface removal of major nutrients was observable. Considering the offshore location of the examined area, the particulate Fe concentration was very low and apparently little Fe had been released from the recently melted ice. The Fe deficiency was also proved by one enrichment experiment, i.e., the nitrate uptake rates increased by one order of magnitude after the addition of Fe to the samples.

Another experiment carried out by Martin et al. in the Atlantic sector of the Southern Ocean showed that the highly productive neritic Gerlache Strait waters have an abundance of Fe (7.4 n*M*) which facilitates phytoplankton blooming and major nutrient removal (133). The results of the investigation carried out in low productivity offshore Drake Passage waters showed low levels of dissolved Fe (0.16 n*M*); the concentration was so low that the phytoplankton was able to use less than 10% of the major nutrient available to them.

The effect of phytoplankton bloom on Fe distribution during the austral summer in coastal areas was studied by Frache et al. (130). Measurements of dissolved and particulate Fe along vertical profiles in the Wood Bay (Ross Sea) were carried out on samples collected during the summer of 1993–1994. The authors did not present the result of each single profile, but reported the mean concentration of Fe through the water column before and after the ice pack melted (Figure 5.11). The metal concentration in samples collected in the first 10 m was 16 n*M* when pack ice was present; the profile presented a minimum concentration of 6 n*M* at a depth of 50 m. After the ice melted the dissolved concentration in the first 10 m was reduced to a mean value of 8.4 n*M*. At the same time as the depletion of Fe in the dissolved phase an increase in Fe was detected in the particulate phase. The mean Fe concentration in the particulate in the first 10 m before the ice melt was 1.6 $\mu g\ g^{-1}$; after pack melt the mean value increased to 20 $\mu g\ g^{-1}$.

Westerlund and Öhman presented the results of Fe concentration in the Weddell Sea and the shelves; they tried to determine whether there are fluxes of Fe from the continent and shelf, hypothesizing that they might represent an important supply of Fe for the offshore waters (132). The Weddell Sea is rich in nutrients and no pronounced oxygen minima are found; thus, on the basis of the observations made by Martin and co-workers, the authors assumed that in the studied

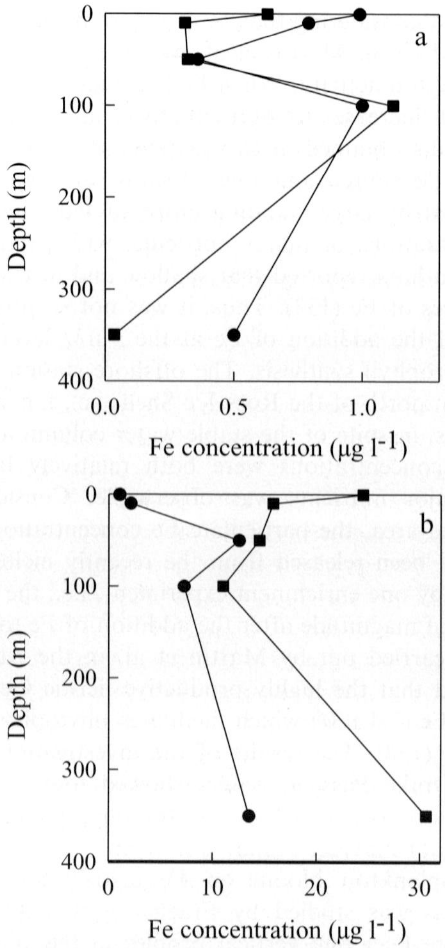

Figure 5.11. Vertical distribution of Fe concentration in the Wood Bay: ●, samples collected before the ice pack melt; ■, samples collected after the ice pack melt; (a), dissolved concentration; (b), particulate concentration. Adapted from Frache et al. (130).

area Fe could be the limiting factor for primary production. The average value for concentration of dissolved Fe was found to be 1.2 n*M*, with somewhat higher values at the Filchner Ice Shelf. The total Fe concentration was found to be considerably higher, with a range between 1 and 6 n*M* in the central Weddell Sea and between 1 and 25 n*M* at the shelves. Results showed some high values of the metal concentration in the top layer, perhaps due to the presence of many icebergs and large amounts of sea-ice in the area studied. Considering the large concentration gradient between the shelves and the main sea the authors demonstrated the transport of Fe from the shelves into the Weddell Sea basin.

Other studies in the Weddell Sea, the Scotia Sea and the intermediate Weddell/Scotia Confluence were carried out by de Baar et al. (162). They found that phytoplankton growth was stimulated by Fe, although they concluded that this element was not the only limiting factor for productivity. Nolting et al. determined dissolved Fe in the same areas, the levels ranging between 2 and 8 n*M* in the surface waters, with analogous levels in deep water; some relative maxima were observed at 200 and 500 m and close to the bottom (126). Over the South Orkneys shelf, the dissolved Fe was about one order of magnitude higher than the other regions examined (about 60 n*M*). They concluded that the levels are adequate to sustain biological growth and that the shelf sediments, together with transport of weathered material by icebergs, appear to be a major source for both dissolved and particulate Fe.

Considering the results of Westerlund and Öhman and those presented by de Baar and co-workers for the same area, it is evident that more knowledge on Fe in the marine environment is necessary to determine whether this metal is a limiting factor for primary production or not (126, 132, 162). In particular, it is necessary to know the bioavailability of the different forms of Fe for producers as well as the composition of the dissolved Fe in oceanic areas.

Studies carried out to evaluate the uptake of Fe by phytoplankton showed that only the dissolved metal is bioavailable and that a thermal or photochemical treatment is necessary for the colloidal Fe to become bioavailable (163). Moreover, the chemical form in which Fe is present can also affect its availability for plankton. The distribution of Fe(II) in the euphotic layer of the equatorial Pacific Ocean was examined by O'Sullivan et al. (164). Its concentration is regulated by the balance between production and removal; Fe(II) can be produced by microbial and chemical reduction, while the loss in surface water is controlled by biological uptake and by oxidation to Fe(III), subsequent hydrolysis, ageing and settling. The results showed maximum concentration near the surface and at the depths with higher chlorophyll *a* levels, the concentration ranging between 0.12 and 0.53 n*M*. Laboratory experiments carried out by the same authors showed that photoreduction can be an important source of Fe(II). Considering the different chemical speciation observed at various depths, different bioavailability can be expected in the examined zone.

6. Conclusions

As highlighted by the studies discussed above, trace metals play an important role in sea water chemistry: they can affect the processes taking place in water and their distribution can give useful information about the processes and characteristics of particular areas.

There is evidence that physical, chemical and biological processes strongly affect trace metals concentration in the Southern Ocean. It was reported that ice melting and glacial till can supply new trace metals to the euphotic layer. Anoxic phenomena can change the oxidation state and consequently the concentration of elements such as Fe or Mn, which, when settling in association with biogenic particles, can affect the distribution of other microelements. The biological uptake or the complexation by organic ligands originated by biological systems can change the distribution of bioavailable forms of micronutrients. For components present at the lowest concentrations, the composition of water masses entering the Southern Ocean can be changed by local chemical and biological processes. Therefore, use of metal tracers could represent a powerful tool in describing the dynamics of the Southern Ocean taking into account also chemical and biological processes. Another consideration is that the processes occurring in the Southern Ocean affect the composition of abyssal waters of all the oceans.

It is evident that the dominant processes in different parts of the Southern Ocean can be very different; detailed studies must therefore be carried out to understand the role of trace metals in biochemical cycles and their utility as tracers. The investigations carried out so far show that the processes changing the characteristics of water masses inside the Southern Ocean have a strong seasonal dependence; the differences are much larger than those observed in tropical or equatorial ocean waters. It is therefore important that the effect of the seasonal processes be evaluated.

A general request from chemical oceanographers, and in particular from those studying the properties of the Southern Ocean, is an improvement in chemical analysis methods with the introduction of sensors that can operate *in situ* for long periods or at least methods that can be used on board to produce large data sets. This request is related to the need to describe large regions in sufficient detail on both spatial and temporal scales. The involvement of trace elements in the biological activity is strongly related to the chemical forms in which they are present. Therefore, a further challenge to the analytical chemist is the improvement of analytical methods with the capacity for better differentiation and measurement of individual species at natural levels.

Acknowledgments

The authors gratefully acknowledge useful discussions with C. Turetta and the technical assistance of V. Cester. This study was carried out in the framework of the "Environmental Contamination Project" supported by the Italian National Programme of Researches in Antarctica (PNRA).

References

1. W. D. Nowlin Jr., *Mar. Chem.*, 35 (1991), 1–7.
2. T. Whitworth III, *Oceanus*, 31 (1988), 53–58.
3. E. G. Carnack, Water characteristics of the Southern Ocean south the polar front. In: *A Voyage of Discovery*, George Deacon 70th Anniversary Volume. M. Angel (Ed.), Pergamon, Oxford, (1977), 15–41.
4. A. L. Gordon, Oceanography of antarctic waters. In: *Antarctic Oceanology* I. *Antarctic Research Series*, J. L. Reid (Ed.) Am. Geophys. Union, Washington, Vol. 15, (1971), 205–221.
5. W. D. Nowlin Jr., J. M. Klink, *Rev. Geophys. Space Phys.*, 24 (1986), 469–491.
6. A. W. Mantyula, J. L. Reid, *Deep-Sea Res.*, 30 (1983), 805–833.
7. E. Fahrbach, M. Knoche, G. Rohardt, *Mar. Chem.*, **35** (1991), 25–44.
8. The Organizing Committee, Conclusion of the Symposium *Biochemistry and Circulation of Water Masses in the Southern Ocean*, Brest 2–6 July 1990, *Mar. Chem.*, **35** (1991), 615–619.

9. S. W. Chisholm, F. M. M. Morel (Eds.), What control phytoplankton production in nutrient-rich areas of the open sea? *Limnol. Oceanogr.*, 36 (1991), 1507–1965.
10. J. H. Martin, R. M. Gordon, S. Fitzwater, W. W. Broenkow, *Deep-Sea. Res.*, 36 (1989), 649–680.
11. E. D. Goldberg, *Mar. Poll. Bull.*, 12 (1981), 225.
12. K. W. Bruland, Trace elements in sea water, In: J. P. Riley, R. Chester, Eds., *Chemical Oceanography*, Vol. 8, Academic Press, London, (1983), 157–220.
13. C. C. Patterson, D. Settle, The reduction of orders of magnitude errors in lead analysis of biological materials and natural waters by evaluating and controlling the extent and sources of industrial lead contamination introduced during sample collection, handling, and analysis, In: P. E. La Fleur (Ed.), *Accuracy in Trace Analysis: Sampling, Sample Handling and Analysis*, US National Bureau of Standards Special Publication, 422 (1976), 321–351.
14. K. W. Bruland, R. P. Franks, G. A. Knauer, J. H. Martin, *Anal. Chim. Acta*, 105 (1979), 233–245.
15. L. Mart, *Fresenius' Z. Anal. Chem.*, 296 (1979), 350–357.
16. L. Mart, *Fresenius' Z. Anal. Chem.*, 299 (1979), 97–102.
17. L. Mart, H. W. Nürnberg, P. Valenta, *Fresenius' Z. Anal. Chem.*, 300 (1980), 350–362.
18. B. K. Schaule, C. C. Patterson, *Earth Planet. Sci. Lett.*, 54 (1981), 97–116.
19. E. A. Boyle, S. S. Huested, S. P. Jones, *J. Geophys. Res.*, 86 (1981), 8048–8066.
20. J. M. Bewers, H. L. Windom, *Mar. Chem.*, 11 (1982), 71–86.
21. P. M. Harper, *Mar. Chem.*, 21 (1987), 183–188.
22. K. W. Bruland, K. H. Coale, L. Mart, *Mar. Chem.*, 17 (1985), 285–300.
23. M. L. A. M. Campos, C. M. G. van den Berg, *Anal. Chim. Acta*, 284 (1994), 481–496.
24. J. R. Donat, K. A. Lao, K. W. Bruland, *Anal. Chim. Acta*, 284 (1994), 547–571.
25. L. S. Cutter, G. A. Cutter, M. L. C. San Diego-McGlone, *Anal. Chem.*, 63 (1991), 1138–1142.
26. I. T. Urasa, Development of new methods of speciation analysis, In: S. Caroli (Ed.), *Element Speciation in Bioinorganic Chemistry*, J. Wiley and Sons, New York, (1996), 121–154.
27. C. Sarzanini, G. Sacchero, M. Aceto, O. Abollino, E. Mentasti, *Anal. Chim. Acta*, 284 (1994), 661–667.
28. A. M. Mota, M. L. Simões Gonçalves, Direct methods of speciation of heavy metals in natural waters, In: S. Caroli (Ed.), *Element Speciation in Bioinorganic Chemistry*, J. Wiley and Sons, New York, (1996), 21–96.
29. I. Ružić, *Anal. Chim. Acta*, 140 (1982), 99–113.
30. M. Plavsic, D. Krznaric, M. Branica, *Mar. Chem.*, 11 (1982), 17–31.
31. C. J. M. Kramer, *Mar. Chem.*, 18 (1986), 335.
32. K. H. Coale, K. W. Bruland, *Limnol. Oceanogr.*, 33 (1988), 1084–1101.
33. K. W. Bruland, *Limnol. Oceanogr.*, 34 (1989), 269–285.
34. G. Capodaglio, K. W. Coale, K. W. Bruland, *Mar. Chem.*, 29 (1990), 221–233.
35. G. Capodaglio, G. Scarponi, G. Toscano, P. Cescon, *Ann. Chim.*, 81 (1991), 279–296.
36. C. M. G. van den Berg, *Mar. Chem.*, 11 (1982), 307–322; correction, *Mar. Chem.*, 13 (1983), 83–85.
37. C. M. G. van den Berg, *Mar. Chem.*, 15 (1984), 1–18.
38. C. M. G. van den Berg, *Mar. Chem.*, 11 (1982), 323–342.
39. L. A. Miller, K. W. Bruland, *Anal. Chim. Acta*, 284 (1994), 573–586.
40. K. S. Johnson, K. H. Coale, H. W. Jannasch, *Anal. Chem.*, 64 (1992), 1065A–1075A.
41. J. Buffle, K. J. Wilkinson, M. L. Tercier, N. Parthasarathy, *Ann. Chim. (Rome)*, 87 (1997), 67–82.
42. M.-L. Tercier, J. Buffle, *Electroanalysis*, 5 (1993), 187–200.
43. M.-L. Tercier, J. Buffle, A. Zirino, R. R. De Vitre, *Anal. Chim. Acta*, 237 (1990), 429–437.
44. A. Daniel, A. R., Baker, C. M. G. van den Berg, *Fresenius J. Anal. Chem.*, 358 (1997), 703–710.
45. E. P. Achterberg, C. M. G. van den Berg, *Anal. Chim. Acta*, 284 (1994), 463–471.
46. Kh. Z. Brainina, R. M. Khanina, V. M. Forshtadt, E. A. Vilchinskaya, G. L. Gaponenko, Proceedings of the J. Heyrovský Centennial Congress on Polarography, 41st Meeting of International Society of Electrochemistry, Prague, (1990), 175.
47. C. M. Sakamoto-Arnold, K. S. Johnson, *Anal. Chem.*, 59 (1987), 1789–1794.
48. T. P. Chapin, K. S. Johnson, K. H. Coale, *Anal. Chim. Acta*, 249 (1991), 469–478.
49. E. A., Boyle, S. D. Chapnick, G. T. Shen, M. P. Bacon, *J. Geophys. Res.*, 91 (1986), 8573–8593.
50. B. J. Bergen, W. G. Nelson, R. J. Pruell, *Environ. Sci. Technol.*, 27 (1993), 938–942.
51. G. Gillain, C. Brihaye, *Anal. Chim. Acta*, 167 (1985), 387–391.

52. G. Capodaglio, G. Toscano, P. Cescon, G. Scarponi, H. Muntau, *Ann. Chim. (Rome)*, 84 (1994), 329–345.
53. G. Capodaglio, G. Scarponi, G. Toscano, C. Barbante, P. Cescon, *Fresenius J. Anal. Chem.*, 351 (1995), 386–392.
54. C. N. Murray, S. Meinke, *J. Oceanogr. Soc. Jpn.*, 30 (1974), 216–221.
55. G. E. Batley, *Trace element speciation: analytical methods and problems*, Batley, G. E., (Ed.), CRC Press, Boca Raton, Florida, (1989), 43–76.
56. D. P. H., Laxen, I. M., Chandler, *Anal. Chem.*, 54 (1982), 1350–1355.
57. G. Capodaglio, G. Toscano, G. Scarponi, P. Cescon, *Ann. Chim. (Rome)*, 79 (1989), 543–559.
58. L. Mart, H. W. Nürnberg, D. Dyrssen, Low level determination of trace metals in arctic sea water and snow by differential pulse anodic stripping voltammetry, In: C. S. Wong, E. Boyle, K. W. Bruland, J. D. Burton, E. D. Goldberg (Eds.), *Trace Metals in Sea Water*, Plenum Press, New York, (1988), 113–130.
59. J. Wang, *Stripping Analysis*, VCH, Deerfield Beach, Fl, (1985), 160.
60. M. Kopanica, F. Opekar, In: *Electrochemical Methods in Chemical and Environmental Analysis*, R. Kalvoda (Ed.), Pergamon, New York, (1987), 58–84.
61. A. Zirino, In: *Marine Electrochemistry*, M. Whitfield, D. Jagner (Eds.), Wiley, New York, (1981), 421–503.
62. C. M. G. van den Berg, In: *Chemical Oceanography*, J. P. Riley, R. Chester (Eds.), Vol. 9, Academic Press, (1988), 197–245.
63. G. E. Batley, T. M. Florence, *J. Electroanal. Chem.*, 61 (1975), 205–211.
64. J. E. Bonelli, H. E., Taylor, R. K. Skogerboe, *Anal. Chim. Acta*, 118 (1980), 243–256.
65. G. Gillain, G. Duyckaerts, A. Disteche, *Anal. Chim. Acta*, 106 (1979), 23–37.
66. I. Gustavsson, *J. Electroanal. Chem.*, 214 (1986), 31–36.
67. H. W. Nürnberg, P. Valenta, Potentialities and applications of voltammetry in chemical speciation of trace metals in the sea, In: C. S. Wong, E. Boyle, K. W. Bruland, J. D. Burton, E. D. Goldberg (Eds.), *Trace Metals in Sea Water*, Plenum Press, New York, (1983), 671–697.
68. W. Lund, Electrochemical methods and their limitations for the determination of metal species in natural waters, In: M. Bernhard, F. E. Brinkman, P. J. Sadler (Eds.), *The Importance of Chemical "Speciation" in Environmental Processes*, Springer-Verlag, Berlin, (1986), 533–561.
69. G. Capodaglio, G. Toscano, G. Scarponi, P. Cescon, *Int. J. Environ. Anal. Chem.*, 55 (1994), 129–148.
70. J. W. McLaren, J. W. H. Lam, S. S. Berman, K. Akatsuka, M. A. Azaredo, *J. Anal. At. Spectrom.*, 8 (1993), 279–286.
71. V. Porta, C. Sarzanini, E. Mentasti, O. Abollino, *Anal. Chim. Acta*, 258 (1992), 237–244.
72. J. W. McLaren, A. P. Mykytiuk, S. N. Willie, S. S. Berman, *Anal. Chem.*, 57 (1985), 2907–2911.
73. K. E. Jarvis, J. G. Williams, E. Alcantara, J. D. Wills, *J. Anal. At. Spectrom.*, 11 (1996), 917–922.
74. L. C. Azaredo, R. E. Sturgeon, A. J. Curtius, *Spectrochim. Acta*, 48B (1993), 91–98.
75. D. Beauchemin, S. S. Berman, *Anal. Chem.*, 61 (1989), 1857–1862.
76. K. Akatsuka, J. W. McLaren, J. W. Lam, S. S. Berman, *J. Anal. At. Spectrom.*, 7 (1992), 889–894.
77. I. Rodushkin, T. J. Ruth, *Anal. At. Spectrom.*, 12 (1997), 1181–1185.
78. G. J. Batterham, N. C. Munksgaard, D. L. Parry, *J. Anal. At. Spectrom.*, 12 (1997), 1277–1280.
79. C. Barbante, M. Krachler, G. Cozzi, Scarponi, G. M. Capodaglio, P. Cescon, XVI Congresso Nazionale di Chimica Analitica, Numana 10–13 June 1998, Abstr. p. 31.
80. O. Abollino, M. Aceto, G. Sacchero, C. Sarzanini, E. Mentasti, *Anal. Chim. Acta*, 305 (1995), 200–206.
81. E. Mentasti, V. Porta, O. Abollino, C. Sarzanini, *Ann. Chim. (Rome)*, 79 (1989), 629–637.
82. V. Porta, O. Abollino, E. Mentasti, C. Sarzanini, *J. Anal. At. Spectrom.*, 6 (1991), 119–122.
83. E. A. Boyle, F. R. Sclater, J. M. Edmond, *Earth Planet. Sci. Lett.*, 37 (1977), 38–54.
84. J. M. Bewers, H. L. Windom, In: *Trace Metals in Sea Water*, C. S. Wong, E. Boyle, K. W. Bruland, J. D. Burton, E. D. Goldberg (Eds.), Plenum Press, New York, (1983), 143–154.
85. C. S. Wong, K. Kremling, J. P. Riley, W. K. Johnson, V. Stukas, P. G. Berrang, P. Erikson, D. Thomas, H. Petersen, B. Imber, In: *Trace Metals in Sea Water*, C. S. Wong, E. Boyle, K. W. Bruland, J. D. Burton, E. D. Goldberg (Eds.), Plenum Press, New York, (1983), 175–195.
86. M. J. Spencer, P. R. Betzer, S. R. Pietrowiz, *Mar. Chem.*, 11 (1982), 403–410.
87. A. R. Flegal, K. Itoh, C. C. Patterson, C. S. Wong, *Nature*, 321 (1986), 689–690.

88. G. Capodaglio, C. Barbante, C. Turetta, G. Scarponi, P. Cescon, *Mikrochim. Acta*, 123 (1996), 129–136.
89. General Services Administration, Federal Standard No. 209D, Washington DC, June 1988.
90. P. J. Paulsen, E. S. Beary, D. S. Bushee, J. R. Moody, *Anal. Chem.*, 60 (1988), 971–975.
91. E. W. Wolff, D. A. Peel, *Ann. Glaciol.*, 7 (1985), 61–69.
92. C. F. Boutron, *Fresenius' Z. Anal. Chem.*, 337 (1990), 482–491
93. NASS-4, Open Ocean sea water Reference Material for Trace Metals, National Research Council of Canada, Division of Chemistry, MACSP, Ottawa, Canada K1A 0R6, 1992.
94. Ph. Quevauviller, C. J. M. Kramer, E. M. van der Vlies, K. Vercoutere, B. Griepink, *Mar. Pollut. Bull.*, 24 (1992), 33–38.
95. S. S. Berman, V. J. Boyko, ICES 16th Round Intercalibration for Trace Metals in sea water, JMG 6/TM/SW, (1987).
96. G. Capodaglio, G. Toscano, P. Cescon, G. Scarponi, H. Muntau, Collaborative sampling error assessment study on sea water analysis: lead, cadmium and copper, *AQUACON MedBas Project Report no. 3*, Joint Research Centre Ispra, Commission of the European Communities, Ispra, Italy (1992).
97. K. W. Hanck, J. W. Dillard, *Anal. Chem.*, 49 (1974), 404–409.
98. B. Imber, M. G. Robinson, F. Pollehne, In: *Complexation of Trace Metals in Natural Water*, C. J. M. Kramer, J. C. Duinker (Eds.), Nijhof/Junk, The Hague, (1984), 429–440.
99. Y. K. Chau, R. Gächter, K. Lum-Shue-Chan, *J. Fish. Res. Board Can.*, 31 (1974), 1515–1519.
100. G. E. Batley, T. M. Florence, *Mar. Chem.*, 4 (1976), 347–363.
101. J. C. Duinker, C. J. M. Kramer, *Mar. Chem.*, 5 (1977), 207–228.
102. R. E. Truitt, J. H. Weber, *Anal. Chem.*, 53 (1981), 337–342.
103. W. G. Sunda; P. J. Hanson, In: ACS Symposium Series, No. 93, *Chemical Modelling in Aqueous Systems*, E. A. Jenne (Ed.), (1979), 147–180.
104. G. Scarponi, G. Capodaglio, C. Barbante, P. Cescon, The anodic stripping voltammetric titration procedure for study of trace metal complexation in sea water, In: *Element Speciation in Bioinorganic Chemistry*, S. Caroli (Ed.), J. Wiley and Sons, New York, (1996), 363–418.
105. J. Buffle, *Complexation Reactions in Aquatic Systems: An Analytical Approach*, Ellis Horwood Ltd., Chichester, (1988), 692.
106. A. Ringbom, *Complexation in Analytical Chemistry*, Wiley, New York, (1979), 395.
107. J. Lee, *Water Res.*, 17 (1983), 501–510.
108. K. W. Bruland, *Limnol. Oceanogr.*, 37 (1992), 1008–1017.
109. J. W. Moffett, R. G. Zika, L. E. Brand, *Deep-Sea Res.*, 37A (1990), 27–36.
110. K. H. Coale, K. W. Bruland, *Deep-Sea Res.*, 37 (1990), 317–336.
111. H. W. Nürnberg, P. Valenta, L. Mart, B. Raspor, L. Sipos, *Fresenius' Z. Anal. Chem.*, 282 (1976), 357–367.
112. J. R. Tushall, P. L. Brezonik, *Anal. Chem.*, 53 (1981), 1986–1989.
113. M. Betti, P. Papoff, *CRC Crit. Rev. Anal. Chem.*, 19 (1988), 271–321.
114. P. del Castillo, *Anal. Proc.*, 28 (1991), 253–254.
115. J. R. Donat, K. W. Bruland, *Mar. Chem.*, 28 (1990), 301–323.
116. L. E. Brand, G. Sunda, R. R. L. Guillard, *J. Exp. Mar. Biol. Ecol.*, 96 (1986), 225–250.
117. G. Sunda, *Biol. Oceanogr.*, 6 (1988–1989), 411–442.
118. M. Whitfield, D. R. Turner, The role of particles in regualting the composition of sea water. In: *Aquatic Surface Chemistry*, W. Stumm (Ed.), Wiley Interscience, (1987), 457–493.
119. K. W. Bruland, *Earth Planet Sci. Lett.*, 47 (1980), 176–198.
120. E. A. Boyle, S. D. Chapnick, G. T. Shen, M. P. Bacon, *J. Geophys. Res.*, 91 (1986), 8573–8593.
121. K. W. Bruland, R. P. Franks, "Mn, Ni, Cu, Zn and Cd in the western North Atlantic", In: C. S. Wong, E. Boyle, K. W. Bruland, J. D. Burton (Eds.), *Trace Metals in Sea Water*, Plenum Press, New York, (1983), 395–414
122. J. D. Burton, P. J. Statham, Trace metals in sea water. In: *Heavy Metals in the Marine Environment*, P. S. Rainbow, R. W. Furness (Eds.), CRC Press, Boca Raton, (1990), 5–25.
123. A. K. Hanson, C. M. Sakamoto-Arnold, D. L. Huizenga, D. R. Kester, *Mar. Chem.*, 23 (1988), 181–203.
124. E. Boyle, J. M. Edmond, *Nature*, 253 (1975), 107–109.

125. R. F. Nolting, H. J. W. de Baar, *Mar. Chem.*, 45 (1994), 225–242.
126. R. F. Nolting, H. J. W. de Baar, A. J. Van Bennekom, A. Masson, *Mar. Chem.*, 35 (1991) 219–243.
127. S. Westerlund, P. Öhman, *Geochim. Cosmochim. Acta*, 55 (1991), 2127–2146.
128. M. J. Orren, P. M. S. Monteiro, Trace elements geochemistry in the Southern Ocean. In: *Antarctic Nutrient Cycles and Food Webs*, W. R. Siegfried, P. R. Condy, R. M. Laws (Eds.), Springer-Verlag, Berlin, (1985), 30–37.
129. G. Bordin, P. Appriou, P. Treguer, *Ocenol. Acta*, 10 (1987), 411–420.
130. R. Frache, F. Baffi, C. Ianni, F. Soggia, *Ann. Chim. (Rome)*, 87 (1997), 367–374.
131. O. Abollino, M. Aceto, G. Sacchero, C. Sarzanini, E. Mentasti, *Ann. Chim. (Rome)*, 86 (1997), 229–243.
132. S. Westerlund, P. Öhman, *Mar. Chem.*, 35 (1991), 199–217.
133. J. H. Martin, R. M. Gordon, S. Fitzwater, *Nature*, 345 (1990), 156–158.
134. G. Capodaglio, C. Turetta, G. Toscano, A. Gambaro, G. Scarponi, P. Cescon, Cadmium, lead and copper complexation in Antarctic coastal sea water. Evolution during the Austral summer. *Int. J. Environ. Anal. Chem.*, 71 (1998), 195–226.
135. P. M. Saager, H. J. W. de Baar, D. L. Huizenga, *Deep-Sea Res.*, 39 (1992), 9–35.
136. G. Scarponi, G. Capodaglio, C. Turetta, C. Barbante, M. Cecchini, G. Toscano, P. Cescon, *Int. J. Environ. Anal Chem.*, 66 (1997), 23–49.
137. G. Scarponi, G. Capodaglio, C. Barbante, G. Toscano, M. Cecchini, A. Gambaro, P. Cescon, Concentration changes of cadmium and lead in Antarctic coastal sea water (Ross Sea) during the austral summer and their relationship with the evolution of biological activity. In: *Ross Sea Ecology*, L. Ianora, Guglielmo (Eds.), Springer-Verlag, Berlin, (2000), 585–594.
138. R. D. Frew, K. A. Hunter, *Nature*, 360 (1992), 144–146.
139. E. A. Boyle, *Paleoceanography*, 3 (1988), 471–489.
140. A. R. Flegal, H. Maring, S. Niemeyer, *Nature* 365 (1993), 242–244.
141. I. H. De Gregori, D. D. Delgado, H. C. Pinochet, *Int. J. Environ. Anal. Chem.*, 34 (1988), 315–331.
142. W. G. Sunda, P. A. Gillespie, *J. Mar. Res.*, 37 (1979), 761–777.
143. D. M. McKnight, F. M. M. Morel, *Limnol. Oceanogr.*, 24 (1978), 823–837.
144. R. F. Srna, K. S. Garrett, S. M. Miller, A. B. Thum, *Environ. Sci. Technol.*, 14 (1980), 1482–1486.
145. G. R. Di Tullio, E. A. Laws, *Deep-Sea Res.*, 38 (1991), 1305–1329.
146. G. A. Knox, in: *Antarctic Ecosystems. Ecological Change and Conservation,* K. R. Kerry, G. Hempel (Eds.), Springer Verlag, Berlin, (1990), 115–128.
147. V. Bruni, M. L. C. Acosta Pomar, T. L. Maugeri, E. Crisafi, R. La Ferla, R. Zaccone, In: *Oceanographic Campaign 1989–90. Data Report. Part II.* (National Scientific Commission for Antarctica, Genova, 1992), 107–122.
148. W. M. Landing, K. W. Bruland, *Geochim. Cosmochim. Acta*, 51 (1987), 29–43.
149. R. M. Gordon, J. H. Martin, G. A. Knauer, *Nature*, 299 (1982), 611–612.
150. P. M. Saager, H. J. W. de Baar, P. H. Burkill, *Geochim. Cosmochim. Acta*, 53 (1989), 2259–2267.
151. L.-G. Danielson, B. Magnusson, S. Westerlund, *Mar. Chem.*, 17 (1985), 23–41.
152. J. L. Symes, D. R. Kester, *Mar. Chem.*, 17 (1985), 57–74.
153. N. M. Price, B., Ahner, F. M. M. Morel, *Limnol. Oceanogr.*, 39 (1994), 520–534.
154. R.M. Gordon, K. H. Coale, K. S. Johnson, *Limnol. Oceanogr.*, 42 (1997), 419–431.
155. S. Emerson, R. E. Cranston, P. S. Liss, *Deep-Sea Res.*, 26A (1979), 859–878.
156. L. Balistrieri, P. G. Brewer, J. W. Murray, *Deep-Sea Res.*, 28 (1981), 101–121.
157. H. J. W. de Baar, C. R. German, H. Elderfield, P. Van Gaans, *Geochim. Cosmochim. Acta*, 52 (1988), 1203–1219.
158. K. W. Bruland, J. R. Donat, D. A. Hutchins, *Limnol. Oceanogr.*, 36 (1991), 1555–1577.
159. K. W. Bruland, K. J. Orians, J. P. Cowen,*Geochim. Cosmochim. Acta*, 58 (1994), 3171–3182.
160. D. A. Hutchins, G. R. Di Tullio, K. W. Bruland, *Limnol. Oceanogr.*, 38 (1993), 1242–1255.
161. J. H. Martin, R. M. Gordon, S. Fitzwater, *Limnol. Oceanogr.*, 36 (1991), 1793–1802.
162. H. J. W. de Baar, A. G. J. Buma, R. F. Nolting, G. C. Cadee, G. Jacques, P. J. Treguer, *Mar. Ecol. Prog. Ser.*, 65 (1990), 105–122.
163. H. W. Rich, F. M. M. Morel, *Limnol. Oceanogr.*, 35 (1991), 652–662.
164. D. W. O'Sullivan, A. K. Hanson, W. L. Miller, D. R. Kester, *Limnol. Oceanogr.*, 36 (1991), 1727–1741.

Environmental Contamination in Antarctica
A Challenge to Analytical Chemistry
S. Caroli, P. Cescon and D.W.H. Walton, editors

Chapter 6

Trace metals monitoring as a tool for characterization of Antarctic ecosystems and environmental management. The Argentine programme at Jubany Station

Cristian Vodopivez, Patricia Smichowski and Jorge Marcovecchio

1. Introduction

1.1. Antarctica: a continent to preserve

The natural balance of the environment has been seriously affected by man in many parts of the planet. However, there is universal consensus that the fate of Antarctica has to be different and that its natural resources, scientific values and beauty must be preserved. Although most inner zones of the Antarctic continent remain unexplored and have minimum human presence, the coastal zone has been thoroughly navigated, with permanent human settlements since the beginning of the twentieth century. The coast of the Antarctic Peninsula, in particular, has been one of the most extensively explored zones, attracting an important human presence. The closeness of the peninsula to South America, its less severe climate and its icefree accessibility in summer account for the numerous research stations and the commercial exploitation of the marine resources. The remnants of human activity such as abandoned stations, field dumps of fuel, rubbish dumps, etc., are still visible.

At present, the principal human activities in Antarctica are scientific research and tourism. Summer population is estimated at 4000 persons, while the winter population is about 900 persons. From 1944 until the mid 1950s there was only limited activity by a few nations, but this increased substantially during the International Geophysical Year (1957–1958). The ratification of the Antarctic Treaty in 1961 consolidated political and scientific interest, which have since then grown substantially. This is reflected by the increase in the number of permanent stations, from 31 belonging to 12 countries in 1966 to more than 40 (18 countries) in 1996 (1). A half of these stations are located in the Antarctic Peninsula region and eight of them are on King George Island (2).

In the last 15 years tourism in Antarctica has shown continuous growth, which has been particularly marked in the area of the Antarctic Peninsula (3–7). About 10,000 persons now visit Antarctica each year. The accessibility of the Peninsula as well as the richness and diversity of its wild life provide a strong attraction for those who want to discover these mysterious and unknown lands. Antarctica has usually been included among the few remaining pristine regions of the planet, primarily because of its isolation from large industrial centres. Detection of pollutants in different Antarctic matrices was originally ascribed to the global transport of atmospheric aerosols (8–11). This is an acceptable explanation for most of the continental area. However, the increasing activity of scientific stations, the improvement of the logistic facilities, as well as the activity of supply and tourism vessels, have also contributed to some extent to the contamination and modification of the Antarctic environment, especially on the local scale.

Recent studies have demonstrated the occurrence of a contaminated halo around scientific stations, where hydrocarbon residues and metals at trace and ultratrace levels were detected. In most cases pollutants have been found within an area of a few hundred meters from the stations, rapidly decreasing with distance from the emission focus (12). Quite often the concentrations of the detected contaminants have been very low and far below the levels deemed to be toxic. The presence of the halo of pollution around stations is a typical indication of human activities and contrasts sharply with the pristine nature of most of Antarctica.

The Antarctic Treaty and the Madrid Protocol provide the necessary framework for environmental management and have obliged all the nations with an active presence in the continent to reduce their impacts on the antarctic environment. The Treaty's aim is to guarantee the peaceful use of Antarctica and to ensure conservation of flora, fauna and the natural environment. Through more adequate environmental impact assessment and management as well as environmental monitoring it is expected the Antarctic will remain the cleanest place on earth despite an increase in human presence.

1.2. Environmental monitoring in Antarctica as a management tool

Since the Madrid Protocol was signed in October 4, 1991, the international community has showed an increasing awareness of the value of environmental monitoring in the preservation of Antarctica. This is reflected in the number of studies that are being carried out by different countries. The need for environmental monitoring in Antarctica was briefly stated in the SCAR/COMNAP Report on Environmental Monitoring on Antarctica – A Discussion Document (13):

> *Environmental monitoring is a fundamental element of basic research, environmental management, and conservation. The organised and systematic measurement of selected variables provides for the establishment of baseline data and the identification of both natural and human-induced change in the environment. Monitoring data are important in the development of models of environmental processes, which in turn facilitate progress towards a predictive capability to detect environmental impact or change. The collection and evaluation of monitoring data is essential for the detection of human perturbation within the natural variability of ecosystem processes. Since all environmental monitoring must be based on testable hypotheses it can contribute to advancement in both basic and applied research.*

Since the 1950s, several monitoring programmes have been undertaken and the results obtained have been of interest not only in the evaluation of environmental pollution on a global scale, but also in investigating the impact produced by research stations on a local scale. These programmes were orientated towards basic research. More recently, Antarctic operators began to consider the importance of environmental monitoring as a fundamental tool for the environmental management of all the activities which are carried out within the area of the Antarctic Treaty (14).

Three principal objectives have been highlighted for environmental monitoring in Antarctica (15);

i) to protect the scientific value of the Antarctic;
ii) to contribute to the continuous improvement of Antarctic environmental management;
iii) to implement the legal requirements of the Protocol and national legislation.

The Environmental Monitoring Programme (EMP) implemented at Jubany Station attempts to accomplish these basic objectives while keeping the basic research activities which provide important information for management decisions.

1.3. The determination of trace elements for characterization of the ecosystems and environmental management

Over the last few decades the interest in the detection and quantification of trace elements in natural matrices (waters, sediments, biota) has increased noticeably as a consequence of concern about anthropogenic activity. This is to a large extent probably due to an increased awareness of the consequences of environmental pollution at all spatial scales. In addition, the importance of the presence or absence of certain metals at trace ($\mu g\ g^{-1}$) or ultratrace ($ng\ g^{-1}$) levels in living organisms is now much better known for a range of temperate species. An improved knowledge of the presence of trace metals in Antarctica will permit not only a better understanding of global distributions, but also provide a baseline against which any potential adverse biological effects can be assessed. Trace elements are also useful for the detection of pollution by local activities in Antarctica. Monitoring trace elements can be a very difficult task as it is crucial to define a strategy by which contamination and losses of the analyte at the different steps can be avoided. The full incorporation of quality control and assurance criteria in all the preanalytical and analytical steps is mandatory in dealing with trace and ultratrace levels. Special attention must be paid to the following aspects:

i) the major source of error is an inadequate sampling strategy. Erroneous data can easily result, especially due to the low concentrations found in environmental samples. If this step is carried out wrongly the remainder of the analysis will be irrelevant;
ii) clean room, or clean glove box conditions, and laminar air-flow cabinets are required for the reduction of environmental contamination in the laboratory. For ultratrace analysis, a class-100 clean laboratory is highly desirable. In a class-100 laboratory a particle count below 100 particles of size above $0.3\ \mu m\ ft^{-3}$ of air is specified (16);
iii) sample preservation is critical. Proper containers, cleaning procedures and

storage conditions are crucial for high quality data. Adsorption, desorption and volatilization may occur so that adequate controls are essential. For sediments with biogenic materials, bacterial degradation may take place during the storage. It is advisable to store the samples in polyethylene bottles and freeze them as soon as possible. In the case of waters, the permissible storage time before analysis in a cool (4°C) dark place after addition of stabilizing agents (HNO_3, pH 2) varies for a number of trace elements from 1 to 6 months (17). The acidification of the sample is undesirable when speciation is required;

iv) in the grinding of solid samples an agate mortar assures homogeneity of the dried samples;

v) in the sample treatment, special attention must be paid to the possibility of contamination caused by the reagents used;

vi) as for all chemical determinations, quantification of the analytes will depend upon the technique used, the blanks, the precision and the validation of the results. Size and fluctuation of the blank must be reduced as much as possible because the accuracy of the measurements is inversely related to the variability of the blank (18);

vii) the selection of a technique mainly depends on the matrix to be analyzed and the laboratory facilities. A general statement about the best technique for the determination of trace metals in a specific environmental matrix is not possible. The instrumental cost, sensitivity and detection limits, accuracy, precision, interferences and the necessary sample volume are important parameters to take into account. In this context, it is self-evident why methods based on atomic spectrometry have been so successful, in many cases in combination with other techniques or a preconcentration step. Inductively Coupled Plasma Mass Spectrometry (ICP-MS) is a technique of choice in trace and ultratrace analysis because it offers powerful detection power. Selectivity, multielemental capabilities, high sample throughput and wide dynamic range are other major advantages. On the other hand, Anodic Stripping Voltametry (ASV) is the electrochemical technique most widely applied to trace analysis because Limits of Detection (LoDs) in the ng g^{-1} range can be reached;

viii) for the validation of the methods developed and for verifying the accuracy of the experimental measurements, two new Certified Reference Materials (CRMs) based on antarctic marine sediment and Antarctic krill are now available (ISS, National Institute of Health, Rome, Italy). (19).

2. The environmental monitoring programme at Jubany Station

2.1. A brief history of Argentina in Antarctica

Argentina has a long history in Antarctica which dates back to 1904, when the first permanent scientific observatory on the South Orkney Islands was established (20). Six permanent stations have been maintained by Argentina over recent years as well as field camps during the austral summers when all activities are noticeably increased (21). Antarctic stations (abandoned and active ones) have been indicated as the principal focus of localized chronic contamination (12, 22, 23). As part of the Argentine response to the Protocol, the Instituto Antártico Argentino (Argentine Antartic Institute, IAA) has promoted studies involving environmental monitoring (24, 25).

After the sinking of the Bahía Paraíso ship in Arthur Harbour in 1989 a series of studies based on the monitoring of trace elements was also performed (26–29). In 1991 an Environmental Monitoring Programme (EMP) was designed to assess the occurrence, concentration and distribution of several trace elements in a coastal ecosystem at Jubany Station (King George Island, South Shetland Islands). Since 1992, the EMP has been carried out by researchers from the IAA, the Argentine Institute of Oceanography (IADO, Bahía Blanca) and the Mar del Plata National University (UNMdP, Mar del Plata). At present other institutions such as the Laboratory of Geological and Edaphic Chemistry (LAQUIGE-CONICET), the Naval Hydrographic Service (SHN, Buenos Aires) and the Atomic Energy National Commission (CNEA, Buenos Aires) are also collaborating to the Programme.

2.2. Environmental monitoring programme objectives

The environmental monitoring programme implemented at Jubany Station attempts to contribute to a better understanding of biogeochemical processes in the costal environment. Systematic evaluation of selected trace elements could be useful to identify natural and anthropogenic changes in the Antarctic environment. Monitoring provides information for an adequate environmental management. The EMP at Jubany Station has a number of targets grouped under two major objectives:

i) *management objective*: providing information from which management decisions can be made; assessing pollutant levels at impacted sites; providing an early warning of environmental deterioration; identifying the activities most responsible for environmental deterioration;
ii) *scientific objective*: providing a better understanding of biogeochemical processes; establishing the baseline of trace elements in the Potter Cove marine environment; identifying biomonitors; assessing bioaccumulation and biomagnification processes; assessing biogeochemical cycles of key elements.

2.3. Activities at Jubany Station

Jubany Station (62° 14′ S, 58° 38′ W) was chosen as a focus for monitoring studies because:

i) King George Island has the largest human population in the Maritime Antarctic;
ii) the station is in an area of environmental value, especially with respect to biodiversity. In addition, it is close to the Site of Special Scientific Interest (SSSI) No. 13;

iii) valuable information about the physical and biological environment in the area is already available;
iv) several other research projects are being carried out in the area and the interchange of information and samples is therefore simplified whilst researchers can also share logistic support and laboratory instruments;
v) the station provides the infrastructure necessary for preliminary treatment and storage of samples.

Working with other specialists, especially those studying coastal ecology, allowed the use of available logistic support with other projects to be optimized (i.e., vessels, communication equipments, diving operation time). In addition, the collection of biotic and abiotic samples was assigned to the corresponding specialists on each subject (i.e., sampling of lichens was assigned to lichen specialists). In the case of higher animals, it was decided to use only tissues of specimens sacrificed for other research purposes (it was decided not to sacrifice higher animals during the pilot stage of the programme). In all cases, the programme focused on monitoring of heavy metals by making the best use of available logistic and human resources and minimizing disturbance to the native flora and fauna.

2.3.1. Investigated area

Potter Cove is a tributary inlet close to the entrance of Maxwell Bay, one of the two big fjords at King George Island (Figure 6.1). The cove is divided into a mouth and an inner part. The mouth area is bordered by steep slopes in the north and by a broad intertidal platform in the southeast. The bottom of the mouth area lies between 100 and 200 m. The inner part is not deeper than about 50 m and barred by a sill of a depth of about 30 m (30).

The glacier reaches the cove in the north and in the east, while the southern shore is a sandy beach. The coast mainly consists of crumbly volcanic andesite interspersed with intrusions of resistent basaltic dykes, which form protruding reefs and promontories (31). Uneven spacing of these structures creates a close neighborhood of protected, pocket-like bights and exposed, open single beaches and rock platforms (32).

The areas of interest in Potter Cove can be summarized as follows: (33–35);

(i) the long term mean current describes a cyclonic gyre (clockwise) around the cove, with the waters entering by the north sector and exiting by the south sector;
(ii) the marked E–W bidirectionality of the wind leads to a two-layer vertical circulation cell. In presence of west winds, an entry of water by the surface and exit in the depth with sinking in the interior of the cove can be noted. The opposite case, with upwelling in the interior, occurs with east winds;
(iii) the tidal current is characterized by low intensity (in comparison to the long term mean current and wind-drift current). Although the contribution of the semidiurnal component to the tide amplitude overcomes that of the diurnal, greater values of current intensity are observed for diurnal periods;

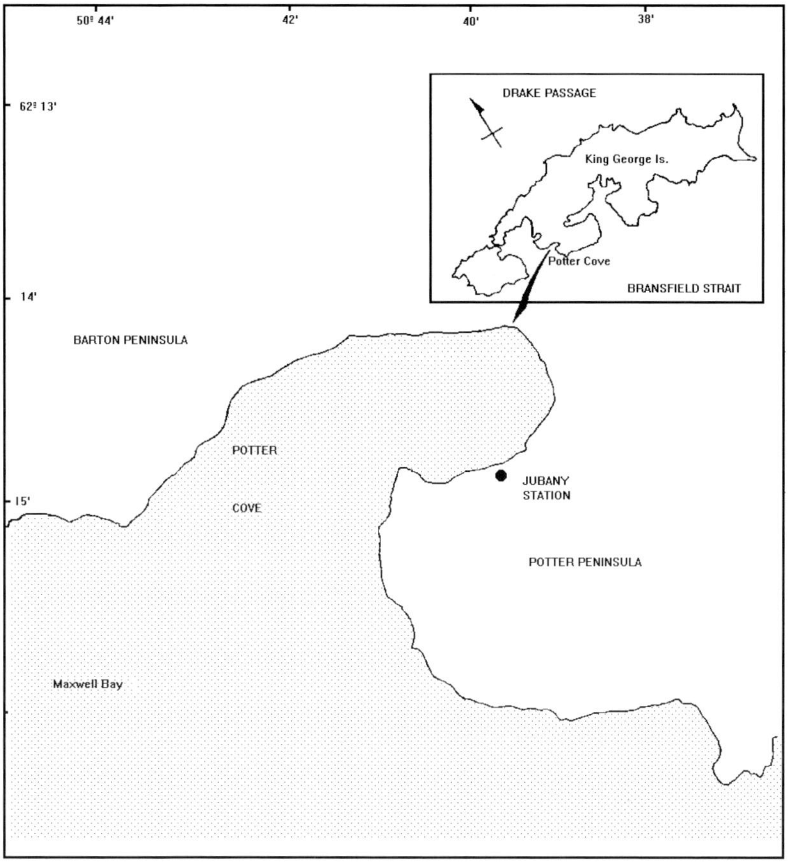

Figure 6.1. Location of Jubany Station, King George Island, South Shetland Islands.

(iv) the wave field in the southern coast of the cove generates an entering littoral current by a narrow coastal strip (10 to 15 m), leading to a west-to-east sediment transport and also causing the spreading of the material supplied by meltwater creeks;

(v) the comparison between summer and winter (with frozen cove) current with different intensities shows slightly greater values in the first case, presumably as a consequence of the inhibiting effect that the ice cover has on the surface wind stress.

3. Materials and methods

3.1. Sampling procedures and treatment

Careful sampling and storage procedures, as described above, were followed in order to assure the validity of the results obtained.

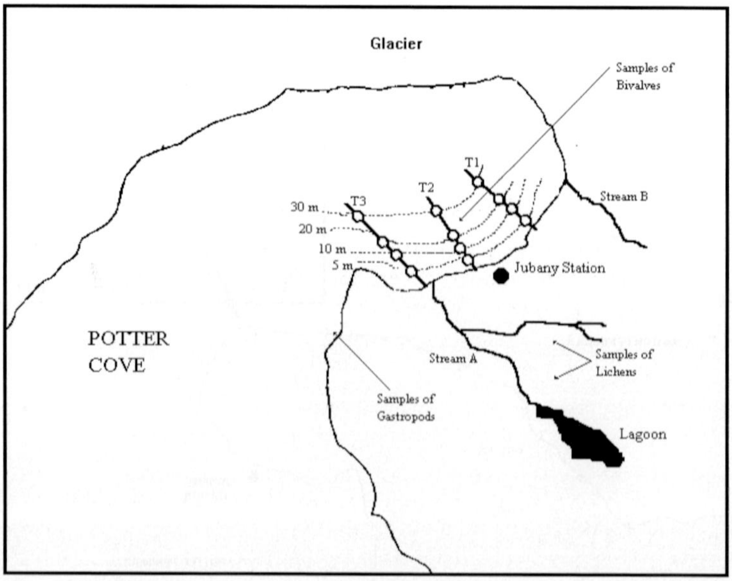

Figure 6.2. Area of sampling around Jubany Station.

The biotic and abiotic samples were collected during the 1992–1993, 1993–1994, 1994–1995 and 1995–1996 austral summer seasons around Potter Cove, King George Island and South Shetland Islands (Figure 6.2). Samples collected during the mentioned austral summers included:

(i) *marine surface sediments*: samples of surface sediments were collected in polyethylene containers by autonomous diving at twelve sites located in three transects within the Cove (Figure 6.2), after removing the upper gravel. Sediment samples were frozen at –20°C, then dried at 40 ± 5°C for 48 hr to constant weight. The samples were then divided into two batches, one for determination of grain size distribution using sieves and the other for chemical analysis.

(ii) *freshwater sediments and suspended particulate matter*: samples of freshwater and sediments were collected in streams A and B and lagoons using polypropylene sampling bottles. Special care was taken to avoid the possible resuspension of sediments. The water samples were vacuum-filtered through a 0.45 μm cellulose acetate filter and the Suspended Particulate Matter (SPM) was frozen at –20°C until their analysis in the laboratory. The same treatments used for marine sediments were followed for fresh water sediments.

(iii) *molluscs*: samples of *Laternula elliptica* were hand-picked by scuba divers at 20–25 m depth, while those of *Nacella concinna* were collected in the intertidal area during low tide. Samples were stored at –20°C until their treatment in the laboratory.

(iv) *lichens*: samples of the lichen species *Usnea aurantiacoatra* and *Usnea antarctica* were hand-collected close to "Jubany", stored at –20°C, then washed with distilled water and dried at 50 ± 5°C to constant weight. Lichens were homogenized in a porcelain mortar and stored in acid washed-glass containers in a dessicator.

3.2. Heavy metals determination

For the analytical determination of metals (Cd, Cu, Fe, Mn, Pb and Zn) in surface sediments, suspended particulate matter and biological matrices, digestion with a 3:1 HNO_3-$HClO_4$ mixture under controlled temperature was used (36). Analysis of sediments and suspended particulate matter were made by Flame Atomic Absorption Spectrometry (FAAS) with air-acetylene flame and deuterium background correction. The analysis of metals in lichens and molluscs were performed by ICP-AES. The operating conditions for FAAS and Inductively Coupled Plasma Atomic Emission Spectrometry (ICP-AES) analysis are shown in Tables 6.1 and 6.2, respectively.

Total Hg in biological matrices was determined using the Cold Vapor Atomic Absorption Spectrometry (CV-AAS) methodology (37). Samples were predigested in a 4:1 HNO_3-H_2SO_4 mixture at 60°C. The digestion was completed with 6%

Table 6.1. Operating conditions for the determination of trace metals in sediments and particulate matter by FAAS

Spectrometer	Shimadzu AA 640–13 (Cd, Cu, Fe, Mn, Pb, Zn) and Buck Scientific 200 A (Hg)						
Element	Cd	Cu	Fe	Hg*	Mn	Pb	Zn
Wavelength (nm)	228.8	324.8	248.3	253.7	279.5	283.3	253.7
Slit width (nm)	0.7	0.7	0.2	0.7	0.2	0.7	0.7
Lamp current (mA)	6	15	15	5	15	8	10
Air flow rate (l min^{-1})	8	8	8	–	8	8	8
Acetylene flow rate (l min^{-1})	4	4	4	–	4	4	4

*Hg was determined by CV-AAS

Table 6.2. Operating conditions for the determination of trace metals in molluscs and lichens by ICP-AES

Spectrometer	Baird ICP 2070
Software	Baird ICP 2070, Version 1.06
Electromagnetic field frequency	40 MHz
Outer gas flow rate (l min^{-1})	8.5
Intermediate gas flow rate (l min^{-1})	1.0
Sample gas flow rate (l min^{-1})	1.1
Observation heigh above load coil (mm)	14
Wavelength (nm)	Cd(I), 228.8; Cu(I), 324.8 (I); Fe(II), 259.9; Mn(II), 257.6; Pb(I), 283.3; Zn(I), 213.9

Table 6.3. Results of the analysis of CRMs to assess analytical accuracy

	Concentration of element ($\mu g\ g^{-1}$)				Concentration of element ($\mu g\ g^{-1}$)		
CRM	Element	Certified	Found	CRM	Element	Certified	Found
NIES No. 2 (Pond Sediment)	Cd	0.82 ± 0.06	0.75 ± 0.07	NIES No. 6	Cd	0.82 ± 0.03	0.78 ± 0.05
	Cr	75 ± 5	72 ± 4	(Mussel)	Cu	4.9 ± 0.3	4.9 ± 0.4
	Cu	210 ± 12	197 ± 15		Fe	158 ± 8	160 ± 8
	Fe (%)	6.53 ± 0.35	6.71 ± 0.38		Hg*	0.05	0.07
	Mn*	770	754		Mn	16.3 ± 1.2	16.0 ± 1.3
	Pb	105 ± 6	104 ± 6		Pb	0.91 ± 0.04	0.90 ± 0.04
	Zn	343 ± 17	349 ± 21		Zn	106 ± 6	107 ± 5

NIES: National Institute for Environmental Studies, Japan
* Qualified value

potassium permanganate solution. The excess of permanganate was eliminated using 20% hydroxylamine hydrochloride solution. Reduction of Hg(II) to Hg^0 was accomplished by a Sn(II) chloride solution. Determinations were made with a Buck Scientific 200 A CVAAS instrument.

All reagents were of analytical grade unless otherwise mentioned. CRMs (marine sediment and mussel tissue), supplied by the National Institute for Environmental Studies (NIES), Tsukuba, Japan were used. Results obtained in the analysis of these CRMs are shown in Table 6.3. The results obtained were compared through one-way analysis of variance (ANOVA).

4. Results and discussion

4.1. Trace metals in marine surface sediments

The distribution pattern of heavy metals in surface marine sediments is regulated not only by their concentrations, but also by their physical-chemical characteristics, mineralogical composition, grain size distribution, organic matter contents, etc. Several environmental conditions such as marine currents, wind, and continental runoff must also be considered (38). Some data on trace metals in Antarctic sediments have been published, but the information available for Potter Cove is limited (27, 39–42).

During the 1994–1995 austral summer, samples of sediments fron Potter Cove were collected in order to assess the presence and origin of Cd, Cr, Cu, Fe, Mn, Pb and Zn. Numerous studies have shown that finer fractions of sediments and finer textured sediments in estuaries and marine zones have higher levels of Fe and Mn oxides and trace metals than those found in coarse materials (43). The water movement in the cove is usually by clockwise currents with the littoral current on the southern coast favouring the sedimentation of the finer particles ($<62\ \mu m$) from a 10 m depth. Table 6.4 and Figure 6.3 show the granulometric composition

Table 6.4. Granulometric composition of the transects 1 and 2

Transect	Grain size fraction (μm)	Depth (m)			
		5	10	20	30
	> 500	1.42	1.97	4.06	0.82
	250–500	5.78	1.64	1.78	2.35
1	125–250	73.87	18.37	11.26	14.23
	62–125	13.75	24.60	18.91	19.11
	< 62	5.18	53.42	63.99	63.49
	> 500	26.48	10.4	8.12	0.71
	250–500	4.78	1.65	1.21	1.02
2	125–250	13.88	8.70	9.16	14.75
	62–125	22.26	15.84	25.43	28.56
	< 62	32.60	53.01	56.08	54.96

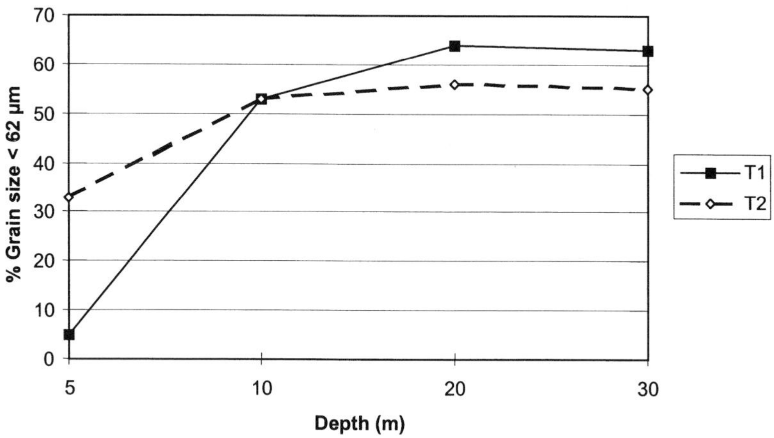

Figure 6.3. Percentage of grain size (fraction < 62 μm) in transects 1 and 2.

of the transects 1 and 2. Total Cd, Cr, Cu, Fe, Mn, Pb and Zn contents of surface marine sediments collected in the transects 1, 2 and 3 are given in Table 6.5.

Chromium, Cu, Mn, Pb and Zn presented a similar distribution trend along the studied area, even though their concentrations are completely different (Figures 6.4–6.8). This fact suggests that these metals have a similar behaviour in the evaluated environment, probably because they are controlled by similar processes and conditions. Cadmium contents in the analysed samples are in all cases below the LoDs of the method employed (<0.09 μg g^{-1} dry weight). Iron showed a particular distribution pattern (Figure 6.8), which is probably related to its abundance (38). No evidence of trace metals in surface sediments of the Potter Cove was observed that could be attributed to contamination caused by scientific or logistic activities carried out in the Jubany Station.

Table 6.5. Metal concentrations (± standard deviation, three replicate) in surface sediments from Potter Cove

Transect	Depth (m)	Cd(*)	Cr(*)	Cu(*)	Fe(**)	Mn(**)	Pb(*)	Zn(*)
	5	<0.09	6.4 ± 0.5	128.5 ± 5.3	12.0 ± 0.8	0.99 ± 0.07	3.3 ± 0.2	46.0 ± 3.1
1	10	<0.09	6.7 ± 0.6	140.1 ± 6.8	12.3 ± 0.7	1.01 ± 0.07	3.1 ± 0.2	53.3 ± 2.1
	20	<0.09	7.7 ± 0.6	156.3 ± 7.1	16.3 ± 0.8	1.13 ± 0.06	5.5 ± 0.3	63.0 ± 2.9
	30	<0.09	6.0 ± 0.4	139.0 ± 5.4	21.4 ± 0.6	1.08 ± 0.05	3.8 ± 0.1	52.5 ± 2.3
	5	<0.09	5.7 ± 0.3	119.1 ± 4.8	16.3 ± 0.7	0.93 ± 0.04	3.5 ± 0.1	48.0 ± 2.2
2	10	<0.09	8.1 ± 0.6	137.6 ± 5.0	18.6 ± 0.7	0.98 ± 0.04	4.8 ± 0.1	59.1 ± 2.6
	20	<0.09	5.1 ± 0.4	109.8 ± 4.6	6.6 ± 0.4	0.94 ± 0.04	3.1 ± 0.1	48.8 ± 2.3
	30	<0.09	6.5 ± 0.3	126.1 ± 5.0	10.3 ± 0.5	0.95 ± 0.04	4.0 ± 0.2	56.0 ± 2.0
	5	<0.09	4.1 ± 0.2	73.4 ± 4.4	9.9 ± 0.7	0.79 ± 0.06	2.3 ± 0.1	45.0 ± 3.0
3	10	<0.09	4.9 ± 0.3	83.5 ± 5.0	5.2 ± 0.3	0.85 ± 0.07	3.9 ± 0.3	49.9 ± 3.4
	20	<0.09	7.3 ± 0.5	110.6 ± 6.3	14.0 ± 0.8	0.95 ± 0.06	3.6 ± 0.2	56.7 ± 2.9
	30	<0.09	6.3 ± 0.4	109.8 ± 6.0	11.0 ± 0.5	0.89 ± 0.05	4.6 ± 0.2	55.6 ± 3.5

(*) in μg g^{-1}, dry weight. (**) in mg g^{-1}, dry weight.

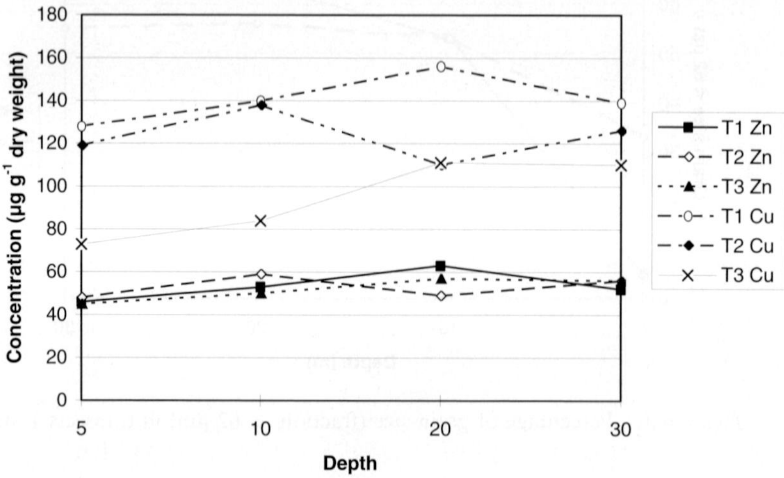

Figure 6.4. Distribution of Zn and Cu in transects 1, 2 and 3.

4.2. Trace metals in freshwater sediments and suspended particulate matter

During the summer glacier meltwater feeds two streams (Figure 6.2) that debouch into the cove. Stream A has its origin in the lagoon by the glacier and debouches in the cove between the sampling transects 2 and 3 (T2 and T3). Stream B has its origin in the glacier itself where melt from its upper section carries down volcanic material and glacier sediments on to a sand beach and then debouches close to the first sampling transect (T1). These streams input sufficient amounts of freshwater and suspended sediments to drastically modify the hydrographic conditions of the

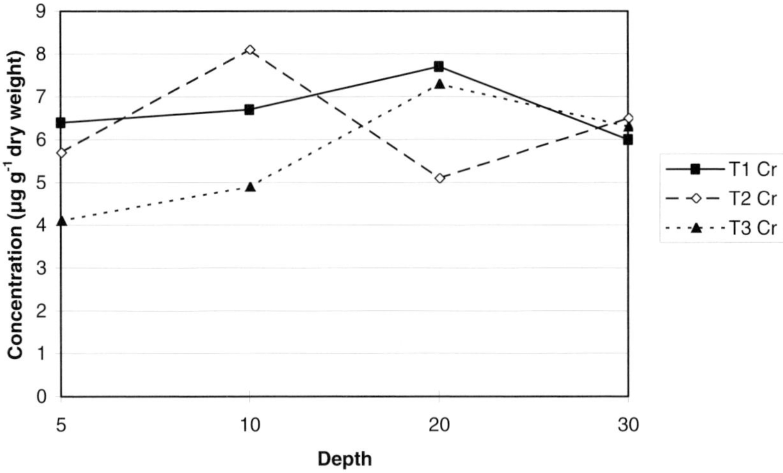

Figure 6.5. Distribution of Cr in transects 1, 2 and 3.

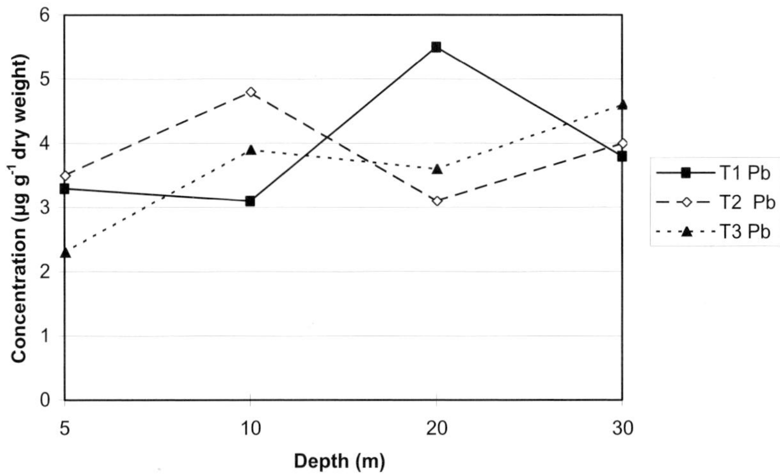

Figure 6.6. Distribution of Pb in transects 1, 2 and 3.

cove (35, 44–46). Table 6.6 shows the water volume and the suspended particulate matter that are introduced by the streams. Total Cr, Cu, Fe, Pb, Mn and Zn contents determined in suspended particulate matter samples collected in the streams are shown in Table 6.7, while those in samples of freshwater sediments of the lagoon and in the streams are given in Table 6.8.

Sediments and suspended particulate matter in freshwater originate from glacial erosion of volcanic rocks, mainly basaltics and andesitics that are primarily composed of olivine and pyroxene and of plagioclase and pyroxene, respectively (31).

Metal contents observed in the sediments from the inner part of the cove coincide with the values obtained in freshwater sediments (Tables 6.5, 6.7 and 6.8),

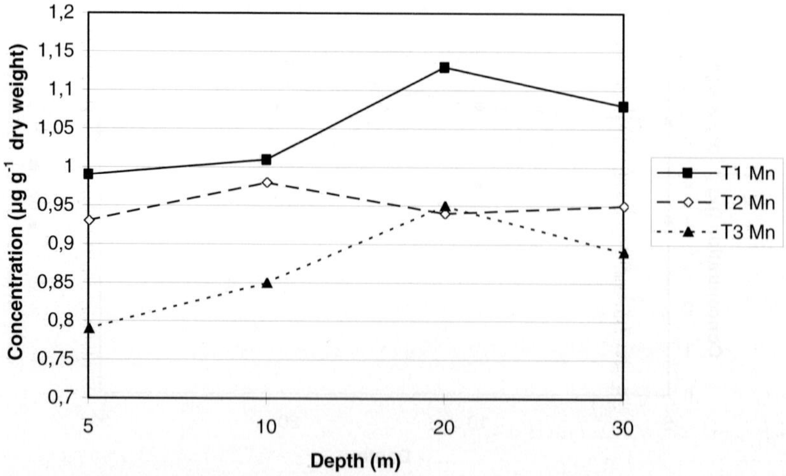

Figure 6.7. Distribution of Mn in transects 1, 2 and 3.

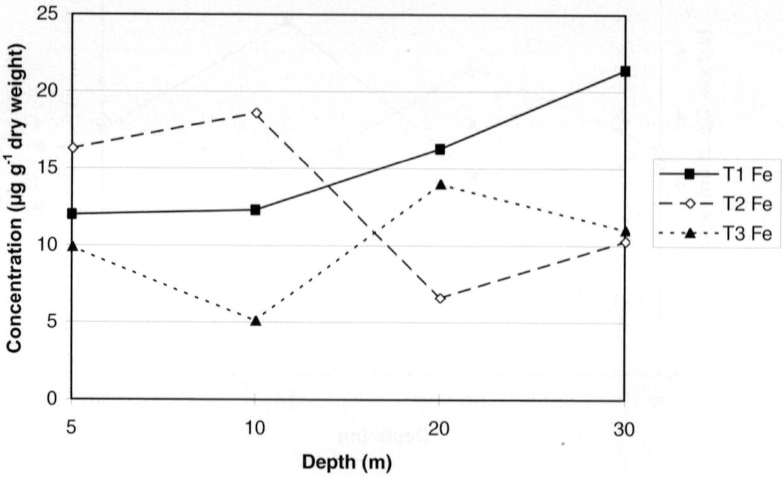

Figure 6.8. Distribution of Fe in transects 1, 2 and 3.

indicating that the contribution of suspended particulate matter from the streams is the principal component of the in-shore surface marine sediments. The metal contribution of stream B is approximately ten times greater than stream A. Metals are dispersed by the dynamic mixing conditions of the cove and, as a consequence, the sedimentation of the finest particles is favoured in the central and southern zones. These studies suggest that the metals analyzed are autochthonous of this environment and consequently the occurrence of heavy metals in sediments of Potter Cove is not related to activities performed in the scientific stations in the area.

Table 6.6. Freshwater and Suspended Particulate Matter (SPM) input from the streams to Potter Cove, King George Island

Stream	Summer input of water to the cove (hm^3)	SPM ($mg\ l^{-1}$)	Reference
A	0.5	No data	44
		25–488	46
B	5.7	100–15000	45
		600–900	46

Table 6.7. Metal concentrations in SPM (± standard deviation) from the two streams studied close to Jubany Station (μg l^{-1})

Stream	Cd	Cr	Cu	Fe	Mn	Pb	Zn
A	<0.09	0.11 ± 0.09	5.6 ± 1.6	2764 ± 158	61.9 ± 0.9	0.20 ± 0.08	17.1 ± 0.4
B	<0.09	2.12 ± 0.23	60.8 ± 3.9	24,613 ± 1302	1293 ± 87	1.82 ± 0.06	26.9 ± 0.8

Table 6.8. Metal concentrations (± standard deviation) in surface freshwater sediments from Potter Cove, King George Island (μg g^{-1} dry weight)

Site	Cd	Cr	Cu	Fe	Mn	Pb	Zn
Lagoon	<0.09	3.0 ± 0.8	82.7 ± 6.4	30,526 ± 415	1418 ± 77	5.8 ± 0.5	63.4 ± 6.3
Stream A	<0.09	3.6 ± 0.8	58.6 ± 4.9	30,504 ± 1020	1292 ± 99	6.2 ± 0.5	78.8 ± 6.1
Stream A (debouch)	<0.09	3.4 ± 0.6	42.7 ± 4.2	28,400 ± 712	1041 ± 85	2.6 ± 0.4	58.6 ± 6.3
Stream B	<0.09	5.0 ± 0.7	59.3 ± 3.3	27,771 ± 995	1308 ± 115	3.8 ± 0.4	67.9 ± 3.0
Stream B (debouch)	<0.09	2.6 ± 0.4	48.5 ± 2.8	23,461 ± 874	935 ± 69	4.9 ± 0.5	52.6 ± 3.2

The values after the ± signs are standard deviations of three replicate measurements

4.3. Trace metals in Antarctic lichens

Lichens have been used as good bioaccumulators of atmospherically transported heavy metals by different researchers (47–53). Lichens have neither roots, nor other adsorptive structures or a waxy cuticle and for mineral nutrition are largely dependent on atmospheric depositions. Lichens act as bioaccumulators via their capacity to adsorb metals from the environment. Three mechanisms have been proposed: (a) intracellular uptake via an exchange process; (b) intracellular accumulation; and (c) trapping of metal-rich particles (54). In spite of that, there are still some gaps in the understanding of the overall process responsible for the metal uptake and accumulation in lichens.

Table 6.9. Trace elements contents in *Usnea aurantiacoatra* and *Usnea antarctica* from Potter Cove, King George Island (μg g^{-1} dry weight)

Species		Cd	Cu	Fe	Mn	Pb	Zn
U. aurantia-	mean ± sd	0.08 ± 0.01	4.2 ± 2.1	751 ± 289	25.1 ± 10.2	1.0 ± 0.9	11.1 ± 3.1
coatra	maximum	0.35	6.9	1264	37.3	2.4	15.9
n = 12	minimum	<0.02	2.0	425	7.4	<0.04	4.1
U. antarctica	mean ± sd	0.07 ± 0.01	6.2 ± 2.8	700 ± 255	32.5 ± 9.3	0.9 ± 1.9	12.4 ± 4.1
	maximum	0.30	10.9	1015	43.9	1.9	26.0
n = 12	minimum	<0.02	4.2	286	19.2	<0.04	7.2

n = number of specimens analyzed.

Data on the occurrence of trace metals in lichen species from different continents have been reported, but this kind of information is scarce for Antarctic species (29, 48, 52–56).

Most lichens growing in the King George Island seem to prefer semi-hydrophilic or dry conditions such as crees with local dry plains and ever drier habitats like small hills or peaks where snow disappears early in summer. Lichens grow well at sites where an adequate moisture is maintained because of snow. Wind-blown sea spray is one of the unfavourable factors for lichens; on the other hand high precipitation in the area seems to dilute the salinity. About 198 species of lichens from King George Island and Nelson Island were found and classified into three major groups based on the distribution pattern within Antarctica: (a) species known only from the maritime Antarctic; (b) species often reported from the maritime Antarctic, but hardly known from continental Antarctic; and (c) species known to occur in the maritime and continental Antarctic (57).

The objective of this preliminary study was to assess levels of Cd, Cu, Fe, Mn, Pb and Zn in two species of epilithic macrolichens: *Usnea aurantiacoatra* (only found in maritime Antarctica) and *Usnea antarctica* (found in maritime and continental Antarctica) (57). *Usnea antarctica* has an antarctic circumpolar distribution: continental and subantarctic regions, New Zealand and Tierra del Fuego (58). Table 6.9 shows the concentrations of Cd, Cu, Fe, Mn, Pb and Zn in specimens of the Antarctic lichens *Usnea aurantiacoatra* and *Usnea antarctica* in areas close to Jubany Station. Two different trends of trace metals accumulation have been identified, namely:

i) lichen species presented a similar capacity for Cd, Fe, Pb and Zn accumulation and no significant differences among them were found. This fact is reflected in the variance study ($p < 0.01$). A comparison of the obtained results was done through ANOVA. Cadmium contents determined in selected lichen species were significantly higher than those previously reported by Poblet *et al.* (mean ≤ 0.01 μg g^{-1}), but similar to those reported by Bargagli and Focardi for *Usnea antarctica* from the Admiralty Bay, King George Island (55, 56). On the other hand, levels of Cd found in lichens from King George Island were lower than those detected in species from continental Antarctica (see Table 6.10). Lead

Table 6.10. Cadmium and Pb contents (± standard deviation) in lichen species from Antarctica (μg g^{-1}, dry weight)

Lichen species	Area	n(*)	Cd	Pb	Reference
U. antarctica	King George Island – Admiralty Bay	3	0.07 ± 0.02	0.35 ± 0.06	55
U. antarctica	Deception Island Whaler Bay	3	0.05 ± 0.01	0.36 ± 0.07	55
U. antarctica	Graham Land – Paradise Harbour	3	0.52 ± 0.04	24.8 ± 5.5	55
U. antartica	N. Victoria Land	22	0.16 ± 0.09	2.63 ± 2.13	55
U. decussata	N. Victoria Land	25	0.22 ± 0.14	0.44 ± 0.25	55
U. antarctica	King George Island – Potter Cove (sampling station 1)	3	0.03	2.85	56
U. antarctica	King George Island – Potter Cove (sampling station 2)	3	0.02	1.10	56
U. aurantia-coatra	King George Island – Potter Cove (sampling station 2)	3	0.01	1.01	56

(*) n = number of specimens analyzed

levels detected in *Usnea antarctica* were almost three times higher than those reported by Bargagli and Focardi (55) for samples from King George Island (see Table 6.10), but the present results are in good agreement with those reported by Poblet *et al.* for Potter Cove (55, 56). The levels are significantly lower than those reported by several authors for different lichen species collected in other environments (48, 52). If one takes into account that lichens incorporate trace elements from the environment where they grow, the comparison provide just a general indication because the data are referred to lichen species from different ecosystems;

ii) a different trend has been identified for Cu and Mn, taking into account that *Usnea antarctica* showed higher ability to accumulate these metals than *Usnea aurantiacoatra.* A highly significant difference ($p < 0.01$) in Cu and Mn contents in both species was observed. Moreover, metal levels in lichen species are similar to those previously reported by other authors (51, 59).

A better understanding of the different trends in the accumulation of metals observed in *Usnea antarctica* and *Usnea aurantiacoatra* demand more studies, such as the estimation of the age of lichens, their growing rate and the composition of the soil and rocky substrates in contact with them.

4.4. Trace metals in antarctic molluscs

The study of certain organisms as pollutant bioaccumulators has several advantages over the chemical analysis of the abiotic compartments because they only accumulate the biologically available forms of the contaminants and they can be used for continued monitoring. Fluctuations in the pollutant concentrations are time-integrated; magnification afforded by bioaccumulation may be advantageous with respect to the accuracy and costs of analysis of trace pollutants close to the analytical LoDs (60). The use of an aquatic organism as biomonitor of metals in coastal water is useful only when the resulting picture of environmental contamination truly reflects ambient metal bioavailabilities (61).

Only certain organisms that accomplish a number of prerequisites prove to be adequate biomonitors, i.e.:

i) organisms should not exhibit significant movement or migration;
ii) organisms should be abundant and representative of the studied area;
iii) organisms should be easy to identify and collect;
iv) organisms should tolerate a wide range of salinity;
v) organisms should accumulate high levels of pollutants and a simple correlation should exist between the pollutant concentrations detected in their tissues and the average bioavailable pollutant concentrations in the studied area;
vi) kinetic studies of pollutants should be possible at the laboratory scale.

Most of the studies published on the suitability of organisms as pollution biomonitors have been carried out on invertebrates, mainly molluscs or crustaceans (62–64). Bivalve molluscs are known to accumulate trace metals in their tissues and have been frequently employed for monitoring coastal waters around the world (62). As a matter of fact, several authors have proposed coastal Antarctic molluscs as suitable biomonitors for the detection of environmental changes (65–67).

Even though some papers related to the occurrence of trace metals in molluscs from East Antarctic have been published, this kind of information is scarce for molluscs from the Antarctic Peninsula (26, 28, 65–70).

At Potter Cove, there are some molluscs that meet the prerequisites to be considered as a useful biomonitor, such as the clam *Laternula elliptica* and the gastropod *Nacella concinna*.

Laternula elliptica, one of the largest bivalves in the Antarctic marine environment, has been proposed as a biomonitor of trace metals in coastal waters (67, 70). Although gastropod molluscs have been less extensively used for monitoring trace metals pollution that mollusc bivalves, they show characteristics useful for these studies (61). The limpet *Nacella concinna* is the most abundant and often the only invertebrate found in the littoral zone of the Antarctic peninsula.

Keeping in mind the above scenario, trace elements were determined in Antarctic molluscs (bivalves and gastropod) from Potter Cove. Table 6.11 shows total Cd, Cu, Hg, Pb, and Zn contents in whole tissue of the two Antarctic molluscs species *Laternula elliptica* and *Nacella concinna* collected in Potter Cove. Mercury and Pb were not detected either in *Laternula elliptica* or in *Nacella concinna,* while Cd, Cu and Zn showed similar distribution trends in both species (their concentrations were significantly different). Trace metal concentrations in *Nacella concinna* detected in these studies have been significantly lower than those reported for gastropod from contaminated coastal environment, but higher than those found in limpets collected in the Antarctic Peninsula (see Table 6.12) (63, 71).

Table 6.11. Trace elements contents in molluscs from Potter Cove, King George Island (μg g^{-1}, wet weight)

Species	Specimen size		Trace metals contents				
			Cd	Cu	Hg	Pb	Zn
	small	Mean ± sd(*)	0.7 ± 0.4	3.6 ± 1.4	<0.04	<0.5	12.8 ± 2.8
L.	<50 mm	CV %	54	39	–	–	22
elliptica	medium	Mean ± sd(*)	0.7 ± 0.4	4.2 ± 1.4	<0.04	<0.5	14.8 ± 2.8
n = 10(**)	<100 mm	CV %	60	33	–	–	18
	large	Mean ± sd(*)	0.6 ± 0.3	4.2 ± 2.0	<0.04	<0.5	14.5 ± 2.6
	> 100 mm	CV %	52	48	–	–	18
	small	Mean ± sd(*)	3.0 ± 0.6	10.3 ± 2.0	<0.04	0.5	11.2 ± 1.1
N.	< 5 mm	CV %	19	19	–	–	10
concinna	medium	Mean ± sd(*)	3.3 ± 0.8	12.0 ± 2.4	<0.04	<0.5	11.8 ± 1.3
n = 10 (***)	<10 mm	CV %	24	19	–	–	11
	large	Mean ± sd(*)	3.5 ± 1.0	6.5 ± 1.8	<0.04	<0.5	10.0 ± 1.6
	>10 mm	CV %	28	27	–	–	16

(*) sd, standard deviation; CV %, coefficient of variation;
(**) n, number of samples analysed analyzed; (***) pooled sample (10 specimens per sample)

Table 6.12. Trace metals contents in *Nacella concinna* from Antarctic Peninsula

Area	n	Tissue	Cd	Cu	Zn	Reference
King George Island	2	foot	2.28	1.15	8.50	26
Fildes Bay			1.65–2.91	0.85–1.45	8.48–8.51	(**)
King George Island	2	foot	1.79	1.41	8.59	26
Potter Cove			1.57–2.00	1.08–1.74	8.08–9.09	(**)
Anvers Island	8	foot	1.21	0.63	11.05	26
Arthur Harbour			0.21–2.51	0.50–0.81	6.4–13.87	(**)
Anvers Island	14	foot	1.47	0.99	7.91	26
Biscoe Bay			0.74–2.6	0.55–1.36	4.23–11.70	(**)
Anvers Island	11	whole	13.4	8.37	67.2	23
Arthur Harbour						(*)
Anvers Island	13	whole	9.65	18.4	68.4	23
Arthur Harbour						(*)
Close to Palmer Station						

(*) μg g^{-1}, dry weight
(**) μg g^{-1}, wet weight

If one compares these results with those reported by Moreno et al., it is evident that the Cu contents are significantly higher, although they are coincident with the Cu values found in marine surface sediments from Potter Cove (26).

The relatively high level of Cd found in *Nacella concinna* with respect to *Laternula elliptica* may be due to its longevity or to the presence of metallothioneins involved in the mechanism of Cd detoxification in the muscle tissue (26, 72, 73).

Table 6.13. Trace metals contents in whole tissue of *Adamussium colbecki* and *Laternula elliptica* from Antarctica (μg g^{-1}, dry weight)

Specie	Area	Cd	Cu	Zn	Reference
A. colbecki n = 27	Syowa Station	34.0	10.6	83.7	68
A. colbecki n = 30	Terra Nova Bay	31.6 ± 7.0	5.6 ± 1.4	80.7 ± 20.8	65
A. colbecki n = 3	Explorers Cove	13.6 ± 1.9	13.8 ± 1.6	147.3 ± 31.5	66
L. elliptica n = 10	Potter Cove	3.2 ± 0.5	18.9 ± 3.2	67.6 ± 18.4	this study

n = number of specimens analyzed

Mean Cd, Cu and Zn contents in whole tissue of Antarctic bivalves are compared in Table 6.13. Copper levels detected in bivalves from Potter Cove were higher than those observed in bivalves from East Antarctica. The levels of Cu detected in *Laternula elliptica* are consistent with the high values of Cu observed in limpet, marine surface sediments and suspended particulate matter. While Zn levels were in agreement with the values reported for *Adamussium colbecki*, those of Cd were significantly lower in *Laternula elliptica*.

Cadmium concentrations reported for *Adamussium colbecki* from Terra Nova Bay and Syowa Station were almost one order of magnitude higher than levels found in *Laternula elliptica* from Potter Cove. The marked bioaccumulation of Cd found in molluscs from East Antarctica was related to two main reasons: *i)* the high bioavailability of this metal in the Antarctic marine environment; *ii)* the high potential of these organisms for metal bioaccumulation (67, 70, 74).

The high bioavailability of Cd in the marine environment from Terra Nova Bay was ascribed to upwelling of deep nutrient rich water nearshore (70). Cadmiun has a nutrient-like concentration gradient in the water column, lower close to the surface and increasing with depth (75). In the water column, Cd would be absorbed by the phytoplankton, during the summer bloom, thus becoming available for benthic molluscs in which the Cd is bioaccumulated. Ecological studies undertaken in the inner and outer part of Potter Cove showed that after seasonal ice retreat no phytoplankton bloom occured (35). In spite of the low planktonic production, rich benthic animal comunities are found in the area (29).

An important question to be answered as regards the marine ecological processes at Potter Cove is whether phytoplankton production provides the base for the benthic animal nutrition or if an alternative food source exists. Recent studies suggest that carbon sources may change in origin and composition along the year. With microphytobenthic primary productivity, input of terrigenous material and bacterial carbon are assumed to be the main sources during the spring-summer seasons, while resuspension of detritic organic carbon is suggested to play a key role fuelling secondary production during the fall and winter periods (35).

The low Cd levels observed in *Laternula elliptica* from Potter Cove might be due either to the low phytoplankton biomass available for consumption or to the possible utilization of alternative food sources. Likewise, alternative food sources

Table 6.14. Trace metals contents (mean ± sd or range) in *Adamussium colbecki* and *Laternula elliptica* from Antarctica

Specie	Area	Tissue	Cd	Cu	Zn	Reference
A. colbecki	Terra	Gill	6.8 ± 1.1	6.5 ± 2.2	114 ± 33.0	65
65–80 mm	Nova Bay	d.g.	142 ± 57	12.6 ± 3.3	74.9 ± 25.4	(*)
		Kidney	11.6 ± 3.3	4.0 ± 1.7	199 ± 89.0	
A. colbecki	Syowa	Whole	34.0	10.6	83.7	68
20–70 cm	Station		26.0–49.0	5.5–14.6	40.0–105	(*)
L. elliptica	Terra	Gill	12.5 ± 3.8	7.8 ± 0.8	207 ± 63.6	67
65–90 mm	Nova Bay	d.g.	48.2 ± 15.6	33.3 ± 5.1	119 ± 10.9	(*)
		Kidney	360 ± 150	8.0 ± 1.2	3300 ± 1400	
L. elliptica	King George	Whole	0.7 ± 0.4	4.2 ± 1.4	14.8 ± 2.8	28
50–100 mm	Island					(**)

d.g., digestive gland; whole, all soft parts combined.
(*) $\mu g\ g^{-1}$, dry weight
(**) $\mu g\ g^{-1}$, wet weight

such as resuspended benthic material and terrigenous material showed a low Cd content (see Tables 6.5 and 6.7).

Metal contents in bivalves usually depend on biological factors, such as body size (related to age) and sex (76). In *Adamussium colbecki* the contents of Cd, Mn and Zn reported by Mauri et al. decreased steadily with body weight in all cases (65). Capelli et al. observed a significant correlation between total Hg contents and shell length for *Adamussium colbecki* (69). In Figure 6.9 and 6.10, it is clearly shown that no correlation between metal contents and shell length was observed neither in *Laternula elliptica* nor in *Nacella concinna*. Similar results were reported by Honda et al. for *Adamussium colbecki* from Syowa Station (68).

It is well known that trace metals are taken up by specific tissues and/or organs and are thus accumulated selectively. Table 6.14 highlights the important differences found when gills, digestive glands and kidneys or whole tissues are analyzed. In consideration of this, studies employing gills, digestive glands, muscle and shells of *Laternula elliptica* and foot and middle intestinal gland of *Nacella concinna* are being undertaken.

5. Concluding comments

Results obtained during the first step of the project carried out at Potter Cove provide clear evidence that the EMP could be a good tool to characterize this system from the environmental point of view. Moreover, the occurrence of heavy metals in this environment has been verified, and their role in the involved biogeochemical cycles is being assessed. The characterization of the environmental condition of surface sediments from the Potter Cove is an important contribution if one

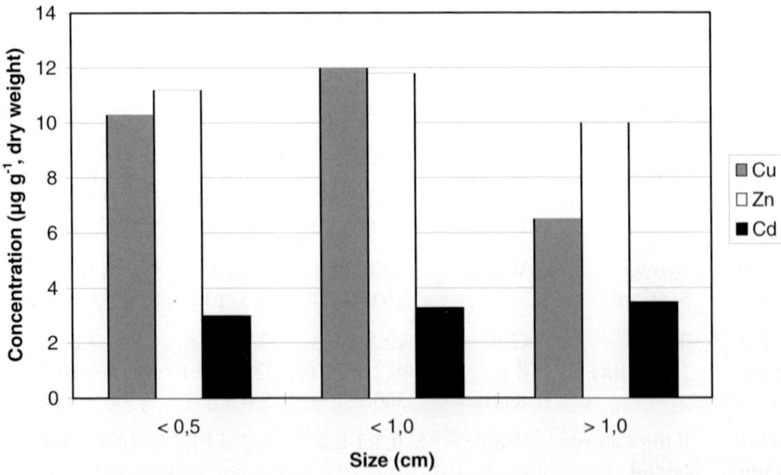

Figure 6.9. Trace elements contents (µg g^{-1} wet weight) in *Nacella concinna.*

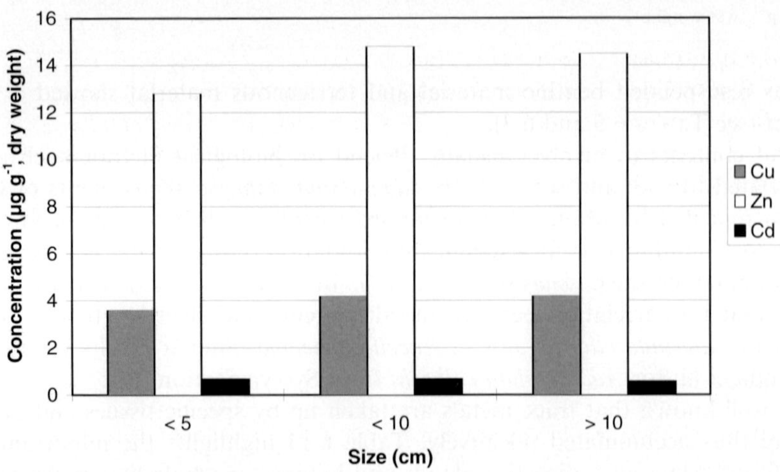

Figure 6.10. Trace elements contents (µg g^{-1}, wet weight) in *Laternula elliptica.*

keeps in mind the paucity of information available on this issue for the mentioned area.

The contents of Cu, Cr, Fe, Mn, Pb and Zn found in marine surface sediments of Potter Cove constitute natural background levels of the system and no evidence of metals from anthropogenic activities was gained. Metals detected in the marine surface sediments are consistent with the values measured in freshwater sediments and suspended particulate matter, thus suggesting that the contribution of metals from the streams is the main component of the inshore sediments. It was demonstrated that Stream B is the most important carrier of trace metals from the island to Potter Cove. Nevertheless, additional research will be necessary to confirm this assumption. Future studies about the relationship between metals concentration and granulometric composition of marine sediments as well as the analysis of soil and rock from Barton and Potter Peninsulas may elucidate the baseline geochemical pattern from the cove.

Lichens (*Usnea aurantiacoatra* and *Usnea antarctica*) could be useful as monitors to evaluate the Antarctic ecosystem and to identify patterns of the deposition of atmospherically transported heavy metals. Thus, it would be possible to study variations in the spatial distribution of these elements not only at the local, but also at the regional scale. Knowing the effect of the different local and regional growth rates and evaluating metal contents of the soil, it would be possible to perform metal biomonitoring in different zones of Antarctica and of the southern hemisphere.

The use of molluscs as biomonitors might provide a useful tool for environmental evaluation and diagnosis. However, reliable baseline data and a better knowledge of the variations in the different natural processes are needed for an appropriate use of these organisms. Analysis of metals contents in sea water, marine suspended particulate matter and phytoplankton and studies of feeding behaviour (alternate food sources such as resuspended organic matter and organic terrigenous input) are needed to evaluate the bioaccumulation in the species investigated.

All the information given above shows that the application of the EMP designed for the Potter Cove ecosystem was successful in this first step, although further work is needed to accomplish the objectives of the programme. Research in Antarctica is an interdisciplinary task that needs joint efforts among the different groups involved. The availability of an international data base is mandatory to compare results, disseminate and obtain information and avoid duplication of efforts at a continent-wide scale. Antarctica is important for scientific purposes and needs cooperative effort to preserve it uncontaminated for the use of future generations.

References

1. SCAR Bulletin No. 24. (1996), published by the SCAR at the Scott Polar Research Institute, Cambridge, England, 429–438.
2. SCAR Bulletin No. 123. (1996), published by the SCAR at the Scott Polar Research Institute, Cambridge, England, 1–8.
3. D. J. Enzembacher, Tourists in Antarctica. Number and trends, *Pol. Rec.*, 28 (1992), 17–22.
4. V. Smith, A Sustainable Antarctic: science and tourism, *Ann. Tour. Res.*, 21 (1994), 221–230.
5. R., Headland, Historical development of antarctic tourism, *Ann. Tour. Res.*, 21 (1994), 269–280.
6. D. J. Enzembacher, Tourism at Faraday Station. An Antarctic case study, *Ann. Tour. Res.*, 21 (1994), 303–317.
7. K. White, Tourism and the Antarctic economy, *Ann. Tour. Res.*, 21 (1994), 245–268.
8. R. W. Risebrough, W. Walker, T. T. Schmitdt, B. W. DeLappe, C. W. Connors, Transfer of chlorinated biphenyls to Antarctica, *Nature*, 264 (1976), 738–739.
9. H. Hidaka, R. Tatsukawa, Review, Environmental pollution by chlorinated hydrocarbons in the Antarctic., *Antarct. Rec.*, 71 (1981), 151–164.
10. E. W. Wolff, Review, Signals of atmospheric pollution in polar snow and ice, *Antarct. Sci.*, 2 (1990), 189–205.
11. E. W. Wolff, The influence of global and local atmospheric pollution on the chemistry of antarctic snow and ice, *Mar. Pollut. Bull.*, 25, (1992), 274–280.

12. M. C. Kennicutt II, S. McDonald, J. Sericano, P. Boothe, J. Oliver, S. Safe, B.J. Presley, H. Liu, D. Wolfe, T. L. Wade, A. Crockett, D. Bockus, Human contamination of the marine environment – Arthur Harbor and McMurdo Sound, Antarctica, *Environ. Sci. Technol.*, 29 (1995), 1279–1287.
13. SCAR/COMNAP, *Environmental Monitoring in Antarctica – A Discussion Document*, (1992), p. 37.
14. FMEEMA, *First Meeting of Experts on Environmental Monitoring in Antarctica, Buenos Aires*, (1992), 37.
15. SCAR/COMNAP, *Monitoring of Environmental Impacts from Science and Operations in Antarctica*, (1996), 43.
16. G. E. Batley, Collection, preparation and storage of samples for speciation analysis, in G. E. Batley (Ed.), *Trace Element Speciation: Analytical Methods and Problems* (1990), CRC Press, Boca Raton, Florida, 1–24.
17. K. R. Huibregtse, J. H. Moser, *Handbook for Sampling and Sample Preservation of Water and Wastewater*, US Department of Commerce, National Technical Information Service, PB-259946, (1976), 257–299.
18. M. Zief, J. W. Mitchell, Basic Aspects of Quantitative Ultratrace Analysis, 12–45, in P. J. Elving (Ed.), *Contamination Control in Trace Element Analysis* (1976), John Wiley & Sons, NY.
19. S. Caroli, O. Senofonte, S. Caimi, P. Pucci, J. Pauwels, G. N. Kramer, A pilot study for the preparation of a new reference material based on antarctic krill, *Fresenius J. Anal. Chem.*, 360 (1998), 410–414.
20. R. Capdevila, S. Comerci, *Historia Antártica Argentina*, Dirección Nacional del Antártico (Ed.) (1986), Buenos Aires, pp. 126.
21. Annual SCAR Report on Argentine Antarctic Scientific Activities. *Progress Report No. 38*, Dirección Nacional del Antártico-Instituto Antártico Argentino (Ed.) (1996), Buenos Aires, pp. 25.
22. M. C. Kennicutt II, T. J. McDonald, G. J. Denoux, S. J. McDonald, Hydrocarbon contamination on the Antarctic Peninsula, I, Arthur Harbor-subtidal sediments, *Mar. Poll. Bull.*, 24, (1992), 499–506.
23. M. C. Kennicutt II, S. J. McDonald, Marine disturbance-contaminants, foundations for ecological research west of the Antarctic Peninsula, *Antarct. Res. Ser.*, 70 (1996), 401–415.
24. R. Comes, L. Ventajas, S. Kocmur, Hidrocarburos polinucleares en el Mar de Weddell, Primer Symposium Español de Estudios Antárticos, Consejo Superior de Investigaciones Científicas (Ed.), (1987), 286–290.
25. L. Ventajas, Determinación de residuos de petróleo en aguas antárticas, Instituto Antártico Argentino, Report No. 315 (1985), pp. 12.
26. J. Moreno, J. M. Gerpe, V. Moreno, C. Vodopivez, Heavy metals in Antarctic organisms, *Polar Biol.*, 17 (1997), 131–140.
27. M. Scagliola, A. Poblet, C. Vodopivez, A. Curtosi, J. Marcovecchio, Distribución de metales pesados en sedimentos marinos superficiales de la Caleta Potter, Isla 25 de Mayo (Shetland del Sur), Antártida, *Terceras Jornadas de Comunicaciones sobre Investigaciones Antárticas.* Instituto Antártico Argentino (1994), Buenos Aires, 401–408.
28. C. Vodopivez, A. Curtosi, La utilización de *Laternula elliptica* y de *Nacella concinna* como indicadores de presencia y distribución de metales pesados en la Isla 25 de Mayo, (Antártida), *Terceras Jornadas de Comunicaciones sobre Investigaciones Antárticas* (1994), Buenos Aires, 393–396.
29. A. Poblet, M. Scagliola, C. Vodopivez, A. Curtosi, J. Marcovecchio, *Usnea aurantiacoatra*: un organismo integrador de metales peasados transportados atmosféricamente en ecosistemas de altas latitudes?, *Terceras Jornadas de Comunicaciones sobre Investigaciones Antárticas* (1994), Buenos Aires, 397–400.
30. H. Klöser, G. Mercuri, G. Laturnus, L. Quartino, C. Wiencke, On the competitive balance of macroalgae at Potter Cove (King George Island, South Setland), *Polar Biol.*, 14 (1994), 11–16.
31. N. Fourcade, Estudio geológico-petrográfico de Caleta Potter, Isla 25 de Mayo, Islas Shetland del Sur, Instituto Antártico Argentino, No. 8, (1960), pp. 15.
32. H. Klöser, G. Ferreyra, I. Schloss, G. Mercuri, F. Laturnus, A. Curtosi, Hydrography of Potter Cove, a small fjord-like inlet on King George Island (South Setland), *Estuar. Coast. Shelf Sci.*, 38 (1994), 523–537.
33. M. Roese, J. Speroni, M. Drabble, C. Pascucci, Medición de corrientes en Caleta Potter, Antártida, *Resúmenes de las Jornadas Nacionales de Ciencias del Mar*, Universidad Nacional de la Patagonia,

(1993), Puerto Madryn, pp. 210.

34. M. Roese, M. Drabble, Wind-driven circulation in Potter Cove. in C. Wiencke, G. Ferreyra, W. Arntz, C. Rinaldi (Eds.) *The Potter Cove Coastal Ecosystem. Synopsis of Research* (in press).
35. I. Schloss, H. Klöser, G. Ferreyra, A. Curtosi, G. Mercuri, E., Pinola, Factors governing phytoplankton and particulate matter variation in Potter Cove, King George Island, Antarctica, in B. Battaglia, J. Valencia, D. W. H. Walton (Eds.), *Antarctic Communities: Species, Structure and Survival* (1997), Cambridge University Press, Cambridge, 135–141.
36. J. E. Marcovecchio, V. J. Moreno, A. Pérez, Determination of heavy metal concentrations in the biota of Bahía Blanca, Argentina, *Sci. Tot. Environ.*, 75 (1988), 181–190.
37. J. F. Uthe, F. A. J. Armstrong, M. P. Stainton, Mercury determination in fish samples by wet digestion and flameless atomic absorption spectrometry, *J. Fish. Res. Bd. Can.*, 27 (1970), 805–811.
38. W. Salomons, U. Förstner (Eds.), *Metals in the Hydrocycle*, (1984), Springer-Verlag, Heidelberg, pp. 350.
39. H. S. Lenihan, J. S. Oliver, J. M. Oakden, M. A. Stephenson, Intense and localized benthic marine pollution at McMurdo Station, Antarctica, *Mar. Pollut. Bull.*, 21 (1990), 422–430.
40. H. S. Lenihan, Benthic marine pollution around McMurdo Station, Antarctica: a summary of findings, *Mar. Pollut. Bull.*, 25 (1992), 318–323.
41. B. Cosma, R. Frache, A. Mazzucotelli, F. Soggia, Trace metals in sediments from the Terra Nova Bay-Ross Sea, Antarctica, *Ann. Chim. (Rome)*, 81 (1991), 371–382.
42. I. A. Alam, M. Sadiq, Metal concentrations in antarctic sediment samples collected during the Trans-Antarctica 1990 Expedition, *Mar. Pollut. Bull.*, 26 (1993), 523–527.
43. S. Luoma, Processes affecting metal concentration in estuarine and coastal marine sediments, in P. Furness, P. Rainbow (Eds.), *Heavy Metals in the Marine Environment* (1990), CRC Press, Boca Raton, Florida, 51–66.
44. L. Varela, Estudio sobre el escurrimiento fluvial de arroyos de deshielo, Instituto Antártico Argentino, Report No. 419 (1994), 31–35.
45. L. Varela, Estimación del caudal sólido en el arroyo Potter, Caleta Potter, Isla 25 de Mayo, Shetland del Sur, *Cuartas Jornadas de Comunicaciones sobre Investigaciones Antárticas,* Dirección Nacional del Antártico, Instituto Antártico Argentino, Buenos Aires (in press).
46. S. Andrade, A. Poblet, A. Curtosi, C. Vodopivez, J. Marcovecchio, A. Pucci, Metales pesados en material particulado en suspensión de Caleta Potter, Isla 25 de Mayo, Shetland del Sur, *Cuartas Jornadas de Comunicaciones sobre Investigaciones Antárticas,* Dirección Nacional del Antártico, Instituto Antártico Argentino, Buenos Aires (in press).
47. F. D. Tomassini, K. J. Nieboer, E. Richarson, B. Grace, Determination of copper, iron, nickel and sulphur by X-ray fluorescence in lichens from the Mackenzie Valley, Northwest Territories and Sudbury District, Ontario, Canada, *Can. J. Bot.*, 54 (1976), 1591–1603.
48. J. Garty, M. Galun, S. Fuchs, N. Zisapel, Heavy metals in the lichens *Caloplaca aurantia* from urban, suburban and rural region in Israel (a comparative study), *Water, Air & Soil Pollut.*, 8 (1977), 171–184.
49. E. Nieboer, D. H. Richardson, L. Tomassini, Mineral uptake and release by lichens: an overview, *The Biologist*, 81 (1978), 226–246.
50. E. Nieboer, D. H. Richardson, Lichens as monitors of atmospheric deposition, 339–388, in S. J. Eisenreich (Ed.), *Atmospheric Pollutants in Natural Water* (1981), Ann. Arbor, Ann Arbor Science.
51. L. Folkenson, Interspecies calibration of heavy metal concentrations in nine mosses and lichens: applicability to deposition measurement, *Water, Air & Soil Pollut.*, 11 (1979), 253–260.
52. Y. Olmez, C. M. Gulovali, G. E. Gordon, Trace element concentrations in lichens near coal fired power plants, *Atmos. Environ.*, 19 (1985), 1663–1669.
53. R. Bargagli, Determination of metal deposition patterns by epiphytic lichens, *Toxicol. Environ. Chem.*, 18 (1989), 249–256.
54. M. L. Antonelle, P. Ercole, L. Campanella, Studies about the adsorption on lichens *Evernia prunastri* by enthalpimetric measurements, *Talanta*, 45 (1998), 1039–1047.
55. R. Bargagli, S. Focardi, Preliminary data on heavy metals in surface soil and macrolichens of Northern Victoria Land, in B. Battaglia, P. M. Bisol, V. Varotto (Eds.), *Proceedings of the Meeting on Antarctic Biology* (1994), Edizioni Universitarie Pavatine, Padova, 227–234.
56. A. Poblet, S. Andrade, M. Scagliola, C. Vodopivez, A. Curtosi, A. Pucci, J. Marcovecchio, The use

of epilithic antarctic lichens (*Usnea aurantiacoatra* and *Usnea antartica*) to determine deposition patterns of heavy metals in Shetland Islands, Antarctica, *Sci. Tot. Environ.*, 207 (1997), 187–194.
57. M. Inoue, Floristic notes on lichens in the Fildes peninsula of King George Island and Harmony Cove of Nelson Island, South Shetland Islands, The Antarctic, *Proceedings of the NIPR Symposium on Polar Biology*, 6 (1993), 106–120.
58. F. J. Walker, The lichen genus *Usnea* subgenus *Neuropogon*, *Bull. Brit. Mus. Nat. History (Bot.)*, 13 (1985), 65–73.
59. M. Saeki, K. Kunji, T. Seki, K. Sugiyama, T. Suzuki, S. Shishido, Metal burden of urban lichens, *Environ. Res.*, 13 (1977), 256–266.
60. W. G. Jones, K. F. Walker, Accumulation of iron, manganese, zinc and cadmium by the Australian freshwater mussel *Velesunio ambiguus (Phillipi)* and its potential as a biological monitor, *Austr. Mar. Freshw. Res.* 30 (1979), 741–751.
61. D. Phillips, Use of macroalgae and invertebrates as monitors of metal levels in estuaries and costal waters, in R. Furness, P. Rainbow (Eds.), *Heavy Metals in the Marine Environment* (1990), CRC Press, Boca Raton, Florida, 81–89.
62. E. D. Goldberg, V. T. Bowen, J. W. Farrington, G. Harvey, J. H. Martin, P. L. Parker, R. W. Risebrough, W. Robertson, E. Schneider, E. Gamble, The mussel watch, *Environ. Conserv.*, 5 (1978), 101–125.
63. G. Bryan, G. W. Potts, G. R. Forster, Heavy metals in the gastropod mollusc *Hiliotis tuberculata (L)*, *J. Mar. Biol. Assoc., UK*, 57 (1977), 379–385.
64. S. L. White, P. S. Rainbow, Heavy metal concentrations and size effects in the mesopelagic decapod crustacean *Systellaspis debilis*, *Mar. Ecol. Prog. Ser.*, 37 (1987), 147–151.
65. M. Mauri, E. Orlando, F. Regoli, Heavy metals in the Antarctic mollusc *Adamussium colbecki*, *Mar. Ecol. Prog. Ser.*, 7 (1990), 27–33.
66. P. A. Berckman, M. Nigro, Trace metal concentrations in the scallops around Antarctica: extending the Mussell Watch Program to the Southern Ocean, *Mar. Pollut. Bull.*, 24 (1992), 322–323.
67. M. Nigro, F. Regoli, R. Rocchi, E. Orlando, Heavy metals in Antarctic molluscs, in B. Battaglia, J. Valencia, D. W. H. Walton (Eds.), *Antarctic Communities: Species, Structure and Survival* (1997), Cambridge University Press, Cambridge, 409–412.
68. K. Honda, Y. Yamamoto, R. Tatsukawa, Distribution of heavy metals in Antarctic marine ecosystem, *Proceedings of the NIPR Symposium on Polar Biology*, 1 (1987), 184–197.
69. R. Capelli, R. V. Minganti, F. Fiorentino, R. Pellegrini, Mercury and selenium in *Adamussium colbecki* and *Pagothenia bernacchii* from the Ross Sea (Antarctica), *Ann. Chim.* (*Rome*), 81 (1991), 357–369.
70. M. Nigro, M. Mauri, F. Regoli, E. Orlando, Ecology and metal distribution in Antarctic molluscs, in B. Battaglia, P. M. Bisol, V. Varotto (Eds.), *Proceedings of the Meeting on Antarctic Biology* (1994), Edizioni Universitarie Pavatine, Padova, 235–255.
71. G. Nickless, R. D. Stenner, N. Terrile, Distribution of cadmium, lead and zinc in the Bristol Channel, *Mar. Poll. Bull.*, 3 (1972), 188–194.
72. G. B. Picken, The distribution, growth and reproduction of the Antarctic limpet *Nacella (Patinigera) concinna* (Strebel, 1908), *J. Exp. Mar. Biol. Ecol.*, 42 (1980), 71–85.
73. F. Nöel-Lambot, J. M. Bouquegneau, F. Frankenne, A. Distéche, Cadmiun, zinc and copper accumulation in limpets (*Patella vulgata*) caught off the Bristol Channel with special reference to metallothioneins, *Mar. Ecol. Progr. Ser.*, 2 (1980), 81–89.
74. A. Viarengo, L. Canesi, M. Pertica, G. Mancinelli, E. Ponzano, A. Mazzucotelli, M. Orunesu, Valutazione del contenuto in metalli (K, Na, Mg, Ca, Mn, Fe, Cu, Cd, Zn) in differenti tessuti di *Adamussium colbecki*: possibile ruolo delle metallotioneine nei processi di omeostasi e detossificazione dei metalli pesanti (Cu, Cd, Zn), in B. Battaglia, P. Bisol, V. Varotto (Eds.), *Atti del Primo Convegno di Biologia Antartica* (1991), Edizione Universitarie Pavatine, Padova, 335–342.
75. K. W. Bruland, R. P. Franks, Mn, Ni, Cu, Zn and Cd in the western North Atlantic, in C. S. Wong, E. Boyle, K. W. Bruland, J. D. Burton, E. G. Goldberg (Eds.), *Trace Metals in Sea Water,* (1983), Plenum Press, New York, 395–414.
76. C. R. Boyden, Trace element content and body size in molluscs, *Nature*, 251 (1974), 311–314.

Environmental Contamination in Antarctica
A Challenge to Analytical Chemistry
S. Caroli, P. Cescon and D.W.H. Walton, editors

Chapter 7

Biomethylation in the Southern Ocean and its contribution to the geochemical cycle of trace elements in Antarctica

Klaus Gustav Heumann

1. Introduction

Biomethylation is known as an important chemical process in the ocean, which can lead to volatile alkylated compounds of elements. Whereas biomethylation is the only substantial alkylation process for some heavy metals in the environment, in the case of halogens other Volatile Halogenated Organic Compounds (VHOCs) are also produced. Even if bioalkylation takes also place in fresh water systems, soil and sediments, the oceans must be assumed to be the major source of naturally alkylated elements, as investigations about VHOCs (1–5) and dimethyl-Hg (6, 7) have indicated. Volatile compounds, dissolved in the ocean, can usually be transferred into the atmosphere, this being an important mechanism in the geochemical cycle of elements and often contributing significantly to the global distribution of individual substances. Most of the methylated compounds are not very stable in the atmosphere because they can be decomposed, e.g., by OH radicals or photodissociation. It is possible that naturally produced iodinated and brominated VHOCs are able to reach the stratosphere, which would have a significant influence on ozone concentrations since iodine or bromine radicals formed act much more efficiently as a catalyst in the decomposition of ozone than chlorine radicals formed by Freons (8).

Biomethylation is therefore an important natural process. However, to follow natural biomethylation without any influence from anthropogenic sources, a remote environment must be chosen. Antarctica and its surrounding polar ocean is one of the few areas on earth not directly influenced by anthropogenic sources. This means, for example, that total heavy metal concentrations, which are very often of anthropogenic origin, do not dramatically exceed the small portion of biogenic methylated metal compounds, which normally prevents identification and quantification of biomethylated metal fractions. It is of general interest to determine how far biomethylation is involved in the geochemical cycle of elements and this can best be studied under the natural conditions of Antarctica. In addition, typical fingerprints of bioalkylated elements produced by polar marine species and their contribution to the global atmospheric concentration of trace substances is another important aspect needing investigation.

Biomethylation, or alternative bioalkylations of halogens and other non-metals, usually leads to permethylated or peralkylated products, which are relatively volatile and not very soluble in sea water. In the case of metals only Hg forms the permethylated compound Me_2Hg, where Me stands for the methyl group (7, 9). In the case of Cd and Pb only the bioproduction of ionic methylated substances ($MeCd^+$, Me_2Pb^{2+} and Me_3Pb^+) has been established so far (10, 11). These ionic methylated compounds are much more soluble in sea water than the permethylated ones and are therefore not so volatile. However, bioproduction takes place mainly in the upper layers of the ocean and it is also physically well known that substances with both hydrophobic (methyl group) and hydrophilic (positive charge) properties can be enriched at the surface layer of a water/gas (here ocean/atmosphere) interface (12). Such an enrichment allows sea spray to transfer these compounds into the atmosphere from where they can also be transported over land. Decreasing Cd concentrations, measured in surface snow of the Filchner Ice Shelf in Antarctica, are related to distance from the ice edge which, together with extremely high enrichment factors of Cd in these snow samples compared with the terrestrial Cd abundance, indicate that such a sea spray effect plays an important role in the distribution of some heavy metals over Antarctica (13).

There is no doubt that biomethylation of bromine and iodine takes place in the ocean (3–5). There is also significant evidence that chlorine is biomethylated in the ocean (14, 15), but bioalkylation of fluorine seems to be relatively unlikely. Other important non-metals, which were found to be biomethylated in the ocean are S and Se (16, 17). With respect to the chemical stability in sea water the following methylated heavy metal compounds can be expected: $MeCd^+$, $MeHg^+$, Me_2Hg, Me_2Pb^{2+}, Me_3Pb^+, Me_4Pb and Me_2Tl^+. Biogenic formation has now been demonstrated for both methylated Hg compounds as well as for the two ionic methylated Pb compounds (6, 7, 9). More recently, the first evidence of monomethyl-Cd in polar and non-polar ocean water as well as in Arctic fresh water samples has been also obtained (11). So far, Me_4Pb has not been shown to be biogenically formed, but Schedlbauer and Heumann recently obtained preliminary results which indicate the presence of very low concentrations of dimethyl-Tl in the ocean (18).

The transfer of volatile compounds from the ocean into the air can be described by two-phase models for the substance flux F from the aqueous into the gaseous phase. For example, by a simplified model described by Liss and Slater (19):

$$F = K\,(c_l - c_g/H) \tag{1}$$

where K is the substance specific exchange constant, c_l and c_g are the concentrations in the ocean and air, respectively, and H is Henry's constant. Equation (1) shows that analytical measurements of the volatile compound in the ocean surface and in the corresponding atmosphere over the ocean are necessary for flux determinations. However, the K value as well as Henry's constant must also be determined for conditions identical to the real ocean-air system. K and H values calculated for iodomethane under Antarctic polar sea water conditions come out as 0.102 and 0.117 m hr^{-1} (20), respectively, whilst for Me_2-Hg under non-polar conditions they are 0.113 and 0.15 m hr^{-1} (21, 22).

One major goal of this chapter on biomethylation in the Southern Ocean is to demonstrate that biomethylation takes also place under polar conditions and therefore contributes substantially to the concentration of VHOCs and heavy metals in the atmosphere. Preliminary calculations of the contribution of Me_2-Hg, iodomethane and bromofrom from polar oceans to the global atmospheric budget of Hg and VHOCs show the importance of biomethylation in polar regions. It can also be demonstrated that individual biological species, e.g., those of polar algae and bacteria, produce characteristic fingerprints of VHOCs and methylated heavy metals which can explain the complex distribution of these compounds in the Southern Ocean.

2. Analytical methods

2.1. Sampling and sample pretreatment

During expeditions with the German research vessel *Polarstern* sea water was continuously pumped under clean conditions from the front of the ship bow at a depth of 10 m into a laboratory, where samples were collected into precleaned PE bottles. Subsequently, the samples were filtered using a 0.45 μm pore sized filter and then analyzed as soon as possible. For the analysis of VHOCs a 10 l precleaned steel bucket was also used in some cases. Only ultra pure chemicals were used for the pretreatment of samples.

2.2. Determination of brominated and iodinated VHOCs and of methylated Hg

50–100 ml of the filtered sea water sample were introduced into the purging unit of the purge and trap/GC system (Figure 7.1). The volatile compounds were then stripped from the sea water sample by a He carrier gas and transferred into a cold trap. This works without additional sample pretreatment for the brominated and

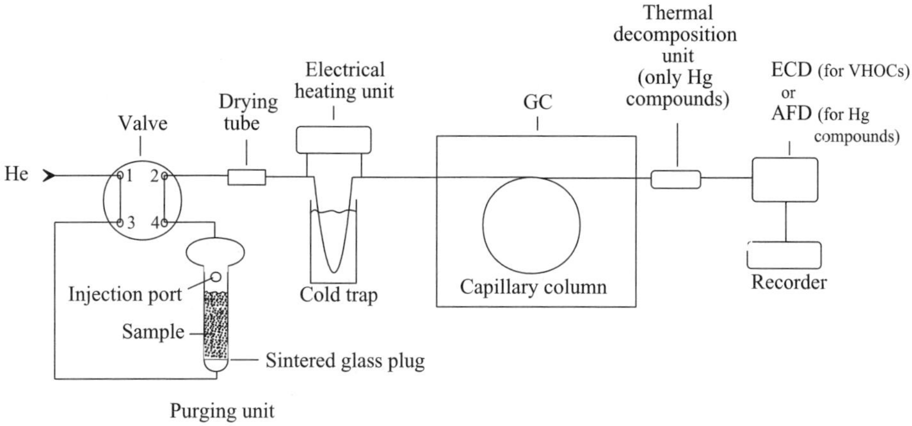

Figure 7.1. Schematic figure of the purge and trap-GC system for the determination of VHOCs and methylated Hg in sea water samples (7, 23).

iodinated VHOCs and for Me_2Hg as well. For the determination of $MeHg^+$ *in situ* derivatization by tetraethylborate was carried out, which forms methylethyl-Hg (MeEtHg). During the purging process, the drying tube, filled with K_2CO_3, prevented clogging of the cold trap (liquid N_2) by moisture. The trapped substances were then transferred into the capillary column of the GC system by thermodesorption and separation was carried out by using different capillary columns for VHOCs and alkylated Hg. The separated VHOC fractions were measured by an Electron Capture Detector (ECD), whereas the alkylated Hg compounds were first decomposed thermally at a temperature of about 830°C and then determined by an Atomic Fluorescence Detector (AFD). A more detailed description of these analytical procedures is presented elsewhere (7, 23).

Atmospheric VHOC concentrations were obtained by sucking air through an adsorption tube filled with Carbosieve. The adsorbed substances were then desorbed by stepwise heating of the tube up to 400°C and connecting the tube directly with the cold trap of the system shown in Figure 7.1 (23). Both methylated Hg compounds, $MeHg^+$ and Me_2Hg, could also be adsorbed on Carbosieve and, afterwards, desorbed at about 160°C. During desorption by a He gas flow $MeHg^+$ was adsorbed in a bottle filled with distilled water and Me_2Hg in a subsequently installed cold trap (24). The analysis of both Hg compounds could then be carried out in the same way as described for sea water samples.

Calibration was always performed by using standard solutions of the substances to be determined in degassed synthetic sea water. The detection limit for VHOCs depends on the substance because the ECD has different sensitivities for different halogenated compounds. The detection limits lie in the range of 0.01–0.25 ng l^{-1} for sea water and of 0.004–0.08 pptv for air samples (CH_3I–CH_2I_2), respectively. The detection limit for Me_2Hg and $MeHg^+$ determinations in sea water is 5 pg l^{-1} (as Hg) for both species.

2.3. *Determination of methylated Cd and Pb by DPASV*

Methylated Cd and Pb compounds were determined by Differential Pulse Anodic Stripping Voltammetry (DPASV). A polarographic analyser equipped with a rotating Hg film electrode, plated *in situ* on a glassy carbon substrate as the working electrode and an Ag/AgCl reference electrode were applied. 50 ml sea water samples were always used. In the case of the determination of methylated Pb, inorganic Pb^{2+} ions must first be separated by coprecipitation with $BaSO_4$. Using a potential of –1 V at pH 2, Me_3Pb^+ and Me_2Pb^{2+} were deposited, whereas at a deposition potential of –0.7 V only Me_2Pb^{2+} was collected at the electrode. This allows for both Pb species to be determined by subtraction. For Me_4Pb determination this compound can specifically be converted by ICl into Me_2Pb^{2+} and then also determined by DPASV. The procedures used for the determination of different methylated Pb species are similar to corresponding descriptions in the literature (25, 26). For all methylated Pb species the detection limit (as Pb) was 450 pg l^{-1}.

Electrodeposition of $MeCd^+$ took place at a potential of –1 V using sea water at its natural pH value of about 8. In this case inorganic Cd^{2+} ions must not be separated from monomethyl-Cd because the stripping potentials of both species are different by 112 mV, which allows monomethyl-Cd to be determined in the presence of the inorganic Cd^{2+}. The detection limit for $MeCd^{+}$ (as Cd) is 470 pg l^{-1}. A more detailed description of $MeCd^{+}$ determinations by DPASV is given elsewhere (11). Sampling of methylated Cd and Pb in the atmosphere was carried out by adsorption on Porapak material, which was also applied by other authors for alkyl-Pb compounds (27). Desorption took place at 150°C, followed by an adsorption of the methylated heavy metal compounds in water where they could be determined by DPASV as described above.

2.4. Determination of dimethyl-Tl

Dimethyl-Tl (Me_2Tl^{+}) was determined in 500 ml sea water samples by combining Isotope Dilution Mass Spectrometry (IDMS) with a species-specific extraction method (18). A ^{203}Tl-enriched spike solution was applied for the isotope dilution step and the production of positive thermal ions was used for mass spectrometric measurements. After species-unspecific Tl enrichment at an anion exchanger, inorganic Tl was oxidized to Tl(III) by a Br_2 solution and then separated by extraction with methyl *iso*-butyl ketone, whereas Me_2Tl^{+} remained in the aqueous phase. The detection limit for Me_2Tl^{+} is 440 pg l^{-1} (as Tl).

3. Bioalkylation of halogens and heavy metals in the polar ocean

3.1. Bioalkylation of Br and I by polar macroalgae

It has been known for some time that brominated and iodinated VHOCs, found in sea water and the corresponding marine atmosphere are mostly released from various biological species in the ocean (3, 28, 29). This is supported, e.g., by exceptionally high values of VHOCs in marine air samples from highly bioactive tropical and polar regions (30, 31). However, only by selective experiments with polar macroalgae could Schall et al. relate, for the first time, the release of distinct VHOC substances to definite polar organisms (32).

The experiments of Schall et al. were carried out with three different species of macroalgae (*Laminaria saccharina, Desmarestia aculeata, Fucus distichus*), which were collected at the shore site of the Kongsfjord on Spitsbergen at about 79° N, 12° E. *Fucus distichus* was the main species found in the eulitoral area (0–2 m depth) and the other two macroalgae were collected from a depth of 2–15 m in the fjord. Different samples of every macroalga were cultivated in incubation vessels (see Figure 7.2) under polar conditions with respect to the temperature and the irradiation at the place of collection. The release rates of volatile organo-Br and organo-I compounds were determined by using the purge and trap-GC-ECD system described in Section 2. Bromoform ($CHBr_3$), dibromomethane (CH_2Br_2), dibromochloromethane ($CHBr_2Cl$), dichlorobromomethane ($CHBrCl_2$), iodomethane (CH_3I, often also called methyl iodide), diiodomethane (CH_2I_2), 1-iodopropane ($CH_3CH_2CH_2I$) and 2-iodopropane (CH_3CHICH_3) were investigated in all experiments.

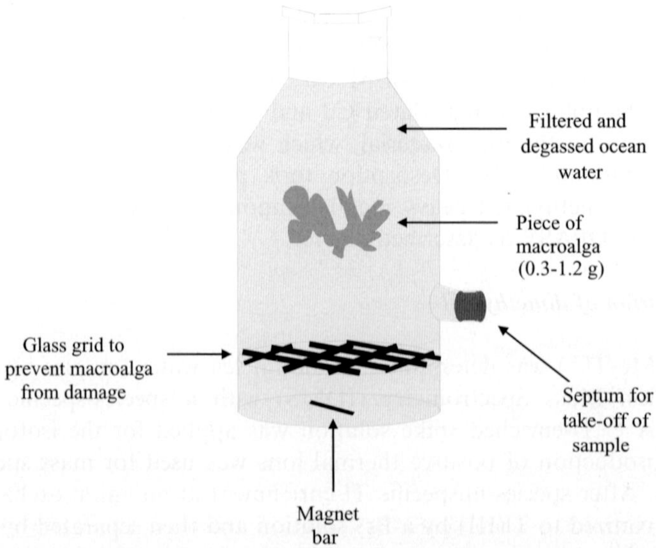

Figure 7.2. Incubation vessel (400 ml) used for selective experiments with polar macroalgae to determine the release of organo-Br, organo-I and methylated heavy metals (10, 32).

Representative results are shown in Figures 7.3 and 7.4 for brominated methanes and organo-I compounds, respectively, given in ng of substance released by 1 g of wet alga per day. As Figures 7.3 and 7.4 demonstrate, the different macroalgae show substantial differences in the release of the various brominated and iodinated

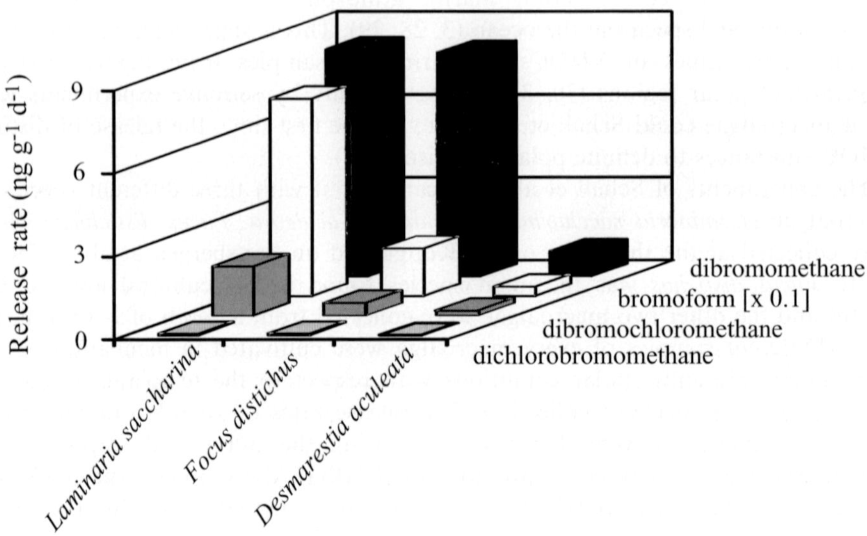

Figure 7.3. Mean releasing rates of brominated methanes by polar macroalgae (note that the bromoform concentration is ten times higher than represented in the figure) (32).

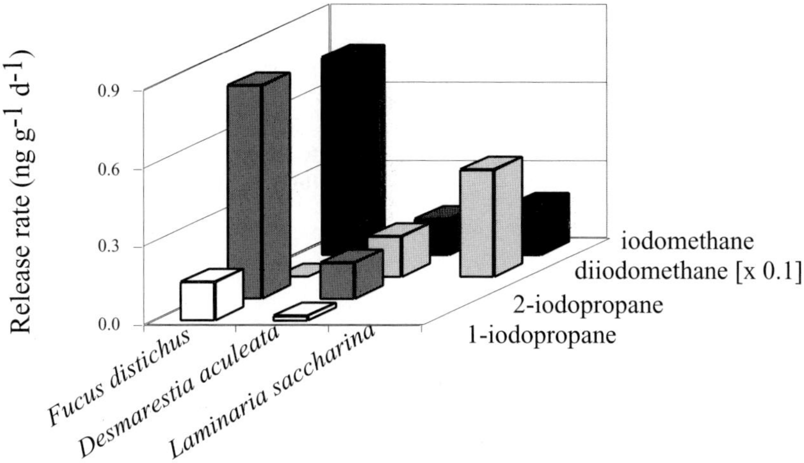

Figure 7.4. Mean release rates of volatile organo-I compounds by polar macroalgae (note that the diiodomethane concentration is ten times higher than represented in the figure) (32).

organic compounds. In the case of the bromomethanes the difference is due to the production rate, but for a single type of macroalga the release rates of the different brominated compounds relate one to another. Bromoform is released as the most abundant organo-Br substance by all macroalgae investigated and the following sequence is found:

$$CHBr_3 > CH_2Br_2 > CHBr_2Cl > CHBrCl_2 \tag{2}$$

Typical fingerprints, with substantial differences in the relative abundance of the various substances, are observed for the iodinated compounds. For example, release of 1- and 2-iodopropane was below the detection limit for *Laminaria saccharina*, but significant amounts of these substances were produced by *Fucus distichus* and *Desmarestia aculeata*. However, the most interesting result is that diiodomethane is the most abundant product from *Laminaria saccharina* and *Desmarestia aculeata*, whereas *Fucus distichus* produces only very low amounts of this iodinated compound. Iodomethane was very often the only iodinated VHOC which was determined in surface sea water and the marine atmosphere in the past (1, 29, 33). It was therefore assumed previously that iodomethane is the most abundant iodinated substance produced by marine biological species. However, the high release rates for diiodomethane from some macroalgae and the additional production of iodopropane show that other iodinated VHOCs must also be taken into account for I marine chemistry.

Further investigations by Laturnus et al. with Arctic macroalgae (34, 35) confirmed most of the results obtained by Schall et al. During these experiments the production of additional halogenated VHOCs, such as bromomethane (CH_3Br), bromoethane (C_2H_5Br), 1,2-dibromoethane (CH_2BrCH_2Br) and iodoethane (C_2H_5I), were proved and it was also shown that the formation of brominated and

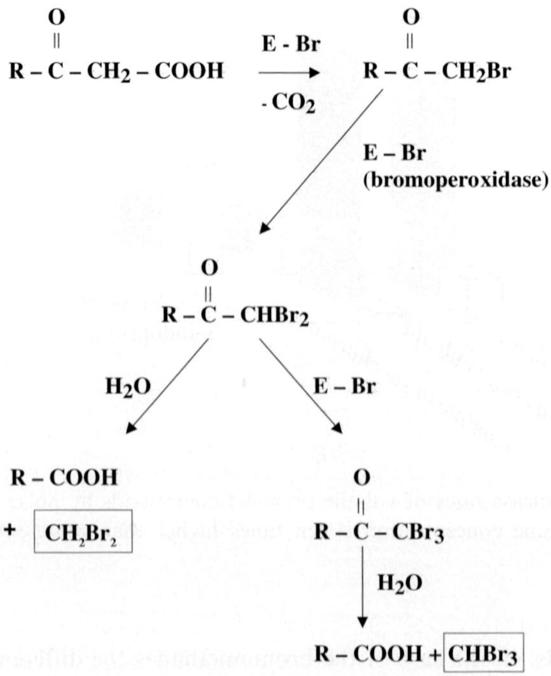

Figure 7.5. Possible way of biosynthesis of bromoform and dibromomethane (38).

iodinated compounds in Antarctic macroalgae is similar to that for macroalgae from the Arctic region (36, 37).

The formation of the bromomethanes $CHBr_3$ and CH_2Br_2 by marine organisms is believed to be an enzymatic bromination of ketone metabolites (38, 39). A possible biosynthesis of bromoform and dibromomethane, postulated by Moore (38), is presented in Figure 7.5. Dibromochloromethane and dichlorobromomethane are possibly directly produced by marine organisms or can also be formed by nucleophilic substitution of bromoform with chloride of sea water according to equations (3) and (4) (3, 23):

$$CHBr_3 + Cl^- \rightarrow CHBr_2Cl + Br^- \quad (3)$$

$$CHBr_2Cl + Cl^- \rightarrow CHBrCl_2 + Br^- \quad (4)$$

Indication that the same biogenesis produces bromoform as well as dibromomethane was obtained both from the good correlation between these substances during the experiments with polar macroalgae and also by the positive correlation (correlation factor 0.92 and 0.91) between the concentrations of these two compounds in Arctic and Antarctic surface sea water samples, respectively (23, 40). This is demonstrated in Figure 7.6 for the correlation of $CHBr_3$ with CH_2Br_2 in Antarctic sea water samples. It is interesting to note that extrapolation of the corresponding compensation curve does not fit the zero point, but shows a significant CH_2Br_2 concentration for $CHBr_3$ at 0 ng l^{-1}, which confirms the subsequent

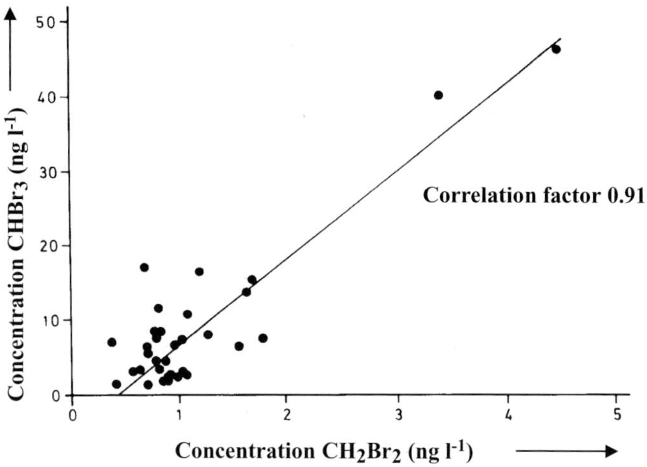

Figure 7.6. Correlation between bromoform and dibromomethane in Antarctic surface sea water samples (40).

bromination steps represented by Figure 7.5. Good correlation also between bromoform and the two different chlorobromomethanes $CHBr_2Cl$ and $CHBrCl_2$, respectively, in Antarctic surface ocean water demonstrates that all these brominated methanes have probably the same or similar pathways of formation (40).

Even if Mehrtens and Laturnus found positive correlation of the brominating and iodinating activity for an Arctic macroalga (41), such a correlation could not be observed between iodomethane and dibromomethane – and also not between iodomethane and the other brominated methanes – in Antarctic surface ocean water, as can be seen from Figure 7.7 (20). A similar negative correlation was also found in Arctic sea water from Kongsfjord on Spitsbergen (23). These results suggest different biogenic geneses for the brominated methanes, on the one hand, and for iodomethane, on the other hand. White assumed that iodomethane could be produced in algae by a reaction of dimethyl sulfonium ions and inorganic iodide, both present in these marine species (42). In this connection it is also remarkable that in north-south concentration profiles of surface sea water of the Atlantic Ocean from Europe to the south polar sea the brominated methanes have shown acceptable positive correlation with chlorophyll-*a* as a general indicator for bioactivity in the ocean, which was not found for iodomethane (see Figure 7.8 and 7.9) (5). The completely different fingerprints of iodinated VHOCs, which have been produced by different polar macroalgae (see Figure 7.4), demonstrate that the genesis is different for different types of macroalgae and it is also different for the various iodinated compounds.

3.2. Production of bromomethanes by polar ice algae

Sturges et al. collected ice algae from the platelet ice layers located at the ice-seawater interface beneath 2–2.5 m of congelation ice at McMurdo Sound during November

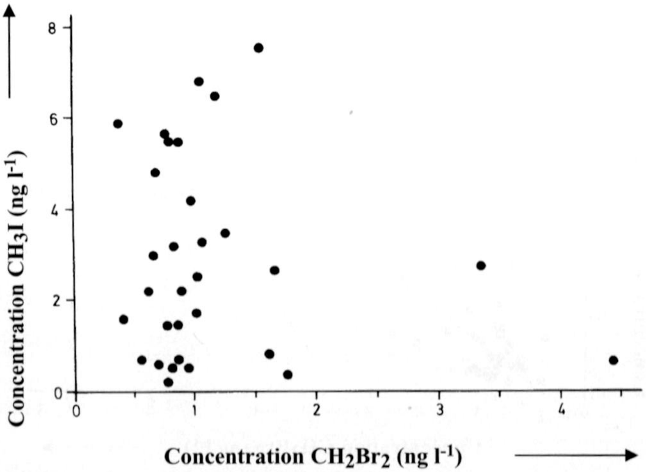

Figure 7.7. Correlation between dibromomethane and iodomethane in Antarctic surface sea water samples (20).

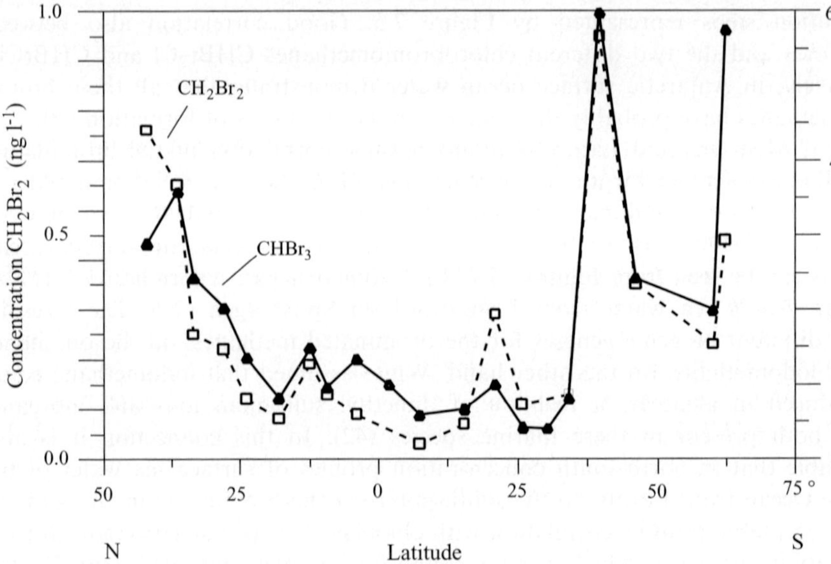

Figure 7.8. North-south profile of dibromomethane and bromoform in surface sea water samples of the Atlantic Ocean from 42° N to the Southern Ocean at 62° S (measurements during November and December) (5).

(late spring in Antarctica) (43). They diluted the samples with filtered sea water and stored them at –1.5°C under irradiation conditions representative for under-ice illumination. *Nitzschia stellata* was the predominant ice alga at two sampling sites, *Porosira pseudodenticulata* was a virtual monoculture at a third sampling site.

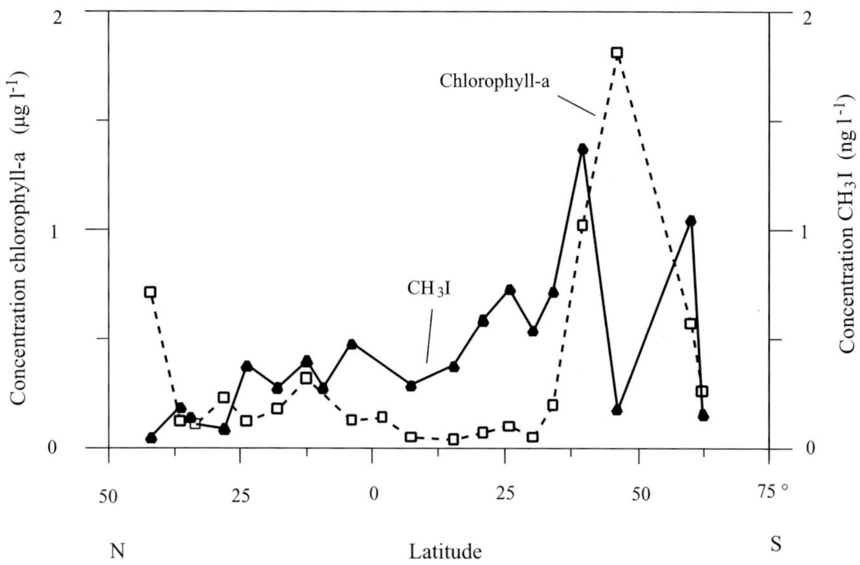

Figure 7.9. North-south profile of iodomethane and chlorophyll-*a* in surface sea water samples of the Atlantic Ocean from 42° N to the Southern Ocean at 62° S (measurements during November and December) (5).

The results of analyzing different bromomethanes are represented in Figure 7.10 together with the amount of chlorophyll-*a* determined in these samples. In addition, concentrations of corresponding compounds from the ice-edge at McMurdo and those determined in the mid-Atlantic by Class and Ballschmiter (3) are also plotted. Comparison of the mid-Atlantic data with the sea water sample from the Antarctic ice-edge show an enrichment of most of the analyzed bromomethanes in the polar ocean water, which indicates a release of these compounds from the ice algae. The different samples of ice algae show significant variations in the bromomethane release, where only the sub-samples NITZ 1.1 and NITZ 1.2, on the one hand, and NITZ 2.1 and NITZ 2.2, on the other hand, agree reasonably well. Contrary to the polar macroalgae (see Figure 7.3), bromoform was not the most abundant brominated methane in all these ice alga samples. In the case of sub-samples NITZ 2.1 and NITZ 2.2 the mixed chlorinated and brominated methanes were analyzed to have the highest concentration. Moreover, sub-samples NITZ 1.1 and NITZ 1.2, with the same type of ice alga, show a quite different sequence with bromoform as the most abundant compound. The distribution of bromomethanes released from *Porosira pseudodenticulata* (PORO 1.1) is again different when compared with the samples from *Nitzschia stellata* and does not agree with the sequence found for macroalgae (equation (2)). It does also not agree with the abundance distribution of bromomethanes in the open Southern Ocean (Table 7.1) as well as with findings at other locations, e.g., in the mid-Atlantic (first set of columns in Figure 7.10).

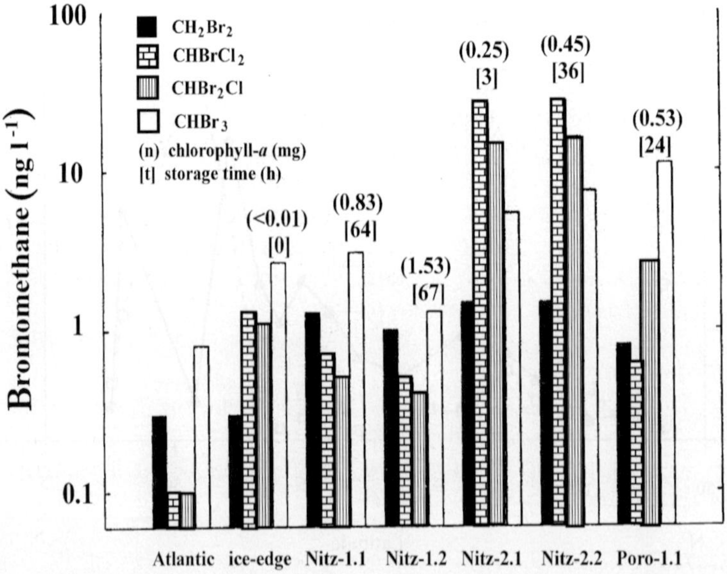

Figure 7.10. Production of bromomethanes by Antarctic ice algae (43) (NITZ 1.1 and 1.2 and NITZ 2.1 and 2.2 are sub-samples; NITZ *Nitzschia stellata*, PORO *Porosira pseudo-denticulata*; Atlantic data from (3)).

Table 7.1. Concentration of bromomethanes and iodinated alkanes in the surface sea water of the open Southern Ocean compared with a coastal site on Spitsbergen (5, 20, 23, 40)

Compound	Concentration range ($ng\ l^{-1}$)			Mean concentration ($ng\ l^{-1}$)		
	Antarctic spring to summer[1)]	Antarctic summer to autumn[2)]	Arctic late summer[3)]	Antarctic spring to summer	Antarctic summer to autumn	Arctic late summer
$CHBr_3$	1.4–47.2	<0.05–15.0	4.7–49.8	6.2	4.0	17.9
CH_2Br_2	0.4–4.4	<0.03–1.1	0.7–6.8	0.9	0.6	2.7
$CHBr_2Cl$	0.1–4.3	0.06–1.5	<0.03–2.4	0.6	0.4	1.1
$CHBrCl_2$	0.2–2.4	<0.03–2.3	0.07–0.6	0.3	0.3	0.2
CH_3I	0.2–7.5	<0.01–1.3	<0.01–1.5	2.6	0.3	0.5
CH_2I_2	n.d.	<0.2	<0.3–3.4	n.d.	<0.2	1.7
CH_2ICl	<0.04–0.2	<0.01–0.3	0.1–0.7	<0.04	0.1	0.3
$CH_3CH_2CH_2I$	n.d.	<0.05–0.3	<0.06–1.0	n.d.	0.2	0.4
CH_3CHICH_3	n.d.	n.d.	<0.07–2.1	n.d.	n.d.	0.6

n.d. = not determined

[1)] Samples collected around the Antarctic Peninsula (60–67° S, 54–69° W) from mid-October to mid-December 1987

[2)] Samples collected on an expedition from Puerto Madryn to Cape Town (62–73° S, 59° W–13° E) from December 1991 to March 1992

[3)] Samples collected in the Kongsfjord on Spitsbergen in September 1992

3.3. Investigations on possible bioalkylation of halogens by Antarctic bacteria

There is little information about possible biomethylation of halogens by polar bacteria. Schall carried out experiments with bacteria obtained from phytoplankton decaying processes (44). Here, a mixture of various diatom species, isolated from surface water of the Weddell Sea was cultivated at 1°C. After the algae had reached the stationary growth stage, they were stored in the dark for one week to increase the number of bacteria, which grow well on the collapsing algal populations. After filtration of the algal material the cells were crushed and again used as substrate for the bacteria growth.

An exponential increase of viable bacteria cells was observed but, even after about 13 days, where the number of bacteria had increased by a factor of approximately 2000, no significant increase in the initial concentrations of bromomethanes and iodinated alkanes (CH_3I, CH_2I_2, CH_2ClI, $CH_3CH_2CH_2I$, and CH_3CHICH_3) in the samples was observed. Further investigations are necessary to determine if this result is characteristic for polar bacteria.

3.4. Biomethylation of heavy metals by polar macroalgae

Biomethylation of the heavy metals Hg, Pb, and Cd by Arctic macroalgae was investigated in similar experiments as described before for brominated and iodinated compounds (10). All these experiments were carried out with the natural total (inorganic) heavy metal content in the corresponding polar sea water, which was approximately 0.8, 9 and 3 ng l^{-1} for Hg, Pb, and Cd, respectively. Representative mean release rates from seven different macroalgae are shown in Figure 7.11. In all cases no production of $MeCd^+$ could be found and only three of the seven macroalgae investigated were able to release trimethyl-Pb. The other possible methylated Pb compounds, Me_2Pb^{2+} and Me_4Pb, were also not detectable. It is remarkable that both methylated Hg compounds, $MeHg^+$ and Me_2Hg, were formed by four macroalgae, but two other macroalgae were only able to release $MeHg^+$. Another macroalga, *Desmarestia aculeata*, only produced Me_2Hg. These results are an important indication of the fact that the existence of $MeHg^+$ in the ocean is not necessarily a consequence of Me_2Hg decomposition in sea water. As in the formation of iodinated compounds the results of Figure 7.11 demonstrate that the mechanism of formation of methylated heavy metals must be different between different macroalgae and is also dependent on the heavy metal type. Otherwise, it cannot be explained how some of the macroalgae are able to methylate Pb or to permethylate Hg and others are not.

The biogenic production of methylated heavy metals was found to be strongly influenced by irradiation (10). The optimum release rate was only achieved in experiments under irradiation conditions comparable with the location where the macroalgae were growing. For example, *Fucus distichus*, which grows in surface sea water down to 5 m, produces methylated Hg only under irradiation conditions comparable with this water depth, and *Laminaria saccharina*, normally found at a depth of 2–15 m, has its highest production rate at irradiation conditions which correspond to 6–10 m. By doubling the natural concentration of Hg^{2+} ions in

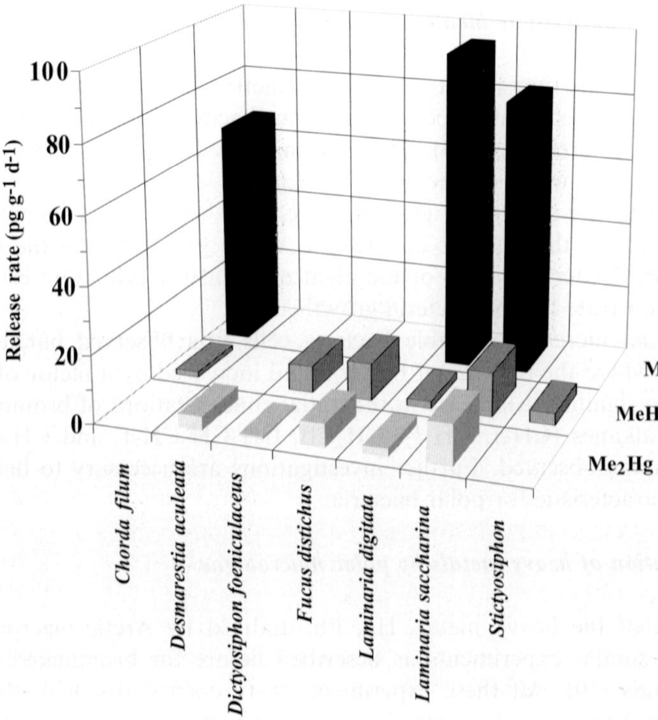

Figure 7.11. Mean release rates of methylated Hg and Pb compounds by seven different polar macroalgae (10).

polar sea water through addition of $HgCl_2$, the releasing rate for $MeHg^+$ and Me_2Hg increased by a factor of 5–10. However, if $MeHg^+$ is added instead of Hg^{2+}, Me_2Hg production is dramatically increased by a factor of 14–33. It must therefore be assumed that, in principle, $MeHg^+$ can be biogenically transformed into Me_2Hg, which further complicates the interpretation of primary sources of methylated Hg compounds by their concentration found in ocean water.

3.5. Biomethylation of heavy metals by polar bacteria

Recently, Pongratz and Heumann have shown that besides methylated Hg and Pb also monomethyl-Cd can be detected in the polar ocean (10, 11, 24). Because the experiments with polar macroalgae did not show any production of methylated Cd, it must be assumed that other marine organisms are responsible for the existence of $MeCd^+$ in the ocean. Incubation experiments with isolated mixed and pure bacterial cultures of polar origin, collected in the Weddell Sea, were therefore carried out and the production of methylated heavy metal compounds was followed during the growth of these bacteria. The total (inorganic) heavy metal concentration of the natural sea water from the Weddell Sea, used in these experiments, was about 1, 9 and 3 ng l^{-1} for Hg, Pb, and Cd, respectively. The concentration of the methylated compounds was always below the detection limit in the beginning of the growth experiments. More experimental details are described elsewhere (45).

The mixed bacterial cultures were obtained after phytoplankton decaying processes and isolation under certain conditions. The production of methylated heavy metals was then followed with time by the increase of organic carbon (C_{org}). Production of Me_3Pb^+ and $MeCd^+$, but no formation of methylated Hg was observed by the mixed bacterial cultures, as can be seen from the representation in Figure 7.12. On the other hand, the pure bacterial cultures, isolated in the form of slant cultures, produced both methylated Hg species, $MeHg^+$ and Me_2Hg, as well as Me_3Pb^+ and $MeCd^+$. This is demonstrated by the results summarised in Figure 7.13 and in Table 7.2. Here, optical density was used to follow the growth of the bacteria. The major product of Hg methylation by these pure bacterial cultures was Me_2Hg, which was preferably formed in the stationary period of growth. In addition, it is interesting that no significant dependence of the production of methylated Hg on the temperature (experiments at 4 and 25°C) was found.

The results of these experiments demonstrate that, in principle, bacteria from the polar ocean are also potential sources for the production of methylated heavy metals. Since methylation of Cd was not observed by macroalgae but by bacteria, monomethyl-Cd may be normally a product of marine bacteria. Bacteria show specific fingerprints in the release of methylated heavy metal compounds, as the results with the mixed bacterial cultures have shown, where methylation of Hg could not be observed. It is also known that bacteria are able to demethylate methylated heavy metals, as was found for organo-Hg (46). Also Schedlbauer followed the change of $MeHg^+$ concentration in sea water samples in the presence of marine

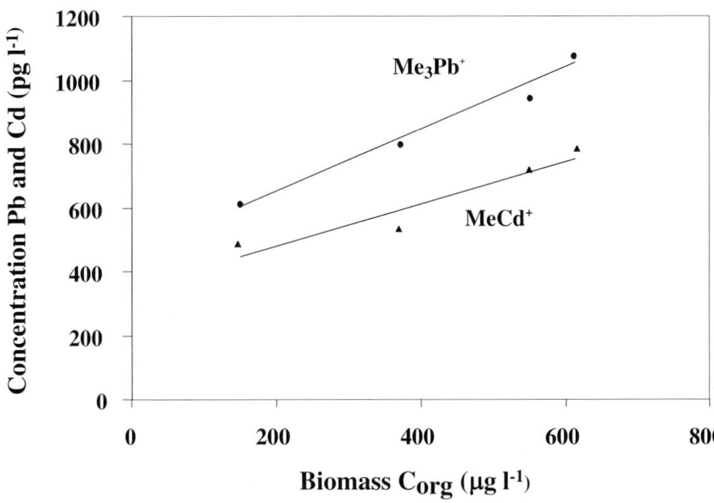

Figure 7.12. Biomethylation of Cd and Pb by isolated mixed cultures of polar bacteria from the Weddell Sea during their growth at 2°C (45).

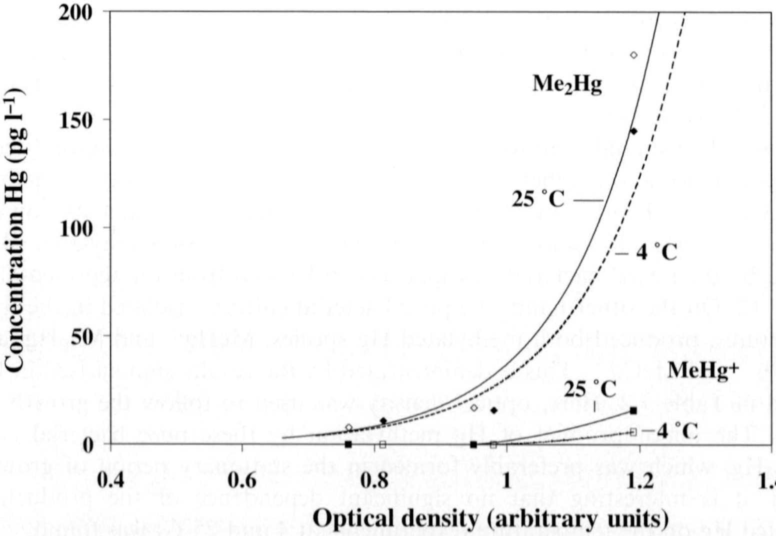

Figure 7.13. Biomethylation of Hg by isolated pure bacterial cultures from the Weddell Sea during their growth at 4 and 25°C, respectively (45).

Table 7.2. Methylation of Cd and Pb by different strains of isolated pure bacterial cultures from the Weddell Sea after a growth for 7 days at 4°C (the optical density, corresponding to the values in arbitrary units in Figure 7.13, was in the range of 1.2–1.9) (45)

Bacterial strain	Concentration (pg l^{-1})	
No.	Me_3Pb^+	$MeCd^+$
1	<LoD	559
2	<LoD	1337
3	<LoD	797
4	710	1065
5	922	948
6	994	799
7	760	494

LoD = Limit of Detection

mixed cultures of algae and bacteria (47). He found a significant decrease of added $MeHg^+$ with increasing growth of algae, as Figure 7.14 demonstrates. However, this took place at a concentration level of methylated Hg, which was a factor of 10^2 to 10^3 higher than the natural concentration in the surface water of oceans (see Figure 7.19–21). The results of these experiments therefore cannot answer the

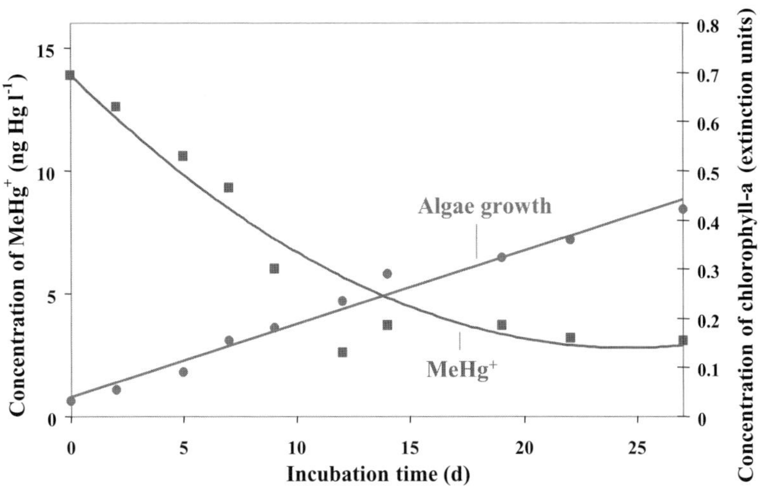

Figure 7.14. Biogenic decomposition and/or incorporation of $MeHg^+$ by incubation experiments with mixed algae and bacterial cultures from the North Sea ($MeHg^+$ and a mixed culture of cultivated algae and bacteria were added to ocean water samples) (47).

question whether decomposition of $MeHg^+$ by bacteria or incorporation into these biological species is responsible for the decreasing concentration in sea water. The concentration curve for $MeHg^+$ in Figure 7.14 suggests a possible equilibrium between decomposition (or incorporation) and formation (or release) of monomethyl-Hg, but this remains to be proved. All these results demonstrate the complexity of the system and shows how complicated the interpretation of measured concentrations of methylated heavy metals in the ocean is.

Another important indication that marine bacteria are essentially involved in the methylation of heavy metals are depth concentration profiles of Me_2Hg, $MeHg^+$, Me_3Pb^+, and $MeCd^+$, which were determined in the South Atlantic (7, 24). The concentration profile for methylated Hg is represented in Figure 7.15 and shows similarities with the one for bacteria measured during the same expedition by Simon et al. (48). Figure 7.16 demonstrates that similar concentration profiles as for methylated Hg are also obtained for trimethyl-Pb and monomethyl-Cd and that, in addition to this, substantial amounts of methylated heavy metals can be found in deeper sea water layers. Contrary to this, the chlorophyll-*a* concentration decreases, as expected, to a value of nearly zero at 200 m depth, even if in the upper sea water layers chlorophyll-*a* shows its maximum at the same depth as found for Me_3Pb^+ and $MeCd^+$. The only biological species able to produce methylated heavy metals, even in the deeper ocean, are bacteria.

3.6. Biomethylation of heavy metals by polar phytoplankton

Different Antarctic ocean water samples were collected in the Bellinghausen Sea at water depths where the chlorophyll-*a* concentration was at a maximum. In these

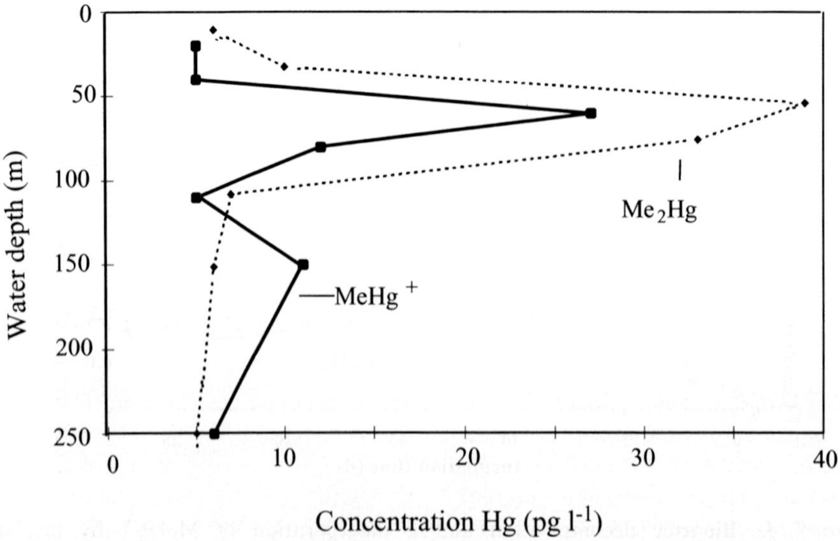

Figure 7.15. Depth profile of $MeHg^+$ and Me_2Hg in the South Atlantic at 53° 30′ S, 9° 00′ E (determined in November 1993) (7).

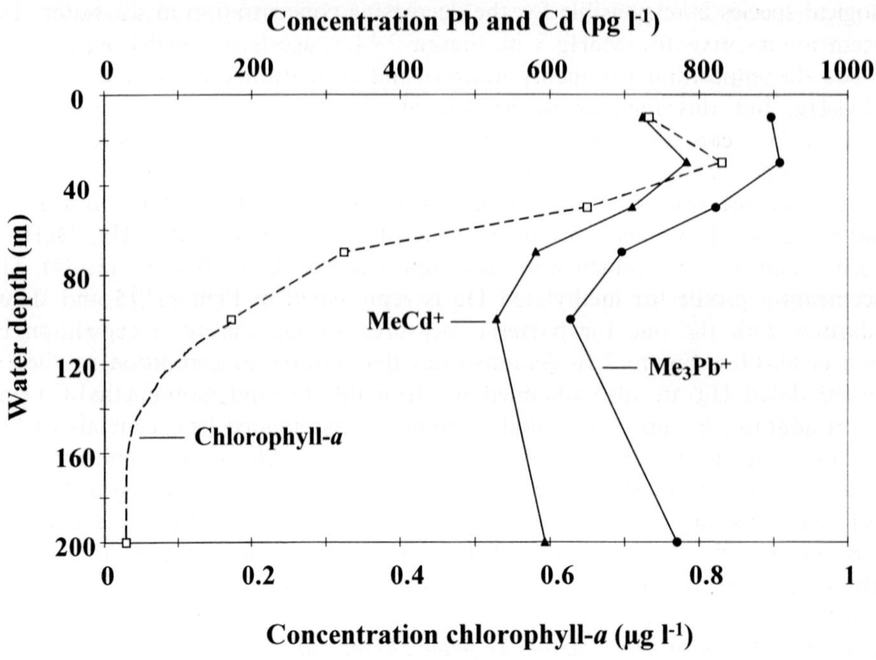

Figure 7.16. Depth profile of Me_3Pb^+, $MeCd^+$ and Chlorophyll-*a* in the South Atlantic at 46° 30′ S, 15° 20′ W (determined in December 1993) (24).

samples, with their natural content of phytoplankton, the production of methylated heavy metals was followed for three days. In seven different sets of samples, collected at latitudes between 58° S and 69° S, five samples showed significant formation of Me_3Pb^+, but only in one sample was there a distinct increase in both methylated Hg compounds and an occurrence of $MeCd^+$ just above the detection limit (24). Further experiments must therefore be carried out to better evaluate the contribution of phytoplankton to the total production of methylated heavy metals in the polar sea.

4. Geochemical cycle of I in Antarctica and its influence on I overabundances in Antarctic meteorites

In 1985 Dreibus et al. reported a mysterious I overabundance in Antarctic meteorites analyzed by radiochemical neutron activation analysis (49). Heumann et al. could confirm these results determining I in different specimens of Antarctic eucrites as well as in high-Fe and low-Fe chondrites by Negative Thermal Ionisation Isotope Dilution Mass Spectrometry (NTI-IDMS) (50). A similarly high enrichment effect was not found for Cl and Br. By analysing different types of Antarctic rocks, a significant enrichment of I was found in surface layers compared with the interior of these samples and, in addition, this tendency could also be observed in Antarctic meteorites (51). From these findings it was assumed that atmospheric I can interact with the surface of Antarctic rocks and meteorites leading to I overabundances in the surface layers of these materials. Moreover, it must be assumed that a volatile I compound, with its primary origin in the polar sea, is responsible for this incorporation of I in the surface layers of rocks and meteorites.

To find the I compound, which may be responsible for this I overabundance in Antarctic meteorites and rock surfaces, I speciation was carried out in the Antarctic marine atmosphere. A system with different filters, which selectively absorbs various I species (particulate I, I_2 and HI, HOI, and organo-I), was applied in connection with NTI-IDMS for I speciation (52). A representative result, using always two filters in sequence for absorption of the same I species, is shown in Figure 7.17. All I species were found in the concentration range of 0.1–0.8 ng I $m^{-3}$of Antarctic air, with I_2 and HI as the most abundant fraction. However, because Antarctic meteorites are preferably found on the blue ice fields in inner Antarctica, the organo-I compounds with their much higher stability compared with the inorganic species I_2, HI, and HIO, must play the most important role for the interaction with the surface of Antarctic rocks and meteorites. The atmospheric life-time of iodomethane was determined to be a few days (53). Due to this relatively long atmospheric life-time it must be assumed that organo-I compounds, and here preferably iodomethane, are transported into the inner of Antarctica where they can be decomposed, e.g., by photodissociation:

$$CH_3I \rightarrow CH_3^{\cdot} + I^{\cdot} \quad (5)$$

The result of a possible photodissociation corresponding to equation (5) is a reactive I radical which can easily react with surfaces of rocks (especially if

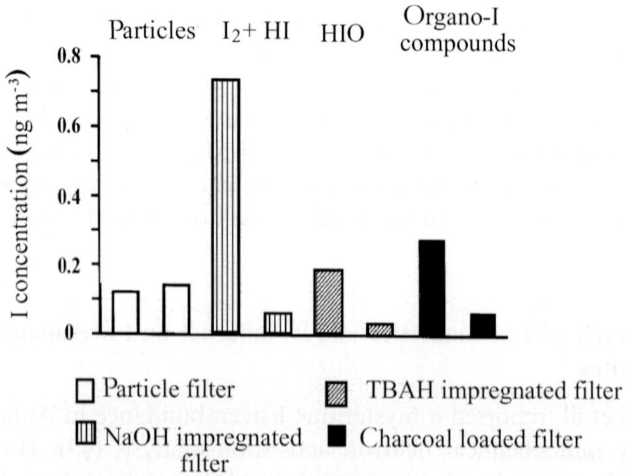

Figure 7.17. Determination of I species in the Antarctic marine atmosphere with a filter system for species-specific absorption (52).

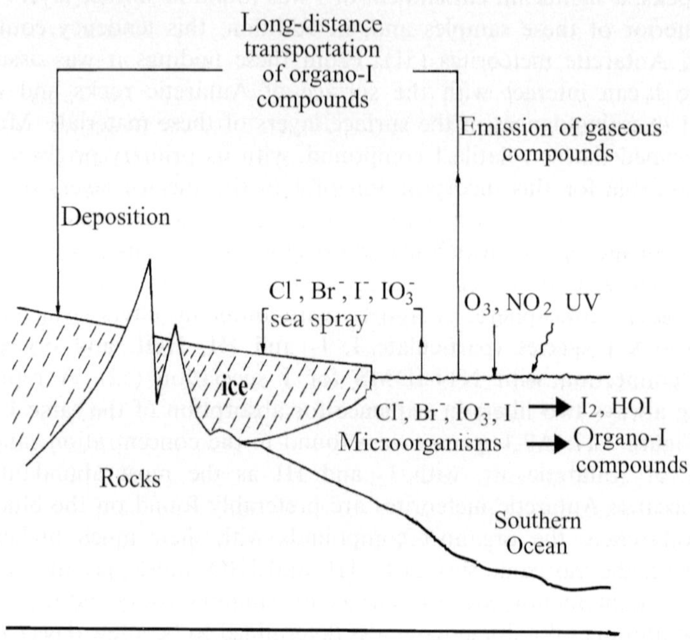

Figure 7.18. Geochemical I cycle for Antarctica with respect to the role of organo-I species (13).

photodissociation is catalyzed by adsorption of CH_3I on solid surfaces) or can be deposited as an inorganic I compound on the ice shield. Figure 7.18 summarizes all these results to give a geochemical I cycle for Antarctica. Sea salt particles from sea spray, which preferentially affect the region near the ice-edge, contain mainly

I^- and IO_3^-, where IO_3^- is more abundant, as further investigations by Wimschneider and Heumann have shown (54). In addition, volatile inorganic and organic I compounds, such as HI, I_2, HIO, and CH_3I, are transferred from the polar ocean into the atmosphere. The volatile inorganic I species have relatively short life-times and should therefore not be transported over long distances. On the other hand, the more stable organo-I compounds can be transported into inner Antarctica.

5. Concentration of bromomethanes and iodinated hydrocarbons in surface sea water of the Southern Ocean

5.1. Concentrations during different seasons and comparison with other regions

Mean concentrations of different bromo- and iodomethanes as well as of the two isomers of iodopropane, measured in surface sea water at a depth of about 10 m at different sites of the open Southern Ocean, are summarized in Table 7.1. Results for coastal sea water of Spitsbergen are also listed for comparison. Even if the measured concentrations at the various locations differ significantly one from the other (in Table 7.1 the minimum and maximum value, measured in the same area, can be seen from the listed range), the mean values of the different bromomethanes exactly follow the abundance sequence obtained by incubation experiments with polar macroalgae (see equation (2)). Unlike these incubation experiments with macroalgae, the total content of bromomethanes in Antarctic sea water is the result of the contribution of the different marine species, which means that, on average, there should be a comparable mechanism of formation for these compounds independent of the biological species involved. The measurements carried out during the period from Antarctic spring to summer (October to December) show higher concentrations compared with determinations carried out during the period from the Antarctic summer to the beginning of autumn (December to March). This could be expected by the higher bioactivity during spring and in early summer compared with later seasons.

The same seasonal variation can also be observed for iodomethane. However, the different iodinated hydrocarbons do not show the same sequence in their abundance for the different areas where the measurements have been carried out. Whereas iodomethane was found to be the most abundant iodinated compound in the open ocean of Antarctica, diiodomethane was more abundant in the coastal water of the Arctic Ocean. This shift in the abundance distribution found in Arctic coastal water may be due to the macroalgae which contribute significantly in coastal areas to the bioalkylated VHOCs. In this connection it is also interesting to mention that in the same samples, for which the north-south concentration profile of iodomethane is shown in Figure 7.9, the CH_2ICl content usually exceeded that of CH_3I (5). From the different abundance distribution of iodinated hydrocarbons at various locations in the ocean it follows that the different biological species may release completely different fingerprints of iodinated compounds. This confirms also the finding that even different types of polar macroalgae show various fingerprints in the production of iodinated hydrocarbons (see Figure 7.4).

Bromomethanes at single locations in the South Atlantic normally show lower concentrations than or similar to those measured for the Southern Ocean. For example, Class and Ballschmiter determined the bromoform and dibromomethane concentration at 6° S, 6° W to be 0.8 and 0.3 ng l^{-1} (3), whereas Abrahamson and Klick measured higher contents of 4.5 and 1.3 ng l^{-1} at 52° S, 6° W (55), respectively. The latter authors also reported low concentrations of iodochloromethane and iodopropane in the range of 0.07–0.13 ng l^{-1} in an Antarctic sea water sample. That diiodomethane is the dominant iodinated compound in coastal waters, especially during spring, was found by Klick on the Swedish coast (56). This confirms the results listed in Table 7.1 for Arctic coastal water.

5.2. North-south concentration profiles

Extended north-south concentration profiles for CH_2Br_2, $CHBr_3$, and CH_3I in surface sea water of the Atlantic Ocean from 42°N southwards to the Southern Ocean at 62° S (expedition from Bremerhaven to south of Cape Town) are represented in Figures 7.8 and 7.9. In Figure 7.9 the Chlorophyll-*a* content is also plotted. A good positive correlation between bromoform and dibromomethane exists (Figure 7.9), which again confirms the assumption that both compounds are produced by the same mechanism (see postulated way of biosynthesis in Figure 7.5). On the other hand, the iodomethane concentrations do not agree well with those of bromomethanes and also do not fit the chlorophyll-*a* curve. This again shows the problem of characterizing the primary sources of VHOCs in ocean water.

The mean concentrations of bromomethanes of this north-south profile are lower by a factor of 1.5–3.6 compared with the concentrations listed in Table 7.1 for the Southern Ocean. On the other hand, the mean values for iodomethane are nearly identical with 0.4 ng l^{-1} for the north-south profile of Figure 7.9 and 0.3 ng l^{-1} for the Southern Ocean, respectively. More investigations must be carried out to better identify the real parameters which result in differences in the production of brominated and iodinated hydrocarbons in the polar regions compared with the other oceans.

5.3. VHOCs in meltwater ponds on the McMurdo Ice Shelf

Bioalkylation does not only take place in the ocean, but also in fresh water systems where biological species exist which are able to produce such compounds. During summer meltwater ponds are formed at many places in Antarctica. The green colour of some meltwater ponds on the McMurdo Ice Shelf indicates high biological activity and cyanobacteria were found to be the dominant organisms in these ponds.

During January 1994 samples from seven of these meltwater ponds were collected at different times and the concentration of iodinated and brominated VHOCs were analyzed (57). For each of the ponds a mean value for the different VHOCs, measured at different times in January, was calculated and the range of the means is listed in Table 7.3. The most spectacular result was that

Table 7.3. Range of mean concentrations of brominated and iodinated hydrocarbons in seven different meltwater ponds on the McMurdo Ice Shelf (57)

Compound	Concentration range (ng l^{-1})
$CHBr_3$	2.5–8.6
CH_2Br_2	0.9–3.0
$CHBr_2Cl$	0.4–1.7
$CHBrCl_2$	0.4–2.0
$BrCH_2CH_2Br$	1.1–9.3
CH_3I	0.6–7.4
CH_2I_2	4.8–19.7
CH_2ICl	0.7–3.4

1,2-dibromoethane, formerly used as additive in leaded petrol, was found to be in the same concentration range as bromoform. This was the first time that biogenic production of this brominated compound could be extensively measured. The biogenic origin of 1,2-dibromoethane was confirmed in experiments with Antarctic macroalgae by Laturnus et al. (36). However, there was no positive correlation in the individual meltwater ponds between 1,2-dibromoethane and bromoform, contrary to the situation between bromoform and dibromomethane. The $CHBr_3/CH_2Br_2$ ratio of mean concentrations varied between 2.5 and 3.5. This is significantly lower than the ratio of 4–10, normally found in polar ocean water samples (23, 40). This difference is probably due to the different dominant biological species producing such brominated compounds in the fresh water ponds compared with the polar ocean.

Diiodomethane was the most abundant iodinated hydrocarbon in all meltwater ponds, which was also observed in coastal waters of the Arctic Ocean (Table 7.1). Iodomethane and iodochloromethane were determined in a comparable concentration, where in three of seven ponds the CH_2ICl concentration exceeded that of CH_3I. This again demonstrates that iodomethane is not the only species to play an important role in the biogeochemical cycle of I in polar regions.

6. Concentration of methylated heavy metals in surface sea water of the Southern Ocean

6.1. East–west concentration profiles

Figure 7.15 represents east-west concentration profiles (from right to left of the graph) of methylated Hg, Pb, and Cd in surface sea water samples, which were determined during an expedition from Cape Town to Punta Arenas from mid-December 1993 to mid-January 1994 (7, 24). In addition, the chlorophyll-*a* concentrations are also plotted, which have been measured by Hirch and Bathmann (58).

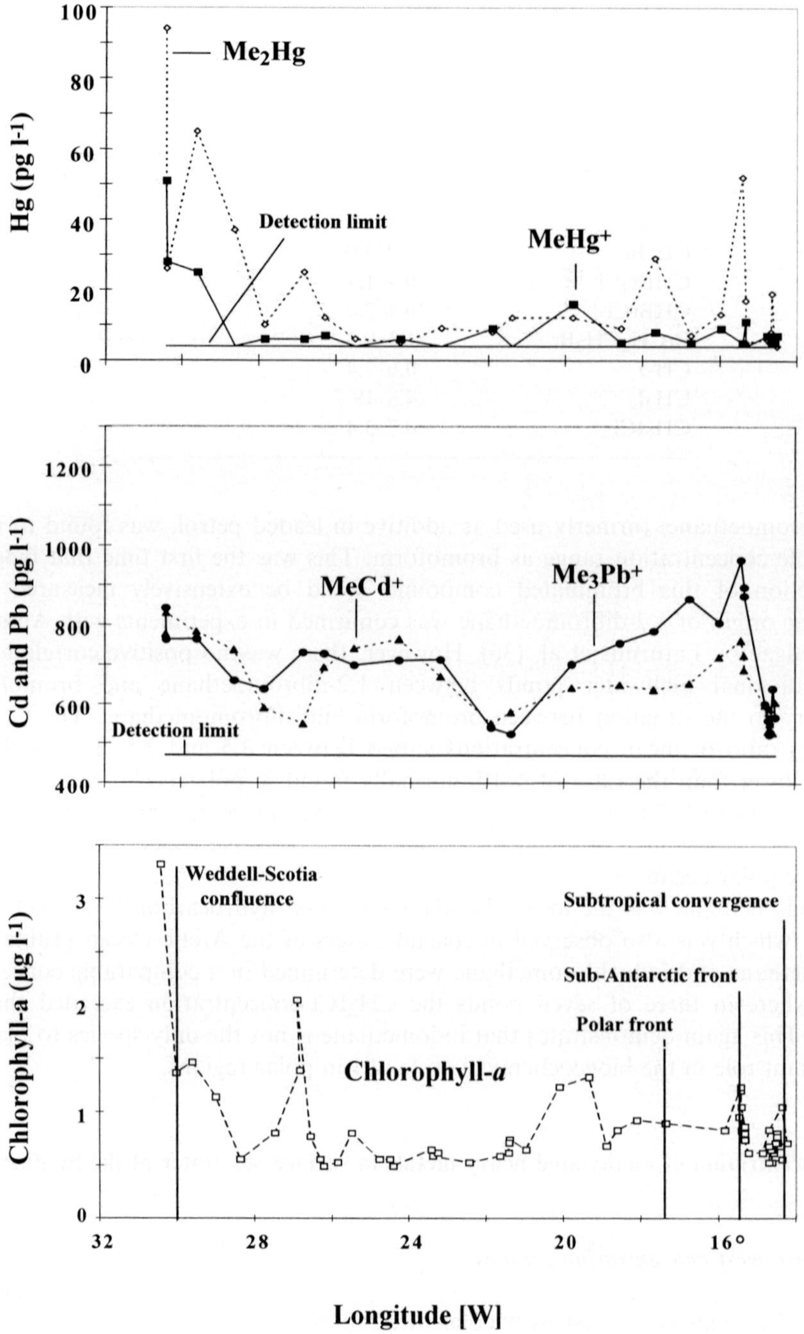

Figure 7.19. East-west concentration profiles of methylated Cd, Pb and Hg and of Chlorophyll-*a* in surface sea water samples on an expedition from Cape Town to Punta Arenas (7, 24) (chlorophyll-*a* concentrations from (58)).

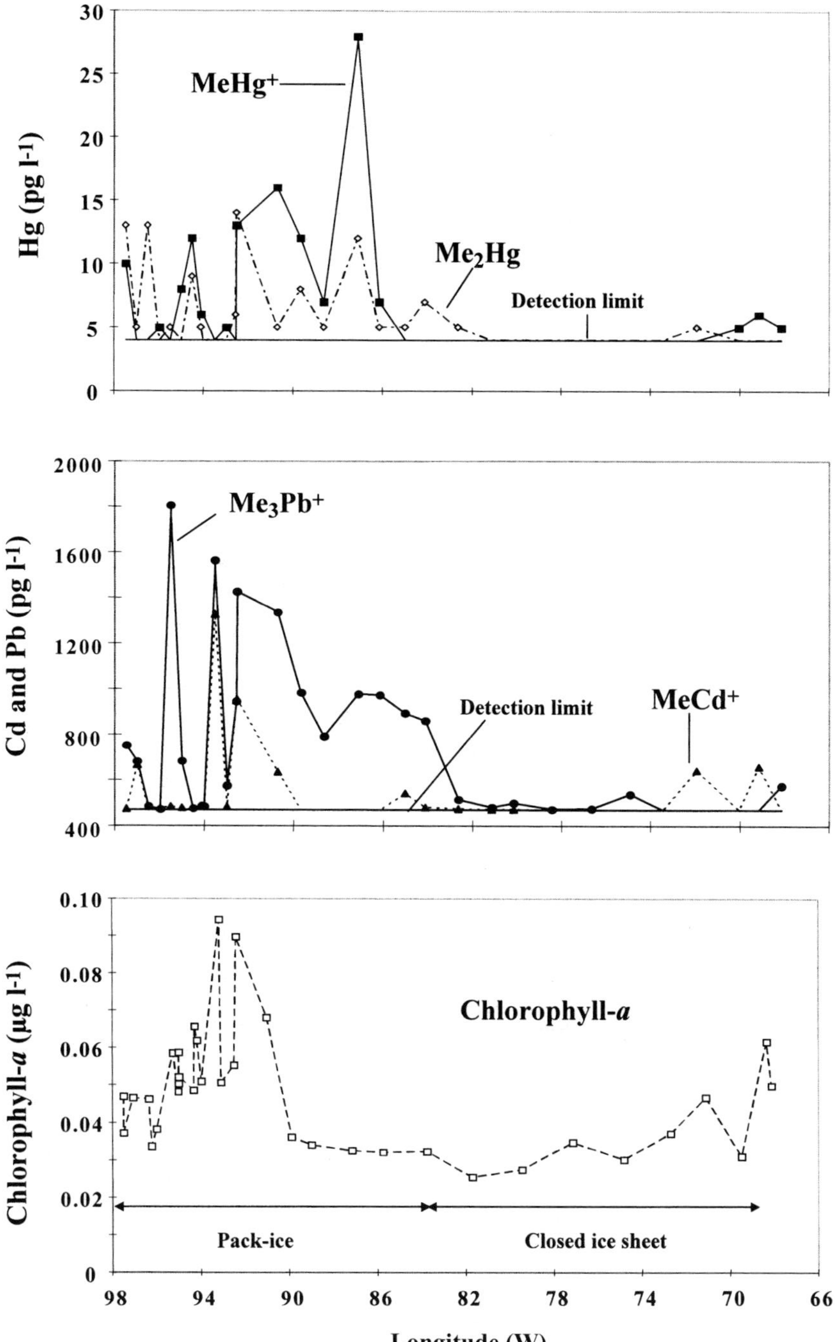

Figure 7.20. West-east concentration profiles of methylated Cd, Pb and Hg and of chlorophyll-*a* in the Southern Ocean west of the Antarctic Peninsula (7, 24) (Chlorophyll-*a* concentrations from (59)).

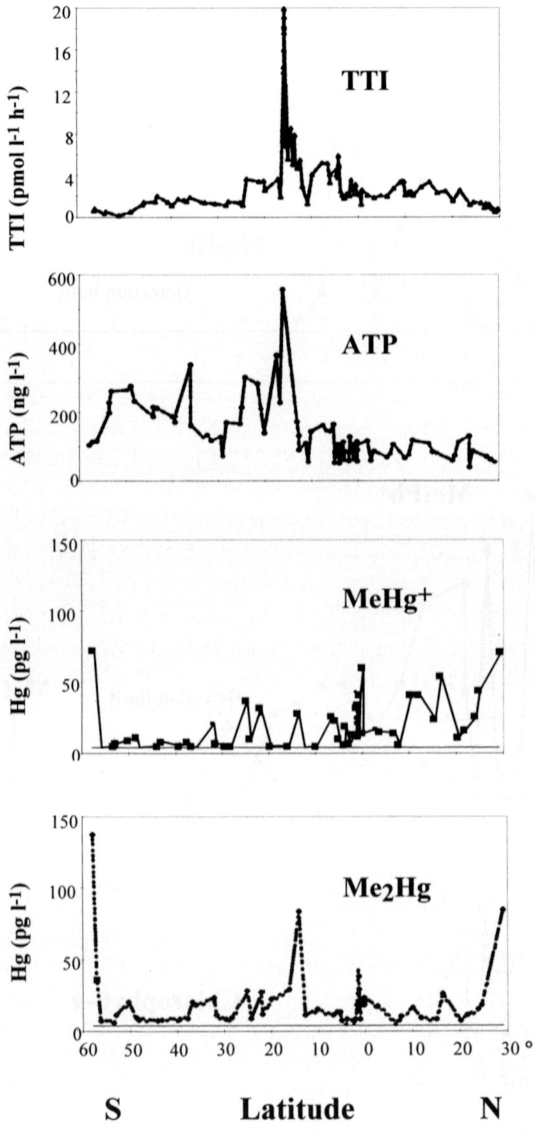

Figure 7.21. North-south concentration profiles of methylated Hg, of adenosine triphosphate (ATP) and of tritiated thymidine incorporation (TTI) determined during an expedition from Europe to the pack-ice border south of Cape Town (7) (ATP and TTI determinations from (60, 61).

The four different oceanic fronts crossed during this expedition are also marked in Figure 7.15. At the subtropical convergence the bioactivity, and therefore also the chlorophyll-*a* concentration, were relatively low compared with the situation at the other oceanic fronts. In some cases, but not in all, peak concentrations of chlorophyll-*a* relate well with high contents of methylated heavy metals. An especially good correlation is observed between Me_2Hg and chlorophyll-*a* at the subantarctic front, the polar front and in the area of the Weddell-Scotia confluence. However, the $MeHg^+$, Me_3Pb^+ and $MeCd^+$ curves do not behave in the same way, which might be again a consequence of the different fingerprints of the biological species producing methylated heavy metals (see Figures 7.11–13). This agrees with the experience described for bromo- and iodomethanes (Figures 7.8 and 9), where chlorophyll-*a* could also not be used as a general indicator for the biogenic production of these VHOCs.

During this expedition from Cape Town to Punta Arenas, the total content of Cd, Hg and Pb was also measured in all samples where methylated species of these heavy metals were found (24). Usually less than 1% of the total Hg was methylated, with a few exceptions, where up to about 5% were present in the monomethylated and dimethylated forms. On the other hand, 8–30% of the total Pb were present as Me_3Pb^+ and 15–30% of the total Cd were present as $MeCd^+$ in samples where methylation of these heavy metals could be detected.

Figure 7.20 illustrates a west-east concentration profile (from left to right of the plot) at a latitude of about 68° S (west of the Antarctic Peninsula) of methylated Cd, Pb and Hg in surface sea water samples (7, 24) analyzed during April 1995. Chlorophyll-*a* was determined by Templin and Bathmann (59). This profile is divided into two different areas, one characterized by pack-ice from 98° W to 84° W, the other one by a totally closed ice sheet from 84° W to 68° W. In the pack-ice section, the methylated heavy metal compounds could normally be determined with concentrations distinctly above the detection limit, whereas in the polar ocean under the closed ice sheet methylated heavy metals were mostly below or at the detection limit, except small concentrations near the shore of the Antarctic Peninsula (right side of Figure 7.20).

6.2. North-south concentration profiles

Figures 7.21 and 7.22 show two different north-south concentration profiles. Figure 7.21 represents the concentration of monomethyl- and dimethyl-Hg in surface sea water during an expedition in October and November 1993 from Europe to south of Cape Town at the pack-ice border at about 60° S (7). The ATP content (60), used as an universal biomass indicator, and the TTI (61), as an indicator for bacterial activities, are also presented in this figure. Whereas in most surface sea water samples the content of $MeHg^+$ exceeds that of Me_2Hg, Me_2Hg concentration is much higher at the pack-ice border as well as at 15° S. The sampling point at 15° S is also the only location where maxima in ATP and TTI clearly correlate with one of the methylated Hg species. The excess of Me_2Hg, compared with $MeHg^+$, and the extremely high TTI rate suggest preferable production of Me_2Hg by bacteria at this location. However, the pack-ice border with its increasing bioactivity during the Antarctic springtime results in the highest concentration of methylated Hg compounds from 30° N southwards.

Figure 7.22 represents a north-south concentration profile in surface sea water samples, which was measured for Me_3Pb^+ and $MeCd^+$ during an expedition from

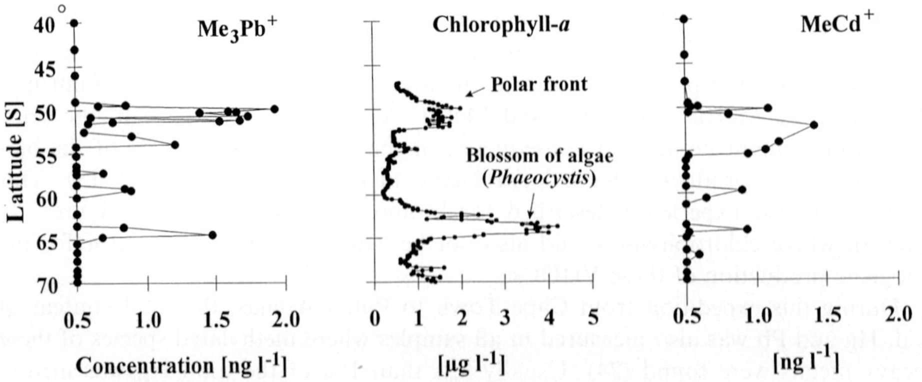

Figure 7.22. North-south concentration profiles of methylated Pb and Cd and of Chlorophyll-*a* during an expedition from Cape Town to the German station Neumayer at 70° S (47) (Chlorophyll-*a* determinations from (62)).

Cape Town to the German Antarctic station Neumayer in December 1995 and January 1996 (47). The chlorophyll-*a* curve, which was measured in parallel samples (62), clearly indicates by its maxima the polar front, on the one hand, and the algal bloom, on the other hand. These two maxima of chlorophyll-*a* agree well with those of the Me_3Pb^+ concentration, but less with the $MeCd^+$ data. However, from 65° S to the ice edge at 70° S methylated Pb and Cd could not be observed above the detection limit of about 0.5 ng l^{-1}.

6.3. Methylated Hg concentrations measured in polar and other oceans

There is very limited knowledge about the concentration of methylated heavy metals in the world's oceans, up to now. Most data are available for $MeHg^+$, but normally only from single locations. A selection of results from polar regions and some other areas is listed in Table 7.4. Single peak concentrations of more than 1000 ng l^{-1} were only found in the English Channel (7), where unknown anthropogenic influences have contributed to these high concentrations. However, as the results in Table 7.4 demonstrate, the natural content in surface sea water samples obviously ranges between <5 and about 200 pg l^{-1}, depending on the biological activity and the various biological species producing methylated Hg in the different areas.

6.4. First evidence of biogenic dimethyl-Tl in the Southern Ocean

During an expedition in December 1995, Schedlbauer and Heumann found significant concentrations of dimethyl-Tl (Me_2Tl^+) in one surface sea water sample of the Southern Ocean at about 64° S, 3° W (18). 1.1 ng l^{-1} Tl as Me_2Tl^+ were determined by IDMS combined with species-specific extraction (see Section 2), which was about 11% of the total Tl found in this sea water sample. During the same

Table 7.4. Concentration ranges of methylated Hg in surface water of the Southern Ocean compared with other areas

Area	Concentration (pg l^{-1})		Reference
	$MeHg^+$	Me_2Hg	
Southern Ocean, Pacific part (51°–71°S)	<5– 28	<5–13	Pongratz and Heumann (7)
Arctic Ocean at 75°N (west-east profile from 11°W to 15°E)	<5–158	<5– 83	Pongratz and Heumann (7)
Kongsfjord on Spitsbergen	<5–60	<5–46	Pongratz and Heumann (10)
Atlantic Ocean (north-south profile from 51°N to 58°S)	<5–3055	<5–1674	Pongratz and Heumann (7)
Coastal site of Washington, USA	—	4–180	Bloom et al. (63)

expedition, Me_2Tl^+ was checked in six other surface sea water samples south of 60° S, but it was always below the detection limit (<0.5 ng l^{-1}), which corresponds to <4% of the total Tl content analyzed. However, evidence that biogenic production of Me_2T^+ was not a single occurrence at the mentioned location in the Southern Ocean could be demonstrated at other regions of the Atlantic Ocean. For example, in a depth profile, determined at 51° S, 10° E, 0.6 ng l^{-1} Tl as Me_2Tl^+ were found in surface sea water, a maximum of 3 ng l^{-1} at a depth of 200 m and even a concentration of 0.7 ng l^{-1} at 4000 m (18), which indicates that bacteria must be involved in the bioproduction of Me_2Tl^+.

7. Transfer of iodomethane and bromoform from the Southern Ocean into the atmosphere

VHOCs play an important role in atmospheric chemistry. Attention has been focused in the past especially on anthropogenic substances of this type. It seems that the importance of biogenic VHOCs in atmospheric chemistry has been underestimated, although high amounts of some of these compounds are transferred from the ocean into the atmosphere. Because of their considerable extent and their relatively high bioactivity the polar oceans substantially contribute to the global atmospheric emission of naturally produced VHOCs.

Iodinated and brominated hydrocarbons can be decomposed in the atmosphere by photodissociation or by reaction with oxidizing agents, such as OH radicals. These degradation products are able, e.g., to decompose tropospheric ozone and it cannot be totally excluded that iodinated and brominated VHOCs also have some

Table 7.5. Atmospheric concentrations of brominated and iodinated VHOCs in Antarctica and comparison with other marine regions

Region	Compound	Concentration (pptv)	Reference
Antarctic Peninsula	$CHBr_3$	6.3 (spring)	Reifenhäuser and Heumann (40)
	CH_2Br_2	3.7 (spring)	Reifenhäuser and Heumann (40)
	CH_3I	2.4 (spring)	Reifenhäuser and Heumann (20)
South Pole	CH_3I	1.8	Rasmussen et al. (33)
Spitsbergen	$CHBr_3$	26 (spring)	Berg et al. (31)
		0.4 (late summer)	Schall and Heumann (23)
	CH_2Br_2	19 (spring)	Berg et al. (31)
		0.4 (late summer)	Schall and Heumann (23)
	CH_3I	1.0 (late summer)	Schall and Heumann (23)
Atlantic Ocean	CH_3I	1.2	Lovelock et al. (1)
South Atlantic	$CHBr_3$	2	Class and Ballschmiter (3)
	CH_2Br_2	2	Class and Ballschmiter (3)

effect on the stratospheric ozone layer. The important question in this connection is whether these substances have any chance within their atmospheric life-time of reaching the stratosphere. The atmospheric life-time of iodomethane, e.g., is about eight days (64). Because the polar ocean is the only source of iodomethane in Antarctica, the comparable concentrations found at the South Pole and at the coast area of the Antarctic Peninsula (see Table 7.5) demonstrate that this compound can be transported over a long distance and also to higher atmospheric levels.

Extended measurements of Rasmussen et al. (33) demonstrated the ubiquitous character of iodomethane in the marine atmosphere with background concentrations of 1–3 pptv. This agrees well with the mean value of 2.4 pptv obtained in spring around the Antarctic Peninsula and also with other marine areas (Table 7.5). Bromoform and dibromomethane were found in slightly higher concentrations in the Antarctic atmosphere compared with iodomethane. However, it seems to be that the emission of these brominated compounds from the polar ocean strongly depends on the season as shown by a comparison of the atmospheric concentrations in spring and late summer, observed on Spitsbergen.

The measured concentrations of VHOCs in surface sea water and in the corresponding air can be used with equation (1) to calculate the flux, and therefore the yearly input, of these substances from the ocean into the atmosphere. Corresponding results for the Southern Ocean and the Arctic Ocean in comparison with the total global input from all oceans into the atmosphere are summarized in Table 7.6 for iodomethane and bromoform. Different assumptions, e.g., the use of representative analytical data or a similar production rate in all oceans, make the calculated transfer into the atmosphere very uncertain. However, the data in Table 7.6 suggest that about 5–10% of the global atmospheric input of iodomethane and probably more than 10% of the bromoform input comes from the Southern

Table 7.6. Transfer of biogenic iodomethane and bromoform from the polar oceans into the atmosphere and comparison with the global input

Compound	Biological source	Area	Calculated input (10^{10} g yr^{-1})	Reference
Iodomethane	all species	Southern Ocean	3.5	calculated from data of Reifenhäuser and Heumann (20)
	all species	all oceans	80	Reifenhäuser and Heumann (20)
	all species	all oceans	50–130	Singh et al. (2), Rasmussen et al. (33)
	all species	Arctic Ocean	0.4	Schall and Heumann (23)
	macroalgae	all oceans	0.0003–0.02	Schall et al. (32)
	macroalgae	all oceans	0.01	Gschwend et al. (28)
Bromoform	all species	Arctic Ocean	5.4	Schall and Heumann (23)
	ice algae	Arctic Ocean	0.5–7	Sturges et al. (65)
	ice algae	Southern Ocean	0.5–8	Sturges et al. (65)
	macroalgae	all oceans	0.4–4*	Gschwend et al. (28)
	macroalgae	all oceans	0.008–0.2	Schall et al. (32)

* Includes also other brominated VHOCs

Ocean. Macroalgae do not contribute much to the global atmospheric concentration of VHOCs, but they can be the dominant contributors in coastal regions. On the other hand, ice algae from polar regions may play an important role in the total transfer of biogenic VHOCs into the atmosphere, as estimations of Sturges et al. (65) suggest.

8. Transfer of dimethyl-Hg from the Southern Ocean into the atmosphere

Permethylated heavy metal compounds, in contrast to ionic methylated species, are the only substances which have a low solubility in sea water and, in addition to this, are volatile enough to be easily transferred from the ocean into the atmosphere. This means that Henry's constant, which is the compound concentration in the gas phase related to that in the aqueous phase, must be high enough for a sufficient air-ocean distribution under Antarctic conditions. For example, Henry's constant was determined to be 0.15 for Me_2Hg, but only about 10^{-5} for MeHgCl (22). Me_2Hg, Me_4Pb, and Me_2Cd are the possible permethylated compounds. From these compounds only Me_2Hg could be determined to be biogenically produced in the polar and other oceans. Me_4-Pb is chemically stable in sea water, so that its absence in the unpolluted Southern Ocean indicates that it is not a biogenic product. Dimethyl-Cd is unstable in sea water and therefore it decomposes after its biogenic formation or it is also not biogenically produced.

From this it follows that calculations of the transfer of methylated heavy metals from the Southern Ocean into the atmosphere on the basis of measured concentrations are only possible for Me_2Hg. However, at different locations in polar regions as well as in other remote areas, distinct concentrations of Me_3Pb^+ and $MeCd^+$ have been determined in the atmosphere (24), which cannot be explained by Henry's law and calculations using equation (1). For example, in marine atmospheric samples of the Pacific part of the Southern Ocean from 60° S to 70° S, west of the Antarctic Peninsula, Me_3Pb^+ concentrations in the range of 350–770 pg m^{-3} and those of $MeCd^+$ in the range of 320-450 pg m^{-3} were determined in eight and five samples, respectively, out of a total number of fifteen samples. It is unclear whether the occurrence of these ionic methylated heavy metal species in the Antarctic atmosphere is due to sea spray transportation or other effects. There is certainly a possibility for the effect by sea spray because biogenic methylation of heavy metals usually takes place in the upper layers of the ocean and such substances with a hydrophilic (ionic) as well as with a hydrophobic side (methyl group) are known to be enriched in boundary layers of an aqueous/gas (ocean/air) system (12).

From Me_2Hg data in surface sea water samples and in the corresponding marine atmosphere, a first estimation of the transfer of this compound from the Southern Ocean, the Arctic Ocean and the Atlantic Ocean was carried out by equation (1) (see Table 7.7) (45). This estimate was calculated under the questionable assumptions that the measured data were representative for the whole area and that no seasonal variation occurs. The great uncertainty in such calculations can also be inferred by the great variability in the data for the total biogenic Hg emission from all oceans, which range from 0.6 ± 10^9 to 7700 ± 10^9 g yr^{-1} (66, 67). However, the calculated input of more than 0.2 ± 10^9 g $yr^{-1}$of Hg as Me_2Hg from each of the polar oceans is more than 10% of the total emission of this heavy metal species from the Atlantic Ocean. In addition, more recent data from Lindqvist et al. and

Table 7.7. First estimation of the transfer of Me_2Hg from polar oceans into the atmosphere compared with estimations of the global input of total Hg

Compound	Region	Source	Calculated input (10^9 g yr^{-1})	Reference
Me_2Hg	Southern Ocean (Pacific part, 51° S–71° S)	biogenic	0.21	Pongratz and Heumann (45)
	Arctic Ocean (64° N–79° N)	biogenic	0.24	Pongratz and Heumann (45)
	Atlantic Ocean (38° N–58° S)	biogenic	1.9	Pongratz and Heumann (45)
	all oceans	biogenic	0.6	Topping and Davies (66)
	all oceans	biogenic	7700	Wood et al. (67)
Total Hg	global emission	natural	3.0	Lindqvist et al. (68)
	global emission	anthropogenic	4.5	Lindqvist et al. (68)
	global emission	anthropogenic and natural	6.1	Nriagu (69)

Nriagu (68, 69) on the Hg emission into the atmosphere by natural and anthropogenic sources demonstrate that the polar oceans contribute significantly to the global content of atmospheric Hg.

Methylated Hg is not very stable in the atmosphere, especially so because OH radicals decompose such species and the atmospheric life-time of Me_2Hg is estimated to be about two days (70). Significant indication that OH radicals decompose methylated Hg was obtained by a first north-south concentration profile of Me_2Hg and $MeHg^+$ in the marine atmosphere over the Atlantic Ocean from Europe to about 55° S (24) (Figure 7.23). If the measured concentrations of methylated Hg are represented by a compensating curve, the minimum of the curve is around the equator, as can be seen from Figure 7.23. This agrees well with the known fact that the OH radical concentration in the marine atmosphere is highest near the equator and decreases with increasing latitudes into northern and southern directions (71). Elementary Hg is the most abundant Hg species in the atmosphere with mean concentrations in the northern hemisphere of about 2 ng m^{-3} and in the southern hemisphere of about 1.3 ng m^{-3} (72). From the results represented in Figure 7.23 it follows that methylated Hg is much less abundant in the atmosphere than elementary Hg by a factor in the range of 10 to >250 (values of >250 correspond to a detection limit of 0.5 pg methylated Hg m^{-3} of air). This unequal abundance can be explained by the fact that methylated Hg compounds have atmospheric life-times of about two days whereas the atmospheric residence time for elementary Hg is estimated to be about one year (72).

A possible geochemical cycle of Hg is discussed in the literature (73). In Antarctica and other remote environments, where analytical data are negligibly influenced

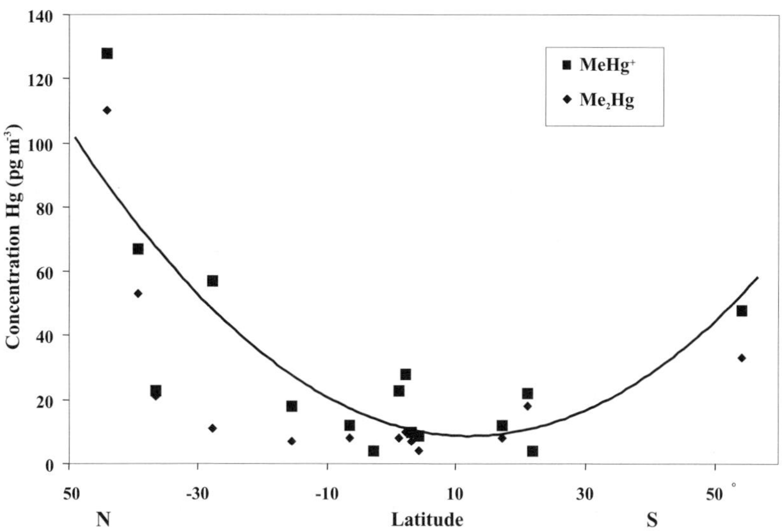

Figure 7.23. North-south concentration profile of methylated Hg in the marine atmosphere over the Atlantic Ocean from Europe to 55° S (24).

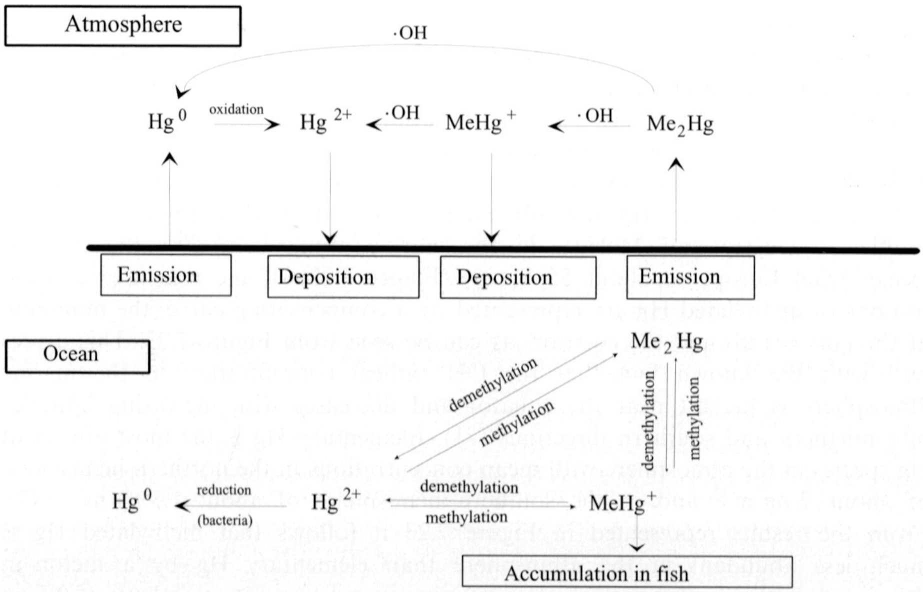

Figure 7.24. Biogeochemical cycle of Hg between the ocean and the atmosphere (following representations in (73)).

by anthropogenic sources, it was possible to prove this geochemical cycle with respect to the biogenic production of methylated Hg compounds, the transfer of methylated Hg from the ocean into the air and its atmospheric decomposition. Following the suggestions of the literature (73), part of the total natural geochemical cycle, which was investigated in the ocean-atmosphere system of Antarctica and other remote areas, is represented in Figure 7.24. Biogenic methylation of inorganic Hg^{2+} and demethylation or incorporation of methylated Hg have been demonstrated for different polar organisms. There is no doubt that Me_2Hg is the compound preferentially transferred from the ocean into the atmosphere where it is decomposed by reactive substances, such as OH radicals into $MeHg^+$ and, ultimately, into zerovalent Hg. Wet and dry deposition from the atmosphere will take place with the ionic Hg species, whereas $MeHg^+$, dissolved in the ocean, is accumulated by fish. Overall, the investigations of the biomethylation in Antarctica have contributed much to the knowledge of the natural biogeochemical cycle of heavy metals.

Acknowledgement

A sincere word of thanks goes to the research students Richard Pongratz, Christian Schall, Oliver Schedlbauer and W. Reifenhäuser who contributed their excellent PhD thesis works and various publications to this review on biomethylation in the polar ocean. Gratitude is also expressed to the Alfred-Wegener-Institute for

Polar and Marine Research, Bremerhaven, to the crew of the German research vessel Polarstern for logistic support and other important assistance and finally to the Deutsche Forschungsgemeinschaft for continuous financial support.

References

1. J. E. Lovelock, R. J. Maggs, R. J. Wade, Halogenated hydrocarbons in and over the Atlantic, *Nature*, 241 (1973), 194–196.
2. H. B. Singh, L. J. Salas, R. E. Stiles, Methyl halides in and over the eastern Pacific (40° N–32° S), *J. Geophys. Res.*, 88 (1983), 3684–3690.
3. T. Class, K. Ballschmiter, Chemistry of organic traces in air. VIII: Sources of bromo- and bromochloromethanes in marine air and surface water of the Atlantic Ocean, *J. Atmos. Chem.*, 6 (1988), 35–46.
4. R. J. Cicerone, L. E. Heidt, W. H. Pollock, Measurements of atmospheric methylbromide and bromoform, *J. Geophys. Res.*, 93 D4 (1988), 3745–3749.
5. C. Schall, K. G. Heumann, G. O. Kirst, Biogenic volatile organoiodine and organobromine hydrocarbons in the Atlantic Ocean from 42° N to 72° S, *Fresenius' J. Anal. Chem.*, 359 (1997), 298–305.
6. R. A. Duce, G. L. Hoffman, B. J. Ray, I. S. Fletcher, G. T. Wallace, J. L. Fasching, S. R. Piotrowicz, P. R. Walsh, E. J. Hoffman, J. M. Miller, J. L. Heffter, Trace metals in the marine atmosphere: sources and sinks, In: H. Windom, R. A. Duce (Eds.), *Marine Pollutant Transfer*, 1976, Heath & Co., Lexington, 77–119.
7. R. Pongratz, K. G. Heumann, Determination of concentration profiles of methyl mercury compounds in surface waters of polar and other remote oceans by GC-AFD, *Int. J. Environ. Anal. Chem.*, 71 (1998), 41–56.
8. M. J. Prather, M. B. McElroy, S. C. Wofsy, Reductions in ozone and high concentrations of stratospheric halogens, *Nature*, 312 (1984), 227–231.
9. R. P. Mason, W. F. Fitzgerald, Mercury speciation in open ocean waters, *Water Air Soil Pollut.*, 56 (1991), 779–789.
10. R. Pongratz, K. G. Heumann, Production of methylated mercury and lead by polar macroalgae – a significant natural source for atmospheric heavy metals in clean room compartments, *Chemosphere* 36 (1998), 1935–1946.
11. R. Pongratz, K. G. Heumann, Determination of monomethyl cadmium in the environment by differential pulse anodic stripping voltammetry, *Anal. Chem.*, 68 (1996), 1262–1266.
12. A. W. Adamson, *Physical Chemistry of Surfaces*, 5th Edition, 1990, Wiley, New York, pp. 69–101.
13. K. G. Heumann, Determination of inorganic and organic traces in the clean room compartment of Antarctica, *Anal. Chim. Acta*, 283 (1993), 230–245.
14. T. Class, K. Ballschmiter, Chemistry of organic traces in air. Part IX: Evidence of natural marine sources for chloroform in regions of high primary production, *Fresenius' Z. Anal. Chem.*, 327 (1987), 40–41.
15. P. D. Nightingale, G. Malin, P. S. Liss, Production of chloroform and other low-molecular-weight halocarbons by some species of macroalgae, *Limnol. Oceanogr.*, 40 (1995), 680–689.
16. R. J. Charlson, J. E. Lovelock, M. O. Andreae, S. G. Warren, Oceanic phytoplankton, atmospheric sulphur, cloud albedo and climate, *Nature*, 326 (1987), 655.
17. D. Tanzer, K. G. Heumann, Gas chromatographic trace-level determination of volatile organic sulfides and selenides and of methyl iodide in Atlantic surface water, *Intern. J. Environ. Anal. Chem.*, 48 (1992), 17–31.
18. O. Schedlbauer, K. G. Heumann, Development of an isotope dilution mass spectrometric method for dimethylthallium speciation and first evidence of its existence in the ocean, *Anal. Chem.*, 71 (1999), 5459–5464.
19. P. S. Liss, P. G. Slater, Flux of gases across the air-sea interface, *Nature*, 247 (1974), 181–184.
20. W. Reifenhäuser, K. G. Heumann, Determinations of methyl iodide in the Antarctic atmosphere and the south polar sea, *Atmos. Environ.*, 16 (1992), 2905–2912.
21. E. M. Prestbo, N. S. Bloom, Frontier Geosciences, Seattle, USA. Personal communication.

22. O. Lindqvist, H. Rodhe, Atmospheric mercury, *Tellus*, 37B (1985), 136–159.
23. C. Schall, K. G. Heumann, GC determination of volatile organoiodine and organobromine compounds in Arctic sea water and air samples, *Fresenius J. Anal. Chem.*, 346 (1993), 717–722.
24. R. Pongratz, Biogene Produktion von neutralen und ionischen Methylschwermetallverbindungen in polaren Gewässern. PhD Thesis work, 1996, University of Regensburg.
25. F. G. Noden, The determination of tetraalkyllead compounds and their degradation products in natural water, In: M. Branica, Z. Konrad (Eds.), *Lead in Marine Environment*, 1980, Pergamon Press, Oxford, 83–91.
26. N. Mikac, M. Branica, Separation of dissolved alkyllead and inorganic lead species by coprecipitation with barium sulfate, *Anal. Chim. Acta*, 212 (1988), 228–230.
27. A. G. Allen, M. Radojevic, R. M. Harrison, Atmospheric speciation and wet deposition of alkyllead compounds, *Environ. Sci. Technol.*, 22 (1988), 517–522.
28. P. M. Gschwend, J. K. MacFarlane, K. A. Newman, Volatile halogenated organic compounds released to sea water from temperate marine macroalgae, *Science*, 227 (1985), 1033–1035.
29. J. E. Lovelock, Natural halocarbons in the air and in the sea, *Nature*, 256 (1975), 193–194.
30. T. Class, R. Kohnle, K. Ballschmiter, Chemistry of organic traces in air. VII: Bromo- and bromochloromethanes in air over the Atlantic Ocean, *Chemosphere*, 15 (1986), 429–436.
31. W. W. Berg, L. E. Heidt, W. Pollock, P. D. Sperry, R. J. Cicerone, Brominated organic species in the Arctic atmosphere, *Geophys. Res. Lett.*, 11 (1984), 429–432.
32. C. Schall, F. Laturnus, K. G. Heumann, Biogenic volatile organoiodine and organobromine compounds released from polar macroalgae, *Chemosphere*, 28 (1994), 1315–1324.
33. R. A. Rasmussen, M. A. K. Khalil, R. Gunawardena, S. D. Hoyt, Atmospheric methyl iodide (CH_3I), *J. Geophys. Res.*, 87 (1982), 3086–3090.
34. F. Laturnus, Release of volatile halogenated organic compounds by unialgal cultures of polar macroalgae, *Chemosphere*, 31 (1995), 3387–3395.
35. F. Laturnus, Volatile halocarbons released from Arctic macroalgae, *Mar. Chem.*, 55 (1996), 359–366.
36. F. Laturnus, C. Wiencke, H. Klöser, Antarctic macroalgae – sources of volatile halogenated organic compounds, *Mar. Environ. Res.*, 41 (1996), 169–181.
37. F. Laturnus, F. Adams, I. Goméz, G. Mehrtens, Halogenating activities detected in Antarctic macroalgae, *Polar Biol.*, 17 (1997), 281–284.
38. R. E. Moore, Volatile compounds from marine algae, *Acc. Chem. Res.*, 10 (1977), 40–47.
39. W. Fenical, Natural halogenated organics, In: E. K. Duutsma, R. Dawson (Eds.), *Marine Organic Chemistry* (1981), Elsevier, New York, 375–393.
40. W. Reifenhäuser, K. G. Heumann, Bromo- and bromochloromethanes in the Antarctic atmosphere and the south polar sea, *Chemosphere*, 9 (1992), 1293–1300.
41. G. Mehrtens, F. Laturnus, Halogenating activity in an Arctic population of brown macroalga, *Laminaria saccharina Lamour, Polar Res.*, 16 (1997), 19–25.
42. R. H. White, Analysis of dimethyl sulfonium compounds in marine algae, *J. Mar. Res.*, 40 (1982), 529–536.
43. W. T. Sturges, C. W. Sullivan, R. C. Schnell, L. E. Heidt, W. H. Pollock, Bromoalkane production by Antarctic algae, *Tellus*, 45B (1993), 120–126.
44. C. Schall, *Identifizierung und Quantifizierung flüchtiger iodierter und bromierter Kohlenwasserstoffe in atlantischen, arktischen und antarktischen Gewässern und der arktischen Atmosphäre sowie Untersuchung der biogenen Quellen dieser Substanzen*, PhD, Thesis work, 1994, University of Regensburg.
45. R. Pongratz, K. G. Heumann, Production of methylated mercury, lead, and cadmium by marine bacteria as a significant natural source for atmospheric heavy metals in polar regions, *Chemosphere*, 39 (1999), 89–102.
46. J. B. Robinson, O. H. Tuovinen, Mechanisms of microbial resistance and detoxification of mercury and organomercury compounds: physiological, biochemical, and genetic analysis, *Microbiol. Rev.*, 48 (1984), 95–124.
47. O. Schedlbauer, *Methylierte Schwermetallspezies im Atlantischen Ozean und der marinen Atmosphäre*. PhD Thesis work, 1999, University of Mainz.
48. M. Simon, University of Oldenburg, Germany, Personal communication.
49. G. Dreibus, H. Wönke, L. Schultz, Mysterious iodine overabundances in Antarctic meteorites, Abstract of the Workshop on Antarctic Meteorites, 1985, Mainz, 11–13.

50. K. G. Heumann, M. Gall, H. Weiss, Geochemical investigations to explain iodine overabundances in Antarctic meteorites, *Geochim. Cosmochim. Acta*, 51 (1987), 2541–2547.
51. K. G. Heumann, J. Neubauer, W. Reifenhäuser, Iodine-overabundances measured in the surface layers of an Antarctic stony and iron meteorite, *Geochim. Cosmochim. Acta*, 54 (1990), 2503–2506.
52. H.-E. Gäbler, K. G. Heumann, Determination of atmospheric iodine species using a system of specificly prepared filters and IDMS, *Fresenius J. Anal. Chem.*, 345 (1993), 53–59.
53. J. E. Lovelock, The production and fate of reduced volatile species, In: E. D. Goldberg (Ed.), *Atmospheric Chemistry*, 1982, Springer, Heidelberg, 199–213.
54. A. Wimschneider, K. G. Heumann, Iodine speciation in size fractionated atmospheric particles by isotope dilution mass spectrometry, *Fresenius J. Anal. Chem.*, 353 (1995), 191–196.
55. K. Abrahamson, S. Klick, Determination of biogenic and anthropogenic volatile halocarbons in sea water by liquid-liquid extraction and capillary gas chromatography, *J. Chrom.*, 513 (1990), 39–45.
56. S. Klick, Seasonal variations of biogenic and anthropogenic halocarbons in sea water from a coastal site, *Limnol. Oceanogr.*, 37 (1992), 1579–1585.
57. C. Schall, K. G. Heumann, S. J. de Mora, P. A. Lee, Biogenic brominated and iodinated organic compounds in ponds on the McMurdo Ice Shelf, Antarctica, *Antarctic Sci.*, 8 (1996), 45–48.
58. S. Hirch, U. V. Bathmann, Alfred-Wegener-Institute for Polar and Marine Research, Bremerhaven, Germany, Personal communication.
59. M. Templin, U. V. Bathmann, Alfred-Wegener-Institute for Polar and Marine Research, Bremerhaven, Germany, Personal communication.
60. T. Bluszcz, O. Schrems, Alfred-Wegener-Institute for Polar and Marine Research, Bremerhaven, Germany, Personal communication.
61. K. Gocke, Institute for Oceanography, University of Kiel, Germany, Personal communication.
62. U. V. Bathmann, V. Smetacek, Alfred-Wegener-Institute for Polar and Marine Research, Bremerhaven, Germany, Personal communication.
63. N. S. Bloom, E. M. Prestbo, J. S. Tokos, E. von der Geest, E. S. Kuhn, Distribution and origins of mercury species in the Pacific northwest atmosphere, *Abstract, World Mercury Conference*, Hamburg, August 1996.
64. W. L. Chameides, D. D. Davis, Iodine: its possible role in tropospheric chemistry, *J. Geophys. Res.*, 85 (1980), 7383–7398.
65. W. T. Sturges, G. F. Cota, P. T. Buckley, Bromoform emission from Arctic ice algae, *Nature*, 358 (1992), 660–662.
66. G. Topping, I. M. Davies, Methyl mercury production in the marine water column, *Nature*, 290 (1981), 243–244.
67. J. M. Wood, F. S. Kennedy, C. G. Rosen, Synthesis of methylmercury compounds by *Methanogenicum bacterium, Nature*, 220 (1968), 173–174.
68. O. Lindqvist, K. Johansson, M. Aastrup, A. Andersson, L. Bringmark, G. Novsenius, L. Hakanson, A. Iverfeldt, M. Meili, B. Timm, Mercury in the Swedish environment – recent research on causes, consequences and corrective methods, *Water Air Soil Pollut.*, 55 (1991), 1–261.
69. J. O. Nriagu, A global assessment of natural sources of atmospheric trace metals, *Nature*, 338 (1989), 47–49.
70. H. Niki, P. D. Maker, C. M. Savage, L. P. Breitenbach, A long-path Fourier transform infrared study of the kinetics and mechanism for OH radical initiated oxidation of dimethyl mercury, *J. Phys. Chem.*, 87 (1983), 4978–4981.
71. U. Platt, J. Rudolph, T. Brauers, G. W. Harris, Atmospheric measurements during Polarstern cruise ANT VII/1, 54° N to 32° S, *J. Atmos. Chem.*, 15 (1992), 203–214.
72. F. Slemr, G. Schuster, W. Seiler, Distribution, speciation and budget of atmospheric mercury, *J. Atmos. Chem.*, 3 (1985), 407–434.
73. P. J. Craig, *Organometallic compounds in the environment – principles and reactions*, 1986, Longman, London.

Environmental Contamination in Antarctica
A Challenge to Analytical Chemistry
S. Caroli, P. Cescon and D.W.H. Walton, editors

Chapter 8

Trace metals in particulate and sediments

Roberto Frache, Maria Luisa Abelmoschi, Franca Baffi, Carmela Ianni, Emanuele Magi, Francesco Soggia

1. Introduction

The Arctic and Southern Oceans are characterized by low temperatures and both regions experience considerable cooling. Ice plays a major role in determining many of their features. The differences between the two oceans are nevertheless significant. The Southern Ocean's boundaries are open with a very large water exchange, whereas the Arctic Ocean is basically an enclosed ocean. The Arctic Ocean is covered by ice most of the year, whereas the Southern Ocean undergoes considerable seasonal variation in ice cover. Finally, the Arctic Ocean is bordered by vast continental shelves with a large input of fresh water from rivers; on the other hand, there is some fresh water input into the seas surrounding Antarctica. The distribution of many chemical species is determined not only directly by the unique characteristics of the polar oceans, but also, indirectly, by how these characteristics control biological processes that in turn influence chemical distribution.

Far from the main anthropogenic sources of the northern hemisphere and separated from the surface circulation of the other major oceans by the subtropical convergence, the Southern Ocean appears to be relatively isolated. But this is merely an appearance: the Southern Ocean, which is the ultimate point reached by deep waters generated in the northern regions, is also a major area of bottom-water formation of the world ocean and is, as mentioned above, broadly opened to the Atlantic, Pacific and Indian Oceans at all depths greater than 200 m. Very large chemical fluxes are involved in these exchanges and therefore the study of this geoecosystem is of basic interest in chemistry (1).

The study of the concentration level and distribution of trace metals appears to be of great interest, because these chemical species, mainly deriving from natural sources, play an important role as micronutrients in the biological cycles.

The understanding of the relationships among phytoplanktonic biomass and nutrients supply is the goal of recent investigations and it has been suggested that micronutrients, as trace metals, may limit phytoplankton growth in nutrient-rich regions as the Antarctic ones (2, 3). It was further hypothesized that nutrients are not removed to depletion in the waters off the continental shelf of Antarctica because of low input rates of atmospheric Fe and the lack of the other sources (e.g., sediments). In other situations the presence of ice edge and island blooms has been related to greater inputs of Fe from ice melting and resuspended sediments (2). Although these hypotheses have not been experimentally proved, they would explain both the geographic and seasonal variations observed in Antarctic productivity. The correct interpretation of previous productivity results from Antarctica can be obtained only by a rigorous examination of trace metal-nutrient-light interactions (4).

Until now the attainment of these results was not possible, as few data are available to describe the trace metal concentrations in the Southern Ocean (5–11). In order to contribute to the knowledge of the distribution of trace elements and of the biological cycles in Antarctica over the past twelve Italian expeditions, attention was focused, among others, on trace metals in solid and in dissolved phases (12–29).

1.1. Trace metals in solid phases

Sediment and particulate matter are formed by various and different geochemical phases (mineralogical and organic fractions) which act as *reservoirs* for trace elements. The natural particulate matter (e.g., clays and oxides of Al, Fe, Mn and Si) plays an important role in the distribution of trace metals because it offers sites for adsorption and it successively deposits, forming a solid sedimentary phase (30). The particulate matter can also be coated with inorganic precipitates and organic macromolecules, which can influence the adsorption of trace elements (31). After deposition, a number of diagenetic reactions take place, which cause a redistribution of trace metals over the various components of the sediment. It is moreover proved that the sedimentation causes a movement towards the surface of the metal-enriched layer (32). Interactions between particulate and metals play an important role in the regulation of dissolved metal concentration (the most bioavailable). In fact, adsorption is the first step in the removal of trace metals from the hydrological cycle, the ultimate sink being the oceanic sediment (33). To understand the correlations between trace metals and biological matrices an evaluation of the interaction of trace metals with particulate matter and sediments is essential. Moreover, sediments indirectly influence the formation of chemical species from dissolved metals because they are a substrate for biogeochemical transformations. Chemical speciation is defined as the identification of the chemical or physical form in which an element is found. The bioavailable fraction is the portion of the total metal concentration in each of the abiotic *reservoirs* which is taken up by organisms. The particulate state of an element can assume an important role in the uptake processes, at least for some kinds of organisms, such as filter feeders and deposit feeders. The bioavailability of trace metals depends on their chemical species as does their toxicity (33).

Although sediments are able to detoxify surface waters by adsorption/precipitation processes, causing a decrease in the dissolved metal concentrations, part of the metals in the sediments is bioavailable, although their quantification in this matrix can be difficult due to the complexity of the system.

Within the general importance of trace metals in particulate matter and sediments for contamination problems and for biogeochemical cycles, there is a particular interest for seeking knowledge of the contents of trace elements in Antarctic marine solid phases.

As previously described, Antarctica plays a critical role in many global phenomena because it is a major site of deposition within a number of biogeochemical cycles. Moreover, there is evidence that 'global environmental problems disproportionately impact on polar regions. In order to understand the influence of human activities in Antarctica, an appreciation of the role of polar oceans is required, as is the response of polar systems to man-induced perturbations' (34). The Antarctic environment also offers very interesting characteristics for the study of the natural phenomena of accumulation and distribution of the suspended load in waters for two reasons: the absence of polluting elements and the presence of terrigenous contributions, essentially of glacial origin. The determination of trace metals in marine solid phases is therefore a very important step to understand geochemical and environmental processes and their possible changes due to anthropogenic activities (35).

1.2. Analytical methods

It is well known that these studies present a number of difficulties and problems due to the possible contamination during sampling, the small amount of particulate samples, the matrix effects which influence the instrumental determination, the different distribution of the metals within the solid phase and so on (36, 37).

Conceptually, the solid material can be partitioned into fractions of different importance which can be extracted selectively (38). The procedures for determining the chemical species of trace metals in solid phases can be grouped into methods which affect the separation between residual and non-residual metals only and more elaborated methods making use of sequential extractions (39).

One of the most frequently used protocols for selective extraction is the one of Tessier et al., with various possible modifications (38). The fractions considered are usually the so-called exchangeable ones, bound to carbonates, bound to Fe and Mn oxides and hydroxides, bound to organic matter and sulphides (not distinguishable) and residual fractions. The differences among the various schemes used by different workers are primarily in the kind of reagents used to solubilize the various fractions.

A simplified version of the scheme of Tessier et al. has been developed during the past years (40, 41). The scheme is made of three sequential extractions called A, B, C, which can leach respectively the exchangeable and carbonatic fraction (A), the reducible Fe-Mn hydroxides phase (B) and the organic matter together with the sulphides (C). The residual fraction is normally disregarded because total dissolution of the same samples is always performed, so that the residual fraction can be calculated as the difference. Moreover, the residual fraction is the less important one from the viewpoint of bioavailability of trace metals.

One of the most important fractions acting as a sink for trace metals is the organic phase, primarily the humic substances, which are complex, polydisperse organic molecules, operationally defined on the basis of their solubility (42, 43).

They are the most important product formed during the decomposition of organic matter and are classified as fulvic acids (completely soluble in aqueous solutions) and humic acids (precipitating below pH 2–3) (44). They can form anionic complexes with metals because of the great variety of functional groups carried on the aromatic backbones, i.e., carboxyl, hydroxyl, carbonyl and amine groups. In recent years, the complexation of trace metals with humic substances in marine environment has been increasingly studied and the importance of this fraction in transport and fixation of trace elements within sediments has been pointed out. The determination of trace metals when bound to humic fraction is not so easy though, because of the lability of the metal-humate species (45–47). A procedure suitable for the extraction of both labile and stable metal complexes is fully described in the experimental section and has been already applied to several Antarctic sediment samples (21, 24).

2. Materials and methods

2.1. Apparatus and reagents

Metal determinations were carried out mainly by Inductively Coupled Plasma Atomic Emission Spectrometry (ICP-AES). The instrument used is a Jobin-Yvon 24 model. It is characterized by an RF-power supplied by a built-in 40.68 MHz generator (JY 2300 model) and a sequential monochromator with a 640 mm focal length, equipped with a highly-dispersive holographic grating (2400 grooves mm^{-1}). The introduction system is a classical Meinhard quartz nebulizer that can be replaced by an ultrasonic nebulizer model Cetac U-5000 AT, with a heater at 140°C and a cryocondenser at –2°C. The ICP-AES conditions are, in general, the following: Ar flow, 13 l min^{-1}; cooling Ar flow, 0.2 l min^{-1}; power, 850 W. Some determinations were carried out with Electrothermal Atomization Atomic Absorption Spectrometry (ETA-AAS). To this end, two atomic absorption spectrometers Varian model AA-300 and AA-400-plus were used, which were equipped with a model GTA 96 graphite tube and with Deuterium and Zeeman background correction, respectively. The analytical wavelengths, checked in order to avoid spectral interferences among major and trace elements, are shown in Table 8.1. Spectrophotometric measurements were performed with a Varian series 634 spectrophotometer, using Fluka Humic Acid as the external standard. All chemical reagents were of analytical or chromatographic grade.

2.2. Particulate matter

2.2.1. Sampling and conservation procedures

The samples of sea water were collected with a Teflon® suction pump or Go-Flo bottles, depending on the sampling depth. Immediately after collection the samples were filtered under N_2 pressure through 0.45 μm pore-size filters (Millipore). Filtration was effected under a laminar flow hood using a closed Teflon® system. The

Table 8.1. ICP-AES and analytical wavelengths (nm)

Element	ICP-AES	AAS-ETA
Cd	226.50	228.80
Cr	267.62	357.90
Cu	327.40	324.80
Fe	259.94	248.30
Mn	257.61	279.50
Ni	352.45	232.00
Pb	220.35	217.00
Zn	213.86	213.86

filters were rinsed with MQ water and stored in individual Petri dishes at –20°C until analysis in Italy. All the equipment used for sampling was prepared by prolonged washing with HCl and rinsing with MQ ultrapure water.

2.2.2. Analytical processes

After drying at 40°C, every filter with particulate matter was weighed and subsequently sonicated in 20 ml of MQ water. After drying as before, the filter was weighed again and the suspension obtained was dried at 40°C and treated with 2 ml of 8*N* HNO_3, at high temperature in a Teflon® bomb, following a standard procedure (47–50). ICP-AES and ETA-AAS were employed, as mentioned above, for metal determination using a standard addition calibration curve.

2.3. *Sediments*

2.3.1. Sampling and conservation procedures

The sediment samples were collected with a stainless steel grab. Immediately after sampling all the samples were frozen at –20°C.

2.3.2. Analytical processes

For total metal solubilization a representative unfrozen sample of 0.5 g, dried at 110°C and finely powdered, is dissolved in a Teflon® bomb by an acid mixture of 5 ml of 40% HF, 1 ml of aqua regia (HNO_3–HCl 1:3 v/v) and 5 ml of MQ water, heated at 180°C for 90 min and then cooled to room temperature. 5 ml of saturated H_3BO_3 solution are then added to the Teflon® bomb and, after dilution to 50 ml in a volumetric flask, the solution is analyzed by ICP-AES. To test the accuracy of the total solubilization method several certified reference materials in sediment matrices are always analyzed along with the unknowns, obtaining results in a good agreement with certified values, as reported in Table 8.2.

Table 8.2. Results ($\mu g\ g^{-1}$) obtained by ICP-AES on two certified reference materials based on sediments from the National Research Council Canada

Element	MESS-1		PACS-1	
	Certified	Found	Certified	Found
Cd	0.59 ± 0.10	0.64 ± 0.12	2.38 ± 0.20	2.09 ± 0.15
Co	10.8 ± 1.9	12.2 ± 1.8	17.5 ± 1.1	18.0 ± 1.7
Cr	71 ± 11	45.7 ± 2.3	113 ± 8	72.4 ± 2.3
Cu	25.1 ± 3.8	28.1 ± 0.9	452 ± 16	432 ± 3
Fe	30.5 ± 1.7 (*)	26.9 ± 1.6 (*)	48.7 ± 0.8 (*)	41.1 ± 1.7
Mn	513 ± 25	460 ± 5	470 ± 12	378 ± 8
Pb	34.0 ± 6.1	38.2 ± 1.7	404 ± 20	408 ± 3
Zn	191 ± 17	197 ± 3	824 ± 22	830 ± 7

(*) $mg\ g^{-1}$

For selective metal solubilization, 5 g of finely powdered sediment dried at 45°C are subsequently shaken with different reagents in order to obtain a selective phase solubilization, as summarized in Figure 8.1. The metal content in the three solutions is determined by ICP-AES. An important difference between this method and all the other sequential schemes is the use, for the exchangeable and carbonatic fraction, of ammonium acetate buffer (pH 5) instead of sodium acetate. The subsequent analysis of the extract is carried out by means of ICP-AES, with serious interference of Na on the signal, whilst ammonium gives negligible interferences (40, 41). Since the ammonium acetate at pH 7 (used for exchangeable fraction only) can also dissolve a small amount of carbonate and since usually the metal content of the exchangeable and of the carbonatic phases is not high and the two fractions have more or less the same importance in speciation models, these are extracted together instead of performing two separate steps (38). In this way the procedure is rather simple and the metal content in this phase (A) is certainly detectable. Another important difference is the use of 8 *N* HNO_3 to solubilize the organic fraction. In several experiments it has been verified that HNO_3 has the same efficiency as H_2O_2, but it is of much easier use.

For metal-humic acids complexes solubilization, 5 g of finely powdered sediment dried at 45°C are shaken with 0.1 *M* acetic acid and the treatment is repeated until the carbonates are completely removed. After centrifugation and washing with water, 0.5 *M* NaOH (degassed with N_2) is added and shaken for 24 hr in order to extract the complexes. The extraction is repeated with a fresh 0.5 *M* NaOH solution and the two extracts are combined. The resulting alkaline solution is filtered through a 0.45 μm Millipore filter and analyzed by ICP-AES. After purification of the extracted humic acids by precipitation and redissolution, they are extracted into chloroform by means of cetylpiridinium chloride and NaCl. The absorbance of the chloroform extract at 450 nm is then measured to determine the concentration of humic acids (51).

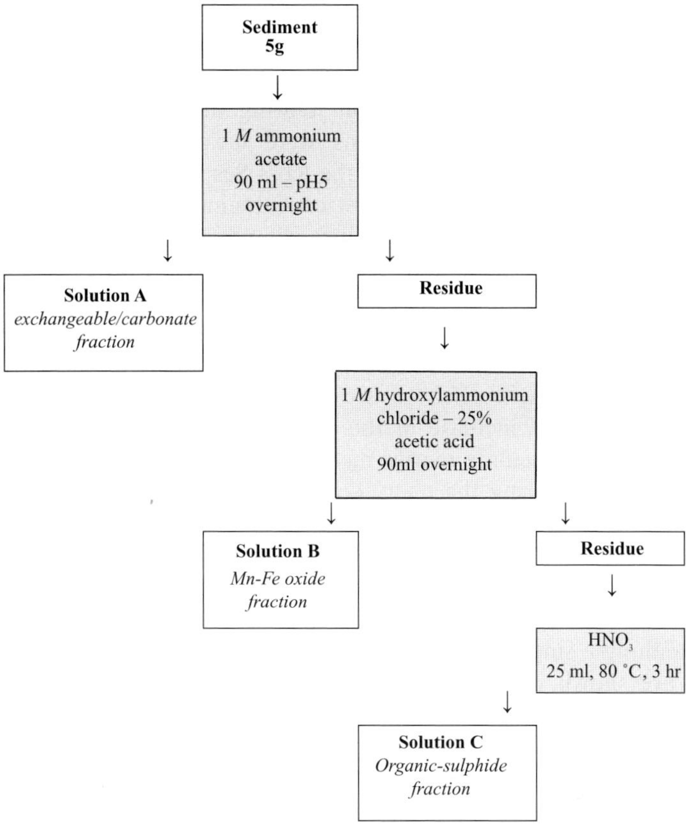

Figure 8.1. Diagram of the selective extraction procedure.

3. Results and discussion

The study of the concentration and distribution of trace metals started with the 1987–1988 expedition and at the beginning it was focused on the quantification of metals in dissolved, particulate and sediments phases. Afterwards, the work was planned in order to obtain data and information useful to understand the role and the fate of trace metals during processes, such as ice melting and biogeochemical cycles.

3.1. Particulate matter

During the 1987–1988 Antarctic Campaign, samples were collected and analyzed from twelve sites at Terra Nova Bay and along the Ross Sea shore. The concentrations of Cr, Cu, Fe and Ni in particulate matter were lower than those found in the Indian sector of the Southern Ocean obtained by a total attack (52). Chromium, Cu and Ni were not site-dependent; the Fe concentration was higher in the area close to Drygalski Glacier than in the remaining sampling stations (12). In the 1988–1989 Campaign, Cd, Cr, Cu, Fe and Ni were analyzed in particulate matter collected in seven stations at the Ross Bay; the data were not enough to characterize the geographical distribution. During the 1989–1990 Campaign, Al, Cd, Cr, Cu, Fe, and Ni concentrations were measured in particulate matter collected in several sampling stations off the Ross Sea shore and in a water column from 20 to 1300 m. This profile is the first one obtained in Antarctica: the content of particulate matter was higher in the surface layer and the metals concentration along the profile showed a maximum concentration value from 20 to 100 m, corresponding to a minimum temperature (17).

A comparison of the data from the different expeditions clearly showed that, for surface waters, Cd, Cu, Ni and, to a lesser extent, Cr and Fe concentrations were higher in nearshore than in offshore areas. A possible explanation was that the main contribution to the increase in the metal concentrations close to the shore might be due to the particulate matter included in the Antarctic pack ice and then transferred to the sea during the melt. To verify this hypothesis in the 1990–1991 and 1991–1992 expeditions some preliminary sampling in the presence and absence of pack were conducted, but the number of samples and the relative data were inconclusive (20). So, in 1993–1994, only one station was chosen at the Wood Bay (Figure 8.2) and both dissolved and particulate metals were determined to obtain detailed data along a single water column (22).

The data are shown in Table 8.3. In the dissolved phase Cu presents quite a smooth profile in the presence and absence of ice and its average concentration practically does not change. On the contrary, the other metals decrease after ice

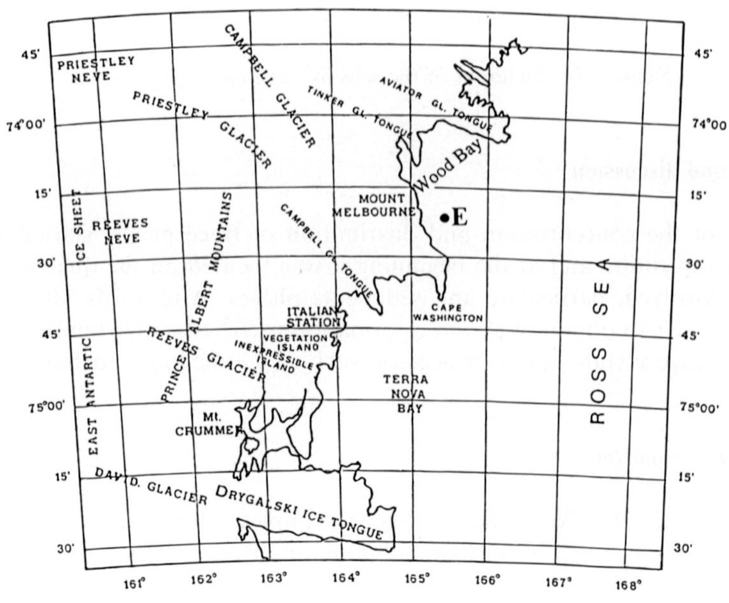

Figure 8.2. Map of the Wood Bay station (E).

Table 8.3. Average concentrations of Dissolved (µg l^{-1}) and Particulate (µg g^{-1}) metals at various depth (m) in the presence (i) and absence (ni) of pack ice

	Cd				Cu				Fe				Ni			
Depth	D / i	P / i	D/ni	P/ni	D / i	P / i	D/ni	P/ni	D / i	P / i	D/ni	P/ni	D / i	P / i	D/ni	P/ni
0.5	0.044	0.005	0.011	0.136	0.28	0.214	0.38	0.449	0.99	1.04	0.63	24.4	0.50	0.044	0.37	0.351
10	0.042	0.008	0.001	0.107	0.33	0.081	0.34	0.348	0.79	2.12	0.31	15.8	0.54	0.453	0.27	0.385
50	0.039	0.010	0.045	0.093	0.25	0.166	0.35	0.430	0.36	12.54	0.32	14.5	0.45	0.216	0.37	0.168
100	0.055	0.006	0.040	0.013	0.42	0.161	0.38	0.306	1.00	7.25	1.12	11.0	0.87	0.130	0.55	0.131
350	0.038	0.014	0.024	0.063	0.35	0.179	0.29	0.546	0.50	13.51	0.035	30.6	0.60	0.028	0.22	0.156

melt. It is interesting to observe that this reduction (especially for Cd and Fe, less for Ni) occurs mainly in the first 50 m. This layer, more or less, corresponds to the photic zone, where, in the absence of ice, the sunlight penetrates and where the phytoplanktonic bloom is centred. In fact, it is quite common for phytoplankton to develop within the marginal ice zones of both northern and southern oceans (1). Therefore, an uptake of these metals from the planktonic mass took place because Cd, Fe and Ni are 'nutrient-type' elements and micronutrients themselves (53). Cadmium, in fact, has no apparent biological function, but it is incorporated because of the lack of discrimination in the uptake mechanism (54). These results clearly show that, after the melting of pack ice, the sea water becomes poorer in those metals with 'nutrient' characteristics. The contribution to dilution of surface water caused by ice melting did not seem very important; in fact, it was found that the surface samples are more concentrated than the 10 m samples. Other authors found the same effect for nutrients, which decreased within the ice edge, but 'the depletion was generally far too pronounced to result from a simple dilution by meltwater' (55, 56).

In the particulate phase it was observed that the amount of all the four metals increases after the pack melting. This increase is very pronounced for Cd (almost ten times) and is significant for Cu and Fe (about two and three times, respectively), while it is not very important for Ni that, moreover, maintains the same profile trend. For Cd and Fe, on the other hand, the phenomenon is rather intense in the first 50 m where the phytoplanktonic bloom occurs, thus indicating an uptake of these two metals, in accordance with the dissolved metal data (Figures 8.3 and 8.4).

As regards Cu, its profiles have a very similar trend, i.e., the increase involves the whole water column as if the overlapping of two phenomena took place: an uptake by planktonic mass in the surface layer and a very fast scavenging in inter-

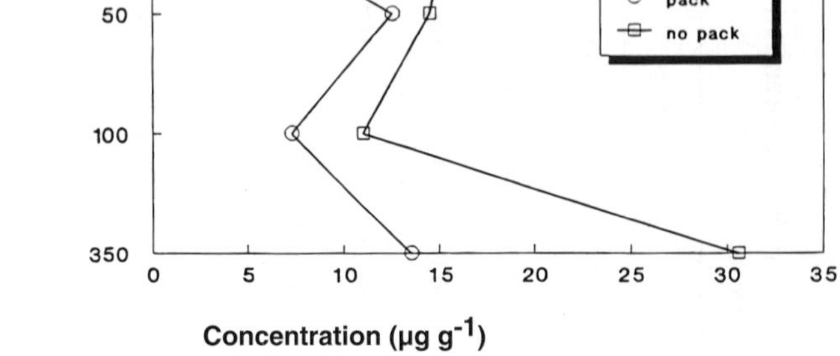

Figure 8.3. Particulate Fe concentration profiles (μg g^{-1}) in the presence and absence of pack ice.

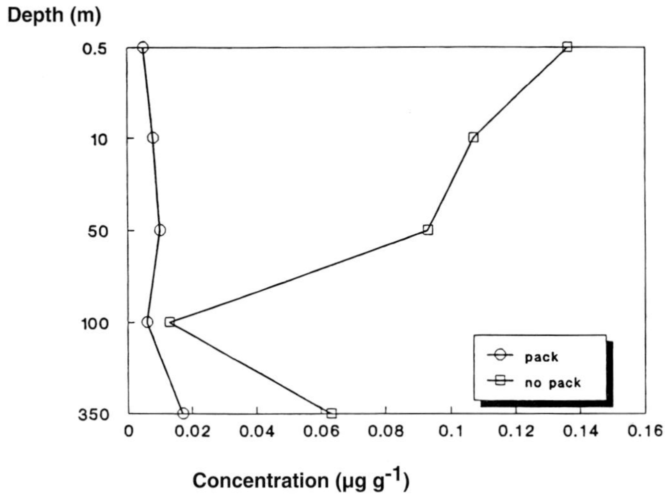

Figure 8.4. Particulate Cd concentration profiles (μg g^{-1}) in the presence and absence of pack ice.

mediate and deep waters. This particular behavior of Cu is not rare: in fact, this metal shows both the characteristics of a 'nutrient type' and a 'scavenging-type' element. Complete hydrological data and nutrient profiles are needed to confirm this preliminary interpretation and more detailed sampling and physical measurements have already been planned.

3.2. Sediments

Using the data obtained in the first expeditions it was easier, with respect to particulate matter, to draw some conclusions on the trace metals concentration in sediments. In general, it was noted that the total concentration of trace elements (Cd, Cr, Cu, Fe, Mn, Ni, Pb and Zn) was lower than that found in marine sediments coming from non-Antarctic seas, especially from those with a strong impact from anthropic areas.

A typical example of the concentration levels is shown in Table 8.4 for samples collected at Terra Nova Bay. In order to obtain a more complete information, a multivariate statistical method, the Principal Component Analysis (PCA), has been applied to all the available data on the total metal concentrations (57, 58).

This procedure was used mainly to achieve a reduction of dimensionality, i.e., to fit a k-dimensional subspace to the original p-variate observations ($p \gg k$). The statistics used to summarize the most important results was the percent of the total variation explained by the first k (usually two or three) components. In the case at hand the interpretation of PCA results was based on the diagrams of coefficients of variables (total concentrations of metals in the samples) and the scatter plot of samples (the stations separated in coastal, intermediate and offshore stations).

Table 8.4. Concentration of metals (μg g^{-1}) in Terra Nova Bay samples (fractions A, B and C, as well as Total)

Station No.	Depth (m)	South latitude	East longitude	Fraction	Cu	Cd	Cr	Fe	Mn	Ni	Pb	Zn
12	0.5	74°18	165°05	A	0.90	0.02	0.05	65	15.0	0.30	1.10	2.50
				B	1.25	0.01	0.50	5.20×10^3	115	2.25	0.80	8.35
				C	7.60	0.04	4.50	3.25×10^4	660	7.40	5.80	50.5
				T	28.5	0.15	14.0	6.05×10^4	1.20×10^3	22.5	8.25	165
21	0.5	74°50	163°55	A	1.15	0.03	0.20	70	6.35	0.60	1.10	2.40
				B	0.80	0.04	0.50	515	8.25	1.15	1.40	1.80
				C	7.55	0.02	20.0	1.71×10^4	240	5.45	10.5	33.5
				T	16.0	0.10	33.0	2.15×10^4	405	16.5	13.0	105
23	23	74°43	164°07	A	0.50	0.04	0.20	45	1.85	0.55	0.30	5.60
				B	0.35	0.02	0.45	445	5.70	0.80	0.80	2.50
				C	5.20	0.02	8.20	1.27×10^4	100	4.70	8.5	24.5
				T	31.0	0.10	17.5	2.32×10^4	450	16.0	10.5	145
29	330	74°45	164°19	A	0.50	0.14	0.25	10	0.85	0.70	0.07	3.45
				B	0.30	0.01	0.45	355	9.00	1.15	0.45	4.00
				C	5.85	0.03	6.00	5.80×10^3	90	3.15	10.5	20.5
				T	30.5	0.20	27.5	1.72×10^4	730	15.0	12.0	155
32	274	74°54	164°11	A	0.60	0.06	025	8	10.5	4.50	0.10	1.85
				B	0.50	0.01	0.60	520	16.0	3.50	0.50	9.35
				C	17.0	0.02	14.50	1.80×10^4	240	8.60	13.5	62.5
				T	33.5	0.15	38.0	2.40×10^4	420	60.0	15.0	120
33	105	74°30	164°11	A	0.95	0.55	0.45	15	5.10	1.50	1.15	12.0
				B	0.60	0.10	0.50	405	7.65	1.75	0.60	5.15
				C	13.0	0.01	16.40	1.29×10^4	190	8.10	10.0	29.0
				T	27.0	0.70	36.5	1.80×10^4	600	40.0	13.5	90.0
34	194	74°39	164°11	A	0.10	0.30	0.30	5	0.70	0.50	0.15	6.30
				B	0.10	0.01	0.63	240	8.70	0.90	0.60	3.60
				C	7.00	0.02	8.60	6.20×10^3	120	3.30	7.35	16.0
				T	22.0	0.35	27.5	1.52×10^4	650	50.0	16.5	82.0
41	454	74°40	164°22	A	0.40	0.15	0.40	10	0.60	0.80	1.40	8.30
				B	0.20	0.02	0.50	405	6.10	1.50	0.20	3.85
				C	7.20	0.02	7.75	6.90×10^3	100	4.50	7.00	21.5
				T	21.0	0.20	26.5	1.60×10^4	505	20.5	12.0	68.5

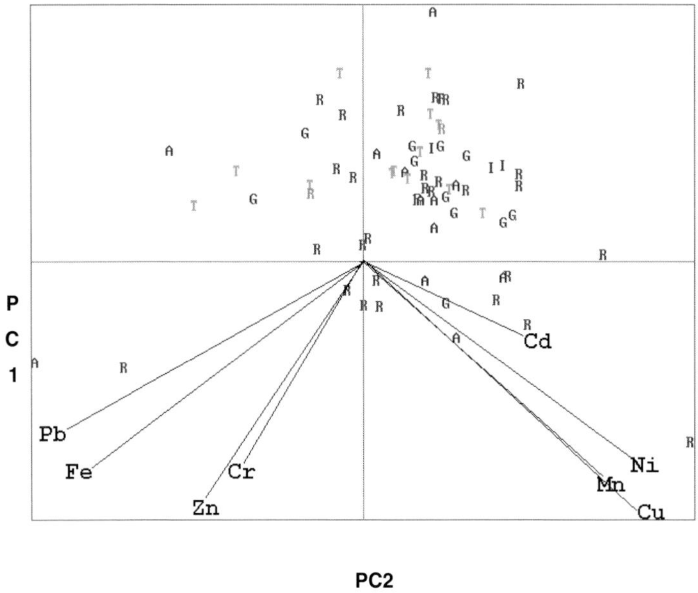

Figure 8.5. The PCA for metal concentrations in Antarctic sediments: *A, G, I* = coastal stations, *T* = intermediate stations, *R* = offshore stations.

The PCA on the data regarding all the Antarctic sediments (1987–1992) gives a significant correlation between the elements (the first three components account for 62.3% of the total variance). In Figure 8.5, which shows the first two principal components, it can be seen that the second component describes the variability of elements. The Cu-Mn-Ni, Fe-Pb and Cr-Zn correlations are particularly evident. The sampling sites instead do not show any appreciable correlation, indicating that there is no geographical diversity between the samples.

Also selective extractions were carried out in order to obtain information about solid speciation of the elements. The results (Table 8.4) show that the metals concentration in the adsorbed, carbonatic, organic and sulphide fractions are in the natural range. The values also show that the extraction sequence is, in general, in the order:

metal concentration in solution A < *metal concentration in solution B* < *metal concentration in solution C*

In terms of percentage, the data showed that a major part of the metals was bound to the silicate or residual fraction. In some samples Cu and Zn were highly concentrated in organic and sulphide fractions. Other studies, based on the extraction with 0.5 *N* HCl, indicate that the samples of this area are not affected by anthropogenic inputs (59, 60).

As said in the Introduction, the humic fraction is one of the most important phase for element speciation and for the pattern followed by the complexed trace metals in the biogeochemical cycles. A preliminary study on marine sediments collected at Terra Nova Bay revealed that the humic fraction consisted only of humic acids, while the absence of fulvic acids had already been observed (21, 61). The sampling stations and humic acid content are reported in Table 8.5, while data on Cr, Cu, Mn, Ni and Zn are shown in Table 8.6 (24). The content of humic acid is variable for all the samples. The amount of the five metals is very different for each sample, but the Cu concentration is about one order of magnitude higher than the others. This is important because Cu is known to have high affinity towards humic acids. Correlation coefficients ($r_{Cr} = 0.81$, $r_{Cu} = 0.85$, $r_{Ni} = 0.84$, $r_{Zn} = 0.67$) show that the concentration of metals measured in the humic fraction actually appears to be representative of the complexed metals, with the exception only of Mn ($r = 0.53$). Each metal has a different partition range (percentile ratio between the metal content in the humic acid fraction and its total content in the sediment): Cr, 0.2 ÷ 1.6; Cu, 4.4 ÷ 39.7; Ni, 0.4 ÷ 4.5; and Zn, 0.1 ÷ 2.8. The value for Mn is less than 0.1, therefore, considering also the 'r' value, one can

Table 8.5. Concentration of humic acids in sediment samples and geographical location of the stations where the sediments were collected

Sample	South latitude	East longitude	Sea depth (m)	Humic acids (μg g^{-1})
12	73°40	171°12	568	73 ± 6
13	74°04	176°49	400	689 ± 34
15	73°44	174°12	383	831 ± 42
24	75°02	169°57	336	273 ± 16
30	74°57	165°47	1113	157 ± 11
31	74°48	164°47	779	842 ± 42
38	74°58	164°48	924	390 ± 23
42	74°51	165°09	753	476 ± 28
46	74°47	165°44	752	561 ± 34

Table 8.6. Content of trace metals in the humic fraction on dried sediment

Sample	Cr (μg g^{-1})	Zn (μg g^{-1})	Ni (μg g^{-1})	Mn (μg g^{-1})	Cu (μg g^{-1})
12	0.064 ± 0.004	<LoD	0.100 ± 0.004	0.259 ± 0.023	1.516 ± 0.089
13	0.367 ± 0.016	0.935 ± 0.004	0.567 ± 0.023	0.226 ± 0.004	7.838 ± 0.141
15	0.255 ± 0.011	0.296 ± 0.012	0.433 ± 0.033	0.319 ± 0.007	8.344 ± 0.116
24	0.204 ± 0.006	0.063 ± 0.003	0.161 ± 0.079	0.118 ± 0.007	2.149 ± 0.118
30	0.079 ± 0.004	<LoD	<LoD	0.046 ± 0.003	2.579 ± 0.121
31	0.686 ± 0.012	1,894 ± 0.017	0.781 ± 0.022	0.303 ± 0.002	6.518 ± 0.078
38	0.217 ± 0.011	0.257 ± 0.008	0.336 ± 0.070	0.332 ± 0.009	5.215 ± 0.125
42	0.234 ± 0.011	0.439 ± 0.014	0.333 ± 0.030	0.299 ± 0.014	3.985 ± 0.518
46	0.336 ± 0.034	0.486 ± 0.003	0.219 ± 0.005	0.206 ± 0.003	2.523 ± 0.030

conclude that Mn measured in the humic fraction is not complexed by humic acids, but is in a different chemical form, which can be extracted by NaOH. For the remaining metals the results obtained in this study seem to reflect the sequence of stability constants that, for the complexes of these metals with humic acids, are in the order Cu > Ni > Zn > Mn (62).

4. Conclusions

The determination of trace metals in Antarctic samples should be characterized by particularly rigorous procedures, especially because of the low levels of concentration and the small amount of sample to deal with. It is therefore necessary to rely on analytical methodologies based on extreme care in sampling and pretreatment to avoid contamination and losses, and to resort to accurate and sensitive instrumental techniques, in order to master as much as possible matrix interferences. During twelve years of participation in the Italian Antarctic Programme, several procedures for the trace metals analysis in the different matrices were developed, starting from the sampling step up to the instrumental determination. These optimized procedures, as described in the experimental section, allow accurate results to be obtained. Data collected up to now offer interesting insights as regards trace metal concentrations and distribution in particulate matter and sediments. It appears that, for particulate matter, pack ice formation and melting processes have a significant effect on the distribution of trace metals, perhaps in part as a consequence of the biological activity linked to this process. The possible effect of the particulate matter entrapped in the pack ice and released during melting has still to be verified. To this aim, during a recent expedition (1997–1998), samples of pack ice have been collected and metals in the dissolved phase and particulate matter are about to be determined.

Metals concentration in sediments show that the Ross Sea and Terra Nova Bay do not suffer from anthropogenic inputs and therefore these values can be assumed to be the natural background levels. The importance of humic acids as regards mobility, concentration and accumulation of trace metals in marine environment, especially of Cu, has been confirmed even for Antarctica. From this point of view, the study of the humic fraction and complexed metals seems to be an interesting opportunity to better understand the role of the sedimentary phase in the biogeochemical cycle of the elements.

The results on the solid phases of Ross Sea, an area where trace metals were poorly investigated in the past, allowed the distribution and the role of these elements in Antarctica to be better assessed. Further expeditions will consider in more detail some peculiar aspects, such as the phenomena at the water-ice interface, and their correlations with biological events.

Acknowledgments

This work was financially supported by the Italian National Research Programme in Antarctica (PNRA).

References

1. E. P. Jones, D. M. Nelson, P. Treguer, Chemical Oceanography, In: W. O. Smith Jr, (Ed.), *Polar Oceanography*, Academic Press, Inc. San Diego, New York, Boston, London, Sydney, Tokyo, Toronto, 1990, 407–476.
2. J. H .Martin, S. E. Fitzwater, Iron deficiency limits phytoplankton growth in the north-east Pacific subarctic, *Nature*, 331 (1988), 341–343.
3. J. H. Martin, R. M.Gordon, Northeast Pacific iron distribution in relation to phytoplankton production, *Deep-Sea*, 35 (1988), 177–196.
4. W. O. Smith Jr, E. Sakshaug, Polar Phytoplankton, In: W. O. Smith Jr, (Ed.), *Polar Oceanography*, Academic Press, Inc. San Diego, New York, Boston, London, Sydney, Tokyo, Toronto, 1990, pp. 491.
5. J. E. Harris, G. J. Fabris, Concentrations of suspended matter and particulate cadmium, copper, lead and zinc in the Indian Sector of the Antarctic Ocean, *Mar. Chem.*, 8 (1979), 163–179.
6. A. Brzezinska, J. C. Van Loon, R. M. Bolicki, Study on the elemental composition of particulate matter from the Southern Baltic, *Proceedings of the International Conference on Heavy Metals in the Environment* (Heidelberg), 1983, 1016–1019.
7. S. Westerlund, P. Ohman, Iron in the water column of the Weddel Sea, *Mar. Chem.*, 35 (1991), 199–217.
8. S. Westerlund, P. Ohman, Cadmium, copper, cobalt, nickel, lead and zinc in the water column of the Weddel Sea, Antarctica, *Geochim. Cosmochim. Acta*, 55 (1991), 2127–2146.
9. R. F. Nolting, H. J. W. Debar, A. J. Vandemrekon, A. Masson, Cadmium, copper and iron in the Scotia Sea, Weddel Sea and Weddel Scotia confluence (Antarctica), *Mar. Chem.*, 35 (1991), 219–243.
10. E. W. Wolff, D. A. Peel, Assessing global and local pollution for heavy metals in Antarctica, *Analusis*, 1994, M41–M43.
11. D. S. Sheppard, J. M. Deely, W. H. L. Edgerley, Heavy metal content of meltwaters from the Ross Dependency, Antarctica, *New Zeal. J. Mar. Freshw. Res.*, 31 (1997), 313–325.
12. F. Baffi, F. Soggia, R. Frache, A. M. Cardinale, Heavy metals distribution in water and suspended particulate matter in Ross Sea, *Ann. Chim.* (Rome), 79 (1989), 607–616.
13. A. Mazzucotelli, B. Cosma, F. Soggia, Trace metals distribution in Antarctic sediments (Terra Nova Bay, Ross Sea) by inductively coupled plasma atomic emission spectroscopy, *Ann. Chim. (Rome)*, 79 (1989), 617–628.
14. R. Frache, F. Baffi, B. Cosma, A. Mazzucotelli, C. Bottinelli, F. Sena, F. Soggia, Trace metals distribution in water, particulate matter and sediments in the Ross Sea and inland lakes (Antarctica), *Proceedings of the Meeting Environmental Impact in Antarctica*, Rome, June 8–9, 1990, 37–46.
15. B. Cosma, R. Frache, A. Mazzucotelli, F. Soggia, Trace metals in sediments from Terranova Bay – Ross Sea, Antarctica, *Ann. Chim. (Rome)*, 81 (1991), 371–382.
16. F. Baffi, A. Cardinale, F. Soggia, R. Frache, Heavy metals distribution in suspended matter in coastal and offshore waters in Ross Sea – Antarctica, *Ann. Chim. (Rome)*, 81 (1991), 523–531.
17. R. Frache, Trace metal distribution in particulate matter and sediments in Ross Sea (Antarctica), *Proceedings of the 2nd Meeting on Environmental Impact – Chemical Methodologies*, Venice, May 26–28, 1992, 60–61.
18. R. Frache, F. Baffi, B. Cosma, A. Mazzucotelli, E. Magi, M. L. Abelmoschi, C. Bottinelli, F. Soggia, Contribution to the study of trace metal concentration in suspended matter and sediments in coastal and offshore waters in Ross Sea (Antarctica), *Proceedings of the 2nd Meeting on Environmental Impact – Chemical Methodologies*, Venice, May 26–28, 1992, 74–82.
19. B. Cosma, F. Soggia, M. L. Abelmoschi, R. Frache, Trace metals distribution in sediments from Terra Nova Bay – Ross Sea, Antarctica, *Intern. J. Environ. Anal. Chem.*, 55 (1994), 121–128.
20. R. Frache, F. Baffi, S. Foggia, B. Cosma, A. Mazzucotelli, E. Magi, P. Rivaro, M. L. Abelmoschi, C. Ianni, C. Bottinelli, M. Ravera, Ricerche sulla distribuzione di metalli pesanti nelle matrici antartiche abiotiche. Ulteriori risultati, *Proceedings of the 3rd Meeting Environmental Impact – Chemical Methodologies*, Venice, March 10–11, 1994, 90–99.
21. E. Magi, T. Giusto, R. Frache, Humic acids in marine sediments: an extraction procedure for the determination of the complexed metals, *Anal. Proc.*, 32 (1995), 267–269.
22. R. Frache, F. Baffi, C. Ianni, F. Soggia, Dissolved and particulate metals in the pack ice melting process in the Ross Sea (Antarctica), *Ann. Chim. (Rome)*, 87 (1997), 367–374.

23. F. Soggia, M. L. Abelmoschi, F. Baffi, M. Grotti, C. Ianni, E. Magi, P. Rivaro, R. Frache, Distribuzione di Fe, Cu, Co e Mn in acque subsuperficiali nel mare di Ross (campagna oceanografica 1994/95), *Proceedings of the 5th Meeting Environmental Contamination*, Venice, February 11–12, 1997, 115–123.
24. E. Magi, Determination of trace metals complexed with humic acids in Antarctic marine sediments. *Chem. Spec. Bio.*, 9 (1997), 67–70.
25. G. Capodaglio, G. Toscano, G. Scarponi, P. Cescon, Lead speciation in the surface waters of the Ross Sea (Antarctica), *Ann. Chim. (Rome)*, 79 (1989), 543–559.
26. G. Capodaglio, G. Scarponi, G. Toscano, P. Cescon, Cadmium complexation in surface sea water of Terranova Bay (Antarctica), *Ann. Chim. (Rome)*, 81 (1991), 279–296.
27. G. Capodaglio, G. Scarponi, P. Cescon, Lead speciation in the Antarctic Ocean, *Proceedings of the Meeting Electroanalysis and Chemometrics of Speciation in Natural Waters*, Liverpool (UK), July, 9–10, 1991, *Anal. Proc.*, 28 (1991), 76–77.
28. G. Capodaglio, G. Toscano, G. Scarponi, P. Cescon, Copper complexation in surface sea water of Terra Nova Bay (Antarctica), *Intern. J. Environ. Anal. Chem.*, 55 (1994), 129–148.
29. G. Scarponi, G. Capodaglio, C. Turetta, C. Barbante, M. Cecchini, G. Toscano, P. Cescon, Evolution of cadmium and lead contents in Antarctic coastal sea water during the austral summer, *Intern. J. Environ. Anal. Chem.*, 66 (1997), 23–49.
30. S. Karlsson, B. Allard, K. Hakansson, Chemical characterization of stream-bed sediments receiving high loadings of acid mine effluents, *Chem. Geol.*, 67 (1988), 1–15.
31. J. A. Davis, Complexation of trace metals by adsorbed natural organic matter. *Geochim. Cosmochim. Acta*, 48 (1984), 679–691.
32. R. O. Hallberg, Diagenetic and environmental effects on heavy metal distribution in sediments: a hypothesis with an illustration from the Baltic Sea. 305–316, A. Kent, F. T. Manheim (Eds.), *Dynamic Environmental Ocean Floor*, 1982, Fanning-Lexington books, Lexington, Massachusetts.
33. W. Salomons, U. Forstner, *Metals in the Hydrocycle*, 1984, Springer-Verlag, Berlin Heidelberg, Germany, pp. 349.
34. W. O. Smith, Preface, (Ed.) W. O. Smith, *Polar Oceanography* Part B, 1990, Academic Press Inc., San Diego, California, pp. 760.
35. U. Forstner, C. Wittman, *Metal Pollution in the Aquatic Environment*, Springer-Verlag, Berlin, 1979, pp. 486.
36. H. Agemian, A. S. Y. Chau, Evaluation of extraction techniques for the determination of metals in aquatic sediments, *Analyst*, 101 (1976), 761–767.
37. R. Chester and F. G. Voutsinou, The initial assessment of trace metal pollution in coastal sediments, *Mar. Poll. Bull.*, 12 (1981), 84–91.
38. A. Tessier, P. G. C. Campbell, M. Bisson, Sequential extraction procedure for the speciation of particulate trace metals, *Anal. Chem.*, 51 (1979), 844–851.
39. B. A. Malo, Partial extraction of metals from aquatic sediments, *Environ. Sci. Technol.*, 11 (1977), 277–282.
40. A. Mazzucotelli, A. Viarengo, G. Martino, R. Frache, Interactions among metals in the spectrochemical determination of trace amounts of cadmium in marine mussels, *Environ. Res.*, 24 (1988), 129.
41. A. Mazzucotelli, F. Soggia, R. Frache, B. Cosma, Electrothermal atomic absorption interactions among trace and major elements in environmental samples: determination of cadmium in the selective and total dissolution of marine sediments, *Appl. Spectrosc.*, 45 (1991), 504–507.
42. R. M. Town, H. K. Powell, Elimination of adsorption effects in gel permeation chromatography of humic substances, *Anal. Chim. Acta*, 256 (1992), 81–86.
43. H. K. Powell, R. M. Town, Solubility and fractionation of humic acid; effect of pH and ionic medium, *Anal. Chim. Acta*, 256 (1992), 47–54.
44. S. Hirata, Phosphorus and metals bound to organic matter in coastal sediments. An investigation of complexes of P, Cu, Zn, Fe, Mn, Ni, Co and Ti by ICP-AES with sephadex gel chromatography, *Mar. Chem.*, 16 (1985), 23–46.
45. P. A. Waller, F. W. Pickering, The lability of copper ion sorbed on humic acid, *Chemical Speciation and Bioavailability*, 2 (1990), 127–138.
46. P. A. Waller, F. W. Pickering, The lability of zinc humate species, *Chemical Speciation and Bioavailability*. 3 (1991), 9–22.

47. P. A. Waller, F. W. Pickering, The effect of pH on the lability of lead and cadmium sorbed on humic acid particles, *Chem. Spec. Bio.*, 5 (1993), 1–22.
48. F. Baffi, M. Fabiano, R. Frache, A. Dadone, Determination des metaux lourds dans la Mer Ligurienne. IV. Cd, Cu, Fe and Ni dans le filtré, dans le particulate and paramètres de l'environnement dans les eaux cotières Liguriennes, *Chem. Ecol.*, 2 (1984), 23–38.
49. M. Fabiano, F. Baffi, P. Povero, R. Frache, Particulate matter in the Genoa Gulf (Summer 1983), *Boll. Oceanol. Teor. Appl.*, 6 (1988), 35–41.
50. M. Fabiano, F. Baffi, P. Povero, R. Frache, Particulate organic matter and heavy metals in Ligurian open Sea, *Chem. Ecol.*, 3 (1988), 313–323.
51. G. Hanschmann, A simple extraction-spectrophotometric method for the estimation of dissolved humic substances in water, *Acta Hydrochim. Hydrobiol.*, 19 (1991), 265–266.
52. J. E. Harris, G. J. Fabris, Concentrations of suspended matter and particulate cadmium, copper, lead and zinc in the Indian sector of the Antarctic Ocean, *Mar. Chem.*, 8 (1979), 163–179.
53. R. Chester, A. Thomas, F. J. Lin, A. S. Basaham, G. Jacinto, The solid state speciation of copper in surface water, particulates and oceanic sediments, *Mar. Chem.*, 24 (1988), 261–292.
54. R. Wollast, M. Loijens, In: J. M. Martin, H. Barth (Eds.), *Water Pollution Research Report*, 13 (1989), 241.
55. D. M. Nelson, W. O. Smith, Phytoplankton bloom dynamics of the Ross Sea ice edge. II. Mesoscale cycling of nitrogen and silicon, *Deep-Sea Res.*, 33 (1986), 1389–1412.
56. D. M. Nelson, W. O. Smith, L. I. Gordon, B. A. Huber, Spring distributions of density, nutrients and phytoplankton biomass in the ice edge zone of the Weddel-Scotia Sea, *J. Geophys. Res.*, 92 (1987), 7181–7190.
57. K. V. Mardia, J. T. Kent, J. M. Bibby, *Multivariate Analysis*, Academic Press, New York, 1979.
58. I. T. Jolliffe, Principal Component Analysis, Springer Verlag, New York, 1986.
59. R. Giordano, G. Lombardi, L. Ciaralli, E. Beccaloni, A. Sepe, M. Ciprotti, S. Costantini, Livelli di elementi in sedimenti marini della Baia di Terranova. Campagna Antartica 1993/94, *Proceedings of the 5th Meeting Environmental Contamination*, Venice, February 11–12, 1997, 108–113.
60. L. Ciaralli, R. Giordano, G. Lombardi, E. Beccaloni, A. Sepe, S. Costantini, Antarctic marine sediments: distribution of elements and textural characters, *Microchem. J.*, 59 (1998), 77–88.
61. L. Campanella, B. Cosma, N. Degli Innocenti, T. Ferri, B. M. Petronio, A. Pupella, Humic compounds in sea water and marine sediments from Antartica, *Int. J. Environ. Anal. Chem.*, 55 (1994), 61–75.
62. S. Hirata, Stability constant for the complexes transition metal-ions with fulvic and humic acids in sediments measured by gel filtration, *Talanta*, 28 (1981), 809–815.

Environmental Contamination in Antarctica
A Challenge to Analytical Chemistry
S. Caroli, P. Cescon and D.W.H. Walton, editors

Chapter 9

Polychlorobiphenyls in Antarctic matrices

Roger Fuoco, Alessio Ceccarini

1. Introduction

Polychlorobiphenyls (PCBs) are a class of non-polar semivolatile organic compounds which includes 209 congeners divided into ten congener classes (Figure 9.1) and are named according to the IUPAC numbering from PCB 1 to PCB 209.

They were synthesized at the end of the 19th century and produced at an industrial level since around 1930. Commercialized under different trade names, i.e., Aroclor, Clophen, Phenoclor, Kanechlor, Fenclor, etc., PCBs have been massively exploited for a variety of purposes. About half of the world production was used as transformer coolants and dielectric fluids in capacitors and the other half as plasticizers, hydraulic and heat transfer fluids, inks, adhesives and many other widely used materials (1–3). The estimated world production in the period 1930–1974 is about 1.2 10^6 tons. Of this *ca.* one third has been released in the environment without any precautions about toxic effects on biota and any care to prevent environmental pollution. More than 60% of them is still in use or deposited in landfills and only 4% has been destroyed or incinerated (3). This has led to the widespread occurrence of PCBs all over the world, even in remote areas (1, 4, 5). PCBs are chemically very stable and their half-life in the environment depends on the number of chlorine atoms in the biphenyl structure. The half-life can be ten to twenty years for higher chlorinated congeners (Figure 9.2), making PCBs one of the most persistent widespread class of environmental pollutants (6). Moreover they are soluble in fatty and lipid-rich tissues and organs of biota where they are accumulated and may act as cancer initiators (1, 2, 7–9). There is also evidence supporting the hypothesis that PCBs may cause reproductive failure in animals (1). The toxicity of a PCB mixture is mainly associated with the presence of non-*ortho* and mono-*ortho* substituted coplanar PCBs (10–13). Biological assay, based on PCB interaction with aromatic hydrocarbon receptors, showed that the most toxic congeners are PCB 15, PCB 37, PCB 77, PCB 81, PCB 126 and PCB 169 (9, 14).

For all these reasons, PCBs have been included in the list of priority pollutants, thus making their monitoring in the environment and studies of their toxic effects on living organisms of prime importance. Many different analytical procedures have consequently been developed, and their diversity relates to the nature of the matrix to be analysed (1, 15, 16).

Congener class	Number of isomers	IUPAC Nos.
Monochloro	3	PCB 1–PCB 3
Dichloro	12	PCB 4–PCB 15
Trichloro	24	PCB 16–PCB 39
Tetrachloro	42	PCB 40–PCB 81
Pentachloro	46	PCB 82–PCB 127
Hexachloro	42	PCB 128–PCB 169
Heptachloro	24	PCB 170–PCB 193
Octachloro	12	PCB 194–PCB 205
Nonachloro	3	PCB 206–PCB 208
Decachloro	1	PCB 209
Total	209	

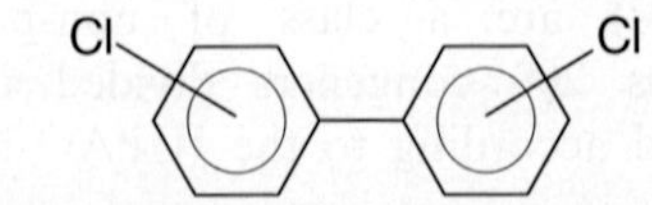

Figure 9.1. Congener classes and number of isomers of PCBs.

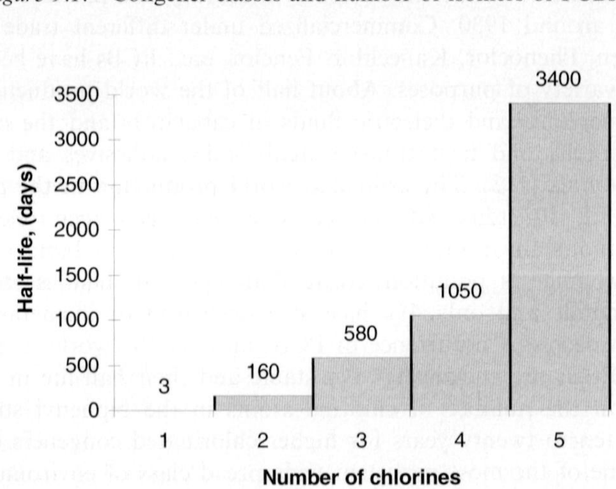

Figure 9.2. Environmental persistence of some PCB congener classes.

A typical analytical procedure can be divided into two parts: the first includes operations generally performed in the field (e.g., sampling, pretreatment if required and storage); the second describes operations generally performed in the laboratory (e.g., extraction of PCBs from the sample, clean-up of the extract, instrumental analysis and data evaluation). In some cases extraction can also be performed in the field. Whenever such complex procedures are used, reliable data can only be obtained if a suitable programme for analytical quality control and quality assurance is run in the laboratory.

Although many papers have been published on the presence of PCBs in several environmental components of different geographic areas, there are very few data on the monitoring of these pollutants in Antarctica. In this respect, it should be noted that the importance of environmental research in Antarctica is mainly related to the very important role that this continent plays in Global Change processes involving the whole terrestrial ecosystem where Antarctica can provide both the global baseline and a valuable historical trend. A substantial contribution to a better understanding of these processes, which is crucial for the survival of all biological species, can be gathered by the observation of suitable chemical parameters in selected geographic areas of our planet, such as Antarctica. It is also worth mentioning that over 70% of the world's fresh water reserves is in the Antarctic ice. Moreover, sediments and ice have acted as recorders of changes over the centuries, thus accumulating very valuable information on the past environmental and climatic events in the depth profile of their chemical composition. Finally, Antarctica is unique in the world and some very peculiar biological species that live there must be protected and managed either because they are not found elsewhere or because they are present in such a huge amount, i.e., krill, that can be considered as a future protein-rich food reservoir. In fact, at the XI Consultative Special Meeting of the Antarctic Treaty held in Madrid in April 1991, Antarctica was defined as a 'natural park and a land of peace and science'. For all these reasons in 1985 the Italian Parliament passed a law on the financial support for several research activities within the Italian National Programme of Research in Antarctica (Programma Nazionale di Ricerche in Antartide, PNRA), including the Environmental Contamination Monitoring Project. The primary aim of this ongoing project is to contribute significantly to a better understanding of the Global Change of the whole terrestrial ecosystem. This can be achieved by monitoring the pollution level of the Antarctic lands and oceans, evaluating its change in the short and long-term by sampling those matrices that retain this information, i.e., sediments, snow and ice, and finally elucidating the diffusion and distribution processes of organic and inorganic pollutants in different environmental compartments.

This chapter includes an overview on the analytical procedures used for the determination of PCBs in environmental matrices and a presentation of the data relevant to the presence of PCBs in the terrestrial ecosystem. Finally, the last section is devoted to the contribution of the PNRA on this matter. A special emphasis is given to the results obtained during several Italian expeditions, particularly in the Victoria Land and the Ross Sea.

2. Determination of PCBs in environmental matrices

In this section a concise overview of the most widely used analytical procedures for the determination of PCBs in environmental matrices (namely, air, sea water, snow/firn/ice, sediment/soil and biota) is given. Regardless of the nature of the sample, the following steps are generally included in an analytical procedure: i) sample collection and storage; ii) sample preparation (extraction of the analytes and cleanup of the extract); iii) instrumental analysis; iv) data evaluation, including analytical quality control.

2.1. Sample collection and storage

Whatever chemical species have to be monitored in a given ecosystem, the first step is to correctly define the information needed; then, the analytical procedures to obtain it can be chosen and tested. Before planning the sampling programme, all the information available on the studied area concerning different chemical, physical and biological parameters which may affect the concentration level of analytes should be carefully considered. Based on this preliminary study, the minimum number of sampling stations, their spatial position and their time frequency can be suitably defined to obtain the information needed.

2.1.1. Atmosphere

The determination of PCBs in the atmosphere is generally performed by sampling large volumes of air (50–4000 m^3) with a rate of uptake ranging from 10 to 800 l min^{-1} (3, 17–22). Depending on the chemical-physical characteristics of the analytes, the content associated with the particulate matter and that present in the gas phase is determined. This can be performed by passing the air through a filtering system which includes a glass fibre filter for the particulate matter and an adsorbing material suitably supported for the volatile fraction. Polyurethane foam, silica gel, Florisil (synthetic magnesium silicate salt) and XAD-2 are the most commonly used adsorbing materials. After sampling, filter and adsorbing material are generally stored at temperature below 0°C in stainless steel or glass containers.

2.1.2. Sea water, fresh water

Surface water samples are generally collected by a Teflon® or stainless steel pumping system without any lubricant or oil. Sampling at different depths can be performed by either Go-Flo or Niskin bottles. The sample volume may vary from 20 to 1000 l (16, 18, 23–26). For large volume sampling, water is passed through a filtering system containing a suitable amount of XAD-2 resin (18, 23, 26). If requested, the water sample is filtered on a 0.45 µm pore size membrane filter and particulate matter is analysed separately. If it is not possible to extract the samples immediately after sampling, they are generally stored at temperature below 0°C in stainless steel containers.

2.1.3. Ice, snow, firn

Ice cores up to about 20 m of depth are generally collected using a metal, hand-operated, ice coring auger with a diameter of about 8 cm. Usually, the auger is painted on the outside with a PCB-free epoxy paint and, before use, it is carefully rinsed with hexane and dichloromethane. For deeper samplings an engine moving system is used. The cores, 10 to 80 cm in length, are wrapped in pre-cleaned aluminium foil and returned to the laboratory, where they are unwrapped, the surface layer scraped with a clean metal scraper, suitably sectioned and finally analyzed (27).

Large quantities of snow are generally collected at each sampling site into either Teflon® bags of 2-mm thickness placed inside cardboard boxes or stainless steel containers. In the case of Teflon bags, they are sealed airtight in the field with a clip sealer. A stainless steel shovel, pre-rinsed with acetone for organic residue analysis, is used to place the snow in the containers, while the operator must use complete clean-room garments to avoid any sample contamination. Samples can be collected on the surface or at different depths down to 4–5 m by excavating a trench. Potential contamination by airplane exhaust on surface snow can be minimized by removing the top 2–3 cm of snow and by sampling upwind of the landing site. The samples are generally stored at –20°C.

2.1.4. Sediment, soil

Surface sediments are generally collected by a stainless steel grab, while a box-corer system is used if depth profile is required. A stainless steel shovel, pre-rinsed with acetone for organic residue analysis, is used to collect soil samples. Also in this case, potential contamination by airplane exhaust on surface soil can be minimized by removing the top 2–3 cm of soil and by sampling upwind of the landing site. Samples are generally stored below 0°C in stainless steel containers.

2.1.5. Biota

This is a very broad matrix category and sampling procedures can vary greatly from each other. In general, analysis is performed on selected tissues and organs, where PCBs are accumulated due to their lipophilic characteristics, or on the whole samples (these, for biota, can be very small in size). Immediately after sampling, tissues, organs or the whole samples should be frozen, homogenised, freeze-dried and finally stored below 0°C in stainless steel or glass containers.

2.2. Sample preparation

2.2.1. Extraction of PCBs

Atmosphere. Extraction of PCBs from both the particulate matter and the adsorbing material used during sampling is generally performed by hexane, dichloromethane and acetone either in a Soxhlet apparatus or in an ultrasonic bath. Petroleum ether, benzene and ethylether are also used (28).

Water. PCBs are non-polar compounds and consequently they are highly soluble in non-polar solvents; this makes their extraction easier from water samples with non-polar immiscible solvents. *n*-Hexane and dichloromethane, or a mixture of them, are the most widely used solvents in liquid-liquid extraction techniques (1, 16, 29–31). Solid Phase Extraction (SPE) and elution with different mixtures of solvents (i.e., acetone/hexane/ dichloromethane) is a very attractive alternative to the liquid-liquid one (29, 32, 33). When large volumes of samples need to be collected, water can be pumped directly through a system containing an adsorbing material suitably supported, i.e., XAD-2, XAD-4, Tenax, C18, C18-NH_2, etc. It can be contained in a column or fixed on a membrane disk (34). SPE has several advantages, i.e., low solvent consumption for the elution of the analytes use in field applications, easy automation and less critical clean-up of the eluate (30, 35). Solid-Phase Microextraction (SPME) is a modified SPE procedure based on the use of a coated fibre, usually made by fused silica, which in many cases eliminates the use of organic solvents (33, 35). Chromatographic stationary phases, such as poly(methylsiloxane), are generally used as chemically bonded coatings of the fibre (36). SPME can be directly coupled with Gas Chromatography (GC) and High Performance Liquid Chromatography (HPLC) (35, 37). However, solid phase extraction has some drawbacks that limit its application to environmental samples, such as low recoveries due to matrix effects, low capacity for samples which have a high content of organic matter and the need of critical calibration procedures for quantitative determinations (35).

Ice/snow/firn. Samples are generally allowed to melt in a clean laboratory and extraction is undertaken as soon as the snow is melted following the same procedures applied to water samples (38).

Sediment, soil. Methylene chloride, petroleum ether, acetone and *n*-hexane, or mixtures thereof, are generally used for the extraction of PCBs from sediment/soil samples. Extraction can be performed either by manual shaking at room temperature or in a more effective way by various apparatuses, i.e., a Soxhlet system, an ultrasonic bath or a microwave oven, at room or higher temperatures. In all cases a wetting agent may be used in the solvent mixture, typically acetone. A very attractive alternative to the classic methods is Supercritical Fluid Extraction (SFE) (15, 39, 40). Sonication at 40–50°C using acetone/hexane/dichloromethane mixtures and wet samples seems to be the fastest and cheapest procedure, though microwave-assisted extraction has higher recoveries (41). Exhaustive comparisons of all these techniques have been reported in the literature (42, 43).

Biological samples. PCBs can be extracted from biological samples by using acetonitrile or cyclohexane/acetone in an ultrasonic homogenizer and pentane/dichloromethane or hexane in a Soxhlet apparatus (23, 44–47).

2.2.2. Clean-up of the extract

The clean-up procedure of the organic extract is very important since it allows for the reduction of any possible interferences from matrix components or other organochlorine compounds which may be present in the sample and are co-extracted with PCBs. Clean-up is generally performed by column chromatography on silica/alumina or Florisil suitably activated (or deactivated) and checked by standard solution in order to find out both the best solvent or mixture of solvents and the optimum volume to be used for selectively eluting PCBs and leaving interferents in the column (1, 5, 30). *n*-Hexane and dichloromethane are the most widely used solvents (1, 5, 30, 48).

In the case of sediments, sulphur should be removed from the organic extract. This can be performed by treating the solution with different desulphurizing reagents (tetrabutylammonium chloride, Cu and/or Hg, $AgNO_3$, *etc.*). Mercury and activated Cu powder are the most widely used (5, 49).

Gel permeation chromatography, sulphuric acid and polyethylene film dialyses have been used for lipid elimination in the organic extract of biological samples (45, 47, 49–53). Lipids can also be eliminated by alkaline alcohol digestion of the sample (54). After lipid elimination the extract can be further cleaned by column chromatography on Florisil or alumina/silica (23, 46–48, 51, 55).

2.3. Instrumental analysis

The choice of the most suitable instrumental technique depends on several factors, such as the physical-chemical characteristics of analytes, the detection limits required, the level and type of interferences, the resolution needed, the identification power required, the accuracy and the precision of the quantitative determination, the availability of instrumentation and finally the cost and the time necessary per each determination. Moreover, extraction and clean-up procedures have to be suitably matched with instrumental analysis. GC coupled with Electron Capture Detection (ECD) or Mass Spectrometry (MS) has been widely applied for the determination of PCBs in organic extracts of environmental samples. In few cases the instrumentation includes the extraction step, such as an SFE system coupled with Supercritical Fluid Chromatography (SFC) or with GC (40).

2.3.1. Gas chromatography

GC on a fused silica capillary column with an MS detector should be used whenever possible for the analysis of organic compounds at trace level in complex mixtures. In fact, it allows the extremely high resolution of GC to be combined with the very high sensitivity and identification power of MS, which makes it possible to determine an analyte at low pg ml^{-1} levels in the final organic extract. However, GC-ECD is very common for PCB determination since it is both the most sensitive and the less expensive technique for chlorinated compounds (5). PCBs can be separated on a 30–50 m fused silica capillary column with 5% phenyl –95% methylpolysiloxane chemically bonded stationary phase (1). On-column injection is very often used, while several oven temperature programmes have been applied for PCB determinations. The initial temperature is generally 10–15°C lower than the boiling point of the solvent and the final one does not exceed 290–300°C.

2.3.2. SFE on-line with GC-MS

Supercritical Fluids (SFs) allow analytes to be extracted from solid samples, i.e., marine sediments, faster and more efficiently since they have lower viscosity and higher diffusivity than liquid solvents (56). CO_2 is the most widely used supercritical fluid with or without a modifier, e.g. methanol and toluene. A very exhaustive discussion on the role of a modifier in the enhancement of the extraction efficiency was recently published (39). Few procedures have been described in the literature based on SFE of organic pollutants from environmental samples, including PCBs and PAHs (39, 41, 56–59). Generally, the extraction is performed off-line the chromatographic-detection system, i.e., GC-MS, and only a few examples of on-line SFE-GC coupling are reported in the literature (40, 41, 57, 60).

2.4. Data evaluation

2.4.1. Identification and quantification

The major drawback when comparing analytical data from different bibliographic sources is mainly related to the method of PCB quantification since there is not yet a standard procedure generally accepted. The following have been the most commonly used: i) total PCB concentration expressed as equivalent quantity of an Aroclor or Clophen mixture taken as reference and calculated on the basis of the area of the most representative chromatographic peaks; ii) individual concentration of only a limited number of selected congeners. The most widely measured congeners are IUPAC Nos. 28, 52, 101, 118, 138, 153 and 180. In this respect, it is worth mentioning that the experimental total PCB concentration, as obtained by measuring a suitable number of the most abundant congeners (about 30 to 60, depending on the concentration level), was generally in a quite good agreement with the value obtained by calculating the sum of the above mentioned seven selected congeners and multiplying it by a factor of 4 for both air and water samples and a factor of 3.5 for both biological and sediment samples, as reported in the literature (7, 19, 61); iii) individual concentration of all identified congeners. In this case, the total PCB content and the congener class distribution can be easily obtained.

In this respect it should be pointed out that the quantification of PCBs either as the sum of PCB congeners or as an estimate to an equivalent quantity of a technical mixture, e.g., Aroclor 1242, Aroclor 1254 or Clophen 60, does not allow valuable information to be gained about the source of contamination and about their environmental fate which is related to the chemical and physical properties of every individual congener (19). As for the determination of individual PCB congeners, they are firstly identified by GC-MS on a standard solution of several Aroclors (i.e., 1221, 1232, 1248 and 1260). The Relative Retention Time (RRT) for each identified congener is then calculated by using one or more Internal Standards (ISs). RTTs are finally applied for chromatographic peak assignment of real samples which can be analyzed by either GC-MS or GC-EC. Experimental Response Factors (RFs) are generally obtained for a limited number of selected PCB congeners (i.e., IUPAC Nos. 13, 28, 35, 52, 81, 101, 118, 127, 138, 153, 169 and 180) and for the ISs, in a suitably selected concentration range. Relative Response Factors (RRFs) to the ISs are then calculated and used in turn to calculate the RRFs for all congeners by extrapolating the values reported by Mullin et al. (62). If an MS detector in the Selected Ion Monitoring (SIM) mode is used, at least three ions should be selected, one as target and two as qualifiers. In this way, it is much easier to evaluate the content of each congener class, from trichloro(Cl_3) – to nonachloro(Cl_9)-biphenyl, and to get a more accurate determination of the total PCB content of the sample. Final extracts of real samples are analyzed after adding a known amount of an IS and quantification can be performed by using commercial computer programmes, which automatically assign chromatographic peaks on the basis of RRTs and calculate the concentration of each congener on the basis of RRFs and the IS concentration/peak area ratio. Moreover, there are some specific data evaluation methods according to the nature of the sample. For instance, in the case of sediment samples, it can be assumed that every particle is coated with a thin layer of organic matter, mainly humic acid, on which organic pollutants are adsorbed. This means that the total amount of organic pollutants is much more likely to be related to the particle surface area per volume unit than to the mass unit of sample (16). The Calculated Specific Surface Area (CSSA) can be obtained by particle size analysis and it is expressed in $m^2\ cm^{-3}$. Comparisons among concentration values of organic pollutants relevant to samples with different particle size distribution may lead to erroneous conclusions if they are expressed in a conventional way, i.e., ng g^{-1} dry weight. For example, a difference up to a factor of four was observed among the total PCB concentrations (ng g^{-1} dry weight) of sediment samples from the Ross Sea (Antarctica), which might indicate a non-uniformity of the pollution level in the studied area (5). These differences became less than 20–30% when PCB concentrations were normalised by dividing them for the calculated specific surface area of each sample (5).

Finally, for biological samples the PCB concentration is generally expressed as g g^{-1} of wet or dry weight of sample and as g g^{-1} of extractable lipids since, as already stated, PCBs are lipophilic in nature and accumulate in lipid-rich tissues and organs. The latter approach allow for better data comparison. It is also worth mentioning that the determination of non-*ortho* and mono-*ortho* substituted PCBs is often performed in these samples, since they are the most toxic congeners (13, 55).

2.4.2. Analytical quality control

The main goal of every procedure for analytical quality control is to allow data within assigned values of accuracy and precision to be obtained. Analytical quality control is primarily achieved by the use of Certified Reference Materials (CRMs) and participation in intercomparison exercises, though calibration solutions and spiked samples can also be used (43, 63, 64).

2.4.3. CRMs

CRMs are the most useful tool for analytical quality control (43, 63, 65). However, spiked samples might be an alternative in many cases, though it must be remembered that spiked analytes generally behave differently from native ones (66). For a correct use of CRMs, their analysis should be scheduled within the time sequence of the analysis of real samples and the results should be reported, for example, on a working analytical control chart (43).

Intercomparison exercises. Participation in intercomparison exercises is a unique opportunity for a laboratory to assess the quality of its analytical capability. Also, they are very useful for estimating the inter-laboratory coefficient of variation for that specific analysis.

Calibration solutions. Calibration solutions are very useful to test an analytical procedure, but their preparation and storage are still one of the main sources of error in the determination of PCBs (43, 63). Certified analytes at purity higher than 99% should be preferred for preparing calibration solutions following suitable procedures, although commercial standard solutions in various solvents are often used (43, 63).

3. The presence of PCBs in the terrestrial ecosystem

The global distribution of a chemical is related to four main processes, namely the local and regional transport (1–100 km), the mesoscale transport (100–1000 km), the hemisphere long range transport (1000–40,000 km) in both the atmosphere and the hydrosphere, where the exchange between environmental components and the multi-phase distribution play a very important role, and finally the inter-hemisphere exchange mechanism. Abiotic and biotic transformation processes should also be taken into consideration. A mesoscale rather than a hemispheric or even global diffusion process seems to be mainly responsible for aerosol transport of semivolatile organochlorines (21).

As far as the persistent anthropogenic organic chemicals are concerned, i.e., PCBs, the main sources are located in the northern hemisphere, where the major industrial nations and most of the global population are concentrated, even though many of these nations banned, or at least severely restricted, PCB usage in the mid-1970s and began several monitoring programmes to evaluate the time-dependent trend of the environmental pollution level. One of the most important findings of these studies is the increasing understanding of the global distribution pattern for the volatile C_1/C_2-halocarbons, though that relevant to the semivolatile halogenated C_6/C_{14}-compounds is still far from being fully elucidated. In particular, PCBs have been found in nearly every environmental compartment regardless of the nature and the geographic provenience of the sample. For this reason they are considered ubiquitous pollutants and have been one of the most studied class of persistent semivolatile organic micropollutants for a long time. Consequently, a very consistent amount of data for many environmental matrices all over the world have been published. In general, their concentrations in the readily exchanged global compartments, i.e., the atmosphere and the ocean, are in the pg m^{-3} or ng m^{-3} range, respectively, owing to their low vapour pressure (10^{-2}–10^{-3} and 10^{-4}–10^{-5} Pa for lower and higher chlorinated congeners) and low water solubility (10^{-3}–10^{-5} and 10^{-6}–10^{-7} mol m^{-3} for lower and higher chlorinated congeners) (21).

In the following sections several environmental matrices are considered, namely, atmosphere, sea water/fresh water, ice/snow/firn, sediment/soil and biota, collected in different geographic areas either highly industrialized, i.e., the Mediterranean basin or remote, i.e., open oceans and the polar regions. A selection of data on the PCB content is presented and compared whenever possible.

3.1. Atmosphere

The main source of atmospheric pollution is the industrial use of PCB technical mixtures obtained by $FeCl_3$-catalyzed chlorination of biphenyl. Thermal dimerization of chlorobenzenes as a results of incomplete combustion processes also might be responsible of the formation of PCB complex mixture, typically rich in higher chlorinated and planar congeners (67, 68). Several studies of both the global baseline PCB air pollution and the relevant short and long range transport mechanisms have been carried out in both the northern and the southern hemisphere, including highly industrialized areas such as the Mediterranean basin, central Europe, the Great Lakes and polar regions as well (3, 4, 17–19, 21, 22, 69, 70). Table 9.1 shows a selection of the published data.

Table 9.1. Total PCB content and individual congener concentration (pg m^{-3}) in the atmosphere

	USA (Great Lakes)			Central Europe (Germany)	Mediterranean Sea	
	Winter	Winter	Summer		Winter	Summer
PCB 28	12–115	20–120	80–450	10	–	–
PCB 52	17–69	3–10	9–32	24	–	–
PCB 101	9–20	–	–	41	–	–
PCB 118	2–7	0.2–1	0.6–5	17	–	–
PCB 138	2–8	–	–	52	–	–
PCB 153	4–7	4–18	1–6	52	–	–
PCB 180	1–4	–	–	23	–	–
Total PCB	272–290[b]	135[b]	1820[b]	830[a]	50–80[b]	500–900[b]
References	21	22	22	21	4,71	4,71

	Southern Atlantic Ocean	Northern Atlantic Ocean	Southern Indian Ocean	Island Arctic Ice	Antarctica		
					Syowa Station	Balleny Island	Antarctic Peninsula
PCB 28	2–10	1–7	2–4	–	–		–
PCB 52	12–41	5–35	3–6	–	–	–	–
PCB 101	10–28	3–23	4–8	–	–	–	–
PCB 118	1–9	0.5–3	–	–	–	–	–
PCB 138	3–21	1–8	2–9	–	–	–	–
PCB 153	3–18	1–6	3–7	–	–	–	–
PCB 180	≤0.5–5	<0.2–1.5	0.2–4	–	–	–	–
Total PCB	136–385[a]	48–313[a]	60–140[a]	9–20[b]	81[a]	96[a]	140–280[b]
References	19	19	69	3	18	18	17

[a] Value calculated as four times the sum of the seven selected congeners
[b] Aroclor 1254 equivalent concentration
[c] Aroclor 1242 equivalent concentration

From all these studies the following conclusions can be drawn: i) the total PCB concentration range is rather ample. Concentration as high as 2000 pg m^{-3} and as low as 20 pg m^{-3} for industrialized and remote areas have been measured, respectively (19, 21); ii) seasonal variations in the tropospheric content of PCBs studied in different geographic areas, namely the Great Lakes and the Mediterranean Sea, showed an increase during summer by a factor of about ten, without changing the congener class distribution (4, 22, 71); iii) the lowest tropospheric total PCB concentration (9–20 pg m^{-3}) has been reported for the Arctic region. This value is consistently lower than that for the North Atlantic Ocean. The values measured in the Antarctic troposphere (70–300 pg m^{-3}) are practically the same of those found in both the Atlantic and Indian sectors of the Southern Ocean. The latter are not so far from those measured in the Mediterranean Sea in winter. It is worth mentioning that a very high content of PCBs was measured in the Antarctic atmosphere at Ross Island-Cape Evans during the 1988 summer. The total PCB content was about 12 ng m^{-3} which is 600 times higher than the geometric mean obtained for the same area in a two-year observation time (1988 and 1989). A contamination accident occurred at McMurdo Base on Ross Island, followed by the volatilization of PCBs from the dumping site and the atmospheric transport could have caused the observed increase (72); iv) the gas-particle distribution depends on the temperature, PCB vapour pressure and the mean aerosol surface. High surface per volume and low temperatures favour the adsorption of PCBs on the particles, whereas low surface per volume and high temperatures transport PCBs in the gas phase. In open ocean atmosphere PCBs are predominantly present in the gas phase (18). In the Great Lakes area, where the content of particulate matter in the atmosphere is in the range of 0.5–3 ng m^{-3} level, more than 90% of the total PCB content was found in the gas phase at a sampling temperature of –20 – –10°C (22). In an urban area of central Europe, where the content of particulate matter in the atmosphere is much higher, more than 60% was found in the particle-phase at a sampling temperature of –4 – –13°C (21); v) the sum tri + tetra + penta- and tri + tetra + penta + esa-chlorobiphenyls account for about 50% and 90% of the total PCB content, respectively; vi) regional, mesoscale and hemispheric PCB transport occurs mainly by air mass movement either in the gas phase or in the particle phase; vii) interhemispheric exchange of semivolatile organochlorines should mainly occur at the ocean surface, where the air-water dynamic exchange equilibrium is established either by water coevaporation, film droplet formation or a combination thereof (21). This conclusion is based on a model which separates the global troposphere into two almost independent hemispheres. This model has been already validated for some chlorinated C_2 compounds and also holds for semivolatile organochlorine compounds (73). As a matter of fact, the exchange time of persistent contaminants in the tropospheres of the two hemispheres is in the range of one year (19, 74); viii) a ratio of about one for the tropospheric total PCB content between the northern and the southern hemisphere is widely accepted, though the major part of the PCB input into the environment is due to human activities in the northern hemisphere.

3.2. Sea water, fresh water

The hydrosphere plays a very important role in the global diffusion mechanism of contaminants, particularly in the inter-hemispheric exchange process, as stated above. Moreover, in the marine environment biomagnification occurs, so that information on the pollution level can contribute to a better evaluation of possible associated risks for organisms, particularly for humans. Moreover, the marine environment is the most important pathway for removing PCBs and thus reducing the pollution level. In the evaluation of the PCB presence, it is very important to characterize the distribution of the total PCB content between the water phase and the suspended particulate matter. Depending on the phase association of organic species, they may be incorporated onto the underlying sediments or transported towards the open sea. It is generally accepted that sediments constitute a sink for the more hydrophobic compounds, including PCBs. Less water soluble chlorinated hydrocarbons with high affinity to suspended materials are readily removed from surface water by sinking particles. Active removal of these chemicals is strongly associated with the primary productivity in the oceans. Table 9.2 shows a selection of the published data.

The highest PCB concentrations in sea water are found in the Mediterranean Sea. The total PCB concentration ranges between 230 and 8600 pg l^{-1} (4). It is clearly demonstrated that these high PCB concentrations are mainly due to the high number of industrialized sites near the coast. Concentrations vary widely in coastal waters, whereas offshore levels are lower and generally more uniform. River waters contribute also to the suspended particulate matter concentration. High PCB concentrations are also due to a very low exchange between Mediterranean and oceanic sea water. Moreover, PCB content of both the water and the particulate phase of surface coastal sea water sampled near Barcelona parallel rather well that found in water samples of the Besos River which flows into the sea in the area studied (75). With few exceptions, the K'_d values were similar in both water systems indicating that the different content of suspended solids and natural colloids do not influence significantly the PCB distribution in the water phase. The composition of the particulate fraction collected near the bottom indicates that sediment remobilization is mostly contributing to this material, which is slightly enriched in PCBs due to their affinity with fast sinking particles. Dilution of urban wastewater into coastal sea water and sediment remobilization can be considered as the dominant physical-chemical processes for the distribution of organic pollutants in these coastal environmental compartments.

Another well studied marine environment with low exchange with the oceanic streams is the Baltic Sea (76, 77). In this area there are only few local sources of PCB pollution, in contrast with the Mediterranean basin, and a lower PCB concentration in sea water was observed, with a rather homogeneous distribution. Winter data seem quite contradictory and vary from 360 pg l^{-1} to 14 pg l^{-1} (76, 77). The total PCB concentration ranges were the following: Baltic Sea 12–71 pg l^{-1} in summer and 14–630 pg l^{-1} in winter; North Sea 61–217 pg l^{-1} in winter and 203–2859 pg l^{-1} in summer. Authors ascribe the anomalous concentration in North Sea (Belt Sea) in summer (exclusively related to the suspended matter) to the human

Table 9.2. Total PCB content and individual congener concentration (pg l^{-1}) in sea water

	Northern Europe							
	Baltic Sea			English Channel		North Sea		
	Winter	Winter	Summer	Winter	Summer	Summer	Winter	Summer
PCB 28	–	–	–	2.8–2.3	1.2–2.0	8.1–39.4	–	–
PCB 52	–	2.5–11.9	0.5–5.3	2.2–1.8	1.0–2.6	3.5–28.0	22.0–43.3	0.1–1.7
PCB 101	–	0.5–6.1	0.3–6.4	1.2–3.0	1.0–4.4	4.5–29.8	9.8–18.7	0.3–1.4
PCB 118	–	0.2–3.1	0.2–1.9	0.9–1.1	0.4–2.3	2.1–11.6	2.5–6.1	0.1–0.6
PCB 138	–	1.4–11.5	0.9–8.5	0.9–3.4	1.2–6.9	5.6–29.1	5.6–11.5	0.1–0.4
PCB 153	–	0.5–2.1	0.8–8.5	0.9–1.1	0.7–5.0	2.4–17.1	5.6–12.6	0.2–0.9
PCB 180	–	0.2–2.9	0.5–4.2	0.9–2.4	0.8–3.9	2.9–22.6	0.7–1.8	0.1–0.3
Total PCB	130–630[a] 50–189[a]	14–74[b]	12–71[b]	28.7–40.0[b]	13.7– 53.7[b]	74–415[b]	133–237 4–7[c]	2–14 203–2850[2]
References	76	77	77	26	26	26	77	26

	Atlantic Ocean		Mediterranean Sea	Adriatic Sea	Antarctica	USA	USA
	Northern	Bermuda	Western		Syowa Station	Great Lakes	Nova Scotia
PCB 28	2.5	11–20	–	–	–	–	
PCB 52	2.6	10–22	–	–	–	–	
PCB 101	1.4	16–57	–	–	–	–	
PCB 118	0.5	9–25	–	–	–	–	
PCB 138	2.1	26–88	–	–	–	–	
PCB 153	0.6	30–88	–	–	–	–	
PCB -180	1.0	17–85	–	–	–	–	
Total PCB	28.1[b]	124–353	230–8600	2000–7500	35–72	470–1600	3100
References	26	88	4,75	5	18	89	51

[a] Sum of 23 congeners
[b] Sum of 31 congeners
[c] Chlorobiphenyls in suspension

activities in this area (i.e., dredging and dumping). Total PCB concentration in this areas were higher in summer than in winter (winter, 98–133 pg l^{-1}; summer, 241–550 pg l^{-1}). The higher values in summer corresponded to the lower salinity observed in that period, which in turn reflected a higher contribution of river water input. No significant differences were evident in congener distribution from winter to summer. In the open North Sea, PCBs were mainly present in the water phase due to a low content of suspended matter (26). In winter, the content of suspended matter was generally lower than 0.1 mg l^{-1}, and the PCB concentration associated with it was below the Limits of Detection (LoDs) of 0.05 pg l^{-1}. In summer, the content of suspended matter increased up to 2.0 ng l^{-1} and a detectable concentration of PCBs was observed.

PCB concentration levels similar to those found in the Mediterranean Sea were observed in water samples from Lake Michigan. A mean concentration of 640 pg l^{-1} was measured in this lake, whilst along the coast from Chicago to Milwaukee (which is the most industrialized area in the southern part of the lake), the PCB concentration reached 1600 pg l^{-1}. Finally, a mean PCB content of 55 pg l^{-1} were measured in sea water samples from Syowa Station (Antarctica) (18). PCB concentrations were nearly the same in sea water samples collected under fast ice and from outer margin of pack ice, though small differences were evident in the congener class distribution. sea water samples collected under fast ice showed a very low percentage of higher chlorinated PCBs as compared with those from snow and fast ice. This result might be explained if one considers that less water-soluble chlorinated hydrocarbons with high affinity to suspended materials are readily removed from surface water by sinking particles, which in turn is strongly associated with the primary productivity in the oceans. In general, it was evident that about 30% of the total PCB content was associated with the particulate matter and also that the sum of tri + tetra- and tri + tetra + penta + hexachlorobiphenyl concentrations accounts for about 40% and 60–80%, respectively, of the total PCB content in the water phase, whereas the particulate one shows a PCB distribution shifted toward higher chlorinated congeners.

3.3. Marine sediments

Data on the PCB content of marine sediments are available for different geographic areas. Very high levels were observed in sediment samples from the Gulf of Bothnia (Northern Europe) (3000–16,000 pg g^{-1}, dry weight), whereas very low concentrations (130–140 pg g^{-1}, dry weight) were observed in samples from Bering Sea and Chukchi Sea (North Sea) (53, 78). Quite high levels were also surprisingly found in the Gulf of Alaska (about 2000 pg g^{-1}, dry weight) (78). PCBs in sediment samples from the Mediterranean Sea vary over a wide range (1000–4000 pg g^{-1} dry weight) and reached very high levels along the coastal line near highly industrialized zones (about 500 and 70 ng g^{-1} dry weight at Rhone river estuary and Ebro river estuary, respectively) (4, 79). Finally, PCB levels as low as 70 ng g^{-1} dry weight were measured in Antarctic sediments, though a very high concentration (up to 4200 ng g^{-1}, dry weight) was found at Winter Quarters Bay (Ross Island) due to a contamination accident, already mentioned in the previous section on Atmosphere (49, 72).

3.3.1. Snow and ice

Snow and ice samples collected at Syowa Station (Antarctica) showed much higher concentrations of PCBs (200–1000 pg l^{-1}) in comparison with water samples. This implies that snow and ice act as transfer media for these pollutants from the atmosphere to the marine environment. However, no significant differences in the PCB content between surface and deep snow samples collected at Mizuho Station were observed, which might support the conclusion that a constant load of these pollutants is present in Antarctica since 1960s (18).

3.3.2. Biological samples

The determination of PCBs has been performed on a variety of biological samples. It is mainly aimed to evaluate the uptake process of these persistent contaminants by the biota in the ecosystem under study and also to elucidate both the biomagnification and the accumulation processes inside the organism. Biota can take up these pollutants by feeding or directly from water, air and the associated particulate matter. Biomagnification is directly related to the trophic transfer in the prey-predator sequence of the overall food chain, i.e., plankton, fish (from smaller to larger), birds and mammals, and may lead to an increase of several orders of magnitude of the PCB concentration. The metabolic pattern and the bioaccumulation process are strongly affected by the relative position of Cl atoms in the biphenyl structure. On this basis, PCBs have been divided into the following classes: i) adjacent hydrogen atoms (H) at the *meta* (*m*) and *para* (*p*) positions only (PCB 52, 44, 101 and 149); ii) adjacent H at the *orto* (*o*) and *m* positions and one *o*-Cl only (PCB 105, 118 and 156); iii) adjacent H at the *o* and *m* positions and two *o*-chlorines (PCB 128, 158, 138 and 170); iv) no *o*-chlorine and substituted in the 3,4 or 3,4,5 positions (PCB 77, 126 and 169); v) none of the above classes (PCB 153, 180 and 194) (80). PCBs of groups 3 and 5 do not have a sufficient planarity to aid the metabolic process, whereas the other groups are metabolised to a grater extent.

Since the accumulation in specific tissues or organs is directly related to the lipid content due to the lipophilic character of PCBs, the concentration is generally expressed as g g^{-1} of extractable lipids. Table 9.3 shows a selection of the data published. Blubber samples are very often analyzed due to the very high lipid content (about 80–90%), even though brain, liver and muscle samples are also taken into consideration. Marine mammals are considered as good indicators of the pollution level since they have a long life-span and a high position in the marine food webs. Sample of dead neonatal and stillborn northern fur seals (*Callorhinus ursinus*) from Pribilof Islands (Alaska) were analyzed to evaluate the PCB level in these organisms at their earliest stage of life (61). Total PCB content on a lipid weight basis in blubber, liver and brain was about 1400, 550 and 60–350 ng g^{-1} extractable lipids, respectively. The benthic fauna is suitable for monitoring regional environmental contamination since it is stationary and it provides the staple diet of many fishes and other predators. The total PCB content in three benthic species from the Gulf of Bothnia (a subarctic industrialized region exposed to high loads of environmental pollutants), namely amphipods, isopods and fourhorn sculpins ranged between 400 and 2200 ng g^{-1} extractable lipids (53). The PCB content in blubber samples from eastern harp seal (*Phoca groenlandica)* collected in the White Sea (1993) ranged from 800 to 5500 ng g^{-1} extractable lipids (45). Very high level of PCBs were found in Scottish marine mammals. In particular, 9020, 19,700, 21,700 and 49,700 ng g^{-1} extractable lipids were measured in blubber samples of whale (*Physeter macrocephalus*), common seal (*Phoca vitulina*), bottle-nosed dolphin (*Pursiops truncatus*) and harbour porpoise (*Phocoena phocoena*), respectively (80). In brain samples of an adult dolphin which was caught near the coast of Massachusetts the total PCB content ranged between

Table 9.3. Total PCB content and individual congener concentration (ng g^{-1} extractable lipids) in organisms

	Sub-Arctic			Northern Europe					
	Alaska			Gulf of Bothnia	Scottish coast				The Netherlands
	Brain (a)	Blubber (a)	Liver (a)	Whole	Blubber (c)	Blubber (d)	Blubber (e)	Blubber (f)	Blood (g)
PCB 28	1–4	8	6	4–17	–	–	–	–	–
PCB 52	3–13	22	14	3–17	115	276	1579	127	–
PCB 101	2–17	14	12		369	351	848	187	–
PCB 118	2–12	78	26	4–47	369	322	516	84	–
PCB 138	4–28	101	39	40–380	354	1287	3040	1131	–
PCB 153	4–25	147	48	40–580	900	2831	6203	3262	–
PCB 180	1–9	26	7	20–260	431	1050	1594	822	–
Total (h)	60–368	1386	532	400–2200	9020	21700	49700	19700	4300 (i)
Lipid (%)					13	88	79	85	
References	61	61	61	53	80	80	80	80	84

Northen Europe		USA		Atlantic Ocean					Mediterranean Sea
The Netherlands (Urban)		Massa-chusetts	Cali-fornia	Falkland islands					
	Adipose (g)	Brain (i)	Brain	Muscle (i)	Liver (n)	Brain (n)	Whole	Muscle (p)	Muscle (q)
PCB 28	8	10	1.9	<3	4.2	4.2	3.5	7.5	14.3
PCB 52	2	83–100	10.2	<3	20.8	4.2	7.1	18.9	22.9
PCB 101	7	75–95	109	<3	8.3	2.5	11.7	17.0	57.1
PCB 118	68	65–80	188	8.0	25.0	8.3	12.9	26.4	165.7
PCB 138	190	242–320	235	18.7	33.3	11.7	30.6	58.5	363.8
PCB 153	260	260–377	43.1	13.0	50.0	19.2	32.9	84.9	366.1
PCB 180	155	81–121	2055	7.3	16.7	6.7	17.6	30.2	208.0
Total PCB[a]	2400	2856–3787	51	175.0	554.0	198.7	407.2	852.0	4190
Lipid (%)				6.4	2.4	1.2	0.85	0.53	1.75
References	84	61	61	46	46	46	46	46	81

(a) Northern Fur seal (*Callorhinus ursinus*); (b) benthic species; (c) whale (*Physeter macrocephalus*); (d) bottle-nosed dolphin (*Tursiops truncatus*); (e) harbour porpoise (*Phocoena phocoena*); (f) common seal (*Phoca Vitulina*); (g) human; (h) calculated value as 3.5 times the sum of the seven selected congeners; (i) as Clophen A60; (l) adult dolphin (*Delphinus delphis*); (m) northern elephant seal (*Mirounga angustirostris*); (n) Gentoo penguins (*Pygoscelis papua*); (o) Argentinian hake (*Merluccius merluccius hubbsi*); (p) flying fish (*Cypselurus cyanopterus*); (q) red mullet (*Mullus barbatus*)

2700 and 3700 ng g^{-1} extractable lipids (61). PCBs were determined in muscle tissue samples of red mullet (*Mullus barbatus*) from the Mediterranean coast of Spain. A mean total PCB content of about 4000 ng g^{-1} extractable lipids was found (81). Gentoo penguins (*Pygoscelis papua*) have a high position in the food chain in the Southern Atlantic and Antarctic waters. Moreover, they are suitable bioindicators for monitoring the baseline pollution of the sub-Antarctic region since they are only influenced to a minimal extent by industrial areas due to their circumpolar occurrence and their restricted migration pattern. The most northely position in which Gentoo penguins are only very exceptionally found is 43° S on the coasts of Argentina (82). The highest PCB concentration in Gentoo penguins caught near the Falkland Islands was found in liver (about 550 ng g^{-1} extractable lipids, 2.4% in lipids), even though brain (about 175 ng g^{-1} extractable lipids) had the highest fat content (6.4% in lipids) (46). The low PCB content in Gentoo penguins may be explained by both the low fat content and a diet mainly based on squid (*Illecebroscus argentinus*) which had no detectable amount of PCBs. This diet is not likely to change very much due to the limited foraging range of these penguins. PCBs were also measured in muscle tissue of Argentine hake (*Merluccius merluccius hubbsi*) (about 400 ng g^{-1} extractable lipids) and flying fish (*Cypselurus cyanopterus*) (about 850 ng g^{-1} extractable lipids) caught in the same area. These values are also very low when compared with fish from more polluted areas of the North-East Atlantic (up to 5000 ng g^{-1} extractable lipids), but are higher than that of Gentoo penguin by a factor of two and four, respectively. PCBs were also determined in biological samples from Antarctica. A mean value of 150 ng g^{-1} (dry weight) was observed in krill samples (23). Assuming a lipid content of about 15%, this corresponds to about 1 ng g^{-1} extractable lipids. The total PCB content in blubber tissues of Weddell seals (*Leptonychotes weddelli*) from the Ross Sea, McMurdo Sound (Antarctica) ranged between 20 and 50 ng g^{-1} extractable lipids (assuming a mean lipid content of 85%) (83). A very high PCB content (more than 6000 ng g^{-1} extractable lipids) was found in south polar skua (*C. maccormicki*). This level was about three times higher than that found in penguins since skuas, unlike penguins, are long-distance migrants. They spend only 4–5 months per year in Antarctica, then migrate across the Southern Ocean and may reach the Northern Pacific and the Northern Atlantic (47). For the sake of comparison, in Table 9.3 is also reported the PCB content in abdominal adipose tissue and blood samples of Dutch citizens from an urban area. The total PCB content was 2400 and 4300 ng g^{-1} extractable lipids for adipose and blood samples, respectively (84).

A wide study was carried out on the PCB content in milk samples from marine mammals (Table 9.4). In particular, milk samples from Antarctic fur seals, Australian sea lions, northern fur seals (Arctic), northern elephant seals (California) and California sea lions were analyzed. Total PCB content ranged from about 40 to 2100 ng g^{-1} extractable lipids in milk samples of Antarctic fur seals and California sea lions, respectively (48). The highest value is comparable with the concentration found in human milk (1300 ng g^{-1} extractable lipids).

On the basis of the concentration of the seven selected PCBs, it appears that the sum of penta + hexa- and penta + hexa + heptachlorobiphenyls account for about 80% and 90% of the total PCB content, respectively, for all biological

Table 9.4. Total PCB content and individual congener concentrations (ng g^{-1} extractable lipids in milk)

	USA (California)	Australia	Antarctica	Northern Europe (The Netherlands)
	Sea lion (*Zalophus californianus*)	Sea lion (*Neophoca cinera*)	Fur seal (*Arctocephalus gazella*)	Human
PCB 28	–	–	–	9
PCB 52	5.9	<1	<1	–
PCB 101	66.0	<1	1.9	9
PCB 118	113.0	11.5	1.9	39
PCB 138	140.0	2.2	2.4	120
PCB 153	198.0	5.0	3.0	110
PCB 180	83.0	2.0	<1	68
Total PCB[a]	2121	80	40	1300
Lipid (%)	47	32	37	
References	48	48	48	84

(a) calculated value as 3.5 times the sum of the seven selected congeners

samples regardless of the sampling site. Moreover, the occurrence of PCBs in every analyzed sample support the conclusion that the global PCB contamination does exist and remains fairly stable in the metabolism of biological organisms.

4. The Italian research on the presence of PCBs in Antarctica

The ongoing Italian research activity dealing with the PCB contamination in Antarctica began in 1987 and has been carried out by participants in the Italian expeditions every year since then. It has been mainly aimed at monitoring the pollution level, along with its time-dependent evolution and at better understanding both transport and diffusion process of these contaminants in Antarctica and on a global scale. PCB content has been evaluated in a variety of environmental matrices, i.e., sea water and the associated suspended matter, sediments, snow/firn, pack-ice, soil and biota. Depth profile in sea water, sediments, snow and ice has been also considered. In particular, the following aims were tackled: i) to get a map on the presence of PCBs in a large area of both Ross Sea, including Terra Nova Bay and Wood Bay, and Victoria Land (Figures 9.3–9.7); ii) to evaluate the PCB depth profile in sea water and marine sediments at Gerlache Inlet and Terra Nova Bay, respectively (Figures 9.5 and 9.6); iii) to investigate the effect of pack-ice melting on the PCB content in coastal sea water at Gerlache Inlet and Wood Bay (Figures 9.5 and 9.6); iv) to elucidate both transport and diffusion mechanisms of PCBs in Antarctica and to evaluate the trend of the PCB level during the last ten years by sampling snow at several depths in selected sites

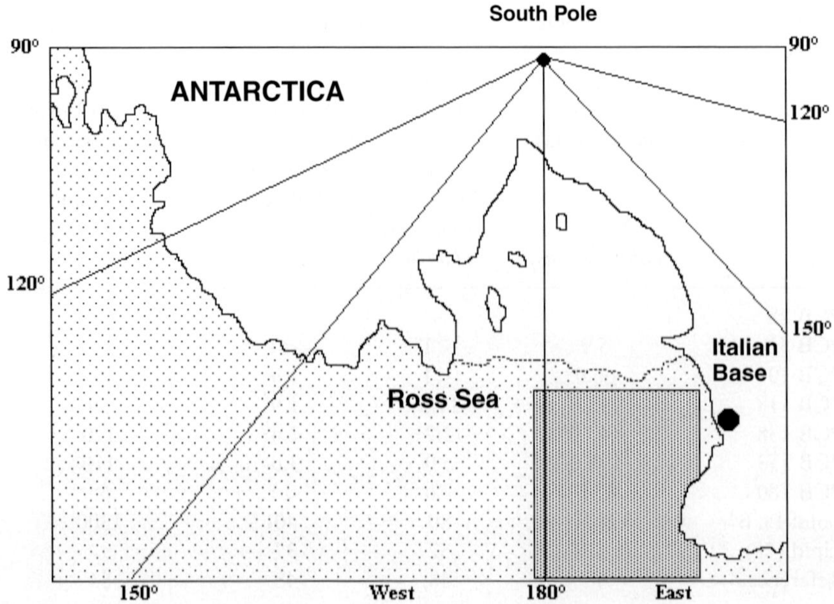

Figure 9.3. Location of marine sediment sampling area at Ross Sea (▒) during the 1990–1991 Italian expedition in Antarctica.

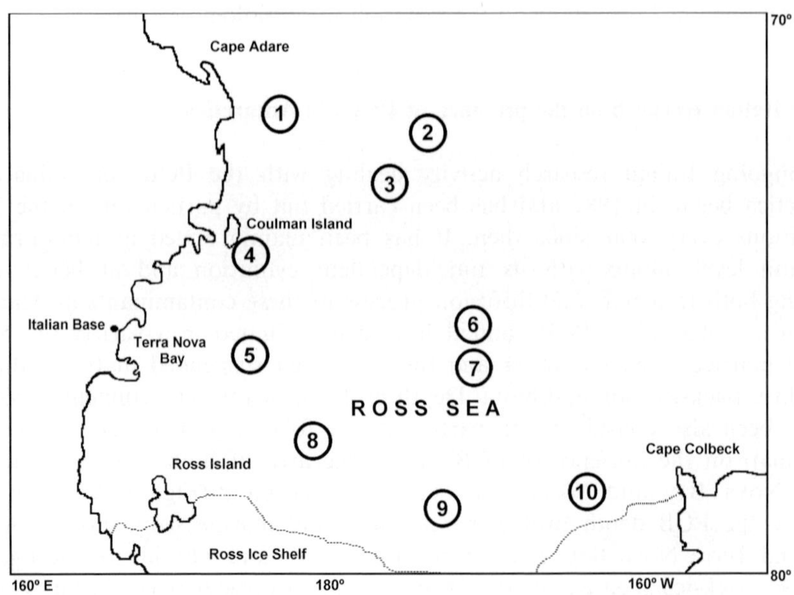

Figure 9.4. Location of sea water sampling stations (1 through 10) at Ross Sea during the 1994–1995 Italian expedition in Antarctica.

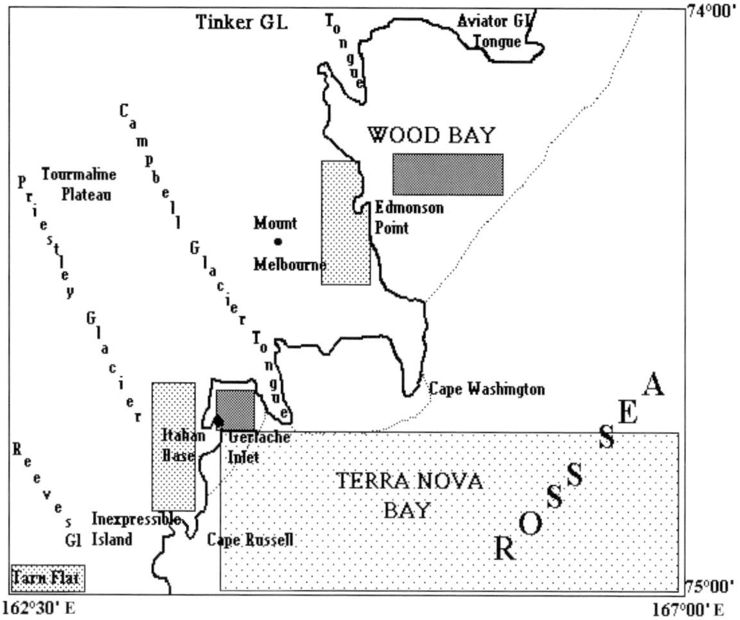

Figure 9.5. Location of sampling areas at Ross Sea and Victoria Land. Lake sediments and soils (Tarn Flat, Inexpressible Island, Edmonson Point) , sea water before and after pack-ice melting (Gerlache Inlet) , surface sea water and marine sediments (Terra Nova Bay) .

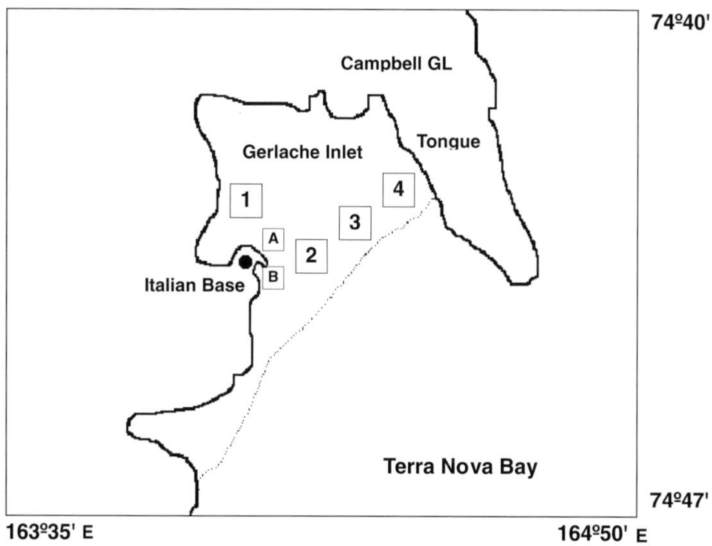

Figure 9.6. Location of sea water and pack-ice sampling stations (1 through 4) at Gerlache Inlet (Terra Nova Bay) during the 1988–1989, 1990–1991 and 1991–1992 Italian expedition in Antarctica. Sampling stations A and B refer to *T. bernacchii* samples collected during the 1993–1994 expedition (see text).

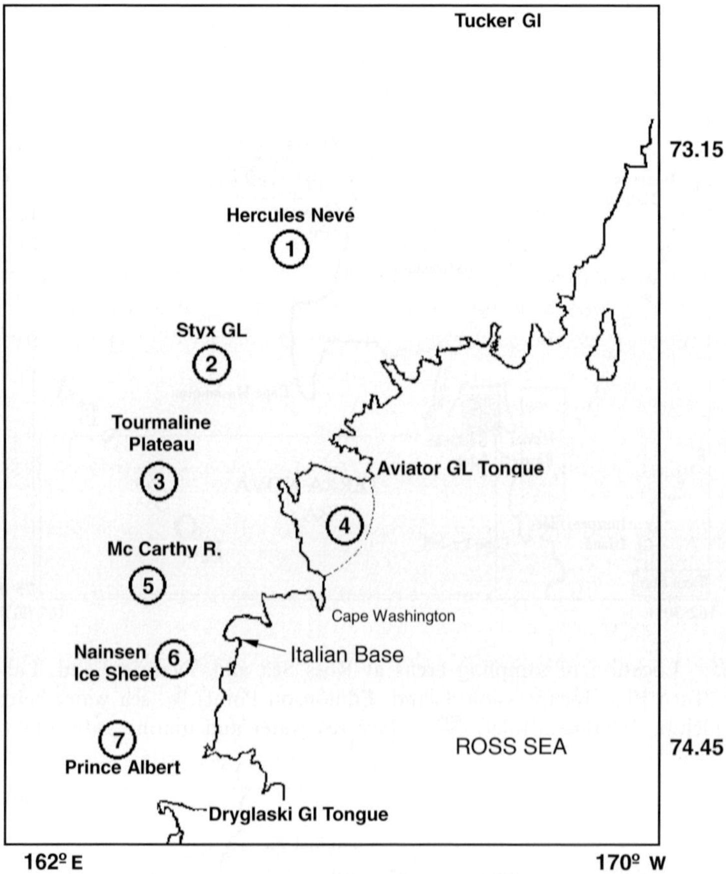

Figure 9.7. Location of snow/firn sampling stations (1 through 7) at Victoria Land during the 1993–1994 and 1994–1995 Italian expeditions.

of Victoria Land (Figure 9.7); v) to provide data on the exposure to PCBs of selected biological species from Terra Nova Bay and to evaluate the environmental impact of human activity at the Italian station (this study was performed by a research team from the University of Siena, Italy).

4.1. Analytical procedure

4.1.1. Sea water, sediment, soil, pack-ice, and snow/firn samples

The analytical procedures for PCB determination in sea water, sediment, soil, pack-ice and snow/firn samples have been described elsewhere (16). Very briefly, all frozen samples were allowed to melt in a clean laboratory. After melting, sea water, pack-ice and snow/firn samples were extracted with *n*-hexane. About 30 g of wet sediment or soil were made homogeneous and then divided into two aliquots of about 20 and 10 g each. The 20 g aliquot was extracted with a 1:1 *n*-hexane/acetone mixture in an ultrasonic bath. After separation from the sample the organic extract was treated for two hours with Cu powder and Hg under magnetic stirring for sulphur removal and finally filtered by a 0.45 μm pore size teflon membrane filter. The 10 g aliquot was used to evaluate the percentage of water in the sample and to perform particle size analysis. As far as water, pack-ice and snow/firn samples are concerned, the extraction of PCBs was performed as soon as the samples were delivered to the laboratory in Italy (about two months after sampling) till the 1990–1991 Italian expedition. Since the 1992–1993 expedition samples were extracted in a clean laboratory at the Italian base in Antarctica by a high efficiency liquid-liquid extraction system (85). In all cases, the extract was cleaned up on a Florisil column (10 mm i.d. × 50 mm length), from which PCBs were selectively eluted with a suitable volume of *n*-hexane. Right before the analysis, this eluate was concentrated at about 100 μl in a conical micro-vial under a mild nitrogen flow, added with an internal standard and finally injected into a gas chromatograph equipped with either EC or MS detection systems. A combined instrumentation based on the on-line coupling of an SFC column clean-up system with GC-MS by a custom-made cryo-trap accumulation cell has been developed in the authors' laboratory and is currently used. By this system up to 100 μl of an organic extract can be injected into the chromatographic system, thus allowing for the determination of organic micropollutants in environmental samples by an MS detector at pg l^{-1} levels or even less (40). Suitable analytical quality control and quality assurance programmes were run in the laboratory during sample analysis in order to get reliable data (43).

4.1.2. Identification and quantification

PCB congeners were identified by GC-MS from a standard solution containing Aroclor 1221, 1232, 1248 and 1260. The RRT for each congener was then calculated by using octachloronaphthalene (OCN) as an IS and applied for chromatographic peak assignment of real samples. RFs were also experimentally obtained for some selected PCB congeners, namely, PCBs 13, 28, 35, 52, 81, 101, 105, 118, 127, 138, 153, 156, 169 and 180 and for OCN in a suitable concentration range. RRFs to OCN were then calculated and used in turn to calculate the RRFs for all congeners by extrapolating the values reported by Mullin (62). Final extracts of real samples were analyzed after the addition of a known amount of OCN and quantification of chromatographic peaks was performed by using the calculated RRFs. In this way, 60 to 80 more abundant congeners were determined. For the sake of comparison with literature data, results relevant to PCBs 28, 52, 101, 118, 138, 153 and 180 were reported along with the total concentration of all congeners determined. The experimental total PCB concentration was generally in a quite good agreement with the value obtained by calculating the sum of the above mentioned seven congeners and multiplying it by a factor of 4 for both air and water samples and a factor of 3.5 for both biological and sediment samples, as reported in the literature, though the calculated value was very often overestimated with respect to the experimental value (61).

4.1.3. Biological samples

The analytical procedures for PCB determination in biological species have been described elsewhere (86). In particular, about 5 g of sample were digested with 1*N* KOH/ethanol solution for 1 hr and the final solution was extracted with hexane. The concentrated hexane was cleaned on 1.5 g silica gel packed in a glass column (10 mm i.d. × 20 cm length). The eluate was concentrated in a rotary evaporator to 6 ml, 3 ml of which were used to determine the total PCB content, while the remaining 3 ml were loaded on a column packed with about 0.12 g of activated carbon and eluted with 100 ml of a benzene-ethyl acetate mixture (50:50, v/v) to separate non-*ortho* coplanar PCB congeners. The determination of total PCBs was performed by a PerkinElmer Autosystem gas chromatograph equipped with a ^{63}Ni ECD system and a SPB-5 fused silica capillary column (30 m long, 0.2 mm i.d., 0.25 μm film thickness). Oven temperature was 100°C, isothermal for 10 min, then up to 280°C at 5°C step min^{-1}. The total content was calculated on the basis of the concentration of about 50 predominant congeners. The determination of coplanar PCBs was performed by a Finnigan MAT GC MS MAGNUM ITD System, operating under the same conditions as for GC-ECD, i.e., in SIM (PCB 77 *m*/*z* 292, PCB 26 *m*/*z* 326 and PCB – 169 *m*/*z* 360). All analyses were accompanied by a suitable analytical quality control programme. Arochlor 1260 was used as an external standard (55).

4.2. Sampling techniques and sampling areas

4.2.1. Sea water, sediment, soil, pack-ice and snow/firn samples

Surface sea water samples were generally collected in a large area of the Ross Sea (Figure 9.4) by a Teflon® pumping system at 0.5 m depth, whenever possible, or by a Go-Flo stainless steel bottle at 20 m depth. Depth profile of the sea water column was performed at Gerlache Inlet and Wood Bay (Figures 9.5 and 9.6) by a Go-Flo stainless steel bottle. Before pack-ice melting, sampling was performed by making a hole in the ice and collecting sea water underneath the inner surface of the pack-ice. Pack-ice samples were collected by a manual ice-coring system. Marine sediments were generally sampled with a stainless steel grab and were collected in a large area of the Ross Sea (Figures 9.3 and 9.5). In a few stations at Terra Nova Bay (Figure 9.5) samples were collected with a box-corer system and aliquots between 0–15 cm and 15–30 cm were obtained, with the exception of one station where aliquots were collected at 0–10, 10–20 and 20–30 cm. Lake sediment samples were manually collected from small lakes, which generally were 30–150 m wide and 0.3–1.2 m deep. Soil samples were also collected manually near these lakes whenever possible (Figure 9.5). Sampling of surface snow samples was performed after having eliminated the first few centimetres in a large area of Victoria Land (Figure 9.7), whereas samples at different depths were collected by digging a trench. In the latter case, sampling sites were selected in collaboration with glaciologists in order to guarantee that snow accumulation was undisturbed, thus preserving the depth profile of its chemical composition the correct sequence of events. In every case, sampling was performed by operators wearing clean-room garments in order to reduce sample contamination. All the samples were stored at –20°C in stainless steel containers pre-cleaned as necessary.

4.2.2. Biological samples

Specimens of two Antarctic fish (*Trematomus bernacchii* and *Chionodraco hamatus*), Adelie penguin (*Pygoscelis adeliae*), south polar skua (*Catharacta maccormicki*) and skin biopsies of Weddell seal (*Leptonychotes weddellii*) from Terra Nova Bay (74°41′42″ S, 164°07′23″ E) were collected in the period 1990–1994 (55). Moreover, two coves near the Italian station at Terra Nova Bay were chosen as sampling sites during the 1993–1994 Italian expedition in Antarctica (Figure 9.6). Cove A was used as a landing place for two small boats for a few weeks each year, while the other one (B) received the waste water of the station (about 10 m^3 day^{-1}). Adult specimens of *T. bernacchii* (body weight range = 210–290 g) were caught using gill-nets in January 1994 in the two coves at a depth of 3 to 8 m and in a control area (on the Gerlache Inlet continental shelf) at a depth of about 80 m. Fish were handled carefully to avoid contamination and the liver samples were immediately removed and frozen in liquid nitrogen. Muscle, kidney, gills and gonads were also isolated and stored at –20°C for transport to Italy. Sex, length and body weight were recorded for each specimen (86).

4.3. Results and discussion

4.3.1. Sea water and pack-ice

Table 9.5 shows the individual concentrations of the seven selected congeners along with the total PCB content of surface sea water samples collected offshore in the Ross Sea. Sea water samples were also collected at Gerlache Inlet and Wood Bay before and after pack-ice melting (Figures 9.5 and 9.6). Table 9.6 shows the total mean PCB content of sea water samples collected at Gerlache Inlet and Wood Bay during four Italian expeditions. The concentration ranges refer to samples collected in all the sampling stations for each expedition. During the 1990–1991 expedition sampling was performed in one station only. Table 9.7 shows individual congener and total PCB mean concentrations of sea water samples collected at Gerlache Inlet and offshore in the Ross Sea during the 1991–1992 and 1994–1995 Italian expeditions, respectively. The mean total PCB concentration in surface sea water samples was 135 pg l^{-1} and the congener class distribution was centred on the tetra- and penta-substituted, as generally found in the open ocean (47). No significant differences in PCB levels were observed between the open sea (Figure 9.4) (mean value 127 pg l^{-1}) and Gerlache Inlet, which is closer to the coast line (Figure 9.5) (mean value before pack-ice melting 130 pg l^{-1}). The 95% confidence intervals of the mean value for open sea samples were calculated and included the concentrations of all the samples analyzed, thus showing a reasonably homogeneous contamination of the area studied. The total PCB content reported in Table 9.5 refers to the experimental values, as obtained by the sum of the concentration

Table 9.5. Individual congener and total PCB concentrations (pg l^{-1}) of sea water samples collected at the Ross Sea during the 1994–95 Italian expedition in Antarctica

	Sampling station										
	1	2	3	4	5	6	7	8	9	10	Mean (CV)*
PCB 28	8	5	4	7	9	8	5	9	10	5	7 (30%)
PCB 52	11	7	9	8	8	14	6	11	14	11	10 (28%)
PCB 101	7	11	5	7	9	8	10	7	9	6	8 (24%)
PCB 118	9	6	4	6	5	5	7	10	6	9	7 (30%)
PCB 138	4	2	6	2	2	2	1	3	1	5	3 (60%)
PCB 153	2	2	1	1	1	1	2	1	1	1	1 (40%)
PCB 180	1	1	2	1	1	2	1	1	2	1	1 (40%)
Total PCBs a)	150	142	113	118	144	139	97	117	135	119	127 (13%)
b)	168	160	124	128	140	160	128	168	172	152	148 (12%)

(*) CV = coefficient of variation
a) Experimental value on the 60-80 most predominant congeners
b) Calculated value as four times the sum of the seven selected congeners

Table 9.6. Effect of pack-ice melting on the total PCB content in surface sea water samples collected at Gerlache Inlet and Wood Bay in different sampling stations during several Italian expeditions in Antarctica (the mean value is reported in brackets)

Expedition	Total PCBs (pg l^{-1})		R(a)
	Before pack-ice melting	After pack-ice melting	
1988–1989	100–160 (120)	150–180 (160)	1.3
1990–1991(b)	140	180	1.3
1991–1992	90–180 (140)	160–230 (205)	1.5
1993–1994(b)	120	160	1.2
Total mean PCBs	130	176	1.3
CV(c)	25%	18%	9%

(a) Ratio between the PCB concentration measured after and before pack-ice melting
(b) Samples collected in only one sampling station
(c) Coefficient of variation

of about 60–80 congeners and to the calculated values, as obtained by the sum of the concentrations of the seven selected congeners multiplied by a factor of four. As already stated, the latter value is very often higher than the former. Moreover, a mean total PCB concentration of 176 pg l^{-1} was observed in samples collected after pack-ice melting and a ratio between the mean PCB content of samples collected after and before pack-ice melting of 1.33 was obtained, which is statistically significant at a 95% confidence level. This increase can be explained if one considers that PCBs are transported into the atmosphere either in the vapour

Table 9.7. Individual congener and total mean PCB concentration of sea water samples collected at Gerlache Inlet, before and after pack-ice melting and in open Ross Sea

	Ross Sea ($pg\ l^{-1}$)	Before pack-ice melting ($pg\ l^{-1}$)	After pack-ice melting ($pg\ l^{-1}$)
PCB 28	7	6	5
PCB 52	10	12	8
PCB 101	8	9	12
PCB 118	6	8	14
PCB 138	3	3	11
PCB 153	1	2	3
PCB 180	1	1	1
Total PCBs	127	133	182

phase or adsorbed on the particulate matter. In both cases the Antarctic climate (low temperatures and dry precipitation) allows these contaminants to be accumulated in snow and ice during their formation. In the case of pack-ice, the melting process which occurs in summer allows all pollutants trapped inside the ice to be transferred to the sea water in a relatively short time. The surface layer is primarily affected in the mixing process. This was confirmed by the measurement of the PCB content in both sea water and pack-ice and also by the depth profiles which were obtained at Gerlache Inlet before and after pack-ice melting (Table 9.8). In particular, the PCB content along the water column ranged in a narrow interval with an increase of about 60% and 30% near the sea bed at 250 m, before and after pack-ice melting, respectively. This was probably due to remobilization processes involving sediments. Moreover, an increase of about 35% confined in a surface layer about 10 m thick after pack-ice melting was also observed. This increase was limited in time and space as already stated. Nevertheless, this phenomenon is very

Table 9.8. PCB depth profile in sea water at Gerlache Inlet before and after pack ice melting (1990–1991 Italian expedition in Antarctica)

	Total PCBs ($pg\ l^{-1}$)	
Depth (m)	Before pack-ice melting	After pack-ice melting
0.5	130	170
10	150	120
25	140	160
250	210	220

dangerous for the Antarctic ecosystem since it occurs in the period of the highest biological activity.

4.3.2. Sediment and soil

Since PCBs are generally adsorbed on the particle surface, the concentration in sediment and soil samples is much more likely to be related to the particle surface area per volume unit than to the mass unit (16). For this reason, the concentration of each sample, expressed in pg g^{-1} (dry weight), was normalized by dividing it by the relevant CS, expressed in square meters of surface per cubic centimetre of dry sample (m^2 cm^{-3}), as obtained by particle size analysis (5). Table 9.9 shows the normalized mean concentration of the seven selected congeners along with the total and the normalized total mean PCB concentration. In this case, the calculated value of the total PCB content was very close to the experimental one (the difference was always lower than 10%). Table 9.10 shows the total mean PCB concentration and the normalized mean PCB concentration for each matrix analysed. In four stations where marine sediments were collected at different depths, a concentration of about 100–200 (pg g^{-1})/(m^2 cm^{-3}) was generally observed in a surface layer of about 10–15 cm, while in deeper layers PCBs were below the LoDs. These results show that the normalized total mean PCB content in marine sediment samples was 150 (pg g^{-1})/(m^2 cm^{-3}) and did not show any significant difference from open sea to the coastal line. Lake sediment and soil samples showed a normalised total mean PCB content of 240 and 130 (pg g^{-1})/(m^2 cm^{-3}) which did

Table 9.9. Normalised individual and total mean PCB concentration of marine sediment, lake sediment and soil samples collected at Ross Sea and Victoria Land, during the 1990-1991 Italian expedition in Antarctica (the coefficient of variation is reported in brackets)

	Marine sediments (pg g^{-1} dry weight)/CS+	Lake sediment (pg g^{-1} dry weight)/CS	Soil (pg g^{-1} dry weight)/CS
PCB 28	2	6	2
PCB 52	3	5	2
PCB 101	6	7	3
PCB 118	11	14	8
PCB 138	10	13	9
PCB 153	13	16	9
PCB 180	3	3	1
Normalized total PCBs			
a)	150 (25%)	240 (27%)	130 (18%)
b)	168	224	119

+ CS is the specific calculated surface area of the sample as obtained by particle size analysis and is expressed in m^2 cm^{-3}

(a) Experimental value on the 60-80 most predominant congeners

(b) Calculated value as four times the sum of the seven selected congeners

Table 9.10. Total and normalised total mean PCB content in all marine sediment, lake sediment and soil samples (the coefficient of variation referred to all samples analysed is reported in brackets)

Sample	Total mean PCBs pg g^{-1} dry weight	Total mean PCBs/CS (pg g^{-1} dry weight)/(m^2 cm^{-3})
Marine sediment	80 (58%)	150 (25%)
Lake sediment	120 (87%)	240 (27%)
Soil	60 (38%)	130 (18%)

not change significantly among sampling stations. Moreover, the normalized total PCB content has a much lower coefficient of variation than the PCB concentration and the 95% confidence intervals of the mean value for each matrix included the concentrations of all the samples analyzed. This led to the conclusion that the spread of concentration was not statistically significant at this level of confidence. It was also shown that PCBs were present in marine sediments only in a surface layer of about 10–15 cm. A 10 cm sediment layer deep should correspond to about 100 years according to the sedimentation rate of the area under study, which was 0.05–0.1 cm $year^{-1}$ as estimated by using the ^{210}Pb method (87). This result agrees quite well with the fact that PCBs were first used at an industrial level about 70 years ago (1). Finally, lake sediment samples presented the highest normalized total PCB concentration. This result might be explained by considering the contributions of atmospheric particulate matter, the primary vehicle of transport and diffusion of PCBs in the environment (1). This contribution seems to be very high and is probably due to the nature of Antarctic lakes, which are formed during the deglacial season from ice melting waters that are rich in atmospheric particulate matter trapped inside the ice matrix during its formation.

4.3.3. Snow, firn

Table 9.11 shows the names, altitudes and distances from the sea of sampling sites. Snow/firn samples were gathered at Hercules Nevé at different depths during two expeditions. Table 9.12 shows the individual concentrations of the seven selected congeners along with both the experimental and the calculated total PCB contents of surface snow samples collected in all the sampling stations. The experimental total PCB content was evaluated on the concentration of 60–80 congeners and ranged between 280 and 730 pg l^{-1}, with a mean value of 540 pg l^{-1} and a coefficient of variation of 28%. The calculated total PCB content was equal to 520 pg l^{-1} with a coefficient of variation of 35%. These results highlight that the area studied presents a reasonably homogeneous PCB level irrespective of the altitude. Since PCBs are mainly transported to Antarctica via the atmospheric aerosol, this implies that the amount deposited on the ground due to dry precipitation is independent of the aerosol particle size distribution. Moreover, the results relevant to the PCB depth profile at Hercules Nevé (Table 9.13) show that the sample

Table 9.11. Altitude and distance from the sea of snow/firm sampling stations at Victoria Land during the 1993–1994 Italian expedition

Sampling station	Sampling site	*Altitude* (m)	Distance from the sea (km)
1	Hercules Nevè (*)	3000	70
2	Styx Nevè	1700	50
3	Tourmaline Plateau	2000	40
4	Wood Bay	0	10
5	McCarthy Ridge	700	40
6	Nansen Ice Sheet	0	15
7	Prince Albert	800	35

(*) Snow/firm samples were collected at Hercules Nevè also during the 1994–1995 Italian expedition

Table 9.12. Individual and total mean PCB concentration (pg l^{-1}) of snow/firn samples collected at Victoria Land (Antarctica) during the 1993–1994 Italian expedition in Antarctica

	Sampling station							
	1	2	3	4	5	6	7	Mean (CV)*
PCB 28	45	51	25	35	25	39	24	35 (31%)
PCB 52	37	51	42	43	29	48	30	40 (21%)
PCB 101	7	20	15	25	14	23	21	18 (35%)
PCB 118	25	18	21	28	15	22	18	21 (21%)
PCB 138	11	6	5	13	6	7	8	8 (37%)
PCB 153	9	7	4	12	3	5	3	6 (52%)
PCB 180	4	2	1	4	1	1	1	2 (71%)
Total PCBs a)	730	510	440	620	280	690	530	540 (28%)
b)	550	620	450	640	370	580	420	520 (20%)

(*) CV = coefficient of variation
(a) Experimental value on the 60-80 most predominant congeners
(b) Calculated value as four times the sum of the seven selected congeners

collected at about 2.5 m has a higher level than the surface one in both expeditions. In order to correctly evaluate this experimental evidence, the following aspects should be considered: analyses were repeated four times, the coefficient of variation was about 20% in all cases and Hercules Nevé was considered as an undisturbed site for snow accumulation, whose accumulation rate was estimated to be about 30–40 cm year^{-1}. Thus, the sample collected at 2.5 m was about 7–8 years old (about 1986–1987) and the difference between its PCB level and that of the surface sample is significant at a confidence level of 90%. In conclusion, if contamination processes at Hercules Nevé from local sources can be excluded, these results might indicate a slight decrease in the PCB amount which reaches the Antarctic continent via the atmosphere and which might be related to the internationally accepted restriction in the usage of PCB-based industrial products.

Table 9.13. Total PCB concentration in snow and firn samples collected at Hercules Nevé at different depths during both 1993–1994 and 1994–1995 Italian expeditions in Antarctica

Sampling depth (m)	Total PCB (pg l^{-1}) 1993–1994	1994–1995
0.1–0.5	710	590
0.5–1.0	590	–
1.0–1.5	780	–
1.5–2.5	1120	910

4.3.4. Biological species

Typical Antarctic species were selected in this kind of investigations, namely *P. bernacchii*, *C. hamatus*, *P. adeliae*, *L. weddellii*, *C. maccormicki* and *T. bernacchii*. In particular, *T. bernacchii* is a bottom feeding fish and is the most abundant species in nearshore waters. Its lifetime is about ten years and it can thus be used for monitoring the environmental impact of human activity at the Italian station. Tables 9.14 and 9.15 show the concentration range and the mean value of both the total PCB content and the non-*ortho* coplanar PCBs in the biological samples collected at Terra Nova Bay. The total PCB content in each species clearly shows the relationship with the trophic level (fish < penguin < seal). Moreover, the comparison of these results with the PCB content in the Antarctic atmosphere (60–180 pg m^{-3}) and hydrosphere (40–130 pg l^{-1}) confirms the biomagnification of these contaminants in the Antarctic food chain (16, 18, 47). The highest total PCB concentration (1162 ng g^{-1}, wet weight) was found in muscle samples from south

Table 9.14. Concentration range and mean value of both total PCBs and non-*ortho* coplanar PCBs in biological species collected at Terra Nova Bay during several Italian expeditions in Antarctica (1990–1994)

Species	Total PCBs (*) (ng g^{-1} wet weight)	PCB 77 (pg g^{-1} wet weight)	PCB 126 (pg g^{-1} wet weight)	PCB 169 (pg g^{-1} wet weight)
P. bernacchii	21 (15–44)	195 (55–312)	100 (21–234)	10 (2–25)
C. hamatus	36 (18–77)	28 (5–20)	6 (1–29)	1 (0.1–4)
P. adeliae	101 (56–188)	250 (115–442)	79 (34–112)	21 (5–48)
L. weddellii	585 (406–750)	1870 (1665–2330)	390 (150–573)	964 (447–1923)
C. maccormicki	1162 (885–1676)	1641 (1220–2300)	2933 (54–552)	560 (134–877)

(*) The concentration range is reported in brackets.

Table 9.15. Mean concentration of both total PCBs and non-*ortho* coplanar PCBs in muscle tissue samples of *T. bernachii* collected near the Italian station at Terra Nova Bay during the 1993–1994 italian expedition in Antarctica

Sampling site	Total PCBs (*) (ng g^{-1} wet weight)	PCB 77 (pg g^{-1} wet weight)	PCB 126 (pg g^{-1} wet weight)	PCB 169 (pg g^{-1} wet weight)
Cove A	17 ± 7	0.3 ± 0.3	0.4 ± 0.4	0.4 ± 0.4
Cove B	4 ± 1.5	0.4 ± 0.3	0.5 ± 0.3	0.5 ± 0.2

polar skua (*C. maccormicki*), which is very likely due to the migration of this bird to more contaminated areas during the southern winter. As far as *T. bernacchii* is concerned, the PCB concentration in muscle tissue samples from species collected at cove A (Figure 9.6) was about four times higher than that of cove B, which in turn was about the same as the PCB level in control species caught at the Gerlache Inlet continental shelf. The non-*ortho* coplanar PCB content was almost the same in all the samples analyzed. The lower total PCB content in species from cove B could be due to the partial detoxification of these compounds by metabolic processes. The latter were related to the observed higher liver microsomal CYP4501A activity for these samples which was evaluated by 7-ethoxyresorufin-O-deethylase (EROD) and benzyloxyrefin-O-deethylase (BROD). This activity might be induced also by other xenobiotics in waste waters from the Italian station (86). In conclusion, the above results show that after more than ten years of scientific activity at the Italian station only a very slight contamination by PCBs took place.

5. Conclusions

Research over the past years shows that PCBs are present in every environmental matrix from Antarctica, though at very low levels. A very peculiar effect of the pack-ice melting process on the PCB content in sea water has been highlighted. This low and quite homogeneous PCB contamination observed in the studied areas may rule out any direct source of PCB pollution in Antarctica, with the exclusion of a very limited area around the Italian station. The snow/firn depth profile also seems to indicate that the amount of PCBs reaching the Antarctica via atmospheric transport has recently decreased. These data clearly need to be confirmed and should encourage governments to continue the financial support of such studies, since they are not only very important for a better understanding of the contamination processes in Antarctica, but are also a source of invaluable information on environmental global changes.

References

1. M. D. Erickson (Ed.), *Analytical Chemistry of PCBs*, Butterworth Pub., Stoneham, MA, USA, 1986, pp. 508.
2. O. Hutzinger, S. Safe, V. Zitko, *The Chemistry of PCBs*, CRC Press, Cleveland, OH, USA, 1974.
3. T. F. Bidleman, G. W. Patton, D. A. Hinckley, M. D. Walla, W. E. Cotham, B. T. Hargrave, *Chlorinated pesticides and polychlorinated biphenyls in the atmosphere of the Canadian Artic*, in D. A. Kurtz (Ed.), *Long Range Transport of Pesticides*, Lewis Pub. Inc., Chelsea, MI, USA, 1990, 347–372.
4. S. W. Fowler, PCBs and the environment: the Mediterranean marine ecosystem, in J. S. Waid (Ed.), *PCBs and the Environment*, Vol. III, CRC Press, Boca Raton, Fla., 1986, 272.
5. R. Fuoco, M. P. Colombini, Polychlorobiphenyls in the environment: analytical procedures and data evaluation, *Microchem. J.*, 51 (1995), 106–121.
6. R. J. Moolenaar, Advances in exposure, health and environmental effects studies of PCBs, *Symposium Proceedings*, R. J. Davenport, B. K. Bernard (Eds.) US EPA, Report No. LSI-TR-507-137B, NTIS PB84-135771, Washington D.C., (1983) 67–96.
7. V. Lang, Polychlorinated biphenyls in the evironment, *J. Chromatogr.*, 595 (1992), 1–43.
8. J. T. Borlakoglu, R. R. Dils, Induction of hepatic cytochrome P-450IA1 in pigeons treated *in vivo* with Aroclor 1254, a commercial mixture of polychlorinated biphenyls (PCBs), *Chem. Brit.*, (1991) 815–818.
9. D. E. Tillitt, J. P. Giesy, G. T. Ankley, Characterization of the H4IIE rat hepatoma cell bioassay as a tool for assessing toxic potency of planar halogenated hydrocarbons in environmental samples, *Sci. Technol.*, 25 (1991), 87–92.
10. J. Mes, D. Weber, Determination of some isomers of polychlorinated biphenyl congeners in fatty food of the Canadian diet, *Chemosphere*, 19 (1989), 1357.
11. C. S. Hong, B. Bush, Determination of *mono-* and non-*ortho* coplanar PCBs in fish, *Chemosphere*, 21 (1990), 173.
12. P. Haglund, L. Asplund, U. Janberg, B. Jansson, Isolation of *mono-* and non-*ortho* polychlorinated biphenyls from biological samples by electron-donor acceptor high performance liquid chromatography using a 2-(1-pyrenyl)ethyldimethylsilylated silica column, *Chemosphere*, 20 (1990), 887.
13. R. Fuoco, M. P. Colombini, E. Samcova, Individual determination of *ortho* and non-*ortho* substituted polychlorobyphenyls (PCBs) in sediments by high performance liquid chromatographic preseparation and gas chromatography/ECD detection, *Chromatographia*, 36 (1993), 65–70.
14. P. De Voogt, D. E. Wells, L. Reutergårdh, U.A.Th. Brinkman, Biological activity, determination and occurrence of planar, *mono-* and di-*ortho* PCBs, in J. Albaigés (Ed.), *Environmental Analytical Chemistry of PCBs. Current Topics in Environmental and Toxicological Chemistry*, Gordon and Breach Science Pub., Amsterdam, 1993, 151–196.
15. R. Fuoco, P. R. Griffiths, Analysis of polychlorobyphenyls (PCBs) in contaminated soil samples by supercritical fluid extraction/supercritical fluid chromatography coupled with FT-IR spectrometry, *Ann. Chim. (Rome)*, 82 (1992), p. 235.
16. R. Fuoco, M. P. Colombini, C. Abete, Determination of polychlorobiphenyls in environmental samples from Antarctica, *Int. J. Environ. Anal. Chem.*, 55 (1994), 15–25.
17. R. R. Weber, R. C. Montone, Distribution of organochlorines in the atmosphere of the South Atlantic and Antarctic oceans, in D. A. Kurtz (Ed.), *Long Range Transport of Pesticides* (Lewis Pub. Inc., Chelsea, MI, USA, 1990) p. 462.
18. S. Tanabe, H. Hidaka, R. Tatsukawa, PCBs and chlorinated hydrocarbon pesticides in Antarctic atmosphere and hydrosphere, *Chemosphere*, 12 (1983), 277–288.
19. J. Screitmüller, K. Ballschmiter, Levels of polychlorinated biphenyls in the lower troposphere of the North- and South-Atlantic Ocean. Studies of global baseline pollution XVII, *Fresenius' J. Anal. Chem.*, 348 (1994), 226–239.
20. U. L. Nilsson, C. E. Östman, Chlorinated polycyclic aromatic hydrocarbons: Method of analysis and their occurrence in urban air, *Environ. Sci. Technol.*, 27 (1993), 1826–1831.
21. K. Ballschmiter, R. Wittlinger, Interhemisphere exchange of hexachlorocyclohexanes, hexachlorobenzene, polychlorobiphenyls and 1,1,1-trichloro-2,2-bis (*p*-chlorophenyl)ethane in the lower troposphere, *Environ. Sci. Technol.*, 25 (1991), 1103–111.

22. J. B. Manchester-Neesvig, A. W. Andren, Seasonal variation in the atmospheric concentration of polychlorinated biphenyl congeners, *Environ. Sci. Technol.*, 23 (1989), 1138–1148.
23. S. R. Gupta, A. Sarkar, T. W. Kureishey, PCBs and organoclhorine pesticides in krill, birds and water from Antarctica, *Deep-Sea Research II*, 43 (1996), 119–126.
24. A. G. Kelly, I. Cruz, D. E. Wells, Polychlorobiphenyls and persistent organochlorine pesticides in sea water at the pg l^{-1} level. Sampling apparatus and analytical methodology, *Anal. Chim. Acta.*, 276 (1993), 3–13.
25. A. Ewen, H. J. Lafontaine, K.-T. Hidesheim, W. H. Stuurmen, Separation and determination of 22 PHAs by minibore HPLC with dual-wavelength UV detection, *Int. Chrom. Lab.*, 14 (1993), 4–8.
26. D. E. Schulz-Bull, G. Petrick, J. C. Duinker, Polychlorinated biphenyls in North Sea water, *Mar. Chem.*, 36 (1991), 365–384.
27. T. J. Murphy, A. W. Schinsky, Net atmospheric inputs of PCBs to the ice cover on Lake Huron, *J. Great Lakes Res.*, 9 (1983), 92–96.
28. J. Stanley, C. Haile, A. Small, E. Olson, Sampling and analysis procedures for assessing organic emissions from stationary combustion sources for exposure evaluation division studies. Methods manual, *U.S. Environmental protection Agency, EPA*-560/5-82-014 (1982).
29. T. Vo-Dinh (Ed.), *Chemical Analysis of Polycyclic Aromatic Compounds* (John Wiley & Sons, Inc., New York, USA, 1989), pp. 494.
30. K. Peltonen, T. Kuljukka, Air sampling and analysis of polycyclic aromatic hydrocarbons, *J. Chromatog.*, 710 (1995), 93.
31. J. D. Millar, R. E. Thomas, H. J. Schattemberg, Determination of organochlorine pesticides and polychlorinated biphenyls in water by gas chromatography, *Anal. Chem.*, 53 (1981), 214–219.
32. J. W. Lintelmann, R. Sauerbrey, A. Kettrup, On-site solid phase extraction and coupled-column high-performance liquid chromatographic determination of six polycyclic aromatic hydrocarbons in water, *Fresenius J. Anal. Chem.*, 352 (1995), 735–742.
33. G. Poole, B. Thibert, H. Lemaire, B. Sheridan, C. Chiu, An assesment of various aprotic solvents to separate PCB from oils, *Organohalogen Comp.*, 24 (1995), 47–50.
34. K. K. Chee, M. K. Wong, H. K. Lee, Optimization of microwave-assisted solvent extraction of polycyclic aromatic hydrocarbons in marine sediments using a microwave extraction system with high-performance liquid chromatography-fluorescence detection and gas chromatography-mass spectrometry, *Anal. Chim. Acta*, 330 (1996), 217.
35. Z. Zhang, M. Yang, J. Pawliszyn, Solid-phase microextraction, *J. Anal. Chem.*, 66 (1994), 844A-853A.
36. Z. Zhang, J. Pawliszyn, Headspace solid-phase microextraction, *Anal. Chem.*, 65 (1993), 1843.
37. J. Chen, J. Pawliszyn, Solid phase microextraction coupled to high-performance liquid chromatography, *Anal. Chem.*, 67 (1993), 2530–2533.
38. D. J. Gregor, W. D. Gummer, Evidence of atmospheric transport and deposition of organochlorine pesticides and polychlorinated biphenyls in Canadian Arctic snow, *Environ. Sci. Tech.*, 23 (1989), 561–565.
39. J. J. Langenfeld, S. B. Hawthorne, D. J. Miller, J. Pawliszyn, Role of modifiers for analytical-scale supercritical fluid extraction of environmental samples, *Anal. Chem.*, 66 (1994), 909–916.
40. R. Fuoco, A. Ceccarini, M. Onor, S. Lottici, Supercritical fluid extraction combined on-line with cold trap gas chromatography/mass spectrometry, *Anal. Chim. Acta*, 346 (1997), 81–86.
41. S. B. Hawthorne, D. J. Miller, J. J. Langenfeld, Quantitative analysis using directly coupled supercritical fluid extraction-capillary gas chromatography (EFE-GC) with a conventional split/splitless injection port, *J. Chromat. Sci.*, 28 (1990), 2.
42. V. Lopez-Avila, R. Young, J. Benedicto, P. Ho, R. Kim, W. Beckert, Extraction of organic pollutants from solid samples using microwave energy, *Anal. Chem.*, 67 (1995), 2096–2102.
43. R. Fuoco, M. P. Colombini, A. Ceccarini, Analitical quality control for the determination of polychlorobiphenyls in environmental matrices, *Mikrochim. Acta*, 123 (1996), 175–183.
44. J. M. O'Neal, W. H. Benson, J. C. Allgood, Analysis of organochlorine pesticides and PCBs in fish and shrimp tissues, in G. K. Ostrander (Ed.), *Techniques in Aquatic Toxicology* (Lewis Pub. Boca Raton, 1996) 561–565.
45. L. Kleivane, O. Espeland, K. A. Fagerheim, K. Hylland, A. Polder, J. U. Skaare, Organochlorine

pesticides and PCBs in the East Ice Harp Seal (*Phoca groenlandica*) population, *Mar Environ. Res.*, 43 (1997), 117–130.

46. J. De Boer, P. Wester, Chlorobiphenyls and organochlorine pesticides in various subantarctic organisms, *Mar. Pollution Bull.*, 22 (1991), 441–447.
47. S. Focardi, R. Bargagli, S. Corsolini, Organochlorines in Antarctic marine chain at Terra Nova Bay (Ross Sea), *Kor. J. Pol. Res.*, 4 (1993), 73–77.
48. C. E. Bacon, D. P. Jarman, D. P. Costa, Organochlorine and polychlorinated biphenyl levels in pinniped milk from the Arctic, the Antarctic, California and Australia, *Chemosphere*, 24 (1992), 779–791.
49. M. C. Kennicutt, S. J. McDonald, L. J. Sericano, P. Boothe, J. Oliver, S. Safe, B. J. Presley, H. Liu, W. Douglas, T. L. Wade, A. Crockett, D. Bockus, Human contamination of the marine environment – Arthur Harbor and McMurdo Sound, Antarctica, *Environ. Sci. Technol.*, 29 (1995), 1279–1287.
50. D. C. G. Muir, R. Norstrom, M. Simon, Organochlorine contaminants in Arctic marine food chains: accumulation of specific polychlorinated biphenyls and chlordane-related compounds, *Environ. Sci. Technol.*, 22 (1988), 1071–1079.
51. G. C. Harding, R. J. LeBlanc, W. P. Vass, R. F. Addison, R. T. Hargrave, S. Pearre Jr., A. Dupuis, P. F. Brodis, Bioaccumulation of polychlorinated biphenyls (PCBs) in the marine pelagic food web, based on a seasonal study in the southern Gulf of St. Lawrence, 1976–1977, *Mar. Chem.*, 56 (1997), 145–179.
52. J. N. Huckins, M. W. Tubergen, J. A. Lebo, R. W. Gale, T. R. Schwartz, Polymeric film dialysis in organic solvent media for cleanup of organic contaminants, *J. Assoc. Off. Anal. Chem.*, 73 (1990) 290.
53. B. N. Bavel, P. A. Bergqvist, D. Broman, K. Lundgren, O. Papakosta, C. Rolff, B. Strandberg, Y. Zebòhr, D. Zook, C. Rappe, Levels of PCBs in the aquatic environment of the Gulf of Bothnia: benthic species and sediments, *Mar. Pollution Bull.*, 32 (1995), 210–218.
54. A. N. Subramanian, S. Tanabe, H. Tanaka, H. Hidaka, R. Tatsukawa, Gain and rates and biological half-life of PCBs and DDE in the bodies of Adelie penguins *Environ. Pollution*, 43 (1987), 39–46.
55. S. Focardi, R. Bargagli, S. Corsolini, Isomer specific analysis and toxic potential evaluation of polychlorinated biphenyls in Antarctic fish, seabirds and Weddel seals from Terra Nova Bay (Ross Sea) *Antarc. Sci.*, 7 (1995), 31–35.
56. T. L. Chester, J. D. Pinkston, D. E. Raynie, Supercritical fluid chromatography and extraction, *Anal. Chem.*, 66 (1994), 106R–130R.
57. S. B. Hawthorne, D. J. Miller, Extraction and recovery of polycyclic aromatic hydrocarbons from environmental solids using supercritical fluids, *Anal. Chem.*, 59 (1987), 1705–1708.
58. S. Reindl, F. Höfler, Optimization of the parameter in supercritical fluid extraction of polynuclear aromatic hydrocarbons from soil samples, *Anal. Chem.*, 66 (1994), 1808–1816.
59. J. J. Langenfeld, S. B. Hawthorne, D. J. Miller, J. Pawliszyn, Effect of temperature and pressure on supercritical fluid extraction efficiency of polycyclic aromatic hydrocarbons and polychlorinated biphenyls., *Anal. Chem.*, 65 (1993), 338–344.
60. M. Lohleit, K. Bächmann, Integrated analysis of solid samples by on-line supercritical fluid extraction-gas chromatography, *J. Chromatogr.*, 505 (1990), p. 227.
61. S. Mössner, I. Barudio, T. S. Spraker, G. Antonelis, G. Early, J. R. Geraci, P. R. Beker, K. Ballschmiter, Determination of HCHs, PCBs, and DDTs in brain tissues of marine mammals off different age, *Fresenius J. Anal. Chem.*, 349 (1994), 708–716.
62. M. D. Mullin, C. M. Pochini, S. McCrindle, M. Romkes, S. H. Safe, High-resolution PCB analysis: synthesis and chromatographic properties of all 209 PCB congeners, *Environ. Sci. Technol.*, 18 (1984), 468–476.
63. D. E. Wells, E. A. Maier, B. Griepink, Calibrants and calibration for chlorobiphenyl analysis, *Inter. J. Environ. Anal. Chem.*, 46 (1992), 255–264.
64. J. Jacob, Method development for the determination of polycyclic aromatic hydrocarbons (PAHs) in environmental matrices, in Ph. Quevauviller, E. A. Maier, B. Griepink (Eds.), *Quality Assurance for Environmental Analysis* Elsevier Science Publ., Amsterdam, 1995, Vol. 17, p. 649.
65. C. Porte, D. Barcelo, J. Albaiges, Quantitation of total *versus* selected polychlorinated biphenyls congeners in marine biota samples, *J. Chromatogr.*, 442 (1988), 564–589.

66. J. J. Langenfeld, S. B. Hawthorne, D. J. Miller, Kinetic study of supercrititcal fluid extraction of organic contaminants from heterogeneous environmental samples with carbon dioxide and elevated temperature, *Anal. Chem.*, 67 (1995), 1727–1739.
67. Å. Bergman, A. Hagman, S. Jacobsson, B. Jansson, M. Ahlman, Thermal degradation of polychlorinated alkanes, *Chemosphere*, 13 (1984), 237–250.
68. K. Ballschmiter, R. Niemczyk, W. Shäfer, W. Zoller, Isomer-specific identification of polychlorinated benzenes (PCBz) and biphenyls (PCB) in effluents of municipal waste incineration., *Fresenius J. Anal. Chem.*, 328 (1997), 583–587.
69. R. Wittlinger, K. Ballschmiter, Studies of the global baseline pollution, XIII, *Fresenius J. Anal. Chem.*, 336 (1990), 193–200.
70. P. Larsson, C. Järnmark, A. Södergren, PCBs and chlorinated pesticides in the atmosphere and aquatic organism of Ross Island, Antarctica, *Mar. Poll. Bull.*, 25 (1992), 281–287.
71. D. L. Elder, J. P. Villeneuve, P. Parsi, G. R. Harvey, Polychlorinated biphenyls in sea water, sediment and over-ocean air of the Mediterranean, Activities of the International Laboratory of Marine Radioactivity, 1976 *Report, International Atomic Energy Agency*, Vienna, 1976, pp. 136.
72. R. W. Risebrough, B. De Lappe, C. Younghans-Haug, PCB and PCT contamination in Winter Quarters Bay, Antarctica, *Mar. Poll. Bull.*, 21 (1990), 523–529.
73. Th. Class, K. Ballschmiter, Global baseline pollution studies, *Fresenius' Z. Anal. Chem.*, 327 (1987), 198–204.
74. J. R. Holton, Global transport processes in the atmosphere, in O. Hutzinger, (Ed.), *Environmental Chemistry – The Natural Environmental and Biogeochemical Cycles*, (Springer-Verlag Pub., Berlin, 1990), vol.1 Part E, 97–146.
75. J. M. Bayona, P. Fernandez, C. Porte, I. Tolosa, M. Valls, J. Albaiges, Partitioning of urban wastewater organic microcontaminants among coastal compartments, *Chemosphere*, 23 (1991), 131–326.
76. D. Dannenberger, A. Lerz, The distribution of selected chlorinated microcontaminants in Baltic waters, 1992 to 1994 *Dtsch. Hydrograph. Z.*, 47 (1995), 301–311.
77. D. E. Schulz-Bull, G. Petrick, N. Kannan, J. C. Duinker, Distribution of individual chlorobiphenyls (PCB) in solution and suspension in the Baltic Sea, *Mar. Chem.*, 48 (1995), 245–270.
78. H. Iwata, S. Tanabe, M. Aramoto, N. Sakai, R. Tatsukawa, Persistent organochlorine residues in sediments from the Chukchi Sea, Bering Sea and Gulf of Alaska, *Mar. Poll. Bull.*, 28 (1994), 746–753.
79. I. Tolosa, J. M. Bayona, J. Albaigés, Spatial and temporal distribution, fluxes and budgets of organoclorinated compounds in northwest Mediterranean sediments, *Environ. Sci. Tech.*, 29 (1995), 2519–2527.
80. D. E. Wells, I. Echarri, Determination of individual chlorobiphenyls (CBs), including non-*ortho*, and mono-*ortho* chloro substituted CBs in marine mammals from Scottish waters, in J. Albaigés (Ed.) *Environmental Analitical Chemistry of PCBS. Current Topics in Environmental and Toxicological Chemistry*, Gordon and Breach Science Pub., Amsterdam, 1993, 197–219.
81. J. Sánchez, M. Solé, J. Albaigés, A comparison of distributions of PCB congeners and other chlorinated compounds in fishes from coastal areas and remote lakes, in J. Albaigés (Ed.), *Environmental Analytical Chemistry of PCBS. Current Topics in Environmental and Toxicological Chemistry*, Gordon and Breach Science Pub., Amsterdam, 1993, 247–262.
82. P. Harrison, Genus Pygoscelis, P. Harrison, (Ed.), *Seabirds: An identification Guide*, Beckenham, Kent: Christopher Helm Pub., UK, 1988, 448.
83. J. O. Strömberg, L. G. Anderson, G. Bjorg, W. N. Bonner, A. C. Clark, A. L. Dick, W. Ernst, D. W. S. Limbert, D. A. Peel, J. Preddle, R. I. L. Smith, D. W. H. Walton, State of marine environment in Antarctica, *UNEP Regional Seas Reports and Studies No. 129*, UNEP 1990.
84. P. A. Greve, P. van Zoonen, Organoclorine pesticides and PCBs in tissues from dutch citizens (1968–1986), in J. Albaigés (Ed.), *Environmental Analytical Chemistry of PCBS. Current Topics in Environmental and Toxicological Chemistry*, Gordon and Breach Science Pub., Amsterdam, 1993, 275–287.
85. L. Zoccolillo, High efficiency liquid-liquid extraction system for determination of organic micropollutants in water samples, private communication.
86. R. Bargagli, S. Corsolini, M. C. Fossi, J. C. Sanchez-Hernandez, S. Focardi, Antarctic fish *Trema-*

tomus bernacchii as biomonitor of environmental contaminants at Terra Nova Bay station (Ross Sea), *Mem. Natl. Inst. Polar Res. Spec.*, 1998, 220–229.

87. M. Frignani, F. Giglio, L. Langone, M. Ravaioli, A. Mangini, Late pleistocene-holocene sedimentary fluxes of organic carbon and biogenic silica in the northwestern Ross Sea, Antartica, *Ann. Glaciol.*, 27 (1998), 697–703.
88. W. Krämer, K. Ballschmiter, Global baseline pollution studies, XII. Content and pattern polychloro-cyclohexane (HCH) and -biphenyls (PCB), and content of hexachlorobenzene in the water column of the Atlanctic Ocean, *Fresenius J. Anal. Chem.*, 330 (1988), 524–526.
89. R. F. Pearson, K. C. Hornbuckle, S. J. Eisenreich, D. L. Swackhamer, PCBs in Lake Michigan water revised, *Environ. Sci. Technol.*, 30 (1996), 1429–1436.

Environmental Contamination in Antarctica
A Challenge to Analytical Chemistry
S. Caroli, P. Cescon and D.W.H. Walton, editors

Chapter 10

Certified reference materials in Antarctic matrices: development and use

Stefano Caimi, Oreste Senofonte, Sergio Caroli

1. Introduction

The increasing complexity of modern societies and, in particular, the mounting awareness of the need for an effective protection of human health and the environment have prompted in recent years, as never before, the international scientific community to focus their attention on the attainment of absolutely valid and comparable experimental information as obtained through research and monitoring. The adoption of quality control and quality assurance systems is central for the achievement of data of proved reliability. This applies all the more to the challenges posed by Antarctic sciences, as illustrated in detail in Chapter 1 of this book, because of the remoteness of the continent, the obvious difficulties in performing all preanalytical steps, and the levels of the determinands in the various environmental and biological matrices, generally much lower than in analogous samples from the more populated northern hemisphere.

Matrix Certified Reference Materials (CRMs) are of primary importance in this context as they offer a powerful means for assessing the accuracy and precision of experimental data, provided that their chemical and physical properties and the concentration ranges of the analytes of interest are as close as possible to those of the unknowns (see Chapter 1 for general concepts and definitions). Unfortunately, it is impossible to produce CRMs for all the matrix-determinand combinations that can be encountered in the real world since their number is virtually endless and the associated costs are beyond imagination (1–14). Hence, a still acceptable compromise is to resort to CRMs that resemble the samples under test in terms of nature of matrix components and levels of the analytes. One major (and inescapable) drawback of this approach is that there is no guarantee that determinands in a given CRM will behave during the analysis exactly as those in the materials under test, primarily because of potentially significant differences in the way the substances to be quantified are bound to the host matrix. To partly circumvent this problem, at least in the long run, it would be highly desirable that effective harmonization of policies and coordination of production programmes be achieved among the principal, tradition-rich bodies involved in the production of CRMs, e.g., the National Institute of Standards and Technology (NIST, formerly National Bureau of Standards, NBS), USA, the Institute for Reference Materials and

Measurements, Joint Research Centre of the European Commission (EC-JRC-IRMM), the National Research Council (NRC), Canada, the Laboratory of the Government Chemist (LGC), UK, and the National Institute for Environmental Studies (NIES), Japan.

For the time being, however, the dismal lack of ad hoc CRMs is keenly felt especially in certain scientific sectors. From this standpoint, Antarctic investigations do suffer from their unavailability, all the more so in consideration of the huge investments such studies entail, whereby the obtainment of unreliable findings will result in an intolerable waste of time and effort.

In full recognition of this unsatisfactory situation, the Italian National Programme of Researches in Antarctica (Programma Nazionale di Ricerche in Antartide, PNRA), starting in the mid 1990s, has invested substancial financial, technical and scientific resources to promote, undertake and produce some CRMs, in close cooperation with other relevant organizations, with the purpose of assisting the environmental analytical chemist in the achievement of reliable research and monitoring data in selected fields of major concern in Antarctica. A full account on the Antarctic CRMs accomplished so far, as well as on those scheduled for the next few years, is given hereafter.

2. The certification project

2.1. General aspects

A thoughtful and meticulous survey of the CRMs already available on the market was first carried out in order to identify aspects of environmental chemistry in Antarctica for which not even the possibility of resorting to satisfactory surrogates existed or was considered appropriate. After this selection, a finer screening was done on the basis of the interest demonstrated by research teams all over the world toward specific activity fields, so as to arrive at a manageable number of cases for which Antarctic CRMs were deemed to be definitely necessary and urgent. As a result, a few hypotheses were forwarded which envisaged the launch of projects centred on the certification of: i) trace and minor elements in Antarctic sediment; ii) trace elements in krill; iii) trace elements in water of the Southern Ocean; iv) trace elements in the scallop *Adamussium colbecki*; and v) polychlorobiphenyls (PCBs) in krill.

After the approval of this general programme by the Italian National Scientific Commission for Antarctica (Commissione Scientifica Nazionale per l'Antartide, CSNA), information on the upcoming start of the certification activities was circulated in all possible ways (announcements in scientific journals, at congresses and meetings, private communications and still other), with the explicit goal of stimulating the interest of all potential, qualified partners. In particular, the objectives and timing of the programme were presented and discussed in the context of the Scientific Committee for Antarctic Research (SCAR), the organization supervising and coordinating the initiatives of countries active in Antarctica (15). Eventually, it was possible to select a number of prominent institutions which adhered to one or more of the certification projects listed above. The Istituto Superiore di Sanità (ISS) of Rome – entrusted with the responsibility and coordination of these activities – managed the collection of the necessary raw masses, their transportation to and storage at the ISS premises, the performance of preliminary tests, the organization and conduct of the certification exercises, the evaluation of experimental data and the flow of information to and from the collaborating centres.

All these steps were always carried out in close cooperation with EC-JRC-IRMM. In general, the obtainment of a sufficient amount of each kind of matrix (in the order of several tens of kilos) poses some operative problems in that chemical contamination by, and/or loss of, the target substances should be avoided or minimized as much as possible. These and other related issues have been already discussed exhaustively in Chapter 1 of this book. Here it will suffice to capitalize on those aspects that more specifically apply to the task at hand. In fact, important as it is, preservation of integrity of the collected materials as well as of the original levels of the analytes of interest is slightly less vital than in the case of samples from which accurate data on the status of the Antarctic environment are directly required (16). A candidate CRM could actually tolerate, to a certain extent, a modest alteration in the original values of the properties under certification and even in the chemical and physical state of the host matrix *vs.* those of the raw material, provided that such variations are still within the naturally-occurring ranges observed for those quantities in samples of the same kind. This means in practice that, under such circumstances, would-be CRMs can still fully accomplish their function because they reflect what the experimentalist is likely to encounter in the real world. On other hand, primarily contamination phenomena are all the more risky even in the case of masses intended for certification purposes for at least two reasons: firstly the extremely low levels of pollutants often found in Antarctic environmental samples, which could be easily by far exceeded if no rigorous measures are taken during the collection process, and secondly the way in which exogenous substances can bind to other matrix components, essentially different as a rule from that typical of endogenous analytes. Both aspects combine and lead to candidate CRMs whose characteristics can become poorly representative of the systems for which they are intended. Of no lesser importance in this context is the physical transformation that the raw mass must necessarily undergo to gain homogeneity and stability, these aspects being mandatory prior to any subsequent action in the characterization and certification process. This is always a rather dramatic step in itself in that pretreatments such as water removal, freeze-drying, grinding, sieving or separation of undesirable components are, in most cases, unavoidable. The native features of the raw materials can thus be profoundly altered to the point that one may wonder whether they still resemble the samples the validity of whose analyses they should verify. Moreover, the above treatments bring about also a change (normally an increase) in analyte concentrations. This effect may add to the already mentioned contamination aspects and render the final product even more distant from the original material. The only pragmatic solution to this complex situation is to fully evaluate pros and cons beforehand and then adopt the procedure which minimizes drawbacks and optimizes the potential uses of the candidate CRMs. This is the philosophy behind the certification projects described in the following.

2.2. The case of the CRM MURST-ISS-A1 Antarctic Sediment

2.2.1. Obtainment and pretreatment of the raw mass

The increasing number of chemical analyses performed worldwide on marine Antarctic sediments for their contents in trace elements did demand the availability of at least one CRM of this kind. In order to foster the production of such a CRM, sediment was sampled in 1994 during the eighth Italian expedition in Antarctica. The sampling site was in Adelie Cove (74°46′ Lat S; 163°57′ Long E), close to Terra Nova Bay where the Italian Base is situated. By means of a dredge, a mass of 100 kg was taken at a depth of 80 m and immediately frozen at –20°C in chemically decontaminated 2-l polyethylene bottles. The whole mass, always kept at this temperature, was then delivered to ISS, where an aliquot of 10 kg was retained for preliminary analytical tests. The remaining mass of 90 kg was forwarded to the EC-JRC-IRMM laboratories for the treatment necessary to transform it into the candidate CRM (17–19).

Chapter 11 reports an exhaustive description of the treatment the sediment was subjected to. Here it may be important to recall that a first sieving of the wet mass was necessary to eliminate all particles larger than 2 mm. These were found to consist mainly of organic debris such as fish scales, fragments of shells as well as other unidentified particles not yet transformed by the process of decomposition and compaction subsequent to sedimentation. Once processed, the resulting ground material turned out to contain mostly particles with an average size of less than 90 μm (97% of the total mass), while particles larger than 150 μm were practically absent. The fraction between 90 and 150 μm was mainly formed by organic material homogeneously distributed over the sediment. A mean moisture content of 0.35 ± 0.05% was measured at bottling. Homogeneity and stability were also found to be satisfactory (for more detail see Chapter 11).

2.2.2. Organization of the certification exercise

Participants were carefully selected among laboratories with a reputation in both the analysis of sediments and active involvement in certification projects undertaken by expert organizations. Representatives of these laboratories were contacted individually to inform them of the planned certification exercise and to encourage their collaboration. In the end, a group sufficiently large was formed that showed much promise for the successful outcome of the project, also because of the wide choice of analytical techniques available. Table 10.1 lists the participating laboratories whose data were accepted along with the techniques used.

Each laboratory was requested to perform analyses in compliance with a detailed set of instructions and to provide information on sample intake, moisture content, digestion procedure, analytical techniques, calibrants, use of CRMs and anything else that could be relevant to the determinations. Five independent analytical

Table 10.1. Laboratories participating in the certification project for trace elements in the *CRM MURST-ISS-A1 Antarctic Sediment* and the instrumental techniques employed

Participant	Technique (*)
Complutense University, Faculty of Chemical Sciences, Department of Analytical Chemistry, Madrid, Spain	FAAS, HG-AAS, Q-ICP-MS
ENEA (National Agency for Alternative Energies), Rome, Italy	HG-AAS, Z-ETA-AAS
Eötvös Loránd University, Institute of Inorganic and Analytical Chemistry, Budapest, Hungary	ETA-AAS, ICP-AES, NAA
Institute for Reference Materials and Measurements, European Commission, Joint Research Centre, Geel, Belgium	ID-MS, NAA, SS-Z-ETA-AAS
Istituto Superiore di Sanità, Applied Toxicology Department, Rome, Italy	ICP-AES, Q-ICP-MS
National Environmental Research Institute, Røskilde, Denmark	NAA
National Institute of Standards and Technology, Gaithersburg, MD, USA	Q-ICP-MS
Pavia University, Department of General Chemistry, Pavia, Italy	NAA

(*) ETA-AAS, Electrothermal Atomization Atomic Absorption Spectrometry; FAAS, Flame Atomic Absorption Spectrometry; HG-AAS, Hydride Generation Atomic Absorption Spectrometry; ICP-AES, Inductively Coupled Plasma Atomic Emission Spectrometry; ID-MS, Isotopic Dilution Mass Spectrometry; NAA, Neutron Activation Analysis; Q-ICP-MS, Quadrupole Inductively Coupled Plasma Mass Spectrometry; SS-Z-ETA-AAS, Solid Sampling Zeeman Atomic Absorption Spectrometry; Z-ETA-AAS, Zeeman Electrothermal Atomization Atomic Absorption Spectrometry

runs (three replicates each) were compulsory for all participants. Once the forms with the experimental data and additional information were returned to the organizers, results underwent basic statistical treatment, and means and standard deviations were calculated for each laboratory.

Results were discussed at a general meeting where all aspects of the analyses done by each laboratory were rigorously perused and checked so as to exclude data in which no complete confidence could be attained. It goes without saying that the same rule had to be followed in the case of data submitted by laboratories which were not able to participate in the final meeting and thus could not defend their findings. The elements for which certified values could be established are shown in Table 10.2. Each of them was obtained by combining means from two or more independent analytical techniques, provided that they were in full compliance with the exclusion criteria touched upon above. Some laboratories quantified additional elements, which though not eligible for certification (e.g., because obtained through one technique only or because the means given by the relevant

Table 10.2. Certified concentrations for trace and minor elements in the *CRM MURST-ISS-A1 Antarctic Sediment*

Element	Mean of means ± confidence interval (95%)	Accepted analytical techniques *
Al (%)	0.671 ± 0.033	FAAS, ICP-AES (2), NAA
As ($\mu g\ g^{-1}$)	4.41 ± 1.06	NAA, Q-ICP-MS (2), Z-ETA-AAS
Cd ($\mu g\ g^{-1}$)	0.538 ± 0.027	ID-MS, Q-ICP-MS (2), SS-Z-ETA-AAS, Z-ETA-AAS
Co ($\mu g\ g^{-1}$)	6.87 ± 0.31	NAA (2), Q-ICP-MS, Z-ETA-AAS
Cr ($\mu g\ g^{-1}$)	42.1 ± 3.4	ICP-AES, NAA (2), Q-ICP-MS, Z-ETA-AAS
Fe (%)	0.244 ± 0.007	FAAS (2), ICP-AES (2), ID-MS, NAA (2), Q-ICP-MS
Mn ($\mu g\ g^{-1}$)	446 ± 19	FAAS, ICP-AES (2), NAA (3), Q-ICP-MS, SS-Z-ETA-AAS
Ni ($\mu g\ g^{-1}$)	9.56 ± 1.05	ICP-AES, NAA, Q-ICP-MS (2), Z-ETA-AAS
Pb ($\mu g\ g^{-1}$)	21.0 ± 2.9	ETA-AAS, ID-MS, Q-ICP-MS, SS-Z-ETA-AAS, Z-ETA-AAS (2)
Zn ($\mu g\ g^{-1}$)	53.3 ± 2.7	FAAS, ICP-AES (2), NAA

* The numbers in parentheses indicate how many laboratories have used the same technique in certifying a given element

laboratories were affected by unacceptably large differences) were thought to furnish informative values. These are listed in Table 10.3.

A total of 500 bottles of ca. 75 g each were obtained, the storage of which requires dry and cool conditions. After opening, the bottles must be placed in a dry desiccator and kept under subdued light. Before each intake the bottle must be shaken for several minutes to rehomogenize the material because segregation phenomena are always possible. The powdered sediment showed a degree of homogeneity for the various elements that permitted intakes as low as 100 mg, albeit 200 mg are recommended to minimize any possible variation in concentration. Prior to analysis, humidity should be determined on a separate aliquot of the material for correction to dry mass, e.g., in an oven at 105°C until constant mass or over P_2O_5 in a desiccator. One of the bottles and its contents are shown in Figure 10.1.

2.3. The case of MURST-ISS-A2 Antarctic Krill

2.3.1. Obtainment and pretreatment of the raw mass

Krill is a small planktonic crustacean primarily living in the Southern Ocean and its total biomass is estimated to be at least 500 million tons (around 500,000 billion individuals). It feeds on phytoplankton and is one of the most important species in the Antarctic food chain. In fact, krill is the basic food for whales, seals, cephalopods, penguins and many other seabirds and is also used for direct human consumption (20–24). Moreover, krill seems to be a promising bioindicator of environmental pollution by trace elements and organic compounds. All these considerations called for the production of a CRM for trace elements based on this matrix (25).

Table 10.3. Informative concentrations for trace elements in the *CRM MURST-ISS-A1 Antarctic Sediment* as obtained by NAA (unless otherwise specified)

Element	Concentration ($\mu g\ g^{-1}$)	Element	Concentration ($\mu g\ g^{-1}$)
Au	0.033 ± 0.005	Nd	33 ± 1
Ba	566 ± 96	Rb	122 ± 2
Br	43 ± 6	S	1215 ± 250
Ca	17,236 ± 1200	Sb	0.25 ± 0.02
Ce	77 ± 4	Sc	9 ± 1
Cl	8125 ± 978	Se	2.2 ± 0.1
Cs	6 ± 2	Si	335,000 ± 35,000
Cu (a, b, c, d)	6 ± 2	Sm	5.2 ± 0.2
Dy	3.3 ± 0.3	Sn (b, c, d)	2.6 ± 0.6
Eu	1.3 ± 0.3	Sr	217 ± 1
Ga	15.6 ± 0.4	Ta	0.94 ± 0.03
Ge	1 ± 0.4	Tb	0.7 ± 0.1
Hf	6 ± 1	Te	1.0 ± 0.1
Hg (e, b)	0.1	Th	11 ± 1
I	39 ± 4	Ti	2976 ± 211
Ir	16 ± 2	Tl (c)	0.29 ± 0.03
K	26,981 ± 1600	U	2.3 ± 0.1
La	36 ± 3	V	47 ± 6
Mg	15,200 ± 1300	W	1.6 ± 0.2
Mo	2 ± 1	Y	19 ± 2
Na	21,390 ± 1500	Yb	2.4 ± 0.2
Nb	12 ± 2	Zr	169 ± 43

a, ICP-AES; b, Q-ICP-MS; c, SS-Z-ETA-AAS; d, Z-ETA-AAS; e, CV-AAS

The total mass of krill necessary for the project (ca. 40 kg) was obtained by combining three different catches sampled during the 1993–1994 Italian expedition in the Ross Sea, the 1994–1995 expedition of the US Office of Polar Program in Marguerite Bay and the 1994–1995 expedition of the Japanese National Institute of Polar Research in Livingston Island. The three masses were at first stored deep-frozen at the ISS and then transferred to EC-JRC-IRMM laboratories for all the pretreatment steps necessary for the production of the material candidate to certification. A preliminary series of interlaboratory tests was undertaken to explore the feasibility of the various preanalytical and analytical approaches necessary to quantify the elements of interest. Thus, a few krill individuals were freeze-dried in order to detect possible operative difficulties in the final preparation of the candidate CRM. By taking into account all pros and cons in terms of homogeneity, hygroscopicity and representativeness of the freeze-dried material, and after thorough consultation with EC-JRC-IRMM, it was decided that the final product should consist of the whole krill organism, without attempting to separate soft from hard tissues. On the basis of the processing problems encountered in the pilot freeze-drying phase, the definitive procedure adopted included the following major phases:

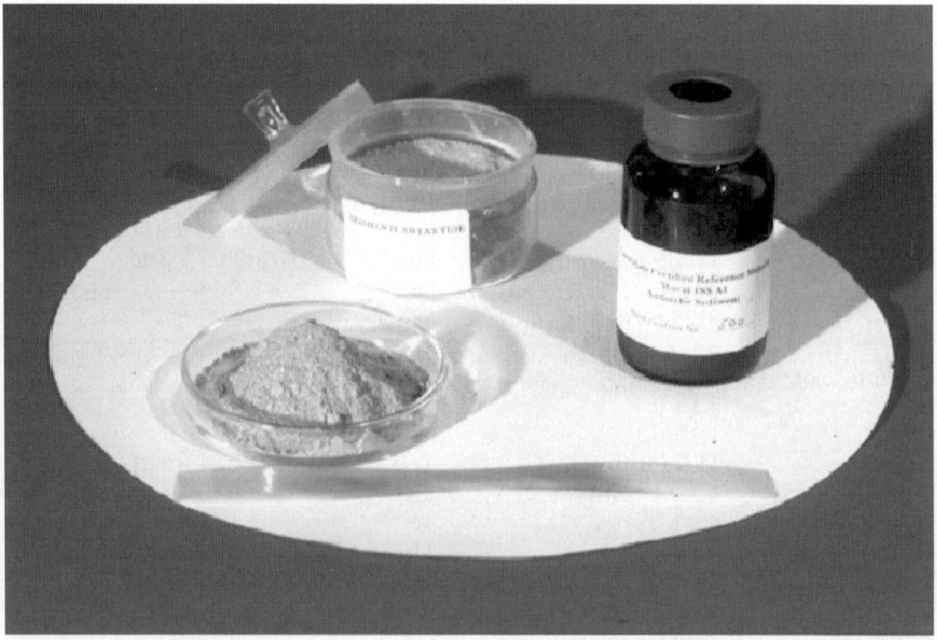

Figure 10.1. A bottle of the *CRM MURST-ISS-A1 Antarctic Sediment* and the powder contained therein.

i) *freeze-drying*. The Italian, US and Japanese krill was freeze-dried separately in an Epsilon apparatus (M. Christ, Osterode, Germany) with a shelf capacity of 5.4 m^2. The programme consisted of a primary drying step of 79 h at increasing temperature (from –20 to 0°C) and of a secondary step of 72 h at 23°C. The final vacuum was 0.9 pa. After freeze-drying, the material was directly transfered to a glove box purged with dry air (moisture content less than 77 mg H_2O m^{-3}) to avoid moisture pick-up during the grinding, sieving and sampling steps;
ii) *pre-crushing, fine grinding*. Each of the three freeze-dried masses was pre-crushed with a Teflon® mortar and pestle. The crushed krill was pulverized using a Pulverisette 5 ball mill (Retsch, Germany) equipped with Teflon® jars and balls. 100 g of material were treated for 10 min at a time;
iii) *sieving and homogenizing*. The three batches of ball-milled powder were sieved separately by means of a polyethylene tool (hole diameter 0.125 mm). Only the fraction smaller than 0.125 mm of the three catches was used for the candidate reference material. The dried, ground and sieved material was first homogenized on a 100-g scale in a Teflon® ball mill, equipped with three 30-mm diameter Teflon® balls, for 5 min. After collection, the 100-g samples were mixed and homogenized on a micro scale in a Turbula mixer for 2 h;
iv) *sampling and packaging*. The final product was sampled under dry air in a glove box, where an accurately weighed amount of krill powder (around 0.5 g) was put into 10-ml amber vials capped with Teflon®-protected rubber stoppers. After a further freeze-drying step, the vials were filled with argon,

sealed and labelled. About 5000 vials of freeze-dried krill were thus produced. Attention is drawn to the fact that the harder components (accounting for chitinous material) of the three krill batches had to be separated from softer parts after a preliminary coarse grinding; otherwise the former components could not be ground further in the presence of the latter. Only at the end of the two separate grinding processes the two fractions were mixed up again and the freeze-dried powders from each catch combined.

The characterization of the total mass thus obtained, performed on representative samples taken during the vial filling procedure, gave additional information on the properties of the material. The average moisture content, measured by Karl-Fisher titration, was lower than 3%. The particle size distribution of the whole mass, measured using the Sympatec analyzer with Helos measuring device, was to peak at 80–90 μm with an upper limit of 435 μm.

2.3.2. Organization of the certification exercise

The overall scheme for conducting the certification project follows the one adopted for the CRM *Antarctic Sediment*. Therefore, the same general considerations hold and will not be repeated here. Only aspects peculiar to this second instance will be highlighted.

Twelve vials of the candidate CRM were distributed to each participant in the certification project. All laboratories were free to use the instrumental approach with which they felt more confident and to pretreat the specimens as necessary to adequately present them to the technique of choice. Table 10.4 lists the participants in this project and the techniques used. For the analysis, laboratories were requested to perform complete and independent dissolutions (for techniques requiring the presentation of samples in the liquid form) of the mass contained in five vials (five independent analytical runs, three replicates each). Digestions were to be carried out on different days for each sample. The dissolution of krill was achieved mostly by microwave-assisted acid digestion.

For ten elements (As, Cd, Co, Cu, Fe, Mn, Ni, Pb, Se and Zn) out of those planned satisfactory results were obtained and certification was possible. Chromium, Hg, and Sn did not meet the preset requirements and could not be certified. Specifically, no agreement could be achieved for Hg among the various sets of data. As regards Cr, the results obtained by most laboratories using Q-ICP-MS were found to be strongly affected by the formation of the double ion ArC^+. Furthermore, unacceptably high SD values (in the order of magnitude of 20 and 30%) characterized the means of Cr and Sn. The concentration values for the accepted elements and the relevant 95% confidence intervals are reported in Table 10.5. Only a relatively small number of measurements were not taken into account in the calculation of these values. As an example of the bar-graphs that present the results obtained in this study, Figure 10.2 shows the pattern followed by three elements (two of which could be certified while the third one was not). Table 10.6 gives additional information regarding some elements not contemplated in the certification project.

Table 10.4. Laboratories participating in the certification project of trace elements in the *CRM MURST-ISS-A2 Antarctic Krill* and the instrumental techniques employed

Participant	Technique (*)
Comisión Nacional de Energia Atómica, (with the cooperation of University of Buenos Aires, Instituto Nacional de Tecnologia Industrial, Instituto Antártico Argentino, Servicio de Hidrografia Naval), Buenos Aires, Argentina	ETA-AAS, ICP-AES, NAA
Complutense University, Faculty of Chemical Sciences, Department of Analytical Chemistry, Madrid, Spain	FAAS, HG-AAS, Q-ICP-MS
Eötvös Loránd University, Institute of Inorganic and Analytical Chemistry, Budapest, Hungary	ETA-AAS, ICP-AES
Graz University of Technology, Institute for Analytical Chemistry, Graz, Austria	CV-AAS, FAAS, Z-ETA-AAS
Institute for Reference Materials and Measurements, European Commission, Joint Research Centre, Geel, Belgium	ETA-AAS, FAAS, NAA
Istituto Superiore di Sanità, Applied Toxicology Department, Rome, Italy	HR-ICP-MS, ICP-AES, Q-ICP-MS
National Institute of Standards and Technology, Gaithersburg, MD, USA	Q-ICP-MS
National Research Council Canada, Ottawa, Ontario, Canada	CV-AAS, ETA-AAS, HG-AAS, ID-MS, Q-ICP-MS
Pavia University, Department of General Chemistry, Pavia, Italy	NAA
Venice University, Environmental Sciences Department, Venice, Italy	HR-ICP-MS

(*) CV-AAS, Cold Vapour Atomic Absorption Spectrometry; ETA-AAS, Electrothermal Atomization Atomic Absorption Spectrometry; FAAS, Flame Atomic Absorption Spectrometry; HG-AAS, Hydride Generation Atomic Absorption Spectrometry; ICP-AES, Inductively Coupled Plasma Atomic Emission Spectrometry; ID-MS, Isotopic Dilution Mass Spectrometry; HR-ICP-MS, Magnetic Sector High Resolution Inductively Coupled Plasma Mass Spectrometry; NAA, Neutron Activation Analysis; Q-ICP-MS, Quadrupole Inductively Coupled Plasma Mass Spectrometry; Z-ETA-AAS, Zeeman Electrothermal Atomization Atomic Absorption Spectrometry

The stability of the material turned out to be acceptable. In fact, no detectable (i.e., exceeding the relevant uncertainty) changes in concentration values, as measured by means of three techniques (ICP-AES, Q-ICP-MS and HR-ICP-MS) were observed for vials stored at –20, 18 and 40°C over a period of six months. The bottling approach chosen, i.e., a small mass of material sealed under Ar in a single-shot vial makes moisture determinations unnecessary. In fact, the amount of each vial (ca. 0.5 g, exactly determined at the moment of bottling with an

Table 10.5. Certified concentrations for trace and minor elements in the *CRM MURST-ISS-A2 Antarctic Krill*

Element	Mean of means ± confidence interval (95%)	Accepted analytical techniques
As (μg g^{-1})	5.02 ± 0.4	HG-AAS, ICP-AES, NAA, Q-ICP-MS
Cd (μg g^{-1})	0.73 ± 0.06	ETA-AAS, FAAS, HR-ICP-MS, ICP-AES, Q-ICP-MS, Z-ETA-AAS
Co (μg g^{-1})	0.11 ± 0.01	HR-ICP-MS, ICP-AES, NAA, Q-ICP-MS
Cu (mg g^{-1})	65.2 ± 2.3	ETA-AAS, FAAS, HR-ICP-MS, ICP-AES, NAA, Q-ICP-MS, Z-ETA-AAS
Fe (mg g^{-1})	56.6 ± 2.3	FAAS, HR-ICP-MS, NAA, Q-ICP-MS, Z-ETA-AAS
Mn (μg g^{-1})	4.12 ± 0.10	ETA-AAS, FAAS, HR-ICP-MS, ICP-AES, NAA, Q-ICP-MS, Z-ETA-AAS
Ni (μg g^{-1})	1.28 ± 0.12	ETA-AAS, HR-ICP-MS, ICP-AES, Z-ETA-AAS
Pb (μg g^{-1})	1.11 ± 0.09	ETA-AAS, HR-ICP-MS, ICP-AES, Q-ICP-MS, Z-ETA-AAS
Se (μg g^{-1})	7.37 ± 1.13	ETA-AAS, HR-ICP-MS, ICP-AES, NAA, Z-ETA-AAS
Zn (mg g^{-1})	66.0 ± 2.0	FAAS, HR-ICP-MS, ICP-AES, NAA, Q-ICP-MS

uncertainty of 0.001 g) must be used at one and the same time. No subsampling is allowed and no shaking is necessary. As manipulation is minimized, the potential chemical contamination of the material is drastically reduced. The vials with their content of freeze-dried krill are shown in Figure 10.3.

2.4. Future activities

2.4.1. Trace elements in Southern Ocean water

The preparation of a third Antarctic CRM for trace elements, *i.e.*, Southern Ocean Water, is under way. The elements to be taken into consideration are As, Cd, Cr, Cu, Fe, Hg, Mn, Ni, Pb and Zn. In view of this, 100 l of water from the Ross Sea (74°46′ Lat S; 163°57′ Long E) were collected during the 1993–1994 Italian Expedition in Antarctica. The water was put into 10-l chemically decontaminated polyethylene containers, immediately frozen at –20°C and shipped to the ISS, where it has been kept in the laboratory at –20°C since. The preparation steps will basically study stabilization and acidification of the water mass and bottling. This aspects are crucial as they entail the long-term storage of the CRM, once prepared, under optimal conditions for stability and minimal leaching of chemical elements from the inner walls of the bottles. For the time being, exploratory work has been done to gain information on the most suitable pretreatments and analytical approaches as well as to identify the concentration range of analytes.

Both water salinity and the generally very low concentrations of the elements considered make analytical determinations rather challenging (26). Removal of the

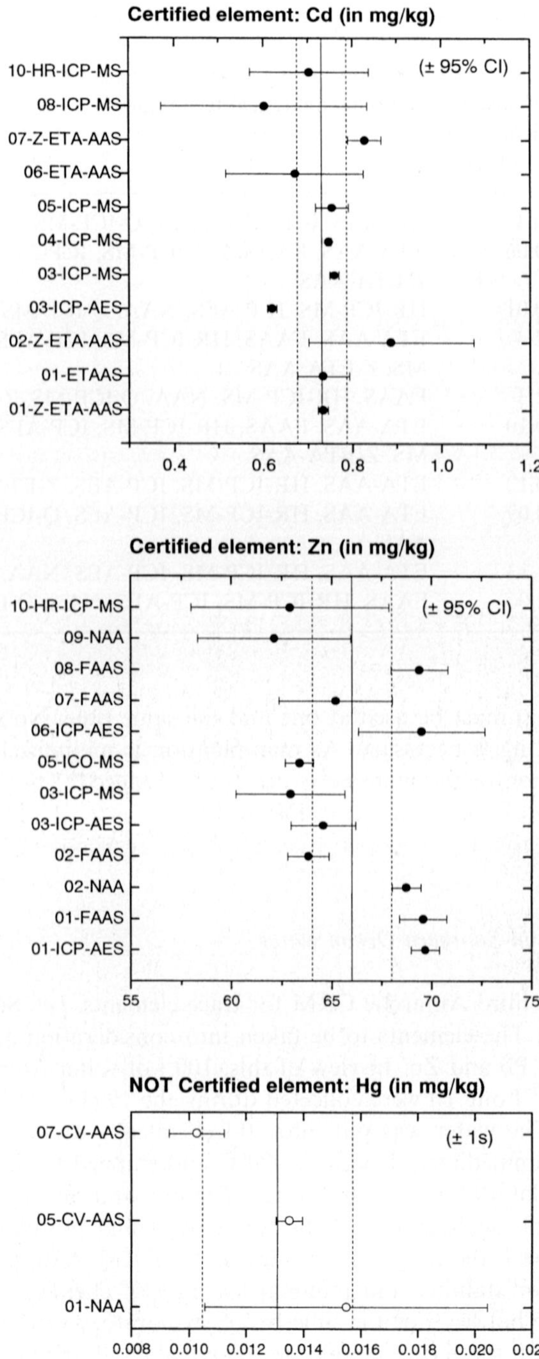

Figure 10.2. Examples of bar graphs obtained in the certification projects of trace elements in Antarctic krill.

Table 10.6. Informative concentrations for trace elements in the *CRM MURST-ISS-A2 Antarctic Krill* as obtained by NAA and other techniques as specified, wherever applicable

Element	Concentration ($\mu g\ g^{-1}$)	Element	Concentration ($\mu g\ g^{-1}$)
Ag (a) (*2*)*	1.38 ± 0.23	Mg	3200 ± 3
Al	23.6 ± 2.2	Mg	5100 ± 200
Br (*2*)*	265.7 ± 1.1	Na	29,400 ± 200
Ca	14,500 ± 400	Rb	4.39 ± 0.34
Cl	48,600 ± 400	Sc	0.0051 ± 0.0003
Cr (a, b, c, d) (*6*)*	0.73 ± 0.14	Sn (f) (*2*)*	0.22 ± 0.01
Hg (e) (*3*)*	0.013 ± 0.003	Sr	242 ± 11
I (*2*)*	7.4 ± 1.0	V (f)	1.21 ± 0.06
K	14,800 ± 300		

a, Q-ICP-MS; b, ICP-AES; c, ETA-AAS; d, Z-ETA-AAS; e, CV-AAS; f, HR-ICP-MS
* Average of values submitted by *n* laboratories

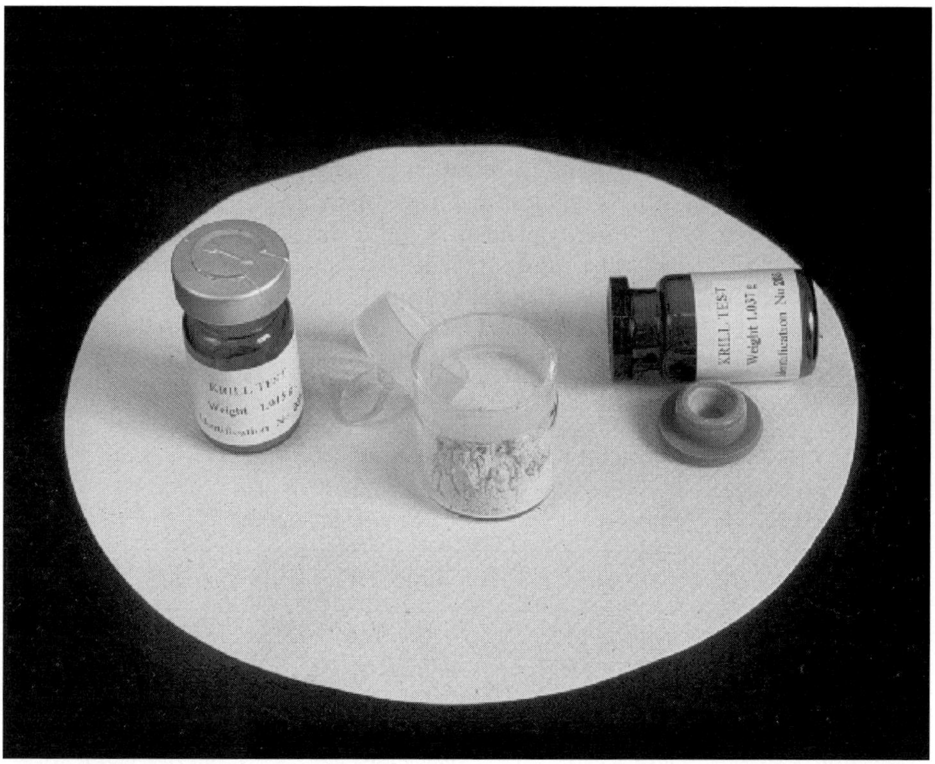

Figure 10.3. Vials of the *CRM MURST-ISS-A2 Antarctic Krill.*

saline matrix may substantially alleviate these difficulties. To explore this possibility, aliquots of ocean water were UV irradiated to disrupt organic compounds, and filtered to remove suspended particulate matter. An on-line preconcentration method was adopted, based on the Flow Injection Analysis (FIA) system and the chelating ion-exchange resin Chelex-100 (Merck, Darmstadt, Germany) (27). Under controlled pH conditions (pH > 5.5, ammonium acetate buffer), and thanks to the scarce affinity of the major components of the saline matrix with the resin, only the elements of interest were retained and could be thus concentrated in the final eluate (2*M* HNO_3). All steps were carried out in a Class 100 clean laboratory, making use of previously decontaminated ametallic containers and tools. Only high purity reagents were employed. The preliminary concentration values obtained by coupling FIA with either ICP-AES or Q-ICP-MS are shown in Table 10.7. Measurement accuracy was checked by including the CRM NASS-4 (NRCC, Canada) in the analytical runs. Results can be considered rather satisfactory and probative of the actual levels to be expected in the ocean water for the elements tested. Contacts are in progress with a number of expert laboratories for their participation in the project which will be hopefully launched in the next few months.

2.4.2. Trace elements in Adamussium colbecki

A preliminary investigation was performed to ascertain the feasibility of producing a CRM for trace elements based on the bivalve *Adamussium colbecki* (28). Samples of this mollusc were collected from shallow waters along the coast of Terra Nova Bay, close to the Italian Base, during the 1999–2000 scientific expedition. In order to test the key steps of the overall certification process, the collected material was freeze-dried (both the entire organism and parts thereof), finely ground and then analyzed by ICP-AES for their content in some trace elements (Cd, Cr, Cu, Fe, Mn, Mo, Ni, Se and Zn). The organs taken into account were the muscle, foot, mantle, gills, kidney, gonads and digestive gland, the latter being of particular interest because of the bioaccumulation of Cd and Zn (29, 30). As regards the

Table 10.7. Concentrations of some trace elements in the Southern Ocean water as determined by ICP-MS ([a]) and ICP-AES ([b])

Element	Concentration in water (μg l^{-1})	Concentration in the CRM NASS-4 (μg l^{-1})	
		Certified	Found
As [a]	1.76 ± 0.12	1.26 ± 0.09	1.3 ± 0.2
Cd [a]	0.074 ± 0.006	0.016 ± 0.003	0.017 ± 0.003
Cu [b]	1.42 ± 0.12	0.228 ± 0.011	0.238 ± 0.024
Fe [b]	0.06 ± 0.002	0.105 ± 0.016	0.106 ± 0.012
Mn [b]	0.24 ± 0.01	0.380 ± 0.023	0.396 ± 0.024
Pb [a]	0.06 ± 0.01	0.013 ± 0.005	0.015 ± 0.006
Zn [b]	0.14 ± 0.01	0.115 ± 0.018	0.117 ± 0.009

Table 10.8. Concentrations of some trace elements in freeze-dried *Adamussium colbecki* as obtained by ICP-AES

Element	Concentration (μg g^{-1})				
	entire organism	muscle	foot	mantle	internal organs
Cd	29.4 ± 1.9	1.49 ± 0.01	4.09 ± 0.31	11.2 ± 0.7	98 ± 1
Cr	2.9 ± 0.9	0.38 ± 0.19	0.93 ± 0.06	4.39 ± 0.45	3.02 ± 0.04
Cu	4.0 ± 0.6	0.889 ± 0.003	3.11 ± 0.63	2.95 ± 0.06	7.78 ± 0.09
Fe	97.2 ± 29.1	37.9 ± 0.5	86.8 ± 1.7	84.9 ± 5.6	264 ± 3
Mn	2.65 ± 0.04	1.10 ± 0.02	2.87 ± 0.06	4.46 ± 0.21	3.59 ± 0.03
Mo	0.69 ± 0.12	0.24 ± 0.01	0.645 ± 0.0004	0.66 ± 0.04	1.77 ± 0.02
Ni	5.11 ± 0.48	1.39 ± 0.08	3.12 ± 0.81	5.05 ± 0.36	7.82 ± 0.06
Zn	68.5 ± 1.0	40.8 ± 0.4	76.3 ± 2.0	157± 9	78.7 ± 1.3

whole organism, the concentrations measured are reported in Table 10.8. These preliminary values agree rather well with those calculated by summing the individual amounts for each organ. As a rule, element concentrations were found to increase in the order muscle-foot-mantle-internal organs. The difference between the lowest and the highest value for a given element spanned up to two orders of magnitude.

As for Southern Ocean water, steps are being taken to select laboratories with high expertise in the analysis of marine molluscs so as to start with the certification project.

2.4.3. Polychlorobiphenyls in krill

The characteristics of krill as a potentially useful organism for environmental monitoring have already been illustrated under Section 2.3. Krill's peculiarity might also be useful to monitor the fate of several organic man-made chemicals, among which polychlorobiphenyls (PCBs) are of particular concern (31). In order to perform a feasibility study, ca. 70 kg of krill were fished in the Ross Sea during the 1999 Italian oceanographic expedition in Antarctica and manipulated as described for the production of the candidate CRM for trace elements, but for a key aspect: no plastic containers or plastic devices were used throughout the process from sampling to bottling to avoid any possible contamination with organic compounds that could eventually mask or alter the original content in the selected PCBs. The fresh material was placed into stainless steel jars soon after catching, stored at –20°C until delivery and pretreated as necessary, always keeping in mind this constraint. Glass vials with approximately 5 g of freeze-dried krill were thus produced and a pilot study is being performed with the participation of a restricted number of highly specialized laboratories. The experimental data obtained so far are shown in Table 10.9. A certification project will be launched in due course.

Table 10.9. Concentration ranges of some PCB congeners in freeze-dried Antarctic krill as obtained by chromatographic techniques

PCB congener	Concentration range and mean (μg g^{-1})	Analytical techniques employed (*)
18	0.009–1.27 (0.43)	GC-ECD, GC-MS
28 + 31	0.12–0.80 (0.51)	GC-ECD, GC-MS
47	0.40–1.23 (0.81)	GC-ECD, GC-MS
52	0.13–5.50 (1.12)	GC-ECD, GC-MS
70	0.05–0.60 (0.23)	GC-ECD, GC-MS
95	0.06–0.40 (0.24)	GC-ECD, GC-MS
99	0.03–0.15 (0.10)	GC-ECD, GC-MS
101	0.10–1.99 (1.10)	GC-ECD, GC-MS
110	0.02–0.20 (0.12)	GC-ECD, GC-MS
118	0.03–0.20 (0.12)	GC-ECD, GC-MS
138 + 163	0.08–0.30 (0.20)	GC-ECD, GC-MS
149	0.09–0.40 (0.23)	GC-ECD, GC-MS
151	0.05–1.67 (0.54)	GC-ECD, GC-MS
153	0.08–0.33 (0.21)	GC-ECD, GC-MS
180	0.07–0.12 (0.09)	GC-ECD, GC-MS
187	0.03–0.08 (0.06)	GC-ECD, GC-MS

(*) GC-ECD, Gas Chromatography with Electron Capture Detection; GC-MS, Gas Chromatography Mass Spectrometry

3. Conclusions

The availability of *ad hoc* CRMs can substantially contribute to the credibility and comparability of experimental information obtained in Antarctica. The global investment made in this remote continent in terms of human and financial resources, and the crucial role played by Antarctic studies in the interpretation of phenomena on a planetary scale do demand that no questionable data be generated or circulated. It would be thus desirable that other countries undertake similar programmes and stimulate the major producers of CRMs to support and participate in such initiatives. International planning and coordination in this context will definitely provide additional evidence of Antarctica as being as a land of peace and science.

References

1. R. Alvarez, Certified reference materials for validating spectroscopic methods and experimental data, *Fresenius' Z. Anal. Chem.*, 324 (1986), 376–383.
2. E. A. Maier, Environmental analysis: is accuracy possible without certified reference materials?, *Anal. Proc.*, 27 (1990), 269–270.
3. H. Muntau, The problem of accuracy in environmental analysis, *Fresenius' Z. Anal. Chem.*, 324 (1986), 678–682.

4. B. Griepink, E. A. Maier, Ph. Quevauviller, H. Muntau, Certified reference materials for the quality control of analysis in the environment, *Fresenius J. Anal. Chem.*, 339 (1991), 599–603.
5. R. J. Mesley, W. D. Pocklington, R. F. Walker, Analytical quality assurance. A review, *Analyst*, 116 (1991), 975–990.
6. S. Caroli, Certified reference materials: use, manufacture and certification. *Anal. Chim. Acta*, 283 (1993), 573–582.
7. G. A. Uriano, C. C. Gravatt, The role of reference materials and reference methods in chemical analysis, *CRC Crit. Rev. Anal. Chem.*, 6 (1977), 361–411.
8. W. E. Van Der Linden, G. De Niet, M. Bos, The role of analytical chemistry in process quality control, *Anal. Chim. Acta*, 216 (1989), 307–319.
9. Ph. Quevauviller, Method performance studies for inorganic analysis: examples of results obtained by plasma spectrochemical techniques in the frame of Community Bureau of Reference (BCR)-certification campaigns, *J. Anal. At. Spectrom.*, 12 (1997), 871–879.
10. R. Dams, Reference materials for trace analysis, *Pure & Appl. Chem.*, 55 (1983), 1957–1968.
11. R. Dybczynski, The contribution of various analytical techniques to the certification of reference materials, *Fresenius J. Anal. Chem.*, 352 (1995), 120–124.
12. *Guidelines for the production and certification of BCR reference materials*. Standards, Measurements and Testing Programme, European Commission, 1994, Doc. BCR/48/93, 54.
13. P. De Bièvre, J. Savory, A. Lamberty, G. Savory, Meeting the need for reference measurements, *Fresenius' Z. Anal. Chem.*, 332 (1988), 718–721.
14. S. B. Adeloju, A. M. Bond, Influence of laboratory environment on the precision and accuracy of trace element analysis, *Anal. Chem.*, 57 (1985), 1728–1733.
15. SCAR/COMNAP Antarctic Environmental Monitoring Workshop Report, College Station, TX, USA, 25–29 March 1996.
16. R. Bargagli, Trace metals in Antarctica related to climate change and increasing human impact, *Res. Environ. Contam. Toxicol.*, 166 (2000), 129–173.
17. S. Caroli, O. Senofonte, S. Caimi, J. Pauwels, G. N. Kramer, Planning and certification of new multielemental reference materials for research in Antarctica, *Mikrochim. Acta*, 123 (1996), 119–128.
18. Papadakis, P. D. P. Taylor, P. De Bièvre, Establishing an SI-traceable copper concentration in the candidate reference material MURST-ISS-A1 Antarctic Sediment using isotope dilution applied as a primary method of measurement, *J. Anal. At. Spectrom.*, 12 (1997), 791–796.
19. S. Caroli, O. Senofonte, S. Caimi, P. Robouch, J. Pauwels, G. N. Kramer, Certified reference materials for research in Antarctica: the case of marine sediment, *Microchem. J.*, 59 (1998), 136–143.
20. D. G. M. Miller, I. Hampton, Biology and ecology of the Antarctic krill (*Euphausia superba*). A review, *Biomass Sci. Res.*, 9 (1989), 166.
21. S. Nichol, Who's counting on krill?, *New Sci.*, 1690 (1989), 38–41.
22. S. Nichol, The age-old problem of krill longevity, *Bioscience*, 40 (1991), 833–836.
23. S. Nichol, W. K. de la Mare, Ecosystem management and the Antarctic krill, *Amer. Sci.*, 81 (1993), 36–47.
24. R. M. Ross, L. B. Quetin, How productive are Antarctic krill?, *Bioscience*, 36 (1986), 264–269.
25. S. Caroli, O. Senofonte, S. Caimi, P. Pucci, J. Pauwels, G. N. Kramer, A pilot study for the preparation of a new reference material based on Antarctic krill, *Fresenius J. Anal. Chem.*, 360 (1998), 410–414.
26. I. Rodushkin, T. Ruth, Determination of trace metals in estuarine and sea-water reference materials by high resolution inductively coupled plasma mass spectrometry, *J. Anal. At. Spectrom.*, 12 (1997), 1181–1185.
27. S. N. Willie, Y. Iida, J. W. McLaren, Determination of Cu, Ni, Zn, Mn, Co, Pb, Cd, and V in sea water using flow injection ICP-MS, *At. Spectrosc.*, 19 (1998), 67–72.
28. M. Nigro, Nearshore population characteristics of the circumpolar Antarctic scallop *Adamussium colbecki* (Smith, 1902) at Terra Nova Bay (Ross Sea), *Ant. Sci.*, 5 (1993), 377–378.
29. M. Mauri, E. Orlando, M. Nigro, F. Regoli, R. Rocchi, Heavy metals in the Antarctic scallop *Adamussium colbecki*, *Mar. Ecol. Progr. Ser.*, 67 (1990), 27–33.
30. A. Viarengo, L. Canesi, A. Mazzuccotelli, E. Ponzano, Cu, Zn and Cd content in different tissues

of the Antarctic scallop *Adamussium colbecki.* Role of matallothionein on heavy metals homeostasis and detoxication, *Mar. Ecol. Progr. Ser.*, 95 (1993), 163–168.

31. R. Sen Gupta, A. Sarkar, T. W. Kureishey, PCBs and organochlorine pesticides in krill, birds and water from Antarctica, *Deep-Sea Res. II*, 43 (1996), 119–126.

Environmental Contamination in Antarctica
A Challenge to Analytical Chemistry
S. Caroli, P. Cescon and D.W.H. Walton, editors

Chapter 11

Preparation and production control of the certified reference material of Antarctic sediment

Jean Pauwels, Gerard N. Kramer, Karl-Heinz Grobecker

1. Introduction

The production of biological and environmental Certified Reference Materials (CRMs) is one of the main activities of the Institute for Reference Materials and Measurements (IRMM), Joint Research Centre of the European Commission, Geel (Belgium) (1). The Reference Materials Unit was therefore requested by the Istituto Superiore di Sanità (ISS), Rome (Italy) to perform the transformation of a batch of ca. 80 kg wet Antarctic coastal marine sediment, collected in the Terra Nova Bay during the XI Italian Expedition to the Antarctic, into a sampled, dry and conservable powder to be certified for the content of a range of elements.

IRMM has established dedicated facilities for the preparation of candidate CRMs aiming at:

i) *being able to handle large quantities of base material in a short time*, which is an important prerequisite to be able to produce CRMs economically. The IRMM facilities are designed to handle typically 200 to 500 kg fresh material in order to produce up to 2000 samples of 20 to 50 g dry powder;
ii) *preparing samples of the correct size*, which often means larger series of smaller samples (to avoid spoilage, contamination or conservation problems in the end-user's laboratory). This requires considerable automation to avoid lengthy and costly bottling and labelling operations;
iii) *preparing samples under controlled conditions*, which means that preparation must take place under a clean atmosphere and low temperature, and that dry products are prepared and packed under inert gas as this is likely to improve stability, and may have a beneficial influence on the cost of both storage and distribution.

Care must also be taken to ensure that particle size distributions are adequate as selected particle size distributions will achieve greater homogeneity in reference materials especially when very low sample intakes are used for measurements. This is extremely important for environmental reference materials such as soils, sediments, etc., and is optimally achieved by jet-milling with ultra-fine classification of particles. With this technique, which is non-contaminating, fast and well-controlled, the sieved fraction below 2 mm of the material is ground by impact of particles accelerated from three high pressure air streams onto each other. It generates a size reduction of material without significant heat generation. Fine powders can be produced, with closely defined particle size distributions and throughputs of 5 to 25 kg hr^{-1}.

Several Bureau Communautaire de Référence (BCR) CRMs of sediments (nos. 462, 535 and 536) have been prepared and certified over the last few years using the above described technique (2–4).

2. Preparation of the candidate CRM

Before starting the preparation of the candidate CRM, a preliminary test was carried out on a small sample of 50 g wet sediment, which was dried for 15 h at 60°C, ground in a porcelain mortar, homogenized for 10 min in a Turbula mixer and subsequently analyzed for various trace metals by Neutron Activation Analysis (NAA) (see Table 11.1) and for its particle size distribution (Figure 11.1), after which the powder was examined microscopically (Figure 11.2). This semi-quantitative check was carried out to verify the initial trace element levels in order to establish whether the preparation steps listed below could be carried out free of contamination and whether the resulting material was of the required quality:

i) drying in Teflon®-protected stainless steel trays;
ii) jet milling in equipment made out of non-ferrous materials;
iii) sampling using a sample divider made out of polyamide and glass.

Subsequently, the main batch was put in production following the flow-sheet given in Figure 11.3.

2.1. Sieving and drying

The frozen material was dispatched to IRMM in polyethylene bottles of 2 to 5 l and subsequently stored in a deep-freezer. Before drying, the unfrozen wet material

Table 11.1. Production control results (in mg/kg) by k_0-NAA on the material used for the 50 g test and on the final product

Element	50 g test	Final product
Al	57,000 ± 5000	65,900 ± 500
As	6.7 ± 0.6	<5
Co	7.0 ± 0.6	7.10 ± 0.14
Cr	42 ± 4	45.0 ± 1.1
Cu	<600	<600
Fe	25,000 ± 3000	24,100 ± 500
Mn	465 ± 40	450 ± 7
Ni	25 ± 5	9.5 ± 0.8
Zn	97 ± 9	<70

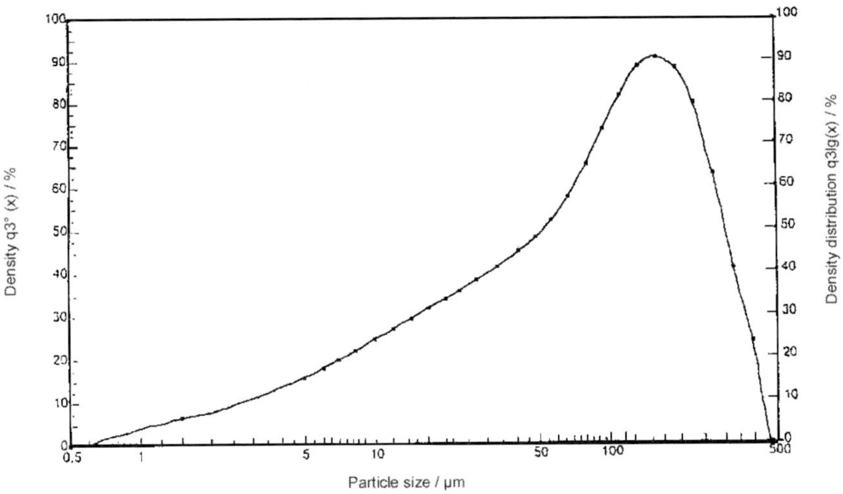

Figure 11.1. Particle size distribution of sample from the 50 g test batch.

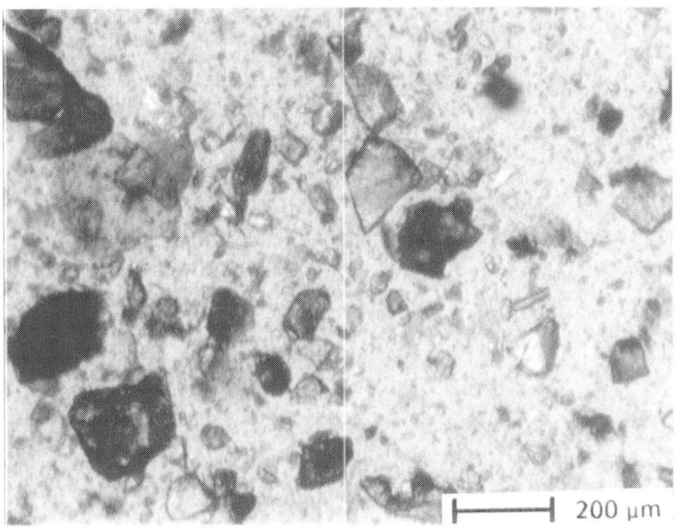

Figure 11.2. Micrograph of sample from the 50 g test batch.

was sieved (<2 mm) using a straining and sieving equipment type Finex 22 (Russell, Belgium) to eliminate foreign coarse materials such as fish, shell and other non-identified biological materials. The sieved fraction was then dried for 90 hr at 60°C in Teflon®-protected stainless steel trays and homogenized for 2 hr using a Turbula mixer. The residual moisture content of the dried material was measured by Karl-Fischer titration (see below), and amounted to 0.42 ± 0.11% (n = 12). The weight of dried material was 44 kg.

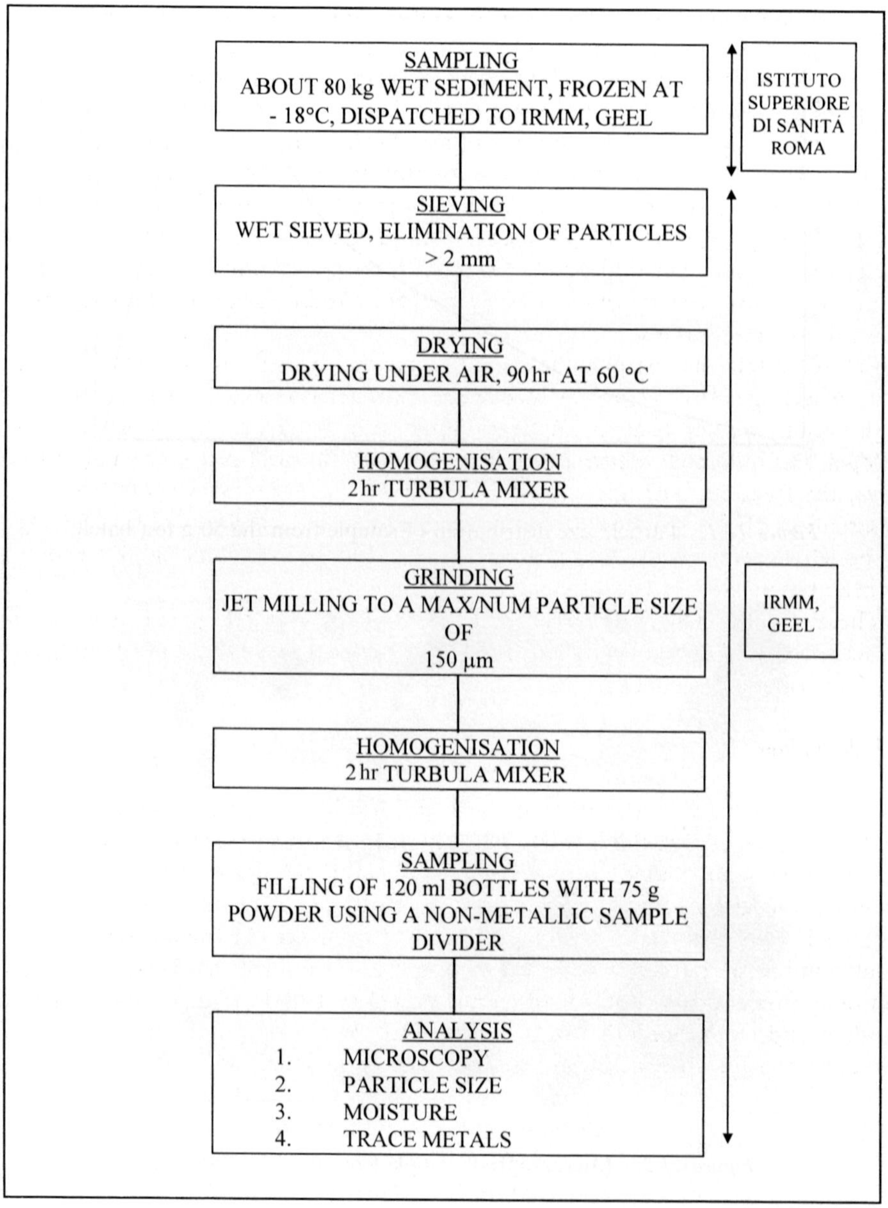

Figure 11.3. Flow-sheet of the preparation of the candidate CRM Murst-ISS A1 Antarctic Sediment.

2.2. Jet milling

Jet-milling with ultrafine classification of particles can be achieved using a Fluidized Bed Opposed Jet Mill 100 AFG (Alpine, Augsburg, Germany). It is a non-contaminating, fast and well-controlled tool for comminution of large amounts (200 kg) of soils, sediments, fly ashes and sludges (2–4).

Fine particles are extracted through an oxide-ceramic classifying wheel on top of the grinding chamber. Depending on the feed material, the pressure of the air, the air volume and the speed of the classifying wheel, particles between 5 and 120 μm, characterized by a closely defined size distribution, can be extracted. Throughput for soils and sediments is 2 to 6 kg hr^{-1}. Segregation of the particles and the air occurs in a cyclone. The major part of the ground material is collected in a bin under the cyclone. The finest particles are extracted and transferred to a filter. The multi-processing system offers closely-defined particle size distribution with an exact top-size limitation.

Before milling the complete batch, a test was performed to verify whether a classifier wheel speed of 3700 rpm was appropriate. It was observed that, due to the high organic matter content of the sediment, a unusually high top particle size of 435 μm was obtained. When a sample of the material was sieved through a 63 μm sieve, the fraction >63 μm corresponded to 3% of the total weight. Microscopic observation of this fraction showed that it contained a high content of biological tissue. However, in order to obtain an "as real as possible" material it was decided to keep the powder as obtained after jet milling.

Therefore, the remaining material (38 rpm 4 kg) was jet milled at a classifier wheel speed of 3700, collected in an 85 l polyethylene container and homogenized for 2 hr using a Turbula mixer.

2.3. Sampling

After homogenization the material was divided into 8 batches of about 4.8 kg each using a non-metallic Fritsch laboratory sample divider. Each of these batches was again divided into 8 batches of about 600 g (total 64 batches). This quantity was used for final sampling of about 75 g of powder into 120 ml well cleaned brown glass bottles with polyethylene inserts and plastic screw caps. The total number of bottles was 512. Representative samples were taken during the sampling procedure to control the moisture content and homogeneity of the final product and to examine the produced powder microscopically and for its particle size. The final product delivered to ISS, Rome, was labelled:

Certified Reference Material
Murst – ISS A1
Antarctic Sediment

3. Production control

3.1. Particle size analysis

The powder was examined microscopically (see Figure 11.4) and particle size measurements were carried out using a Helos particle size analyser (Sympatec,

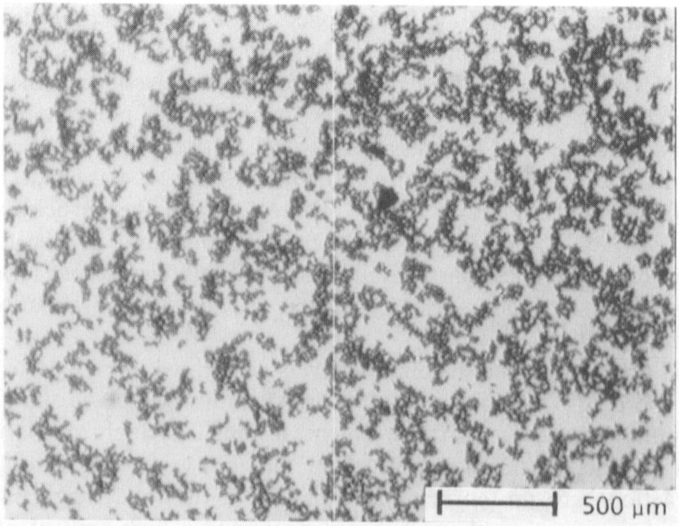

Figure 11.4. Micrograph of jet-milled Antarctic sediment.

Germany). The particle size distribution of a representative sample is shown in Figure 11.5.

3.2. Residual water content

As the drying methods commonly used for the determination of the water content of powders do not selectively measure the water content, but rather the mass lost under specific drying conditions, the Karl Fischer titration method was used

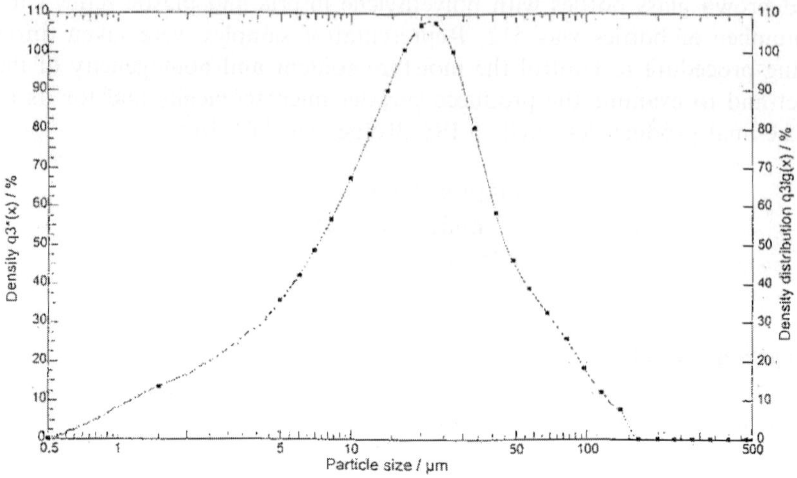

Figure 11.5. Particle size distribution of jet-milled Antarctic sediment.

instead. This method is selective for water and is based upon the following specific two-step chemical reaction:

$$CH_3OH + SO_2 + Z \rightleftarrows ZH^+ + CH_3OSO_2^-;$$

$$ZH^+ + CH_3OSO_2^- + I_2 + H_2O + 2\,Z \rightleftarrows 3\,ZH^+ + CH_3OSO_3^- + 2\,I^-.$$

During the titration, water is extracted by a working medium, usually CH_3OH. Sulphur dioxide and CH_3OH react to methyl sulphite. In a stoichiometric reaction which requires water the ester is consecutively oxidized to $CH_3OSO_3^-$ by the titrating reagent I_2. Z is a base which leads to the practically complete reaction on the right-hand side of the equation by neutralizing the methyl esters.

Nowadays the orginal base pyridine has been replaced by the more efficient and nontoxic imidazole. The first excess of I_2 indicates the end point of the titration (5, 6). The experimental conditions are summarised in Table 11.2.

Ten samples taken during the bottling procedure were analyzed for their residual water content. The mean water content found was 0.35 ± 0.05%, which is fully acceptable to ensure long-term stability.

Table 11.2. Karl Fischer titration parameters. Instrument: 701 K-F-Titrino + 703 Titration Stand (Metrohm, Herisau, Switzerland). Reagents and calibrants were supplied by Riedel de Haën, Seelze, Germany

Bivoltametric end point indication	(= *bipotentiometric*)
I (pol)	50μA
End point voltage	250 mV
Sensitivity range	500 mV
Stop criterion	Time
Delay time	20 s
Maximal titrating rate	2 ml/min
Minimal volume increment	Smallest possible
Start volume	OFF/0 mL
Conditioning	ON
Reagents	Hydranal Composite and methanol
Calibrant (aq. sol.)	Hydranal 5.00 ± 0.02 mg H_2O/ml
Calibrant (solid)	Sodium tatrate 15.66 ± 0.05% H_2O

3.3. Homogeneity control by Solid Sampling Zeeman Atomic Absorption Spectrometry

Representative samples were taken during the bottling procedure to determine a number of trace elements (As, Cd, Cu, Hg, Mn, Pb, Sn, Tl, Zn) in Antarctic coastal marine sediment by Solid Sampling Zeeman Electrothermal Atomization Atomic Absorption Spectrometry (SSZ-ETA-AAS) (see Table 11.3). This technique is particularly suited to homogeneity control because of the usually low sample mass (0.1–10 mg) and the high number of parallel measurements (10–100). Additionally, all measurements are performed without any chemical sample

Table 11.3. Solid Sampling Zeeman Electrothermal Atomization Atomic Absorption Spectrometry parameters

Element wavelength (nm)	Furnace (sheath gas)	Drying (°C/s)	Ashing (°C/s)	Atomization (°C/s)	CRM (supplier)	Sample mass interval (mg)
As 193.7	Graphite (Ar)	300/10	500/20	1800/6	MESS-1 (NRCC)	0.07–0.5
Cd 228.8	Graphite (Ar)	300/10	500/20	2300/3	MESS-1 (NRCC)	0.006–0.07
Cu 324.8	Graphite (Ar)	300/10	500/20	2500/5	MESS-1 (NRCC)	0.03–0.09
Hg 253.7	Nickel (Ar)	–	–	1000/10	CRM 62 (BCR)	30–75
Mn 403.1	Graphite (Ar)	300/10	500/20	2300/5	MESS-1 (NRCC)	0.003–0.013
Pb 368.4	Graphite (Ar)	300/10	500/20	2300/3	MESS-1 (NRCC)	0.1–0.6
Sn 286.3	Graphite (Ar)	300/10	500/20	2500/4	MESS-1 (NRCC)	0.03–0.06
Tl 276.8	Graphite (Ar)	300/10	500/20	1500/4	SRM 2704 (NIST)	0.5–2.2
Zn 307.6	Graphite (Ar)	300/10	500/20	2500/3	MESS-1 (NRCC)	0.1–0.6

MESS-1, Marine Sediment, National Research Council of Canada. CRM 062, Olea Europea, Bureau Communautaire de Référence, EU. SRM 2704, Buffalo River Sediment, National Institute of Standards and Technology, USA.

pretreatment and therefore the relative standard deviation of the parallel measurements is mainly caused by the inhomogeneity of the material (7–9). Calibration and quality control of the determinations were based on CRMs of different origin, but matrix and content of the analytes were chosen to be as similar as possible to the Antarctic sediment (see Table 11.4).

As some of the signals for the element concentrations were outside the linear calibration ranges, the material had to be diluted by adding ultra-pure graphite powder (Schunk–Kohlenstoff, Germany) at a ratio of 21:1 before treatment of both components in a planetary mill (Fritsch Pulverisette 5, Germany) to produce a homogeneous mixture. All trace element measurements were performed by a Grün SM 30 Zeeman-correction AAS instrument (Grün Analytische Meßsysteme, Germany).

Since As measurements suffered from severe matrix interference due to increased sample mass (which caused remarkable signal suppression in the undiluted material), the sample was diluted with graphite. This reduced the signal suppression, although the phenomenon was still present. Therefore, it was only possible to ascertain an indicative value using a linear regression approach and extrapolating to zero mass. However, a homogeneity factor $H_E = 1.25\% \times \sqrt{mg}$ (with H is

Table 11.4. Calibration and quality control of trace element determinations by SSZ-ETA-AAS

Element	CRM used for calibration	Certified value (mg/kg)	Quality control by SSZ-ETA-AAS (mg kg^{-1})
As	MESS-1	10.6 ± 1.2	9.16 ± 1.5 (n = 14)
Cd	MESS-1	0.59 ± 0.1	0.595 ± 0.116 (n = 12)
Cu	MESS-1	25.1 ± 3.8	23 ± 3.3 (n = 4)
Hg	CRM 062	0.28 ± 0.02	0.30 ± 0.03 (n = 2)
Mn	MESS-1	513 ± 25	467 ± 86 (n = 10)
Pb	MESS-1	34 ± 6.1	32.5 ± 7 (n = 4)
Sn	MESS-1	3.98 ± 0.44	4.39 ± 0.65 (n = 10)
Tl	SRM 2704	1.20	1.24 ± 0.09 (n = 4)
Zn	MESS-1	191 ± 17	186 ± 21 (n = 4)

MESS-1, Marine Sediment, National Research Council of Canada.
CRM 062, Olea Europaea, Bureau Communautaire de Référence, EU.
SRM 2704, Buffalo River Sediment, National Institute of Standards and Technology, USA.

homogeneity factor of element E, 1.25% is the relative standard deviation of the measurements of As and mg is the average mass in mg of the samples analyzed) was derived from the results, thus confirming the good homogeneity of the material at the mg level.

For Cd no matrix interferences were observed from the increased sample mass. Nevertheless, all analyses were performed on the diluted sediment, because the Cd content of the sample met exactly the linear range of the Cd spectral line at 228.8 nm. The accuracy of the result was estimated at ±10% (1s), so that a final result of 0.53 ± 0.05 mg kg^{-1} could be given. A homogeneity factor H_E = 1.06% × $\sqrt{mg}$ was derived from these measurements confirming the good homogeneity of the material also for Cd.

Copper measurements were performed on the diluted sediment because the Cu content of the sample fell in the linear range of the Cu spectral line at 324.8 nm. The accuracy of the result was estimated at ±20% (1s), so that the final result of 5 ± 1.2 mg g^{-1} could be achieved. A homogeneity factor H_E = 5.8% × $\sqrt{mg}$ was derived from these results confirming the good homogeneity of the material for Cu.

Mercury measurements were carried out on the original sediment because the original very low Hg content was very close to the limit of detection of the method. No matrix interference was observed, although a slight trend was seen in the results *vs.* sample mass. This is probably due to poor peak evaluation (peaks were only slightly above background). The Hg concentration was estimated as less than 0.01 mg kg^{-1}. Given the very low Hg content, the calculation of a homogeneity factor is meaningless.

Manganese measurements were performed on diluted sediment because the Mn concentration had to be determined on the poorly sensitive spectral line at

403.1 nm. The accuracy of the result was estimated at ±20% (1s), so that a final result of 455 ± 91 mg kg^{-1} could be given. A homogeneity factor H_E = 1.54% × $\sqrt{mg}$ was derived from these results, confirming the good homogeneity of the material for Mn.

Lead measurements were carried out on both diluted and undiluted sediment. No matrix effects were observed, but the standard deviation for the diluted sediment was clearly lower. The accuracy of the result on the diluted sample was estimated at ±10% (1s), so that a final result of 24.6 ± 2.5 mg kg^{-1} was obtained. On the undiluted sample a comparative value of 23 ± 2.2 mg kg^{-1} was produced (n = 11). A homogeneity factor H_E = 2.88% × $\sqrt{mg}$ was derived from these results, confirming the good homogeneity of the material for Pb.

Tin measurements were also performed on both diluted and undiluted sediment. No difference could be observed. The results for Sn in the diluted sample is given in Table 11.5. The accuracy of these results was estimated at ±10% (1s), so that the final result of 2.6 ± 0.3 mg kg^{-1} could be achieved. On the undiluted sample a comparative value of 2.5 ± 0.3 mg kg^{-1} (n = 21) was produced. A homogeneity factor H_E = 1.6% × $\sqrt{mg}$ was derived from these results, confirming the good homogeneity of the material for Sn.

Thallium measurements were performed on the original sediment on the analytical spectral line at 276.8 nm. The accuracy of the result was estimated at ±10% (1s), thus leading to a final result of 0.29 ± 0.03 mg kg^{-1}. A homogeneity factor H_E = 11.7% × $\sqrt{mg}$ was derived from these results. This factor is relatively high due to the relative large sample intake.

Zinc analyses were carried out on both diluted and undiluted sediment. The accuracy of the final result on the diluted sample was estimated at ±10% (1s), thus leading to a final result of 45.9 ± 4.6 mg kg^{-1}. On the undiluted sample a compara-

Table 11.5. Homogeneity control of trace elements in the candidate CRM Antarctic Coastal Sediment MURST-ISS A1 by SSZ-ETA-AAS

Element	Murst-ISS A1 Concentration (mg kg^{-1})	Sample mass (mg)	Homogeneity factor H_E (% × $\sqrt{mg}$)
As	Approximately 10 (n = 10, dil.)	0.015–0.040	1.25
Cd	0.53 ± 0.05 (n = 22, dil.)	0.005–0.07	1.06
Cu	5 ± 1.2 (n = 21, dil.)	0.03–0.09	5.80
Hg	<0.01 (n = 5, ndil.)	30–75	–
Mn	455 ± 91 (n = 26, dil.)	0.003–0.013	1.54
Pb	24.6 ± 2.5 (n = 9, dil.)	0.1–0.6	2.88
Sn	2.6 ± 0.3 (n = 20, dil.)	0.025–0.065	1.60
Tl	0.29 ± 0.03 (n = 10, ndil.)	0.4–2.2	11.7
Zn	45.9 ± 4.6 (n = 12, dil.)	0.1 – 0.6	3.00

(dil = diluted, ndil = not diluted)

tive value of 44.8 ± 4 mg kg^{-1} (n = 13) was produced. A homogeneity factor $H_E = 3\% \times \sqrt{mg}$ was derived from these results, confirming the good homogeneity of the material for Zn.

All results are summarized in Table 11.5.

4. Conclusions

A candidate CRM for trace elements in Antarctic coastal marine sediment was prepared by coarse sieving, drying, jet-milling with ultrafine classification of particles and Turbula mixing. The material was found to be of appropriate particle size distribution, dryness and homogeneity to be certified as a stable and homogeneous CRM.

References

1. G. N. Kramer, J. Pauwels, The preparation of biological and environmental reference materials, *Mikrochim. Acta*, 123 (1996), 87–93.
2. Ph. Quevauviller, M. Astruc, L. Ebdon, G. N. Kramer, B. Griepink, The certification of the contents (mass fractions) of tributyltin and dibutyltin in coastal sediment (CRM 462), EUR 15337 EN, 1994, European Commission, Luxembourg, pp. 36.
3. J. W. M. Wegener, E. A. Maier, G. N. Kramer, W. P. Cofino, The certification of the contents (mass fractions) of polycyclic aromatic hydrocarbons: pyrene, benz(a)anthracene, benzo(a)pyrene, benzo(e)pyrene, benzo(b)fluoranthene, benzo(k)fluoranthene and indeno(1,2,3-cd) pyrene in fresh water harbour sediment CRM 535, EUR 17795 EN, 1997, European Commission, Luxembourg, pp. 51.
4. J. W. M. Wegener, E. A. Maier, G. N. Kramer, W. P. Cofino, The certification of the contents (mass fractions) of chlorobiphenyls IUPAC No 28, 52, 101, 105, 118, 128, 138, 149, 153, 156, 163, 170 and 180 in fresh water harbour sediment, EUR 17799 EN, 1997, European Commission, Luxembourg, pp. 58.
5. E. Scholz, Karl Fischer Titration, Berlin: Springer (1984).
6. K. Schmitt, H.-D. Isengard, Method for avoiding the interference of formamide with the Karl Fischer titration, *Fresenius J. Anal. Chem.*, 357 (1997), 806–811.
7. U. Kurfürst, J. Pauwels, K. H. Grobecker, M. Stöppler, H. Muntau, Microheterogeneity of trace elements in reference materials – determination and statistical evaluation, *Fresenius J. Anal. Chem.*, 345 (1993), 112–120.
8. J. Pauwels, C. Hofmann, C. Vandecasteele, Calibration of solid sampling Zeeman atomic absorption spectrophotometry by extrapolation to zero matrix, *Fresenius J. Anal. Chem.*, 348 (1994), 411–417.
9. K. H. Grobecker, U. Kurfürst, Solid sampling by Zeeman graphite-furnace-AAS, a suitable tool for environmental analyses, in H. Lieth, B. Markert (Eds.), *Element Concentration Cadasters in Ecosystems*, B, VCH Verlag, Weinheim (1990), 121–137.

Environmental Contamination in Antarctica
A Challenge to Analytical Chemistry
S. Caroli, P. Cescon and D.W.H. Walton, editors

Chapter 12

Antarctic Environmental Specimen Bank

Francesco Soggia, Carmela Ianni, Emanuele Magi, Roberto Frache

1. Introduction

The role of Environmental Specimen Banking (ESB) is to provide a long-term, carefully controlled and documented resource to allow changes in toxic and hazardous chemicals in the environment to be accurately assessed. Several countries have already developed projects relating to ESB to provide for the systematic collection and long-term storage of representative environmental specimens for future analyses (1–13). The Antarctic continent, because of its distance from highly populated areas and its very limited biological activity, is an excellent place for research on global change. It can be considered as a key part of the "environmental memory" of the earth and ideal in many ways as a source of baseline data.

The Banca Campioni Ambientali Antartici (the Italian Antarctic Environmental Specimen Bank, BCCA, see logo in Figure 12.1) is an integral part of the Italian Project on the Biogeochemical Cycles of Antarctic Contaminants (Italian National Antarctic Research Programme, PNRA of the Programme Nazionale di Ricerche in Antartide) (14). Research units are co-ordinated by the Department of Environmental Sciences of the University of Venice along with the project concerning the preparation of multielemental Antarctic Certified Reference Materials (CRMs), i.e., sediment, krill, sea water and *Adamussium colbecki*, co-ordinated by the Istituto Superiore di Sanità (National Institute of Health, ISS). They will allow accuracy of analytical data obtained from Antarctic environmental matrices to be constantly checked.

Before establishing an ESB for Antarctica it was essential to undertake pilot studies in three areas, i.e.:

i) choice of the matrices to be sampled as representative of the Antarctic ecosystem (e.g., environmental quality bioindicators);
ii) definition of sampling methods, treatment and preservation of specimens according to their expected future use (chemical characterization of inorganic and organic substance of relevant biochemical and toxicological importance in the different ecological systems);
iii) choice of the sampling areas in the Victoria Land and Ross Sea (Figure 12.2).

Figure 12.1. Logo of the Antarctic Environmental Specimen Bank (Banca Campioni Ambientali Antartici, BCCA).

2. Aims and objectives of the Antarctic Environmental Specimen Bank

Environmental research often needs to evaluate temporal changes, particularly those for chemical parameters. In many cases, monitoring of an ecosystem would be both easier and more reliable if a comparison between past events and present conditions could be easily undertaken. There are difficulties in relying on previous data sets for comparisons due to differing analytical techniques, accuracy and precision. However, the problem can be solved with long-term storage of representative samples characteristic of particular ecosystems to allow the same analyses on temporally separated specimens to be carried out. The establishment of the BCAA for the chemical characterization of changes in ecosystems is an essential part of the evaluation of Antarctic environmental contamination.
The objectives of the BCAA in the medium term can be summarized as follows:

i) retrospective control of the analytical data by ensuring that historical comparisons of data trends can be performed at different times by using previously stored specimens;
ii) study of the pathways, transport and enrichment of chemical elements and compounds in the Antarctic ecosystem. An assessment of the global transport of contaminants is best carried out in Antarctica because of its distance from the other continents and the lack of urban or industrial developments;
iii) opportunity for new research using parameters which are not being studied at present;

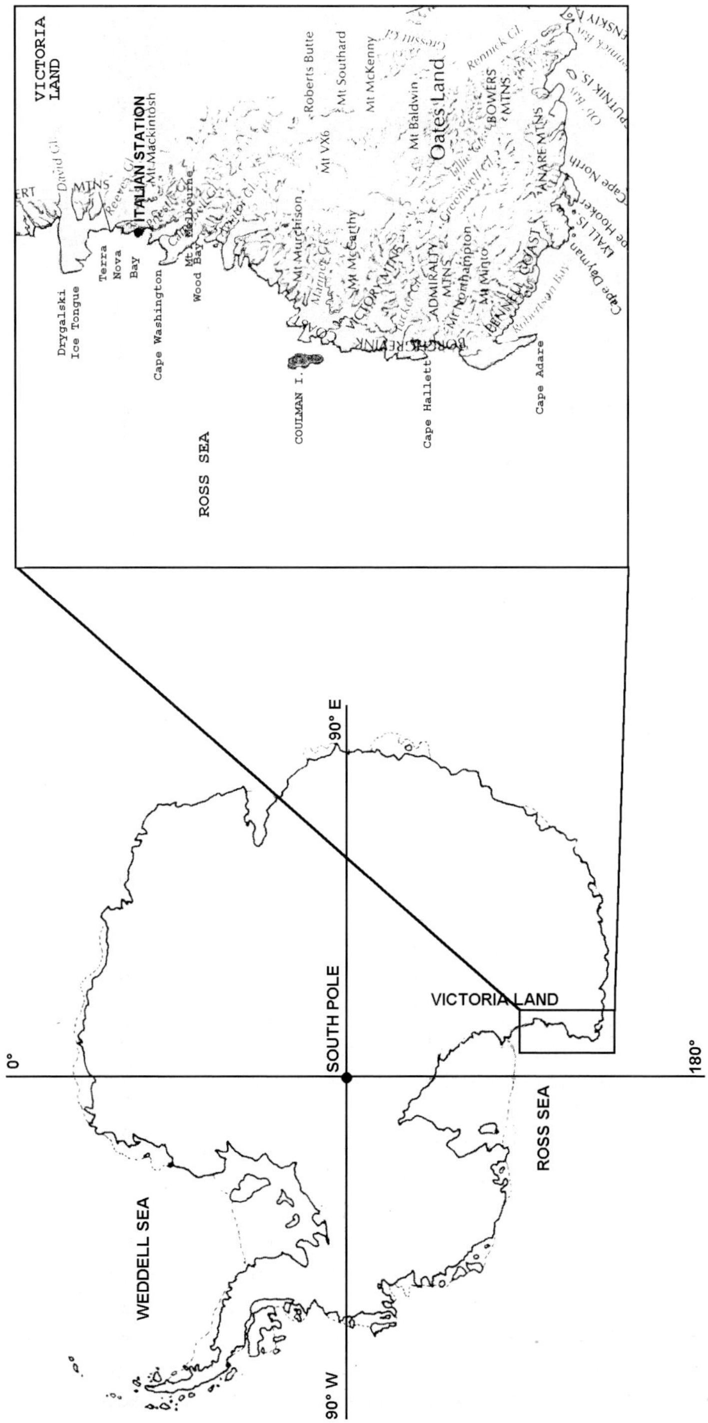

Figure 12.2. Sampling area. Victoria Land and Ross Sea (Antarctica).

iv) maximum accessibility of samples to research, making their original collection most cost-effective by increased use;
v) chemical characterization of samples in collaboration with other research groups relevant to the PNRA;
vi) establishment of a data bank containing information on samples storage, collection, treatment and chemical characterization.

In the international context there are already specimen banks serving biological, ecological, medical and other kinds of projects (15–20). The objectives of the BCAA put an emphasis on environmental chemistry and the establishment of baselines similar to the approaches followed by other ESBs, i.e., long-term storage of representative environmental specimens for future analyses, retrospective control and new research using parameters which are not being studied at present. Chemical characterization of specimens is also a major objective.

In the future, co-operation among national ESBs operating in Antarctica will be very important. In fact, the Antarctic banking projects would be best implemented by a multidisciplinary approach (chemical, biological, oceanographical, geological, etc.) for most effective exchange of information and samples (21).

3. Methodologies for preservation of specimens

3.1. Sampling

The timing, frequency and location of sampling, as well as the type, number and size of specimens to be taken, is usually determined by a combination of factors, i.e., strategy of ESBs, characterization type, sample chemical concentration, distribution, abundance and availability of the population and/or materials to be sampled, seasonal variability, storage room, ease of collection and transport, costs, etc. (13).

From a chemical point of view the terrestrial environment is basically suitable for collecting abiotic matrices (sediment, soil, snow, ice, etc.) and only a small variety of biotic matrices (lacustrine algae, mosses and lichens) because of their limited levels of biological activity. The marine environment has much greater diversity and so offers more organism types. Mammals and birds will be considered in a different perspective, taking into account the experience of other ESBs (22–24). The statistical design has to be determined with respect to the regulations of the Antarctic Protocol (which restricts the sampling of some species), but, at the same time, trying to ensure that the most useful species can be used as bioindicators of the environment quality (25–30).

The differences in sampling techniques between liquid and solid and between abiotic and biotic samples need to be clearly determined by the concentration type and level (major, minor, trace and ultratrace for total and speciation analysis) of analytes to be assessed (31). There is a severe contamination risk during collection from sampling tools, means of transport (helicopters, boats, cars, etc.), and handling operations carried out without using any protection (gloves, masks, boots and decontaminated overalls). Consequently, it is important to develop interdisciplinary Standard Operative Procedures (SOPs) relevant to the sample treatment, pretreatment and preservation (32).

The field sampling methods used for the specimens stored in the BCAA are listed below:

Marine and lake sediment: the marine sediments were collected by a stainless steel grab from a boat or a ship. The lacustrine sediments were collected on the lake shore with a polypropylene scoop. The samples collected were manually pre-homogenised and transferred into 1 l polycarbonated cylindrical jars. The jars were wrapped in double polyethylene bags and immediately stored at –30°C.
Soil: samples of about 15–20 kg were collected with a polypropylene scoop, manually pre-homogenized into a polyethylene container and transferred into 1 l polycarbonate cylindrical jars. The jars were wrapped in double polyethylene bags and immediately stored at –30°C.
Plants: lacustrine algae and mosses were collected by hand, wearing a particle mask and polyethylene gloves. The specimens in polycarbonate cylindrical jars, wrapped in double polyethylene bags, were immediately stored at –80 or –150°C.
Snow: the snow samples were collected from both surface and deeper layers by an operator wearing suitable clothes (mask, polyethylene gloves and overalls) in order to minimize the possible contamination of the specimen and facing against the direction of wind. The snow from the surface layer below 2 cm was transferred (taking it directly with the bottle mouth) into the 1 l wide mouth LDPE bottles. The snow collected from deeper layers (about 1–2 m) was sampled by using the trench technique and inserting the 1 l wide mouth LDPE bottles directly into the horizontal layer. The bottles were wrapped in double polyethylene bags and immediately stored at –30°C.
Pack-ice: the specimens of pack-ice, which was about 2.5 m thick, were collected by coring with a manual steel corer (Model BTC, Ducan, UK). The cores were wrapped in double polyethylene bags and stored in a polypropylene tube at –30°C.
Organisms: the epifaunal scallop *Adamussium colbecki* is one of the most common Antarctic bivalves (33). Since this benthic invertebrate is a filter feeder, it accumulates elements and compounds from the sea providing a potentially useful environmental quality biondicator. The specimens of *Adamussium colbecki* were collected manually by scuba diving at a depth of about 20 m and stored in polyethylene bags at –80 or –150°C.
Marine suspended particulate matter: the subsurface samples (about 0.5 m depth) of inshore sea water were collected before and after pack-ice melting from the prow of a small rubber boat moving upwind, using a teflon-made pump and a 50 l polyethylene tank. The samples were collected under the pack-ice by drilling the surface with an ice corer and by using a Teflon® aspiration tube inserted into the hole, through which sea water was pumped to the surface with a Teflon® diaphragm pump activated by nitrogen in a 50 l polyethylene tank. Pumps, tubes and tanks were acid-cleaned and sea water conditioned. The subsurface samples (about 30 m depth) of offshore sea water were collected by 20–30 l Go-Flo teflonated bottles (General Oceanics, USA) lowered from the ship's side by a winch

provided with a kevlar cable. Pressure filtration of samples was carried out by a Teflon® apparatus (Sartorius SM16540, Germany) using acid-cleaned 0.45 μm pore size cellulose nitrate membrane filters (Sartorius SM11306, 142 mm diameter). The membrane with suspended particulate matter was placed in a polystyrene Petri capsule (150 mm diameter) immediately after filtration and frozen at –80°C. Sample handling was carried out under Class 100 (US Federal Standard 209D) laminar flow hoods (Gruppo Flow, Milan, Italy) installed in clean laboratories at the Terra Nova Bay Station and on an oceanographic ship.

Atmospheric particulate matter: a Sierra Andersen high-volume sampler (Model Sierra 235 Cascade Impactor Grasbery-Andersen PM10 by Tecora, Italy) was used for the characterization of organic compounds. The sampling period was about two weeks and the air volume sampled was about 20,000 m^3 for each glass fibre filter (3.0, 1.5, 0.95, 0.49, 0.3 μm). Each filter was then subdivided into four parts and a quarter of a filter was stored at –80°C in the BCAA. The glass fibre was cleaned with an organic solvent (hexane). The filters with atmospheric particulate matter were placed in an aluminium sheet and frozen at –80°C. Sample handling was carried out under Class 100 laminar flow hoods installed in clean laboratories at Terra Nova Bay.

3.2. pretreatment and sample storage

The samples are placed, usually immediately after being collected, in different types of containers made of plastic, aluminium or other metals, glass, etc., depending on their future use and the storage temperatures. In this preliminary phase the BCAA decided to use polycarbonate containers for soils, sediments, moss and algae; polystyrene containers for filters collecting marine suspended particulate matter; polystyrene Petri capsule for filters with atmospheric particulate matter; low-density polyethylene (LDPE) containers for snows; and polyethylene bags for pack-ice core.

Establishing the protocol to standardize the cleaning of containers according to the sample, type of material and chemical characterization to be effected was seen as critical. The following procedure illustrates the protocol for the cleaning of LPDE bottles for the analysis of trace metals:

i) wash with detergent and rinse with plenty of deionized water;
ii) immerse the bottles completely in a 10% HNO_3 (Carlo Erba RPE, Italy) solution prepared with ultrapure water obtained by a Milly-Q system (Millipore, USA), heated at 40°C, for one week. Repeat this operation three times replacing the washing solution each time;
iii) rinse with ultrapure water and dry the bottles under a laminar flow hood (Class 100);
iv) wash with chloroform in order to remove the possible traces of organic substances that favour the adsorption of trace metals;
v) carry on the washing for at least another week by shaking always with 1% HCl (Merck Suprapur, Germany) prepared with ultrapure water. Replace daily the washing solution with a fresh one;

vi) after cleaning a few elements are determined in order to evaluate the container blank by using Electrothermal Atomization Atomic Absorption Spectrometry (ETA-AAS) and Inductively Coupled Plasma Mass Spectrometry (ICP-MS).

As previously outlined in the Sampling section, sediments and soils were pre-homogenized on-site by subdividing the samples into different sub-samples (which will be similar, but not identical) in order to minimize the possible risk of contamination. Some of these specimens will be accurately homogenized at the BCAA in Italy and delivered to laboratories on request for valid comparison of data and analytical techniques and for the evaluation of the stability of specimens during storage (34, 35).

For all environmental specimens, particularly those of the fauna, it is important that, as soon as possible after being collected, they are frozen and stored at a temperature of –80°C (or less) to inhibit enzymatic and microbial degradation (36). Freezing and storage in liquid nitrogen, if available, would be even better. For the pilot sudies the specimens collected by the BCAA were frozen at either –30, –80 or –150°C (depending on the analytical method used for analysing the specimens) and transported to the BCAA facilities in Italy carefully maintaining the storage temperatures during transport and transfers.

4. Results

4.1. Development of an **ad hoc** *software for data archiving*

All data concerning collection, treatment, storage, archiving and chemical characterization of specimens are in a database running in a Windows environment. The graphics interface allows two modules (*Consultation* and *Updating*) to be used for the information management in order to make data input and search easier for inexperienced users. The application, developed expressly for the data management on the BCAA specimens, consists of an executable file BCAA.EXE (at present available only in Italian, version 1.0, September 1996) and a series of function libraries which allow for the interface connection with the database. This software can be used with development tools for multimedia applications. The information and data on the stored specimens are managed in DBF format that can be controlled by other applications as well.

The graphic images are digitized with 256 colors on a base of 640 × 480 pixels. The minimum hardware configuration required is: PC 486 DX 33 MHz, RAM 4 Mb, Hard Disk with 10 Mb available, SVGA 640 × 480 256 colors, Windows® 3.1.

The BCAA is now managed by using two databases: one focuses on the information about sample collection, treatment and storage, whereas the other deals with data on the chemical characterization. The database field features are listed in Table 12.1. The information and data concerning the chemical characterization of samples are subdivided into six groupings identifying different characterizations, i.e.: Elements, Organic Compounds, Inorganic Compounds, Organometallics, Radionuclides, Others.

The replies to queries for each grouping are displayed and/or printed with the following data:

Table 12.1. Description of the statistical data registered at the BCAA

Statistical data	Description
YEAR	Year when the expedition in Antarctica took place
MATRIX	Initials of the environmental matrix sampled [a]
EXPEDITION	Expedition progressive number
AREA	Code based on the maps by the International Map of the World of the US Geological Survey, Washington (1:250.000) [b]
SITE	Collection site code
SPECIMEN	Specimen univocal code
LATITUDE (S) and LONGITUDE (E o W)	Collection site co-ordinates: – degrees in sixtieths – primes in degree hundredths – seconds in prime thousandths
SAMPLING	Sampling procedures
SAMPLING DATE	Specimen sampling date
ARCHIVE	Specimen archiving code inside the cold-storage facilities
STORAGE	Storage date of the sample in the BCAA
DELIVERY	Delivery date of the sample to a research group for the chemical characterization
TYPE	Specimen typology
CONTAINER	Type of container used for the specimen
TEMPERATURE	Specimen storage temperature (in °C)
QUANTITY	Specimen weight or volume
DELIVERY INFORMATION	Information on the specimen delivery
SPECIMEN NOTES	Notes on individual specimens
Classification of the organisms	
PHYLUM CLASS ORDER FAMILY GENUS SPECIES	Information relevant to the biological classification
Biometry of the organisms	
SIZE WEIGHT AGE SEX OTHER BIOMETRY NOTES	Data relevant to biometry

Table 12.1. Continued

Statistical data	Description
Chemical characterisation	
ELEMENTS	Replies to queries for each grouping are displayed and/or printed with the following data:
ORGANIC COMPOUNDS	research group: responsible for the facilities that carried out the chemical characterisation;
INORGANIC COMPOUNDS	analytical procedure: description of the equipment and analytical procedures used; results: data on the variables that were studied;
ORGANOMETALLICS	publications: possible literature concerning data relevant to the chemical characterisation of the sample received by the BCAA.
RADIONUCLIDES	
OTHER	

[a] See Table 12.2.
[b] See Table 12.3.

i) research group: responsible for the facilities that carried out the chemical characterization;
ii) analytical procedure: description of the equipment and analytical procedures that were used;
iii) results: data on the variables that were studied;
iv) publications: possible literature concerning data relevant to the chemical characterization of the sample received by the BCAA.

4.2. Description of the graphics interface of the programme BCAA.EXE (version 1.0, September 1996)

The software enables the display of information for five groups of environmental matrices presenting similar problems with regard to sampling and storage, as shown in Figure 12.3. The matrices belonging to the five groupings (see Table 12.2) can be recalled individually, after selecting the requested grouping on the next page, as shown in Figure 12.4 for the *Organisms* grouping. After selecting the individual matrix, the data reference on the specimens depends on the collection year and area (see Table 12.3), as shown in Figure 12.5.

After the selection of the year and area data can be recalled in the appropriate window (see Figure 12.6) for each single collection site. This is also graphically displayed on the map. This page displays the collection co-ordinates of each site. It is also possible to select the two active keys *Sampling* and *Archive card* that give information, on methodologies relevant to the sampling and data on individual specimens, respectively.

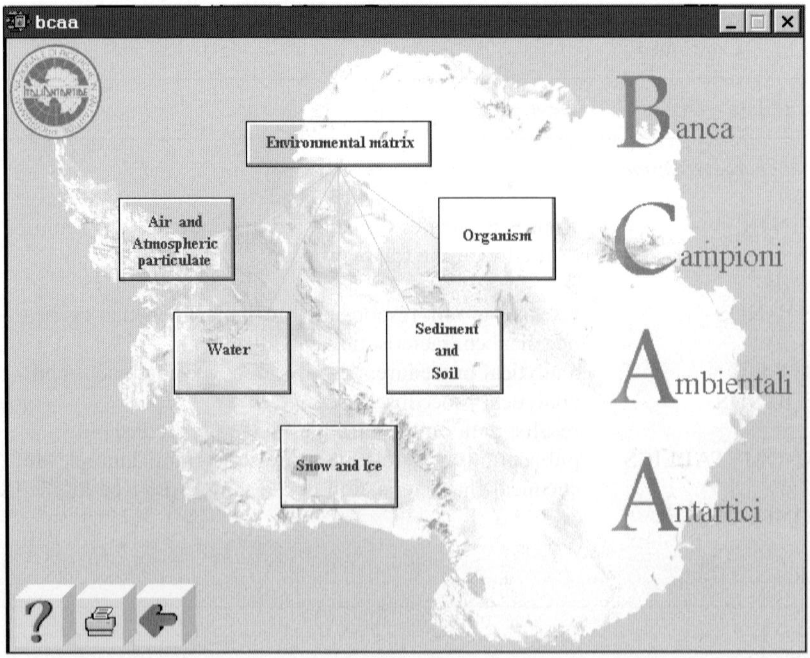

Figure 12.3. Groups of environmental matrices.

For each specimen collected in the selected site the sample archiving database is displayed in a window (see Figure 12.7) showing the following data:

- Sampling Date: day/month/year.
- Quantity: sample weight or volume.
- Type: sample typology (surface, core, etc.).
- Container: type depending on the future use of the sample.
- Storage Temperature: temperature for the sample storage (in °C).
- Archive Code: indicating the immediate location of the sample inside freezers.
- Storage Date: sample storage date.
- Delivery Date: sample delivery date to the BCAA.
- Chemical Characterization: indicating presence or absence of the chemical characterization data.
- Taxonomy: indicating presence or absence of data concerning the classification of the organisms.
- Biometry: indicating presence or absence of data on the biometric parameters of the organisms.

The fields *Specimen*, *Delivery Date, Chemical Characterisation, Taxonomy* and *Biometry* can be selected in order to display other relevant information for a specimen, the name and address of the person holding the specimen, the chemical characterization or, with regard to organisms, data concerning taxonomy and biometry. By selecting the field relevant to the *Chemical Characterisation* (see Figure 12.7) it is possible to recall data for characterization typology and, after

Table 12.2. Environmental matrices of the Antarctic ecosystem codified in the software of the BCAA

Code	Matrix
	Air and Atmospheric Particulate
AI	Air
AP	Atmospheric particulate matter
	Sediment and Soil
LS	Lake sediment
MS	Marine sediment
SL	Soil
	Water
SW	sea water
LW	Lake water
LP	Lake suspended particulate matter
MP	Marine suspended particulate matter
IW	Ice melting water
	Snow and Ice
IC	Ice
SN	Snow
PI	Pack ice
PP	Pack ice suspended particulate matter
	Organisms
PL	Plankton
BE	Benthos
NE	Necton
AL	Algae
MO	Mosses
LI	Lichens

choosing the requested one, data concerning the *Research Group, Analytical Procedure, Results* and *Publications* are displayed (see Table 12.1). The required data can be printed in each single phase of the programme. As regards the *Updating* module the procedure is the same as for the interface on data *Consultation*; the only differences are relevant to the access limitations (using a password) and to the on-line control of input data. This software will be updated in English in the year 2001 and the database will be available on Internet (www.BCAA.Unige.it)

4.3. Balance of stored Antarctic environmental specimens

The collection and storing of abiotic and biotic samples from different environments (atmospheric, marine and terrestrial ecosystems) was begun systematically during the austral summer in 1994–1995.

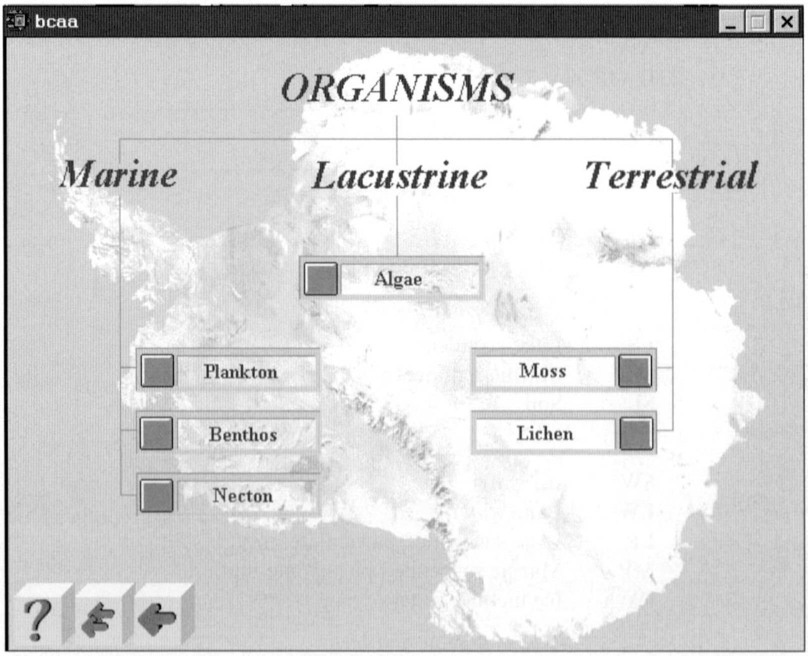

Figure 12.4. Individual matrices related to organisms.

Table 12.3. Maps and geographic coverage of the BCAA sampling areas

Map	Minimum and maximum latitude	Minimum and maximum longitude	Code
Ross Sea	67° S–85° S	160° E –150° W	N.Z.M.S. 135 [a]
Sequence Hills	73° S–74° S	157°30′ E–162° E	SS 55-57/8 [b]
Mount Murchison	73° S–74° S	162° E–166°30′ E	SS 58-60/5 [b]
Coulman Island	73° S–74° S	166°30′ E–171° E	SS 58-60/6 [b]
Reeves Névé	74° S–75° S	157°30′ E–162° E	SS 55-57/12 [b]
Mount Melbourne	74° S–75° S	162° E–166°30′ E	SS 58-60/9 [b]
Mount Joyce	75° S–76° S	157°30′ E–162° E	SS 55-57/16 [b]
Relief Inlet	75° S–76° S	162° E–166°30′ E	SS 58-60/13 [b]

[a] Map from Ross Sea Regions, Department of Lands and Survey, Wellington, New Zeeland. Scale 1:3,000,000, 2nd Edition, January 1970.
[b] Code International Map of the World. Maps from the US Geological Survey, Washington. Scale 1:250,000, 1968.

Table 12.4 shows the typology and number of specimens. In general, the biotic matrices (mosses, marine organisms) are stored in freezers at –80 or –150°C and the abiotic matrices (sediments, soils, ice, snow) are stored in freezers at –30°C and –80°C, depending on the analytical method used for analysing the samples. The

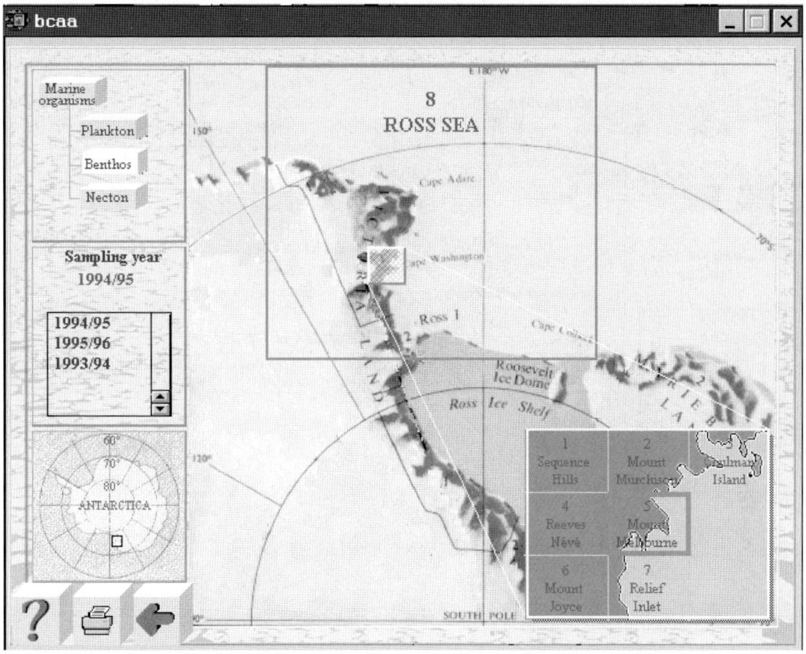

Figure 12.5. Choice of years and sampling zones.

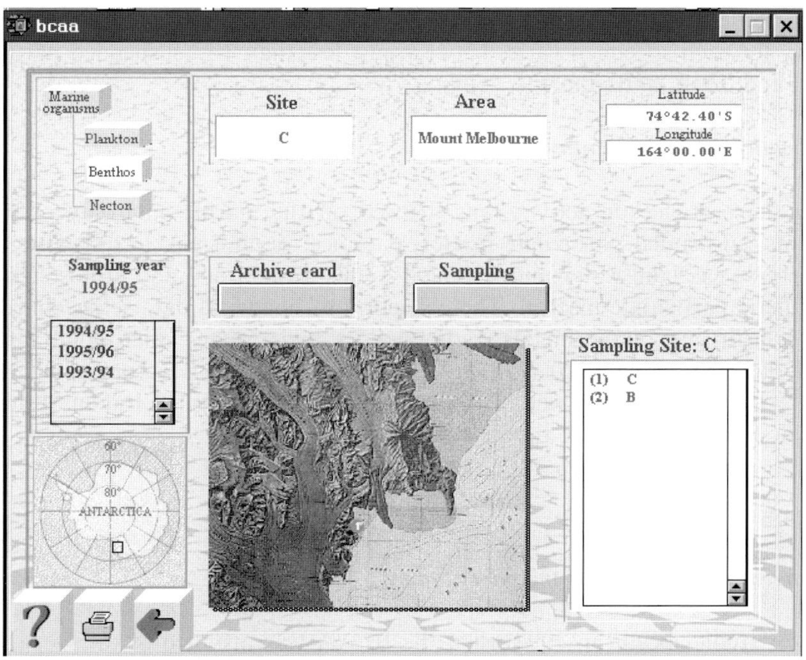

Figure 12.6. Display of collection site.

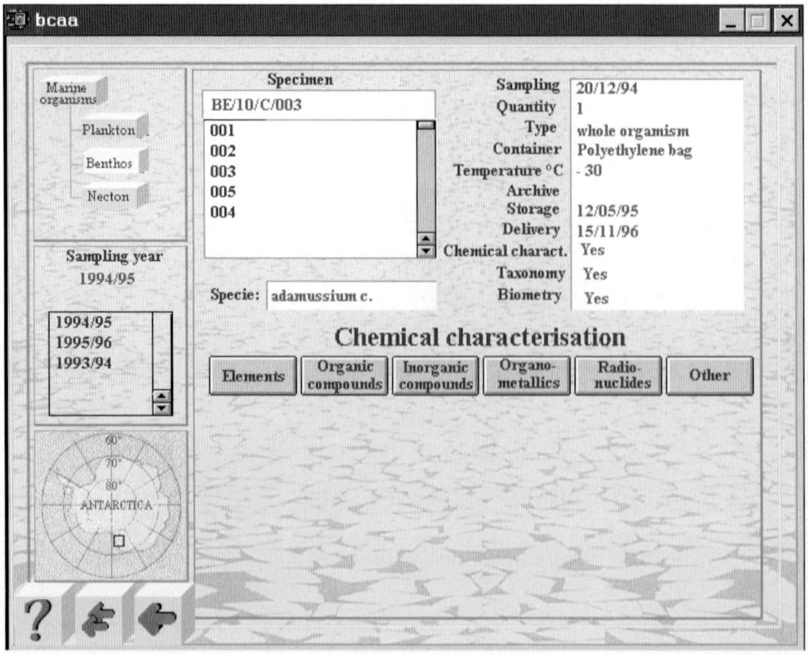

Figure 12.7. List of specimens collected in the selected site and display of data for each single specimen.

quantity for each specimen varies between just less than 1 g (suspended particulate matter) and 0.5–2 kg (sediments, soils).

4.4. Chemical characterization of samples

During the past decade activity was focused on the marine environment (water, sediments, suspended particulate matter, pack ice, organisms), terrestrial environment (snow, ice, lake water and sediments, soils, organisms) and atmosphere (air, atmospheric particulate matter). The collection of historical data series concerning minor constituents, trace metals, radionuclides and organic micropollutants was carried out (see Table 12.5) (37–55). The data collection is co-ordinated by the *Data Centre CHI,* which is one of the Data Centres forming the Italian System for Antarctic Data Exchange of the PNRA.

5. Discussion and conclusions

5.1. Current development of the pilot studies

The pilot studies began in 1994 when the BCAA was installed at the Department of Chemistry and Industrial Chemistry (University of Genoa) (56). These first years allowed the sampling procedures to be defined, the cleaning of containers to

Table 12.4. Type and number of specimens collected during the expeditions in Antarctica and stored at the BCAA

	Expeditions in Antarctica							
Specimen	1987–1988	1988–1989	1990–1991	1993–1994	1994–1995	1995–1996	1996–1997	1997–1998
Marine sediment	4	1	34	28	10	8		379
Lake sediment		5	6	13		8	32	60
Marine suspended particulate matter					21	9		58
Sea water								65
Lake water								2
Snow					8	15	59	33
Soil					1		227	60
Algae					1		7	15
Mosses							5	10
Adamussium colbecki						100		260
Fish								162
Zooplankton								57
Phytoplankton								18
Pack ice							16	24
Atmospheric particulate matter							10	2

be standardized (by carrying out blank analysis), the appropriate materials and the storage temperature to be chosen as a function of the analytes (inorganic or organic total contents or individual chemical species) and a registration protocol for all data and information to be organized as a database.

At present the BCAA facilities include two cold-storage rooms at –30°C with a volume of about 40 m^3, three freezers (capacity of about 1000 l) at –80°C and one freezer at –150°C. All the handling operations of specimens and the cleaning procedures of containers are carried out under laminar flow hoods (Class 100) using the facilities and laboratories of the Department of Chemistry and Industrial Chemistry, Analytical and Environmental Chemistry Group, where the BCAA is located. The information management is carried out through a PC software at the BCAA and in a notebook computer in order to update data on-site. The total number of specimens of the various components of the ecosystem stored at present at the BCAA exceeds 2000. Future expeditions are expected to collect more specimens increasing the total number of samples at the BCAA to a few thousands.

5.2. Future developments

The medium-term objective (the next 5–10 years) envisages the collection of a considerable number of specimens every 1–2 years in order to be able to eventually

Table 12.5. Chemical characterisation of Antarctic matrices carried out by the research units of the project 2c.4 "Biogeochemical Cycles of Contaminants"

Matrix	Chemical characterisation
Sea water	As, Ca, Cd, Co, Cr, Cu, Fe, K, Li, Mg, Mn, Na, Ni, Pb, Sb, Se, Sr, Zn ^{212}Bi, ^{214}Bi, ^{137}Cs, ^{40}K, ^{214}Pb, ^{238}Pu, ^{239}Pu,^{226}Ra, ^{90}Sr, ^{232}Th, ^{208}Tl, ^{238}U, aliphatic hydrocarbons (alkanes, alkenes, terpenoids, cycloalkanes), aromatic hydrocarbons (benzenics, polycyclics), steroids, acids/esters, amides, alcohols, aldehydes, ketones, phenols, phthalates, chlorinated pesticides, humic components
Marine suspended particulate matter	Al, Cd, Co, Cr, Cu, Fe, Mn, Ni aliphatic hydrocarbons (alkanes, alkenes, terpenoids), aromatic hydrocarbons (benzenics, polycyclics), steroids, acids/esters, amides, alcohols, aldehydes, ketones, phenols, phthalates, chlorinated pesticides
Pack ice	aliphatic hydrocarbons (alkanes, alkenes, terpenoids), aromatic hydrocarbons (benzenics, polycyclics), steroids, acids/esters, amides, alcohols, aldehydes, ketones, phenols, phthalates, chlorinated pesticides
Pack ice suspended particulate matter	aliphatic hydrocarbons (alkanes, alkenes, terpenoids), aromatic hydrocarbons (benzenics, polycyclics), steroids, acids/esters, amides, alcohols, aldehydes, ketones, phenols, phthalates, chlorinated pesticides
Marine sediment	Al, As, Ba, Be, Bi, Ca, Cd, Ce, Co, Cr, Cs, Cu, Eu, Fe, Gd, Hf, Ho, La, Lu, Mn, Nd, Ni, Pb, Rb, Sb, Sc, Se, Sn, Sr, Ta, Tb, Th, Tm, U, Yb, Zn, Zr ^{212}Bi, ^{214}Bi, ^{137}Cs, ^{40}K, ^{210}Pb, ^{214}Pb, ^{238}Pu, ^{239}Pu, ^{90}Sr, ^{232}Th, ^{208}Tl, ^{238}U, aliphatic hydrocarbons (alkanes, alkenes, terpenoids), aromatic hydrocarbons (benzenics, polycyclics), steroids, acids/esters, amides, alcohols, aldehydes, ketones, phenols, phthalates, polychlorobiphenyls humic components
Ice melting water	B, Ca, Cd, Co, Cu, Fe, Hg, K, Li, Pb, Mg, Mn, Mo, Na, Ni, SiO_2, Tl, V, Zn Cl^-, F^- , HCO_3^-, NO_3^-, SO_4^{2-}, chlorinated pesticides, chlorinated hydrocarbons
Lake water	Al, As, B, Ba, Ca, Cd, Co, Cr, Cu, Fe, Hg, K, Li, Mg, Mn, Mo, Na, Ni, Pb, Si, Sr, Tl, V, Zn, Cl^-, F^-, HCO_3^-, NO_3^-, SO_4^{2-}, ^{137}Cs, ^{90}Sr, ^{238}Pu, ^{239}Pu, chlorinated pesticides, chlorinated hydrocarbons

Table 12.5. Continued

Matrix	Chemical characterisation
Lake sediment	Al, As, Ba, Ca, Cd, Ce, Co, Cr, Cs, Cu, Eu, Fe, Gd, Hf, Ho, K, La, Lu, Mg, Mn, Na, Nd, Ni, Pb, Rb, Sb, Sc, Se, Sm, Sr, Ta, Tb, Th, Tm, U, Yb, Zn, Zr
	Chlorinated pesticides, humic components
	^{212}Bi, ^{214}Bi, ^{137}Cs, ^{40}K, ^{210}Pb, ^{212}Pb, ^{214}Pb, ^{238}Pu, ^{239}Pu, ^{226}Ra, ^{90}Sr, ^{232}Th, ^{208}Tl, ^{238}U,
Soil	Ca^{2+}, Cl^-, K^+, NO_3^-, Mg^{2+}, Na^+,SO_4^{2-}
	Al, Ba, Ca, Cd, Co, Cr, Cu, Fe, K, Mg, Mn, Na, Ni, Pb, Sr, Zn,
	^{212}Bi, ^{214}Bi, ^{137}Cs, ^{40}K, ^{214}Pb, ^{232}Th, ^{208}Tl, ^{238}U
	polychlorobiphenyls, humic components
Snow	Cl^-, NO_3^-, SO_4^{2-}, F^-, CH_3COO^-, $HCOO^-$, $CH_3SO_3^-$, HCO_3^-, Al^{3+}, Ca^{2+}, H^+, HCHO, H_2O_2, K^+, Mg^{2+}, Na^+, nss SO_4^{2-}
	B, Ca, Cd, Cr, Cu, Hg, K, Li, Mg, Mn, Na, Ni, Pb, Si, Sr,
	^{137}Cs, ^{210}Pb
	aliphatic hydrocarbons (alkanes, alkenes, terpenoids), aromatic hydrocarbons (benzenics, polycyclics), steroids, acids/esters, amides, alcohols, aldehydes, ketones, phenols, phthalates, polychlorobiphenyls, chlorinated hydrocarbons
	humic components
Ice	Al^{3+}, Ca^{2+}, CH_3COO^-, $CH_3SO_3^-$, Cl^-, F^-, H^+, $HCOO^-$, K^+, Mg^{2+}, Na^+, NO_3^-, SO_4^{2-}
	CH_4, CO, CO_2
Air	CCl_4, CFC, CH_4, CO, CO_2, H_2, aliphatic hydrocarbons, phthalates, alkyl phosphates, polyaromatic hydrocarbons
	NO_2, NH_3, SO_2
	^{222}Rn
Atmospheric particulate matter and aerosol	Al, Br, Ca, Ce, Co, Cr, Cu, Fe, Hg, K, La, Mg, Mn, Na, Ni, P, Pb, Rb, S, Sb, Sc, Se, Si, Sr, Th, Ti, V, Zn,
	Na^+, Cl^-, SO_4^{2+}, NO_3^-
Lichens	Chlorinated pesticides
	^{137}Cs, ^{40}K ^{210}Pb
Mosses	^{137}Cs, ^{40}K, ^{210}Pb, ^{214}Pb
Algae	^{137}Cs, ^{40}K, ^{90}Sr, ^{238}Pu, ^{239}Pu
	Al, Ba, Ca, Cd, Co, Cr, Cu, Fe, K, Mg, Mn, Na, Ni, Pb, Sr, Zn
Chionodraco hamatus	^{137}Cs, ^{40}K, ^{238}Pu, ^{239}Pu, ^{90}Sr
	Hg, Se
Adamussium colbecki	Hg, Se, Pb, Sn
	^{137}Cs, ^{40}K, ^{90}Sr, ^{238}Pu, ^{239}Pu
Euphausia superba	^{137}Cs, ^{40}K
	As, Cd, Cr, Cu, Fe, Mn, Ni, Pb, Na, Zn
Trematomus bernacchii	Hg, Pb, Se, Sn
	^{137}Cs, ^{40}K, ^{90}Sr, ^{238}Pu, ^{239}Pu

manage about 5000 samples at different temperatures as a function of the chemical characterization to be carried out. The storage capability will be consolidated down to –150°C, and the sample treatment depending on the sub-sampling will be developed, providing the BCAA with suitable freeze dryers and homogenizers (57).

A few specimens will be regularly analyzed in order to check their stability (58). Use will also be made of the CRMs in Antarctic matrices produced within the Project CRMs for Antarctic Research and Analytical Quality Assurance, thus continuing the collaboration already started with proficiency tests (59, 60). The chemical characterization of specimens will be increased in collaboration with related research groups and all new information will be stored in the BCAA database. Within the information management a protocol concerning the collection, treatment and storage of specimens, according to a statistical design and to the choice of some biondicator samples, will be completed. The software will be updated in English and the database will be made available on Internet. In collaboration with the Data Centre CHI the database will be also interfaced with all the chemical data produced in the last ten years in order to correlate previous data with the specimens present at the BCAA.

As mentioned above, *CHI* is one of the Data Centres forming the *Italian System for Antarctic Data Exchange*. It is housed at the University of Venice within a joint-project between the Ente per le Nuove Tecnologie, L'Energia e l'Ambiente (Agency for New Technologies, Energy and Environment, ENEA) and the Consiglio Nazionale delle Ricerche (National Research Council, CNR), under the appointment of the PNRA of the Italian Ministry of Higher Education, Scientific Research and Technology.

Acknowledgements

This work was financially supported by the Italian National Research Programme in Antarctica (PNRA).

References

1. H. Emons, J. D. Schladot, M. J. Schwugher, Environmental specimen banking in Germany: present state and further challenges, *Chemosphere*, 34 (1997), 1867–1873.
2. P. R. Becker, B. J. Koster, S. A. Wise, R. Zeisler, Biological specimen banking in Arctic research: an Alaska perspective, *Sci. Total Environ.*, 139/140 (1993), 69–95.
3. M. Morita, J. Yoshinaga, H. Mukai, Y. Ambe, A. Tanaka, Y. Shibata, Specimen banking at National Institute for Environmental Studies. Japan, *Chemosphere*, 34 (1997), 1907–1919.
4. R. M. Kiriluk, D. M. Whittle, M. J. Keir, A. A. Carswell, S. Y. Huestis, The great lakes fisheries specimen bank: a canadian perspective in environmental specimen banking, *Chemosphere*, 34 (1997), 1921–1932.
5. E. Kubin, H. Lippo, J. Karhu, J. Poikolainen, Environmental specimen banking of nationwide biomonitoring samples in Finland, *Chemosphere*, 34 (1997), 1939–1944.
6. J. Kucera, I. Obrusnick, J. K. Fuksa, J. Vesely, K. Stastny, J. Hajslova, P. Mader, D. Miholova, J. Sysalova, Environmental specimen banking in the Czech Republic: a pilot study, *Chemosphere*, 34 (1997), 1975–1987.

7. J. Golimowski, K. Dmowski, Standard species from Poland for environmental specimen bank, *Chemosphere*, 34 (1997), 1989–1995.
8. T. Odsjo, A. Bignert, M. Olsson, L. Asplud, U. Eriksson, L. Haggberg, K. Litzen, C. de Wit, C. Rappe, K. Aslund, The Swedish environmental specimen bank – application in trend monitoring of mercury and some organohalogenated compounds, *Chemosphere*, 34 (1997), 2059–2066.
9. M. E. Poulsen, G. Pritzl, The Danish Environmental Specimen Bank: status of establishment, *Sci. Total Environ.*, 139/140 (1993), 61–68.
10. R. Jayasekera, M. Rossbach, Implementation of specimen banking concepts in developing countries. First trial – Sri Lanka, *Sci. Total Environ.*, 139/140 (1993), 139–142.
11. V. A. Borzilov, Development of requirements for environmental specimen banking in ecological monitoring (exemplified by Chernobyl NPP accident area), *Sci. Total Environ.*, 139/140 (1993), 197–201.
12. B. Giege, T. Odsjo, Coordination of environmental specimen banking in the Nordic countries, its mission and strategy, *Sci. Total Environ.*, 139/140 (1993), 37–47.
13. R. A. Lewis, B. Klein, M. Paulus, C. Horras, Environmental specimen banking, in M. Stoeppler (Ed.), *Hazardous Metals in the Environment*, Elsevier, Amsterdam, 1992, 19–48.
14. P. Cescon, Cicli biogeochimici dei contaminanti, 69–71, in "Programma Esecutivo Annuale 1996", 1996, MURST-PNRA, Roma, 139.
15. F. H. Kemper, Human organ specimen banking – 15 years of experience, *Sci. Total Environ.*, 139/140 (1993), 13–25.
16. S. Panico, Gli studi prospettici basati su banche di campioni biologici: una nuova generazione epidemiologica, *Ann. Ist. Super. Sanità*, 28 (1992), 365–370.
17. K. S. Subramanian, Canadian Human Specimen Bank: an emerging Great Lakes health effects program, *Sci. Total Environ.*, 139/140 (1993), 109–121.
18. L. Gerhardsson, D. Brune, N. G. Lundstrom, G. Nordberg, P. O. Wester, Biological specimen bank for smelter workers, *Sci. Total Environ.*, 139/140 (1993), 157–173.
19. F. W. Jekat, R. Eckard, F. H. Kemper, Environmental specimen banking and poisons control—a new challenge, *Sci. Total Environ.*, 139/140 (1993), 507–514.
20. E. Jellum, A. Andersen, P. Lund-Larsen, L. Theodorsen, H. Orjasaeter, The Janus serum bank, *Sci. Total Environ.*, 139/140 (1993), 527–535 .
21. B. Giege, Coordination in ESBs in the Nordic Countries: background, results and future prospects, *Chemosphere*, 34 (1997), 1867–1873.
22. T. I. Lillestolen, N. Foster, S. A. Wise, Development of the National Marine Mammal Tissue Bank, *Sci. Total Environ.*, 139/140 (1993), 97–107.
23. E. Hahn, K. Hahn, M. Stoeppler, Bird feathers as bioindicators in areas of the German Environmental Specimen Bank – bioaccumulation of mercury in food chains and exogenous deposition of atmospheric pollution with lead and cadmium, *Sci. Total Environ.*, 139/140 (1993), 259–270.
24. N. Miyazaki, Contaminant monitoring studies using marine mammals and the need for establishment of an International Environmental Specimen Bank, *Sci. Total Environ.*, 154 (1994), 249–256.
25. *Protocol on Environmental Protection to the Antarctic Treaty, Final Report of the XIth Antarctic Special Consultative Meeting*, Madrid, 3–4 October 1991.
26. A. D. Schladot, F. Backhaus, M. Burow, M. Froning, C. Mohl, P. Ostapczuk, M. Rossback, Collection, preparation and characterisation of fresh, marine candidate reference materials of the German Environmental Specimen Bank, *Fresenius J. Anal. Chem.*, 345 (1993), 137–139.
27. M. Rossbach, M. Stoeppler, Multielement fingerprinting for characterisation: earthworm samples from the environmental specimen bank of the FRG, *Fresenius J. Anal. Chem.*, 332 (1988), 636–639.
28. I. I. Kryshev, I. N. Ryabov, T. G. Sazykina, Using a bank of predatory fish samples for bioindication of radioactive contamination of aquatic food chains in the area affected by the Chernobyl accident, *Sci. Total Environ.*, 139/140 (1993), 279–285.
29. K. Oxynos, J. Schmitzer, A. Kettrup, Herring gull eggs as bioindicators for chlorinated hydrocarbons (contribution to the German Federal Environmental Specimen Bank), *Sci. Total Environ.*, 139/140 (1993), 387–398.
30. P. Ostapczuk, M. Burow, K. May, C. Mohl, M. Froning, B. Süssenback, E. Waidmann, H. Emons, Mussels and algae as bioindicators for long-term tendencies of element pollution in marine ecosystems, *Chemosphere*, 34 (1997), 2049–2058.

31. M. Stoeppler, Sampling and sample storage, in M. Stoeppler (Ed.), *Hazardous Metals in the Environment*, 1992, Elsevier, Amsterdam, 9–18.
32. T. Odsjö, Manual and Standard Operating Procedures in Nordic Environmental Specimen Banking, *Chemosphere*, 34 (1997), 1955–1960.
33. W. L. Stockton, The biology and ecology of the epifaunal scallop *Adamussium colbecki* on the west side of McMurdo Sound, Antarctica, *Mar. Biol.*, 78 (1984), 171–178.
34. A. M. Caricchia, S. Chiavarini, C. Cremisini, R. Morabito, R. Scerbo, Influence of storage conditions on the determination of organotin in mussels, *Anal. Chim. Acta*, 286 (1994), 329–334.
35. Y. Ambe, H. Mukai, Long term stability of benzo[a]pyrene in stored atmospheric particulate matter samples, *Chemosphere*, 34 (1997), 2023–2028.
36. B. Giege, M. Korhonen, T. Odsjö, G. M. Paulsen, M. E. Poulsen, Coordination of Environmental Specimen Banking in the Nordic Countries – Report on forthcoming inter-nordic cooperation in environmental monitoring, *Nordiske Seminar og Arbejds Rapporter*, 1993, 609.
37. G. Capodaglio, G. Toscano, G. Scarponi, P. Cescon, Copper complexation in the surface sea water of Terra Nova Bay (Antarctica), *Int. J. Environ. Anal. Chem.*, 55 (1994), 129–148.
38. PNRA, *Environmental Impact in Antarctica*, Book of Extended Abstracts, National Meeting, Rome (Italy), June 8–9, 1990.
39. G. Saini, C. Baiocchi, D. Giacosa, M. Roggero, Determination of some heavy metals in snow and sea water from Terra Nova Bay (Antarctica), *Ann. Chim. (Rome)*, 81 (1991) 317–324.
40. PNRA, *Settore Impatto Ambientale*, collection of papers January 1986–July 1991, ENEA, Progetto Antartide, Rome (Italy), 1991.
41. PNRA, *Environmental Impact Chemical Methodologies*, Book of extended abstracts, 2nd Meeting, Venice (Italy), May 26–28, 1992.
42. PNRA, *Impatto Ambientale Metodologie Chimiche*, Book of extended abstracts, 3rd National Meeting, Venice (Italy), 10–11 March, 1994.
43. PNRA, *Contaminazione Ambientale*, Book of extended abstracts, 4th National Meeting, Venice (Italy), 6–7 December, 1995.
44. PNRA, *Contaminazione Ambientale*, Book of extended abstracts, 5th National Meeting, Venice (Italy), 11–12 February, 1997.
45. R. Capelli, V. Minganti, F. Fiorentino, R. De Pellegrini, Mercury and selenium in *Adamussium colbecki* and *Pagothenia bernacchii* from the Ross Sea (Antarctica) collected during Italian expedition 1988–89, *Ann. Chim. (Rome)*, 81 (1991), 357–369.
46. P. G. Desideri, L. Lepri, L. Checchini, Identification and determination of organic compounds in sea water in Terra Nova Bay (Antarctica), *Ann. Chim. (Rome)*, 79 (1989), 589–605.
47. P. G. Desideri, L. Lepri, L. Checchini, Identification and determination of organic compounds in antarctic sediments, *Ann. Chim. (Rome)*, 81 (1991), 595–603.
48. F. Baffi, F. Soggia, R. Frache, A. M. Cardinale, Heavy metal distribution in water and suspended Particulate matter in Ross Sea, *Ann. Chim. (Rome)*, 79 (1989), 607–616.
49. B. Cosma, F. Soggia, M. L. Abelmoschi, R. Frache, Determination of trace metals in antarctic sediments from Terra Nova Bay – Ross Rea, *Int. J. Environm. Anal. Chem.*, 55 (1994), 121–128.
50. E. Mentasti, V. Porta, O. Abolino, C. Sarzanini, Trace metal determination in antarctic sea water, *Ann. Chim. (Rome)*, **79** (1989), 629–637.
51. R. Udisti, E. Barbolani, G. Piccardi, Ion Chromatographic separation method of some species and their determination at ppb level in Antarctic snow and ice, *Ann. Chim. (Rome)*, 81 (1991), 325–341.
52. R. Fuoco, M. P. Colombini, C. Abete, Evaluation of pack melting effect on polychlorobiphenyl content in sea water samples from Terra Nova Bay – Ross Sea (Antarctica), *Ann. Chim. (Rome)*, 81 (1991), 383–394.
53. L. Campanella, T. Ferri, B. M. Petronio, A. Pupella, M. Soldani, Organic matter and metals in Carezza lake sediments, *Ann. Chim. (Rome)*, 81 (1991), 417–437.
54. G. Crescentini, M. Maione, F. Bruner, Measurements of tropospheric concentration of halocarbons in Antarctica, *Ann. Chim. (Rome)*, 81 (1991), 491–501.
55. C. Triulzi, F. Nonnis Marzano, A. Mori, A. Casoli, M. Vaghi, Presence of radionuclides in biotic and abiotic matrices collected in the environment around the Italian Base in Antarctica, *Ann. Chim. (Rome)*, 81 (1991), 549–561.

56. F. Soggia, S. Boero, R. Frache, Banca Campioni Ambientali Antartici (BCAA), Proceedings of the 4th National Meeting, Venice (Italy), 6–7 December, 1995, 48–52.
57. R. Zeisler, J. K. Langland, H. Harrison, Cryogenic homogenization of biological tissues, *Anal. Chem.*, 55 (1983), 2431–2434.
58. S. A. Wise, B. J. Koster, J. K. Langland, R. Zeisler, Current activities within the National Biomonitoring Specimen Bank, *Sci. Total Environ.*, 139/140 (1993), 1–12.
59. S. Caroli, O. Senofonte, S. Caimi, J. Pauwels, G. N. Kramer, Planning and certification of new multielemental reference materials for research in Antarctica, *Mikrochim. Acta*, 123 (1996), 119–128.
60. S. Caroli, O. Senofonte, S. Caimi, P. Pucci, J. Pauwels, G. N. Kramer, A pilot study for the preparation of a new reference material based on Antarctic krill, *Fresenius J. Anal. Chem.*, 360 (1998), 410–414.

Environmental Contamination in Antarctica
A Challenge to Analytical Chemistry
S. Caroli, P. Cescon and D.W.H. Walton, editors
2001 Elsevier Science B.V.

Chapter 13

The future role of quality assurance in monitoring and research in the Antarctic

Michael A. Champ, Adriana Y. Cantillo, Gunnar G. Lauenstein

1. Polar monitoring and research

The need for environmental monitoring in polar regions has been identified at scientific and political levels. On 14 June 1981, the eight Arctic circumpolar countries (USA, Canada, Denmark, Iceland, Norway, Sweden, Finland and the former Soviet Union) signed an Arctic Environmental Protection Strategy which among other requirements commits each country to assess on a continuing basis the threats to the Arctic environment and to monitor the levels of, and to assess the effects of, anthropogenic pollution in all components of the Arctic environment. The current focus of this strategy is on persistent organics, heavy metals and radionuclides (1).

The recently completed Protocol to the Antarctic Treaty on environmental Protection (26 countries) is a comparable document that recognizes the unique opportunities in Antarctica for scientific monitoring of, and research on, processes of global as well as regional importance. It also states that regular and effective monitoring shall take place to allow assessment of the impacts of ongoing activities, including the verification of predicted impact. The Antarctic Treaty Consultative Parties have long recognized the need to protect the Antarctic environment and have requested support from the Scientific Committee on Antarctic Research (SCAR) to provide necessary expert scientific advice and the Scientific Committee established by the 1980 Convention on the Conservation of Antarctic Marine Living Resources (CCAMLR) to begin to implement a plan to monitor key components of Antarctic ecosystems. A Quality Assurance (QA) plan is the first step in implementing a monitoring programme.

2. Quality assurance

QA must be an integral part of any multinational multi-year monitoring effort from the inception of any environmental programme. It cannot be done retroactively. The quality of the data must be stated in Data Quality Objectives that will meet the needs of the programme. Therefore, the purpose of the monitoring programme must be known before sample collection and analysis begin. If this is not the case, then the data produced may not be accurate enough for the differences over space or time that the monitoring programme intends to detect. Cost is also an important factor since production of data of higher quality than needed to answer the programme objectives is an unnecessary financial burden. The Management/Coordinators of an environmental programme must take an active lead in and provide support to QA/Quality Control (QC) efforts. Otherwise data of limited use will result.

2.1. QA/QC in an ongoing monitoring programme

The NOAA National Status and Trends (NS&T) Programme determines the current status of, and changes over time in, the environmental health of US estuarine and coastal waters. Concentrations of organic and inorganic contaminants are determined in bivalves, bottom-dwelling fish and sediments. Two projects of the NS&T Programme are the major sources of data: the National Benthic Surveillance Project and the Mussel Watch Project. This long-term monitoring programme (1984 to the present) is an environmental programme that uses numerous aspects of QA and can serve as an example for environmental monitoring programme managers.

In the NS&T Programme, concentrations of organic and inorganic contaminants are determined in bivalves, bottom-dwelling fish (through 1992) and sediments. The analytes include 24 polycyclic aromatic hydrocarbons, 18 polychlorinated biphenyl congeners, DDT and its metabolites, 9 other chlorinated pesticides, organotins, 5 major elements, and 12 trace elements. The quality of the analytical data generated by the NS&T Programme is overseen by the performance-based QA Project (2, 3). This Project has been in operation since 1985 and is designed to document sampling protocols, analytical procedures and laboratory performance and to reduce intralaboratory and interlaboratory variation. In addition, the QA Project facilitates comparisons among different monitoring programmes with similar QA activities and thus extends the temporal and spatial scale of such programmes. It is necessary that sampling sites, sampling protocols and analytical procedures be described in detail, and this has been done for the NS&T Programme (4–6). The NS&T Programme does not prescribe specific analytical methods, but encourages the use of state-of-the-art procedures. This allows the use of new or improved analytical methodology or instrumentation without compromising the quality of the data sets. It also encourages the laboratories to use the most cost-effective methodology while generating data of documented quality. The analysis of reference materials, such as the Certified Reference Materials (CRMs) produced by the National Research Council (NRC) of Canada, the Standard Reference Materials produced by the National Institute of Standards and Technology (NIST), and the control materials generated for use by NS&T laboratories as part of the sample stream, is required. A minimum of 8% of the organic analytical sample string consists of blanks, reference or control materials, duplicates and spike matrix samples. The use of control materials does not entirely replace the use of duplicates and spiked matrix samples. A minimum of 2% of the standard inorganic sample string consists of calibration materials and reference or control materials. Analytical data from all control materials and all matrix reference materials are reported to the NS&T Programme office. These data are stored in the NS&T Programme office.

The Limits of Detection (LoDs) are calculated and reported annually on a matrix and analyte basis. Since 1989, the method used for calculating LoDs is that used by EPA and is described in detail in the 07.01.1988 edition of the Federal Register (Definition and Procedure for the Determination of the Methods Detection Limits (MdLs) – Revision 1.11). If the EPA method is not used or is modified the procedure used for MDL calculation is described in detail. Acceptable limits of precision for organic control materials are ±30% on average for all analytes, and ±35% for individual analytes. These limits apply to those materials where the concentrations of the compounds of interest are at least ten times greater than the LoDs. The application of these guidelines in determining the acceptability of the results of the analysis of a sample is a matter of professional judgment on the part of the analyst, especially in cases where the analyte level are near the LoDs. All NS&T laboratories are required to participate in a continuing series of intercomparison exercises utilizing a variety of solutions and natural matrix materials. The organic analytical intercomparison exercises are coordinated by NIST and the inorganic exercises by NRC. Results of these exercises have been described (7–16).

It has been shown that the performance of laboratories improves with time, as the result of experience gained through participation in intercomparison exercises (10–16). This improvement can only be demonstrated through the continued analysis of a material, such as a CRM, SRM or a control material with known analyte concentrations. The NOAA intercomparison exercises for trace metals for 1991 through 1993 used the NRC CRM sediment BCSS-1 as part of the exercise materials. Typical results reported by a laboratory joining the exercise programme in 1991 are presented in Figure 13.1. The accuracy of the Cr, Se and Zn determinations improved with time, as did the precision of Se analysis.

No CRMs or SRMs are regularly analyzed specifically as part of the trace organic intercomparison exercises, so an evaluation similar to the one done for the trace metal exercises using changes in CRM and SRM results over time is not possible. A measure of improvement of laboratory performance can be made, however, by comparing the performance of a laboratory joining the exercises for the first time and that of a laboratory that has participated for several years (Figure 13.2). Laboratories newly joining the exercises usually have larger percent errors than the veteran laboratories. Within a year or two, however, the performance of the new laboratories typically improves and equals those of the veteran laboratories.

As part of the evaluation of results of the intercomparison exercises for major and trace elements, NRC assigns a performance evaluation criteria based on the number of times results reported by a laboratory fall within acceptable criteria. The percentage of laboratories achieving superior or good performances has increased since 1991 from 46% to 83% (Figure 13.3). Superior-rated laboratories submitted results for most analytes within the 95% confidence intervals; good-rated laboratories submitted many results within the accepted range with a minimum number of outliers (16).

To ensure that high quality environmental data are derived from monitoring

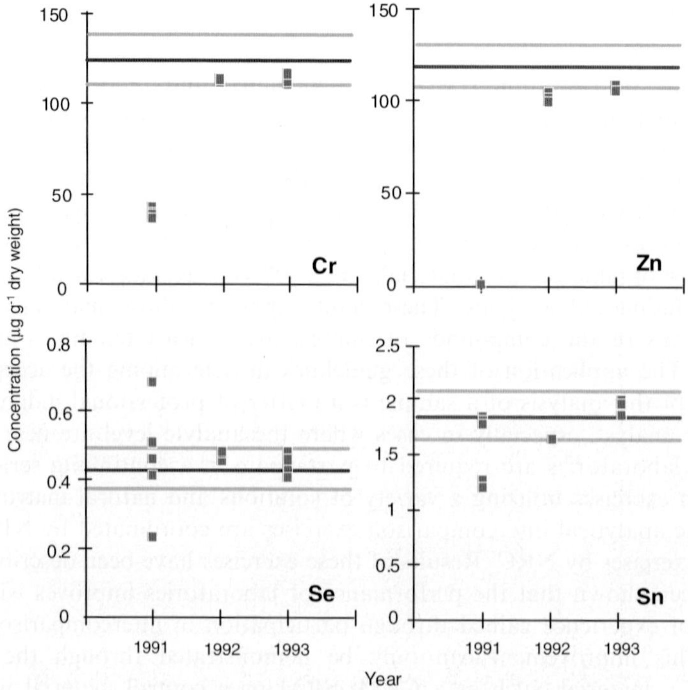

Figure 13.1. 1991–1993 Cr, Se, Sn and Zn intercomparison exercise results of five replicates of BCSS-1 reported by a laboratory participating in the exercises for the first time in 1991. Solid line is the certified value and dashed lines are the uncertainty. All data are in µg g^{-1} dry weight.

programmes, QA must begin even before a contract is awarded. Organizations proposing to perform analyses of large quantities of environmental samples should be required to perform analyses of representative matrix samples provided to them as part of the laboratory selection process. Since this demands considerable expense, the testing should not be required of otherwise unqualified laboratories or in cases where the contract itself is relatively small. Laboratories competing to analyze bivalve mollusks under contract to the NS&T Mussel Watch Project were required to undergo analytical tests of their ability to quantify environmental contaminants as part of the contract evaluation process. In 1994, competing laboratories were tested, but using matrix materials for the quantification of both trace elements and organic contaminants (17). Three laboratory groups participated in the exercises. All laboratories were within the acceptance criteria for the quantification of trace elements and all laboratories performed reasonably well for the quantification of organic contaminants. The laboratories' successes may be the result of the fact that all laboratories participating in the analytical testing had been long-term participants in the NS&T QA project.

Once data are received from the analytical laboratories these data should be reviewed to ensure that the results make environmental sense. Even if all the QA

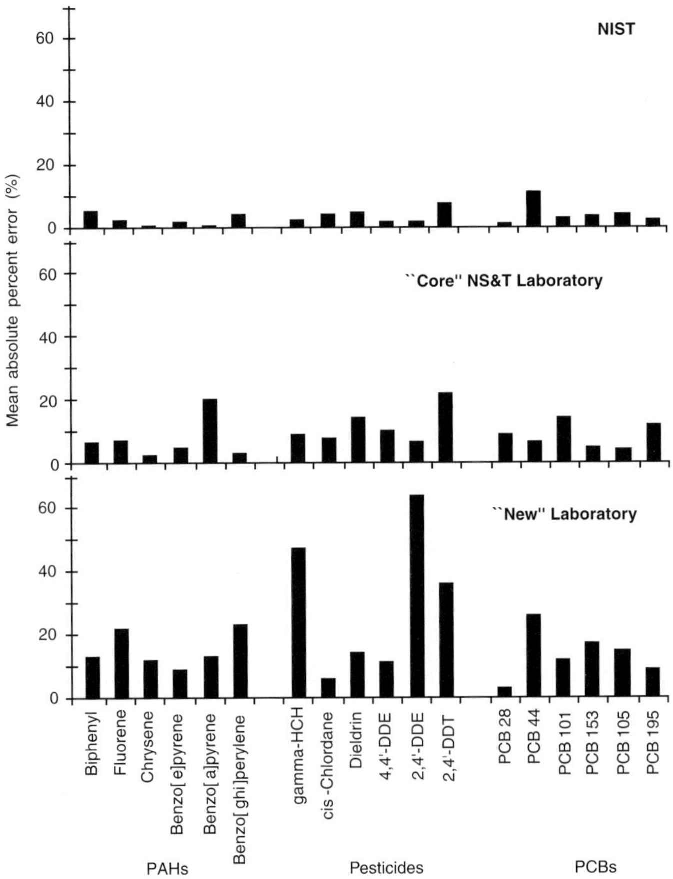

Figure 13.2. 1990 enriched bivalve tissue extract intercomparison exercise results. Mean absolute percent errors of PAHs, pesticides and PCB congeners are shown. Analyses were performed by NIST, an NS&T laboratory, and a non-NS&T laboratory participating in the intercomparison exercise for the first time.

procedures were followed data may appear suspect as compared to earlier sample results or data may not be consistent with what would be expected to result from the environment from which the samples were taken. Confirmatory analyses by another laboratory can be performed on homogenates of suspect samples. Results from these analyses may either resolve the issue or indicate that further action is required.

2.2. *QA/QC in a monitoring programme*

One of the basic components of a monitoring programme is a rigorous QA/QC system that encompasses sampling, analytical processes and data management. Such a QA/QC system must be in place before sampling and data gathering activities start and must continue through the life of the monitoring programme.

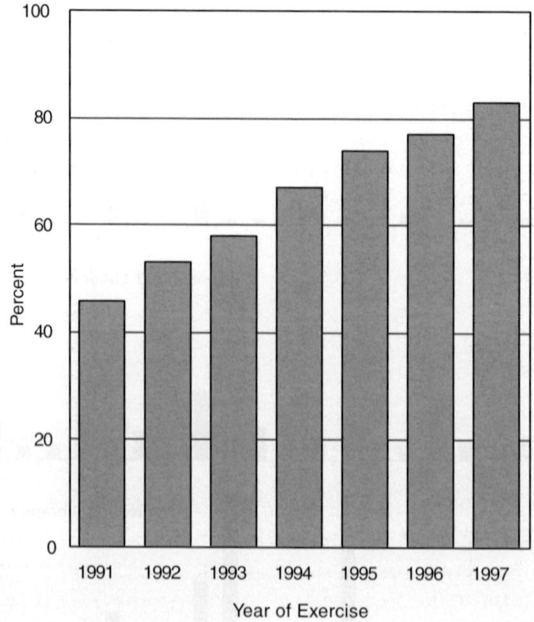

Figure 13.3. Percentage of laboratories participating in the major and trace element intercomparison exercises with performances rated in the superior and good category for the analysis of tissues (16).

Whenever possible, intercomparison exercises should be done to compare and document laboratory/equipment performance and thus extend the range of comparability. The QA/QC programme must be supported by top management and resources must be allocated for it. This is not an area to reduce or exclude for lack of funds.

2.3. Standards and reference materials

The use of Reference Materials (RMs) is part of good QA practices that insure analytical data of documented quality. An RM is a material or substance one or more properties of which are sufficiently well established to be used for the calibration of an apparatus, the assessment of a measurement method or for the assignment of values to materials. A CRM is an RM one or more of whose values are certified by a technically valid procedure accompanied by or traceable to a certificate or other documentation which is issued by a certifying body such as NRC, NIST or others. An SRM is a CRM produced and certified by NIST.

During the last few years, there has been an increase in the number and type of RMs of environmental origin and their use in the environmental analytical community is increasing. At the request of the Intergovernmental Oceanographic Commission/United Nations Environment Programme Group of Experts on Standards and References Materials (GESREM), NOAA has periodically prepared a publication that assembles and updates all information available on RMs for use in marine chemistry and marine pollution research and monitoring (18). This publication was recently expanded to include all aspects of environmental science. The current edition lists more than 1200 RMs from 28 producers and contains information about their proper use, sources, availability and analyte concentrations. RM types included are: ashes, gases, oils, rocks, sediments, sludges, soils, tissues and waters; instrumental performance evaluation RMs; and physical properties RMs. Indices are included for elements, isotopes, and organic compounds. An excellent discussion of various aspects of QA and of the use of RMs can be found elsewhere (19, 20).

3. Conclusions

QA is a necessary component of environmental quality monitoring. It is not only necessary to ensure that reliable data result within a project, but that those data are also comparable to data resulting from other projects. Elements of any QA programme should include: testing of candidate analytical laboratories before any actual field samples are collected; training of field team in the proper way to collect and document the collection of field samples; the adequate use of blanks, spiked blanks, RMs and the performance of duplicate analyses; participation in a QA programme that extends beyond the organization performing the monitoring; and, for monitoring programmes that subscribe to a performance driven QA programme, method documentation with time. An added QA element is that confirmatory analyses should be performed by an exchange of sample splits among laboratories participating in the monitoring effort. When all elements of QA/QC are adhered to, laboratory performance is not only adequate for the purpose of producing reliable data, but laboratory performance actually improves with time.

Appendix

Definitions

There are several concepts and terms that are essential to discussions about QA, even the concept itself. While at a very detailed level any definition can be challenged as being too narrow or too broad, the definitions presented below are useful (20). It should be noted that these terms are quite recent in definition and are not usually given in statistics books. Other terms like pollution have evolved over hundreds of years (21). Some key terms used in the field like reproducibility, standard deviation, standard error, replicate analysis, blanks, spiked samples and blind samples are self-explanatory.

Quality Assurance (QA) is a system of activities whose purpose is to provide to the data user the assurance that the data meet defined standard of quality. It consists of QC and quality assessment. QA applies to field and laboratory practices including collection, identification, storage, preservation, shipment and analysis of samples.

Quality Control (QC) is the overall system of activities whose purpose is to control the quality of the data to meet the needs of the user in a satisfactory, adequate, dependable and economic way.
Quality Assessment is the system of activities whose purpose is to provide assurance that the QC activities are being performed effectively.
Sensitivity is a measurement of the capability of methodology or instrumentation to discriminate between samples having different concentrations of analytes.
The **Limit of Detection** is the smallest concentration/amount of a component of interest that can be measured by a single measurement with a stated level of confidence. This subject is discussed in detail elsewhere (22).
Accuracy is the degree of agreement of a measured value with the true value of the quantity under concern. Inaccuracy results from imprecision (random error) and bias (systematic error) in the measurement process. Bias can only be estimated from the results of measurements of samples of known composition. SRMs are ideal for use in such an evaluation (22).
Precision is the degree of mutual agreement characteristic of independent measurements as the result of repeated applications of the process. Precision is a measure of the level of reproducibility of a given methodology or instrumentation under optimum conditions, while a accuracy is the degree of agreement of a measured value with the true or expected value of the quantity under concern (or simply stated) accuracy is the closeness of a measured or computed value to its true value; precision is the closeness of repeated measurements of the same quantity. (23).
Data Quality Objectives (DQOs) are the stated precision and accuracy ranges that are deemed acceptable for a given measurement. If, for example, data need to have an accuracy of ±1%, then data resulting from a measurement system with an accuracy of ±20 would not meet the DQOs. If, however, only the determination of the presence or absence of a substance is needed, then data with an accuracy of ±20 % may be more than adequate for this purpose.
Reference Material (RM) is a material or substance, one or more of whose values are sufficiently homogeneous and well established to be used for the calibration of an apparatus, the assessment of a measurement method or for assigning values to materials (24).
Certified Reference Material (CRM) is a reference material, accompanied by a certificate, one or more of whose values are certified by a procedure which establishes its traceablility to an accurate realization of the unit in which the values are expressed and for which each certified value is accompanied by an uncertainty at a stated level of confidence (25).

References

1. M. A. Champ, D. A. Flemer, D. H. Landers, C. Ribic, T. DeLaca, The roles of monitoring and research in polar environments, *Mar. Poll. Bulletin.*, 25 (1992), 220–226.
2. A. Y. Cantillo, G. G. Lauenstein, Performance based quality assurance of the NOAA National Status and Trends Program, in M. Parkany (Ed.), *Quality Assurance for Analytical Laboratories*, Royal Society of Chemistry, Cambridge, UK (1993), 34–43.

3. A. Y. Cantillo, G. G. Lauenstein, Performance-based quality assurance – the NOAA National Status and Trends Program experience, *Proceedings of the National Water-Quality Monitoring Council National Conference*, Reno, NV. July 7–9, 1998, III-63 – III-73.
4. G. G. Lauenstein, M. Harmon, B. P. Gottholm, National Status and Trends Program: monitoring site descriptions for the first five years of Mussel Watch and National Benthic Surveillance Projects, *NOAA Technical Memorandum NOS ORCA 70. NOAA/NOS/ORCA*, Rockville, MD, 1993, 360 pp.
5. G. G. Lauenstein, A. Y. Cantillo (Eds.), Sampling and analytical methods of the NOAA National Status and Trends Program National Benthic Surveillance and Mussel Watch Projects 1984–1992: Vol. I–IV. *Technical Memorandum NOS ORCA 71. NOAA/NOS/ORCA*, Silver Spring, MD, 1993.
6. G. G. Lauenstein, A. Y. Cantillo, Sampling and analytical methods of the National Status and Trends Program Mussel Watch Project: 1993–1996 update, *NOAA Technical Memorandum NOS ORCA* 130, 1998, 233.
7. N. Valette-Silver, Elemental analyses in marine sediment and biological tissues, *NOAA Technical Memorandum NOAA/NOS/ORCA* 66, Rockville, MD, 1992, pp. 39 plus appendices.
8. A. Y. Cantillo, Quality Assurance Project intercomparison exercise results 1991–1993, *NOAA Technical Memorandum* 79, NOAA/NOS/ORCA, Silver Spring, MD, 1995a, pp. 219.
9. A. Y. Cantillo, R. M. Parris, Quality Assurance Project trace organic intercomparison exercise results 1986–1990, *NOAA Technical Memorandum NOS ORCA 69*, NOAA/NOS/ORCA, Silver Spring, MD, 1993, pp. 179.
10. S. Willie, S. Berman, NOAA National Status and Trends Program fifth round intercomparison exercise results for trace metals in marine sediments and biological tissues, *NOAA Technical Memorandum NOAA/NOS/ORCA 86*, NOAA/NOS/ORCA, Silver Spring, MD, 1995a, pp. 40 plus appendices.
11. S. Willie, S. Berman, NOAA National Status and Trends Program sixth round intercomparison exercise results for trace metals in marine sediments and biological tissues, *NOAA Technical Memorandum NOAA/NOS/ORCA 85*, NOAA/NOS/ORCA, Silver Spring, MD, 1995b. pp. 44 plus appendices.
12. S. Willie, S. Berman, NOAA National Status and Trends Program seventh round intercomparison exercise results for trace metals in marine sediments and biological tissues, *NOAA Technical Memorandum NOAA/NOS/ORCA 84*, NOAA/NOS/ORCA, Silver Spring, MD, 1995c, pp. 51 plus appendices.
13. S. Willie, S. Berman, NOAA National Status and Trends Program eighth round intercomparison exercise results for trace metals in marine sediments and biological tissues, *NOAA Technical Memorandum NOAA/NOS/ORCA 83*, NOAA/NOS/ORCA, Silver Spring, MD, 1995d, pp. 50 plus appendices.
14. S. Willie, S. Berman, NOAA National Status and Trends Program ninth round intercomparison exercise results for trace metals in marine sediments and biological tissues, *NOAA Technical Memorandum ORCA 93*, NOAA/NOS/ORCA, Silver Spring, MD, 1995e, pp. 52 plus appendices.
15. S. Willie, S. Berman NOAA National Status and Trends Program tenth round intercomparison exercise results for trace metals in marine sediments and biological tissues, *NOAA Technical Memorandum ORCA 93*, NOAA/NOS/ORCA, Silver Spring, MD, 1996, pp. 52 plus appendices.
16. S. Willie, NOAA National Status and Trends Program eleventh round intercomparison exercise results for trace metals in marine sediments and biological tissues, *NOAA Technical Memorandum ORCA 120*, NOAA/NOS/ORCA, Silver Spring, MD, 1997, pp. 51 plus appendices.
17. G. G. Lauenstein, A. Y. Cantillo, Analytical evaluation of laboratories wishing to perform environmental characterization studies, *Environm. Toxicol. Chem.*, 16 (1997), 1345-1350.
18. A. Y. Cantillo, Standard and reference materials for environmental science (Part 1 and Part 2). *NOAA Technical Memorandum 94*, NOAA/NOS/ORCA, Silver Spring, MD, 1995b, pp. 752.
19. J. K. Taylor, *Handbook for SRM Users.* NBS Special Publication 260–100, National Institute of Standards and Technology, Gaithersburg, MD, USA, 1985a, p. 85.
20. J. K. Taylor, *Principles of Quality Assurance of Chemical Measurements.* NBSIR 85-3105. National Institute of Standards and Technology, Gaithersburg, MD, USA, 1985b, p. 71.
21. M. A. Champ, Etymology and use of the term "pollution", *Can. J. Fish. Aqu. Sci.*, 40 (1983), 5–8.

22. L. H. Keith, W. Crummett, J. Deegan, Jr., R. A. Libby, J. K. Taylor, F. Wentler, Principles of environmental analysis, *Anal. Chem.*, 55 (1983), 2210–2218.
23. R. P. Sokal, F. J. Rohlf. 1981, *The principles and practice of statistics in biological research,* 2nd Edition, Freeman and Company. NY, NY. pp. 859.
24. ISO VIM, 1993, 6. 13.
25. ISO VIM, 1993, 6. 14.

Environmental Contamination in Antarctica
A Challenge to Analytical Chemistry
S. Caroli, P. Cescon and D.W.H. Walton, editors

Chapter 14

The Italian environmental policy of research in Antarctica, with special regard to the Antarctic Treaty and the Madrid Protocol

Pietro Giuliani, Milo Kuneshka, Luana Testa

1. Introduction

Italy signed the Antarctic Treaty in 1981 and a National Programme for Scientific Research in Antarctica (Programma Nazionale di Ricerche in Antartide, PNRA) was established in 1985. Funding was provided for five years and this was renewed five years later. At present, the programme is in its sixteenth year of existence.

The PNRA has a broad range of activities, mostly centred at the Research Station Baia Terra Nova (Terra Nova Bay, TNB). The site of the station is in the Ross Sea region, at 74°41′42″S and 164°07′23″E, and was selected by Italian scientists in co-operation with the New Zealand Antarctic Programme. It is located on a rocky, ice-free coastal area. The site has many desirable characteristics, from both a logistic and a scientific point of view.

Logistically, the bathymetry of the area allows for an easy access by ship; furthermore, the station is located near the tip of a small peninsula, which protects a bay, called Thethys Bay, from the open sea. A sea-ice airstrip can be prepared on Thethys Bay for aircraft use. This airstrip has been used in the last six years by an Italian Air Force C-130 aircraft flying from Christchurch and from the US base at McMurdo, carrying personnel, equipment and supplies and allowing the TNB activities to start earlier than in the past. In the first six years of activity the presence of sea ice did not allow the ship to arrive before the first half of December. Using the aircraft the base could be opened in the last decade of October.

From the scientific point of view, the Terra Nova Bay area is in the centre of a region of considerable interest for geological, oceanographic, biological and geophysical research. The PNRA runs a wide spectrum of science research projects, grouped in five major themes:

- geological evolution of the Antarctic continent and of the Southern Ocean;
- global change;
- observatories and geographical information;
- methodologies for environmental conservation;
- technological research.

The TNB station is a summer-only station at present. It could be easily adapted for year-round use should the PNRA decide to implement winter activities.

Italy became a Consultative Party to the Treaty after the start of research activities at TNB in October 1987 and a full member of the Scientific Committee on Antarctic Research (SCAR) in September 1988.

In this chapter the relationship between the Antarctic Treaty and the Protocol on the Environmental Protection to the Antarctic Treaty will be briefly outlined, together with a discussion of the most important points of the Protocol. The implementation of the Protocol by the Italian Antarctic Research Programme will be then described in some detail, including some examples drawn from the experience accrued so for with the monitoring of the impact of human activities around the TNB research station.

2. The Antarctic Treaty and environmental protection

The Antarctic Treaty does not address explicitly the protection of the Antarctic environment; one of the very few references is in art. IX, 1.f: "Measures regarding the preservation and the conservation of living resources in Antarctica", where the Contracting Parties are exhorted to recommend the measures to their governments (1–3).

Later, a number of recommendations on environmental issues were prepared by the Antarctic Treaty Consultative Meetings (ATCM). The first was Recommendation VI-4: "Man's impact on the Antarctic environment" (Tokyo, 1970); in it, the Representatives recommend to their Governments that:

1. they invite SCAR through their National Antarctic Committees
   a. to identify the types and assess the extent of human interference which has occurred in the Treaty area as a result of man's activities,
   b. to propose measures which might be taken to minimize harmful interference and
   c. to consider and to recommend scientific programmes which will detect and measure changes occurring in the Antarctic environment;
2. they encourage research on the impact of man in the Antarctic ecosystem;
3. they take *interim* measures to reduce known causes of harmful environmental interference;
4. they consider including in the agenda for the VII ATCM an examination of this matter in the light of any further available information.

Recommendation VI-4 is important not only because it mentions for the first time man's impact, but also because it prescribes, for the first time, under point 1.c environmental monitoring. It is not surprising that it took until 1970 for these subjects to be mentioned. The environmental sensitivity was developing and growing in the late sixties and times were ripe for the transfer of this awareness to the Antarctic context. At the VII ATCM in Wellington in 1972 the Parties invited their governments with Recommendation VII-1 to take note of the SCAR responses to the Recommendation VI-4 and to discuss it further. Moreover, they should consider adopting them as voluntary guidelines.

Recommendation VIII-11 of the VIII ATCM (Oslo, 1975), quoting Recommendation VI-4 and Recommendation VII-11, asks the governments to observe a code of conduct dealing with waste disposal and environmental impact assessment. Another recommendation of the Oslo ATCM, Recommendation VIII-13, deals with the protection of the Antarctic environment. It prescribes the use of protective measures for the environment, the prediction of possible environmental modifications and the monitoring of changes in the environment. This is the first clear mention of the need to protect the environment, to perform an assessment of possible impacts and to monitor the environmental changes.

At the IX ATCM (London, 1987) the same concepts are expressed in Recommendation IX-5, "Man's impact on the Antarctic environment". This Recommendation is of particular importance, because the Parties recognize as a primary responsibility the protection of the Antarctic environment from harmful human interference, the consideration of the environmental impact of future activities and the continuation of environmental monitoring. Moreover, in the IX ATCM the Antarctic Treaty area was designated as a Special Conservation Area and the "Agreed Measures for the Conservation of Antarctica Flora and Fauna" were adopted; the Parties also decided to recommend to their governments that they reaffirm their commitment to environmental protection.

Environmental monitoring issues were discussed at great length during the long negotiations for the Convention on the Regulation of Antarctic Mineral Resources Activities (CRAMRA). In Art. 4 of CRAMRA the monitoring of key environmental parameters is considered a basic principle governing potential activities. Also in the Convention for the Conservation of Marine Living Resources (CCAMLR) environmental monitoring activities are considered an important management tool. Within CCAMLR a Committee on Ecosystem Monitoring Programme (CEMP) has evolved, developing agreed international protocols of considerable practical value.

At the XII ATCM (Canberra, 1983) Recommendation XII-3 called on the Governments to evaluate scientific activities and related logistics in order to minimize their possible harmful effects on the environment. SCAR was asked to provide advice on the types of research and logistic activities which might be expected to have a significant impact on the Antarctic environment and to suggest procedures to be used in the evaluation of possible environmental impacts. SCAR responded with the report "*Man's Impact on the Antarctic Environment: a Procedure for Evaluating Impacts from Scientific and Logistic Activities*". This report, tabled at the ATCM XIII (Brussels, 1985) proved a milestone in the matter of Antarctic EIAs, until superseded by the Madrid Protocol.

At the ATCM XIV (Rio de Janeiro, 1987) Recommendation XIV-2 was adopted. In it there are two clear references to environmental monitoring in connection with environmental evaluation, which should also cover "identification of measures, including monitoring programmes, that could be taken to minimise or mitigate impacts and detect possible unforeseen effects" and recognize that "key indicators of the environmental effects of activities should be monitored". In the final report another important statement is made about the need for improving the comparability and the accessibility of Antarctic scientific data and the importance of developing an Antarctic data directory.

The XV ATCM (1989, Paris) represented a real turning point for the protection of the Antarctic environment. The parties, in Recommendation XV.1 prompted their governments to undertake as a priority objective the "further elaboration, maintenance and effective implementation of a comprehensive system for the protection of the Antarctic environment and its dependent and associated ecosystems aimed at ensuring that human activity does not have an adverse impact on the Antarctic environment and its dependent or associated ecosystems or compromise the scientific, aesthetic or wilderness values of Antarctica". The parties asked that a Special Antarctic Treaty Consultative Meeting be held in 1990 to explore and discuss proposals relating to the comprehensive protection of the Antarctic environment.

This recommendation was the starting point for the preparation of the Protocol on the Environmental Protection to the Antarctic Treaty. The XI ATSCM was convened in Viña del Mar in Chile in November 1990, followed by a further meeting in 1991 in Madrid, where the Protocol was finally approved and signed on October 4, 1991. Another Recommendation (XV-5) adopted at the XV ATCM makes an important distinction between two different types of environmental monitoring in Antarctica, the "scientific" monitoring, performed in order to study changes in the global environment, taking advantage of the unique conditions of Antarctica, and the "service" monitoring, centred on human activities in Antarctica in order to monitor their impacts.

3. The Protocol on environmental protection to the Antarctic Treaty

The principle behind the preparation of the Protocol was the implementation of a comprehensive approach for the protection of the Antarctic environment and its dependent and associated ecosystems. This implies the adoption and implementation of an integrated and internally consistent set of rules to replace the many measures for environmental protection adopted by ATCMs, to make the system simpler, more consistent with international conservation objectives and less confusing to the scientific and logistic personnel in Antarctica. At the XVI ATCM in Bonn in 1991 the Parties decided to examine all existing recommendations and to reorder them in the light of the just signed Protocol.

The Protocol on Environmental Protection to the Antarctic Treaty designates Antarctica as "a natural reserve, devoted to peace and science" and provides for an indefinite ban on mineral resources exploitation and on mineral prospecting. This ban, and indeed all parts of the Protocol, can be renewed after 50 years from the entry into force of the Protocol at the request of any Consultative Party.

Together with the Protocol four Annexes, forming an integral part of the Protocol, were approved. They deal with environmental impact assessment, conservation of fauna and flora, waste management and disposal and prevention of marine pollution. A few days after the Signing of the Protocol (October 4, 1991) a fifth annex on area protection and management was approved by the XVI ATCM in Bonn on October 17, 1991. Another annex dealing with rules and procedures relating to liability for environmental damage arising from activities taking place in the Antarctic Treaty area and covered by the Protocol is under discussion by a working group of legal experts. This annex is required under Art. 16 of the Protocol. The Protocol also establishes a Committee on Environmental Protection (CEP) whose task is to provide advice and recommendations to the ATCMs on the implementation of the Protocol.

The conditions for entry into force having been completed with the deposit of the instrument of acceptance by Japan on December 15, 1997, and the Protocol with Annexes I to IV entered into force on January 14, 1998. The CEP was therefore inaugurated at ATCM XXII in Tromsoe, Norway (May 1998).

4. The implementation of the protocol by the Italian Antarctic Research Programme at TNB Station

Italy ratified the Protocol with the Law no. 54 of February 15, 1995, and the instrument of ratification was deposited on March 31st, 1995. It is worth recalling that in the Final Report of the XI ATSCM the parties were exhorted to apply the Protocol, and in particular the annexes, even before its entry into force. Most Parties complied with the suggestion and at the various ATCMs following the Madrid meeting, reported on their implementation of the Protocol. Italy also began the implementation of the Protocol a few months after the signing, following in this not only the exhortation of the XI ATSCM, but also an explicit statement to do so from the President of ENEA, the agency responsible for implementing the PNRA.

At the beginning of the Italian Antarctic activities no environmental impact assessment of the station was available. Neither the Italian legislation at the time, nor the Antarctic Treaty required it. Several environmental assessments were prepared instead for other less important activities, such as the installation of a seismographic station. Recently an environmental overview of the base was prepared and tabled at the XXI ATCM.

TNB station has been built and operated in such a way as to keep the environmental impact to a minimum. A strict waste management plan has been enforced from the start; a liquid waste treatment plant and an environmental monitoring programme have been in operation from the second year. An emergency plan for oil spills has been prepared, with a number of people specifically trained to this end. The oil spill plan and a fire protection plan have recently been merged in the station emergency plan. The fuel used at the station is stored in three main fuel tanks, located at about 300 m from the station buildings. The tanks are made of welded steel, with double skins and all welds are X-rayed. The three tanks contain enough fuel for two years of operation, thus reducing the need for refuelling and the risk of oil spills.

5. Waste Management

A formal Waste Management Plan prepared in April 1991 consolidated and updated the operating procedures and instructions developed from 1985–1986 onwards. The objectives of this plan have been:

- implementation of a waste minimization programme both for the bulk and the weight of wastes with special focus on the reduction of packing materials;
- waste recycling;
- storage, transport and disposal in accordance with national legislation;
- waste separation at the origin;
- establishment of a system of documentation for all waste management activities;
- removal of all wastes from the TNB station, field activities, logistic and research vessels from the Antarctic Treaty area.

TNB discontinued any open burning some years ago and now operates a two-chamber high temperature incinerator, mainly for kitchen waste.

6. Sewage treatment plant

At the beginning of the PNRA activities it was decided to install at TNB a sewage treatment plant. The Madrid Protocol allows the discharge of sewage and domestic liquid wastes to be made into the sea for small stations up to 30 persons. For larger stations treatment by maceration is required as a minimum measure.

The biological treatment plant designed for 40–50 persons was installed during the 1986–1987 campaign. In the following summer, in order to verify the effectiveness of the treatment, monitoring of the Biochemical Oxygen Demand (BOD) was begun in the area of effluent discharge. These data were used to adjust the effectiveness of the process on a daily basis. In the following years the scope of monitoring was broadened to include other chemical parameters and surfactants, nitrites, oils, etc. This monitoring indicated that the plant was unable to cope with the increasing load, as the summer population of the station rose to 70 people. In the 1991–1992 campaign a physical-chemical plant was added first in parallel with the biological plant, then in series. In 1995 a completely new and larger physical-chemical treatment plant was installed. In general, the experience with sewage treatment plants at TNB has been good. The mixing of the warm effluent stream is satisfactory and, except very close to the discharge pipe, enhanced levels of nutrients cannot be detected. The sludges resulting from the treatment are returned to Italy.

7. Environmental monitoring

The main purpose of applied environmental monitoring is the detection of local and regional environmental effects caused by specific human activities. In the case of an Antarctic research station these activities are the scientific and logistic support during the operation of the station. At the Italian TNB station an environmental monitoring programme was planned and implemented from 1986–1987. Obviously, environmental monitoring of activities and impacts is only valid and useful if it is performed within the scope of an environmental management strategy. It is perhaps worth mentioning that sound environmental principles have always been at the forefront of Italian activities in Antarctica. It must also be said that the station is new, thus making attainment of good environmental practice simpler than in many of the older stations.

Recommendation XV-5 and the Protocol call for national programmes to establish environmental monitoring programmes for activities that include:

1. waste disposal;
2. contamination by oil and other noxious substances;
3. construction and operation of stations, field camps, ships, aircraft and logistic support in general;
4. implementation of field programmes;
5. recreational activities;
6. activities related to protected areas.

Several of the research projects carried out in Antarctica under the Italian Programme deal with environmental monitoring for basic research purposes. The applied environmental monitoring implemented at TNB over the last 12 years has borrowed ideas and methodologies from these scientific projects. Scientific monitoring has been concerned with sea water, fresh water, soils, particulates, sediments, air, etc, as well as different biota. Applied environmental monitoring has concentrated on only some of these.

In setting up the monitoring programme the following factors were taken into consideration:

1. the type of environment;
2. the equipment and personnel available;
3. the cost of monitoring and its duration;
4. the review of progress;
5. the interference between monitoring and other activities;
6. the relevance of the monitored variables.

In the following a brief outline of the monitoring activities is given.

1986–1987: The monitoring programme was initiated and the collection of baseline samples began before the start of construction of the station buildings. The initial monitoring was centred on airborne particulates and fresh water samples. The collection of samples was done with a low volume air sampler (72 hr sampling time, 20 l/min of air flux) located at about 1200 m from the centre of the station. The first set of samples aimed at the evaluation of 44 elements, including K, Mg, Na (marine origin), Al, La, Sc, Si (crustal origin) and Cd, Hg, V (anthropic origin). Analyses were performed using Instrumental Neutron Activation Analysis (INAA) in the TRIGA nuclear reactor of the Casaccia Research Centre, near Rome. The very low concentrations of some elements, very close and sometimes below the threshold of sensitivity of the instrumentation, has been a significant problem which, at times, has made the measurement a technical challenge. Because of this, the sampling periods have been gradually increased.

1987–1988: Sampling for airborne particulates continued in the vicinity of the station buildings with two new automatic low volume samplers (5 days sampling time, 20 l/min of air flux), at about 100 m NE and at 600 m SW.

1988–1989: Two automatic samplers were used. It appeared that, even with an increased station activity, there was no significant increase in the values of the measured pollutants. On the basis of the results obtained so far it was decided to carry on with the multi-elemental characterization of the environment in order to obtain enough data to perform statistical analyses and to obtain a reliable set of background values.

1989–1990: The choice of low volume samplers at TNB, during the three first campaigns, was simply related to previous experiences and sampler availability. However, analytical requests imposed the use of high volume samplers. In this campaign two new high volume samplers were put in operation. The location of these instruments was selected on the basis of logistic activities and geomorphological considerations. Collected filters were analyzed by the INAA technique for the determination of minor and trace elements and by High Performance Liquid Chromatography (HPLC) for Polycyclic Aromatic Hydrocarbons (PAHs) determinations. This latter determination is important because PAHs are characteristic of combustion products. Biotic accumulation of PAHs was investigated in the marine bivalve *Laternula elliptica*, an organism selected because of its local abundance and low mobility. In this campaign the PAHs sampling methodology had to be adjusted because of the high lipid contents in these bivalves. From the successive campaigns onwards it was possible to collect data regularly on PAHs. Reliable monitoring of PAHs became important because of the installation of more powerful diesel generators, the use of an incinerator and the general increase in logistic activity.

1990–1991: Two additional high volume samplers were installed; at this point the circle of monitoring points around the station was completed. This allowed the determination of the main directions of airborne transport of pollutants to be determined, thus confirming the very low level of contamination from the most important sources, the diesel generators and the various vehicles. The programme of analyses was reviewed, because of difficulties in the operation of the TRIGA reactor used in the neutron activation analyses. The use of Atomic Absorption Spectrometry (AAS) for the detection of Cd and Pb was also considered. As a consequence of this and also in order to simplify the analyses of samples it was also decided to reduce the number of elements investigated to seven. Thus, the attention was directed to the anthropic elements and few natural elements (Cd, Fe, Pb, Zn, etc.). As far as PAHs were concerned, only 11 compounds were selected, among which phenanthrene, anthracene and pyrene.

1991–1992: This was a small campaign, with few participants at TNB station. The weather was very bad for most of the time, with very high winds.

Measured concentrations of all monitored elements, including PAHs, were extremely low.

1992–1993: In this very reduced campaign, concentrating on station maintenance, the only data collected were those relating to airborne particulates and PAHs. The station activity was modest and the measured values were also low.

1993–1994: This was a large campaign with a complex programme. A fifth high volume sampler was installed at Skua Lake. This sampler was also used for spot sampling to verify filter efficiency as a function of sampling time. Single source characterization was also performed. The spectrum of PAHs monitoring was broadened to include components deriving specifically from diesel engine operation, such as coronene, cyclopental [c, d] pyrene, etc. This brought the number of PAH compounds monitored to 23.

1994–1995: Another high volume sampler was added at Campo Icaro because activities tended to expand in that direction.

1995–1996,
1996–1997: The monitoring activity was the same as in the previous years.

In the following sections the experience with the applied environmental monitoring and the analytical techniques used will be described in more detail. Also the impact of man's activities on the environmental state of the atmosphere around the Italian station will be discussed.

8. Materials and methods used in environmental monitoring at TNB

8.1. Sampling sites

In Figure 14.1.a and b the TNB station plan and different sampling sites are shown. During the first campaign (1986–1987) a SEA AIR CAD-M low volume sampler (72 hr sampling with an air flux of 20 l/min) was installed in the site called Skua Lake, about 1200 m SW from the main station building, at an elevation of around 120 m asl. Later, during two successive campaigns (1987–1988 and 1988–1989) two automatic APM1 low volume samplers by Elecos (120 hr sampling with an air flux of 20 l/min) were used. During the 1987–1988 campaign two samplers were installed one in the site called Old Oasi, at about 600 m SW from the main camp and at 45 m asl, and the other close to the main camp, at about 100 m NE from it at 10 m asl.

During the 1988–1989 campaign the samplers were situated one 100 m E of the station at 10 m asl and the other at the Skua Lake (see above). From the 1989-1990 campaign, besides the low volume air samplers previously mentioned, different PM-10, size-selective high volume air samplers by Sierra Andersen were used because of different analytical and data interpretation requirements. During the 1989–1990 campaign the low volume samplers were located as in the previous campaign (see above). Two high volume samplers were installed, one at the site called Eneide, located at about 700 m WSW the main camp and 83 m asl and the

(a)

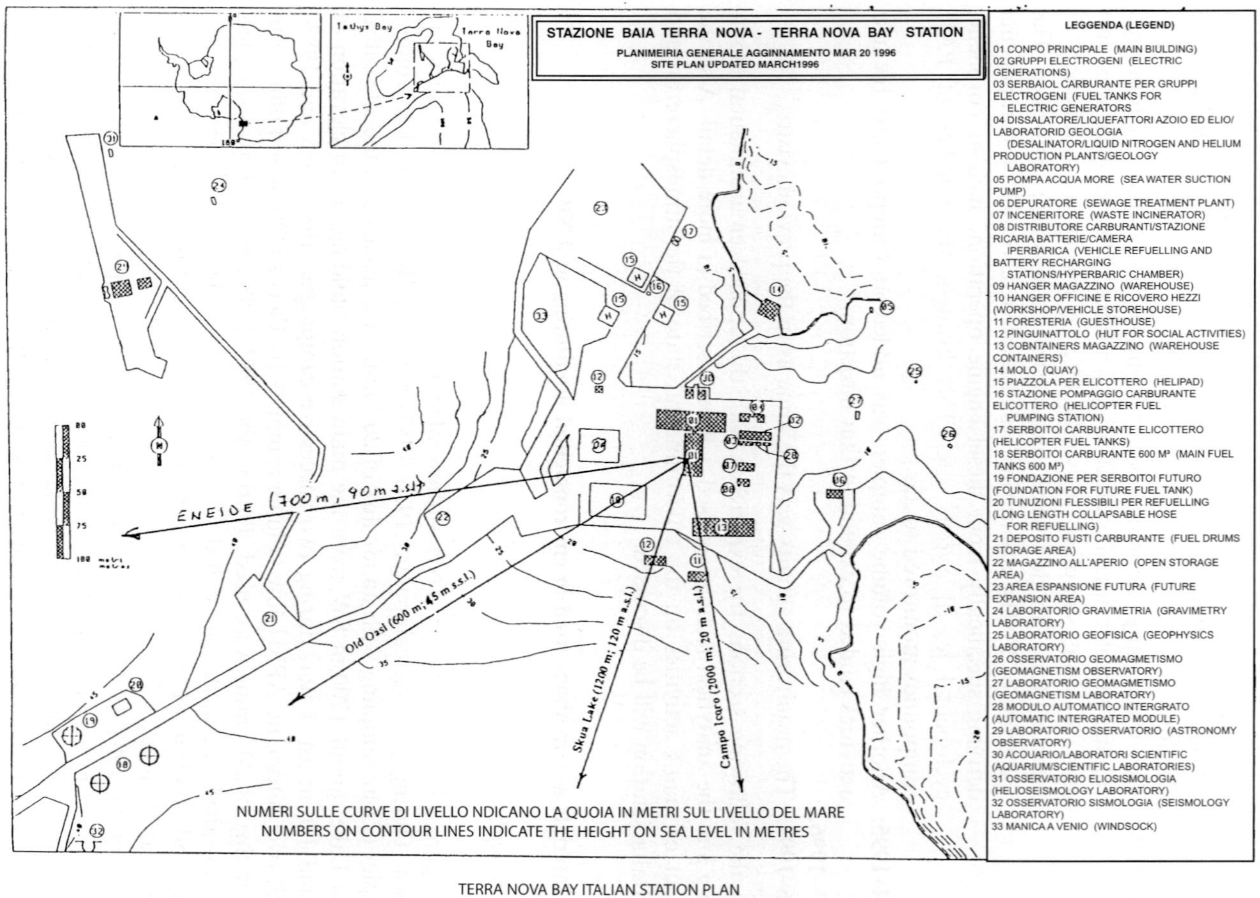

Figure 14.1. (a) Terra Nova Bay Italian station plan; (b) sampling sites around TNB station.

(b)

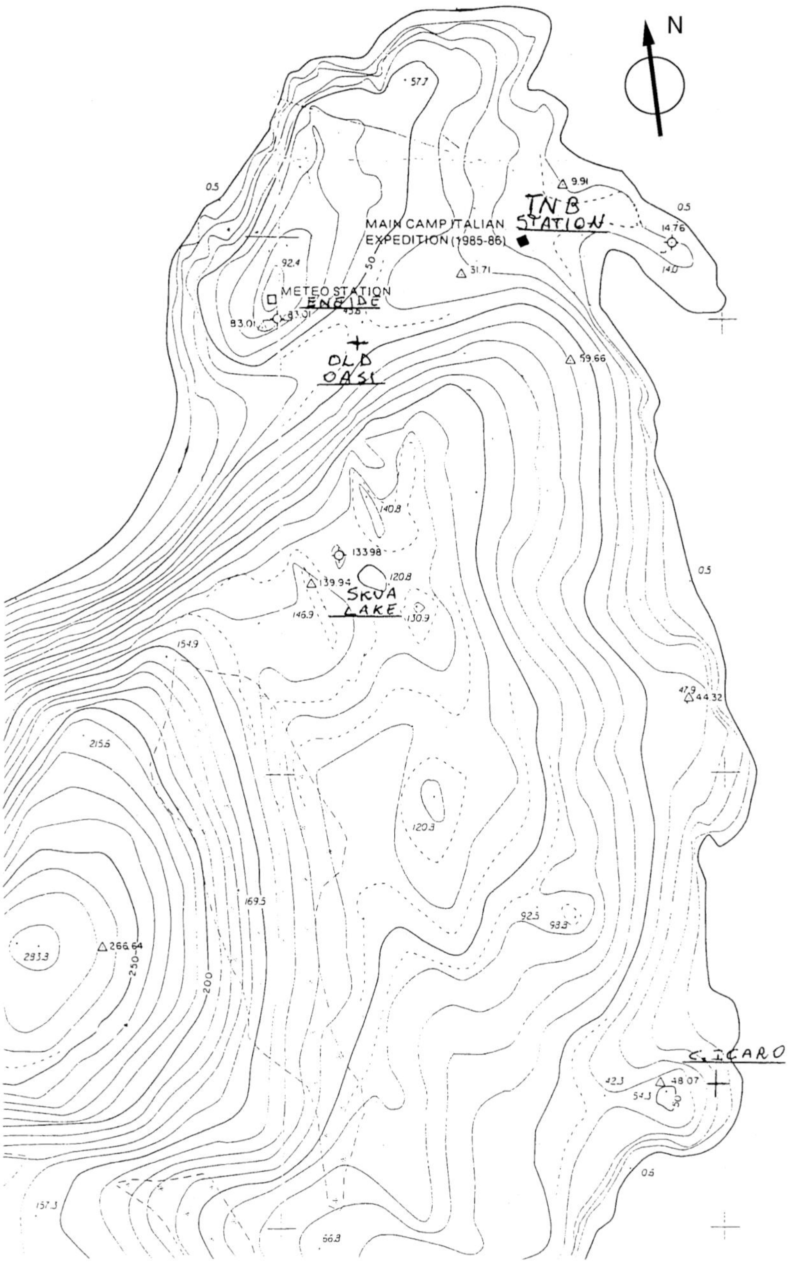

other at 120 m S the main camp and 15 m asl. For the first time, during the 1990–1991 campaign, four high volume air samplers were installed in order to monitor the four cardinal points around the station area. These four sites are described as: North Base (about 100 m North of the main camp and 10 m asl); East Base (about 100 m East of the main camp and 10 m asl); South Base (about 120 m South of the main camp and 15 m asl) and West Base (about 200 m West of the main camp and 15 m asl). All used the same type of filter to make the comparison of all data possible.

Initially, a low volume system was installed at Skua Lake, but this was changed in 1991–1992 to match the high volume air samplers around the station. The data from Skua Lake indicated that the site had become affected by anthropic activities. Thus, it was decided to have a new reference site at Campo Icaro situated about 2000 m S from the main camp and 20 m asl. From the 1994–1995 a high volume air sampler of the same type as all the others was installed there.

8.2. Samplers

The PM-10, size-selective high volume air sampler by Sierra Andersen selects and collects all particles with aerodynamic diameter of less than 10 μm. The symmetrical design ensures insensivity to wind-direction. The inlet design and the internal geometry makes the collection efficiency independent of wind speed from 0 to 36 km/hr^{-1}. The particles are then accelerated through multiple circular impactor nozzles. Those with an aerodynamic diameter less than 10 μm are carried vertically upward by the air flow. The sampler is equipped to control volumetric flow, timing, etc. The air flow rate is 1.1310 m^3 min^{-1} with an accuracy of volumetric flow control of 1% deviation over 24-hr sampling period. A 72-hr sampling period was used.

Several authors confirm that elements related to anthropogenic activities, are concentrated into very small size particles, with a mean aerodynamic diameter less than 2 μm, in the atmosphere. At the same time heavy metals, whose emissions in the atmosphere are related to natural sources, are concentrated into coarse particles with a mean aerodynamic diameter greater than 2 μm (4–10).

Many studies have demonstrated that several toxic metals, including Pb and its compounds, are associated with the fine particulate matter in the ambient air. This is important not only from a health point of view, since fine particles are breathable, but also because fine particles tend to persist in the atmosphere where they can undergo chemical reaction and can be transported from their sources over long distances to pristine areas of the environment (4, 7, 11–13). Furthermore, it is well known that the sub-μm particles are subject to long range transport in the atmosphere (10). Thus, it is believed that using high volume samplers it is possible to collect heavy metals of both natural and anthropogenic origin. Any long range transport component, even if it is present on the filters, will be there as an almost constant background. The variable component will be due to the local sources, both natural and/or anthropogenic.

8.3. Filters

The sampling of atmospheric particulate matter by means of filtration is one of the most commonly used techniques. The low cost and the simplicity of the method are two valuable characteristics, while experience shows that the samples collected on filters may in many cases be stored for subsequent analysis without deterioration.

With an appropriate choice of the samplers and the respective filters, almost any desired amount of sample can be collected in a given interval. Filters remove particles from a gas stream by a number of mechanisms. These include direct interception, internal deposition, diffusion deposition, electrical attraction and gravitational attraction. The mechanism which predominate in a given case will depend on the flow rate, the nature of the filter and the nature of the atmospheric particulate matter. Since a variety of collection mechanisms are involved in filtration, it is not surprising that, for a given aerosol and a given filter, the collection efficiency varies with impact velocity and particle size. The efficiency of a given filter for a given particle size could be high at low flows, due primarily to the effect of diffusion. At increasing velocity, it could first fall off and then, with still higher velocities, begin to rise due to increasing internal deposition.

This pattern has been observed in several experimental penetration tests (5, 13). Since the concentration of the constituents in Antarctica is low, it is necessary to increase the air volume sampled to obtain enough sample mass for analysis. It is important to have a nearly uniform sampling rate over the duration of the sampling period. One should also take into consideration that the flow resistance of all filters increases with increasing loading, but the rate of change is not the same for all filters.

The analytical techniques used for samples collected at the TNB station are INAA, Electrothermal Atomization (ETA) AAS, HPCL and Gas Chromatography–Mass Spectrometry (GC-MS). The two first techniques provide data on heavy metals and other elements, while the others provide data on PAHs. For INAA it would be preferable to collect the atmospheric particulates on polycarbonate membranes, which would also be acceptable for ETA-AAS determinations, but these cannot be used for the determination of PAHs and are unsuitable for high volume samplers because of their considerable resistance to the air flow. Quartz microfibre filters are therefore the best compromise. Advantages are their good mechanical strength, low flow resistance, inertness to all extracting acids (except HF) and organic solvents, as well as being relatively cheap and simple to use. Among the disadvantages are the broad distribution of pore size and density, the high trace element impurities, the impossibility to determine siliceous compounds, etc. Another serious drawback of the quartz fibre filters is the variability of the constituent concentrations. The experimental INAA data show significant variations in concentration for control filters randomly selected from different filter boxes (relative to the same type of filter and from the same producer). This makes reliable analytical determinations difficult especially for low concentrations of atmospheric heavy metals. The high volume air samplers at TNB use Whatman quartz microfibre filter sheets of 20.3 × 25.4 cm dimensions. Tables 14.1 and 14.2 provide some summary characteristics of these filters. Blank filters, before being fitted in the samplers, are submitted to a thermal treatment because half of the sheet is used for PAHs determinations. From the other remaining half, random circular portions with diameter of 43 mm are taken, which are used for INAA and ETA-AAS determinations.

Table 14.1. Summary of air sampling quartz microfibre filter characteristics (2)

Void size μm	Fibre diameter (μm)	Thickness (μm)	Weight/area (mg/cm^2)	Ash content (%)	Maximum operative temperature (°C)	Tensile strength (gm/cm)	ΔP_{100} in H_2O
NA*	NA	450	8.5	100	540	250–300	15.3

* Information not applicable

Table 14.2. Elemental composition and concentrations of air sampling quartz microfibre filters. (Mean values of INAA determinations)

Element	Concentration (μg/g)	Element	Concentration (μg/g)	Element	Concentration (μg/g)
Al	10,400 ± 520*	Hf	0.28 ± 0.03	Sb	0.71 ± 0.05
Ba	47.40 ± 4.65	Hg	0.003 ± 0.002	Sc	0.022 ± 0.003
Ce	1.11 ± 0.16	K	1128 ± 120	Sm	0.21 ± 0.03
Co	0.060 ± 0.010	La	0.62 ± 0.08	Ta	0.018 ± 0.009
Cr	4.32 ± 0.10	Mn	1.50 ± 0.50	Tb	0.061 ± 0.010
Cs	0.034 ± 0.006	Mo	1.09 ± 0.50	Th	0.21 ± 0.02
Eu	0.05 ± 0.02	Na	4398 ± 450	U	0.16 ± 0.05
Fe	56.70 ± 2.50	Rb	0.014 ± 0.002	Zn	23.00 ± 0.20

* Counting uncertainty defined as: (Gross Area Counts + Background Counts)$^{1/2}$ / Net Area Counts

8.4. Analytical techniques

Analytical determinations of heavy metals in the atmospheric particulate matter are performed using INAA and ETA-AAS. INAA is a multi-elemental and non-destructive technique. Using such a technique, it is possible to analyze the sample without any kind of pretreatment, which avoids the possibility of additional contamination and, at the same time, allows for a high degree of sensitivity. However, there are some heavy metals of a great interest, such as Cd, Ni and Pb, that cannot be determined with sufficient precision due to spectral interference or are insensitive to the neutron irradiation (e.g., Pb). These are analyzed by means of the ETA-AAS technique.

Before samples can be analysed by INAA, they must be submitted to neutron irradiation. The interaction between neutrons and different elements in the sample, makes some of them become γ-active and so measurable by a γ-spectrometer. It depends on what kind of heavy metals are present in the sample, on the neutron flux and the neutron flux composition and on the cross-section of the relative n-γ nuclear reactions. The neutron irradiation of the quartz microfibre filters is performed at the TRIGA MARK II nuclear research reactor at the ENEA

Research Centre of Casaccia (Italy). Two different irradiation facilities are used, namely, the pneumatic transfer (so-called Rabbit) and the rotating rank facility (so-called Lazy Susan).

The first yields two groups of isotopes of a half-life of not more than a few days, i.e., Al, Cu, Mg, Mn, Ti and V (irradiation time of 10 s, decay time of 5 min, counting time of 5 min) and As, Br, Ga, K, La and N (irradiation time of 10 s, decay time of 24 hr, counting time of 30 min). In both cases the neutron flux is 1.25×10^{13} n cm^{-2} s^{-1} with a Cd ratio $R_{Cd} = 2.90$.

The second facility was used in the case of isotopes of a half-life more than one week. Again there are two groups of elements, namely, Ba, Br, La, Lu, Mo, Rb, Sb, U and W (irradiation time of 25 hr, decay time of 1 week, counting time of 30 min) and Ba, Ce, Co, Cr, Eu, Fe, Hf, Rb, Sb, Sc, Se, Sr, Ta, Th and Zn (irradiation time of 25 hr, decay time of 1 month, counting time of 3 hr). In both cases the neutron flux is 2.6×10^{12} n cm^{-2} s^{-1} with a Cd ratio $R_{Cd} = 3$. Different γ-spectrometric measuring equipments by EG&G ORTEC were used. The Certified Reference Materials (CRMs) used for the calibration of the measurement systems were "Urban Particulate" and laboratory standards, as detailed in Table 14.3.

For ETA-AAS analyses, the filter portions are submitted first to a 5 ml ultrapure HNO_3 (1:1) mineralization in Teflon® containers and then to a thermal treatment in a Microwave (MW) heating system (CEM81D). After such a treatment, the sample was diluted up to 10 ml with double-distilled water. The instrumental analysis was carried out by a PerkinElmer 5100 ZL spectrometer with a L'vov pyrocoated platform. In Table 14.4 are given all the ETA-AAS analytical parameters for Cd and Pb.

During the analytical procedure, besides different laboratory standards, the "Urban Particulate" CRM was also used.

Table 14.3. Nominal concentrations of the Standard Reference Material NBS 1648 "Urban Particulate" constituents.

Element	Concentration	Element	Concentration	Element	Concentration	Element	Concentration
Nominal concentrations in μg/g, unless otherwise noted.							
Ag	(6)	Cr	403	La	(42)	Sc	(7)
Al	3.42%	Cs	(3)	Mg	(0.8%)	Se	27
As	115	Cu	609	Mn	(860)	Sm	(4.4)
Ba	(737)	Eu	(0.8)	Na	0.425%	Th	(7.4)
Br	(500)	Fe (total)	3.91%	Ni	82	Ti	(0.4%)
Cd	75	Hf	(4.4)	Pb	0.655%	U	5.5
Ce	(55)	I	(20)	Ru	(52)	V	140
Cl	(0.45%)	In	(1.0)	S	(5%)	W	(4.8)
Co	(18)	K	1.05%	Sb	(45)*	Zn	0.476%

* Values in parantheses are not certified, but are given for information only.

Table 14.4. ETA-AAS analytical parameters for Cd and Pb

Parameter	Cd	Pb
Wavelength (nm)	228.8	283.3
T (°C) pretreatment	850	850
T (°C) atomization	1900	1900
Matrix modifier	0.2 mg $NH_4H_2PO_4$ + 0.01 mg $Mg(NO_3)_2$	0.2 mg $NH_4H_2PO_4$ + 0.01mg $Mg(NO_3)_2$

9. Results

The presence of heavy metals in the atmospheric particulate matter in Antarctica can be attributed to different sources, both natural and anthropogenic. Some authors state that almost all natural sources of heavy metals in Antarctica are generally situated in the southern hemisphere (4, 14, 15). The natural sources are normally volcanic activities, erosive processes, continental dusts, marine spray from the ocean, low-temperature biological processes, etc. (7, 10, 16–18). Important local human sources of heavy metal emissions into the Antarctic atmosphere are presumed to be the Antarctic stations and their activities, especially all kinds of transport, power plants, waste burning (incinerators), etc. (10, 12, 15, 19).

Various studies confirm the seasonal behaviour of the heavy metal concentrations in the Antarctic atmosphere (14, 19). The Italian monitoring programme only provides summer data, but this will be relevant when looking for the direct effect of station activities. About 40% of the heavy metal point emissions in the Antarctic atmosphere are deposited not more than 1 km from the station and, at greater distances, only along the down-wind "plume" (21). Snow fall has a major effect on heavy metal measurements, effectively limiting their redistribution by trapping them under the snow sheet. The snow surface generally remains undisturbed for no more than one week on the average (21). Results suggest that during heavy snowfalls, the atmospheric concentration levels for almost all heavy metals are low or very low.

During the first campaigns, a broad view of the environment around the Italian new station was needed. A sampling scheme was adopted for the determination of more than 40 different heavy metals by INAA. Interestingly, concentration levels below the instrumental sensitivity were found. This was due to a sampling protocol (low volume samplers, 72-hr sampling), which provided insufficient material to analyze, and to the still pristine environmental conditions. In Table 14.5 are shown the atmospheric particulate matter elemental composition and the relative concentrations for the campaign 1986–1987. The 1987–1988 campaign was characterised by the building activities in the main camp. The sampler installed at the edge of the main camp was able to signal such a situation. Indeed, the concentration levels for all the heavy metals were significantly higher for the samples collected by the baseline sampler installed at Skua Lake (Table 14.6). The quality of these data relative to the 1987–1988 campaign was confirmed by the data from the successive campaign (1988–1989), during which the building activities were reduced

(Table 14.7). During the 1989–1990 campaign both low and high volume samplers were used. It is clear that the two sets of data cannot be compared because of the different sampling parameters. For the first time data for Pb are available at two sites, i.e., South Base and Eneide, with high volume samplers (Table 14.8). The

Table 14.5. Atmospheric particulate matter elemental composition and the relative concentrations (Skua Lake, 1986–1987)

Element	Concentration (ng/m^3)	Element	Concentration (ng/m^3)	Element	Concentration (ng/m^3)
Al	61 ± 30	Fe	<100	Rb	<0.10
Ce	0.1 ± 0.05	La	<0.1	Sb	<0.1
Co	<0.1	Mn	<0.2	Sc	<0.1
Cr	<0.5	Na	217 ± 46	V	<0.2

Table 14.6. Atmospheric particulate matter elemental composition and the relative concentrations (Skua Lake, Italian station 1987–1988)

Element	Concentration (ng/m^3) Skua Lake	Concentration Italian station	Element	Concentration (ng/m^3) Skua Lake	Concentration (ng/m^3) Italian station
Al	110 ± 20	460 ± 50	Na	230 ± 30	435 ± 45
Ce	0.3 ± 0.1	0.7 ± 0.2	Rb	<0.10	0.6 ± 0.3
Co	0.08 ± 0.01	0.3 ± 0.05	Sb	1.7 ± 0.4	3.4 ± 0.5
Cr	4.0 ± 0.4	6.9 ± 0.7	Sc	0.008 ± 0.001	0.04 ± 0.01
Fe	70 ± 10	315 ± 30	Th	<0.02	0.05 ± 0.01
La	<0.1	0.4 ± 0.1	V	<0.2	0.4 ± 0.1
Mn	1.4 ± 0.4	12 ± 4	Zn	–	23 ± 2

Table 14.7. Atmospheric particulate matter elemental composition and the relative concentrations (Skua Lake, Italian station, 1988–1989)

Element	Concentration (ng/m^3) Skua Lake	Concentration Italian station	Element	Concentration (ng/m^3) Skua Lake	Concentration (ng/m^3) Italian station
Al	32 ± 9	70 ± 10	Na	230 ± 40	127 ± 30
Ce	0.21 ± 0.04	0.4 ± 0.1	Rb	nr	nr
Co	0.06 ± 0.01	0.2 ± 0.02	Sb	<3.0	3.0 ± 0.5
Cr	2.5 ± 0.3	1.0 ± 0.1	Sc	0.006 ± 0.001	0.03 ± 0.01
Fe	42 ± 5	77 ± 10	Th	0.02 ± 0.01	<0.04
La	0.15 ± 0.02	0.1 ± 0.01	V	nr	nr
Mn	nr*	nr	Zn	–	8.1 ± 0.9

nr, not reliable

Table 14.8. Atmospheric particulate matter elemental composition and the relative concentrations (1989–1990 campaign)

Element	Concentration (ng/m^3) Skua Lake	Concentration (ng/m^3) East Base	Concentration (ng/m^3) Eneide	Concentration (ng/m^3) South Base
Co	0.12 ± 0.05	0.20 ± 0.10	0.20 ± 0.10	0.20 ± 0.10
Cs	<0.01	<0.01	0.03 ± 0.01	0.04 ± 0.01
Cu	<20	<20	–	84 ± 16
Fe	90 ± 20	380 ± 50	110 ± 30	390 ± 40
K	190 ± 80	710 ± 160	<300*	330 ± 110
La	0.2 ± 0.1	0.4 ± 0.1	<0.1	0.5 ± 0.1
Mn	1.3 ± 0.8	4.5 ± 2.7	12 ± 2.0	16.4 ± 1.6
Na	450 ± 70	780 ± 90	<500*	<500
Pb	–	–	0.87 ± 0.05	4.87 ± 0.10
Sc	0.03 ± 0.01	0.11 ± 0.01	0.04 ± 0.01	0.12 ± 0.02
V	1.4 ± 0.4	2.4 ± 0.8	4.7 ± 0.5	4.1 ± 0.5
Zn	6.3 ± 0.8	10.8 ± 1.4	18.7 ± 2.5	27.5 ± 2.8

* At Eneide and South Base high volume samplers were used, while at Skua Lake and East Base low volume samplers were still in use (see above). As a consequence, different background levels, due to different filters and different detection limits, were measured.

concentration of V increased in comparison with those of the previous summer with the highest mean value at Eneide (high volume sampler) and was in good agreement with what happened at this site during the campaign (building activities very close to the site). The Pb concentration showed two significant peaks. The first (14–17 January 1990 sampling period) for both sampling sites, of 8.07 and 5.85 ng/m^{-3}, respectively, coincided with significant peaks for Zn and V. The second Pb peak of 9.70 ng/m^{-3} (1–5 February 1990 sampling period) is also supported by a significant Zn peak, but crustal elements like Fe do not show peaks in this period. Throughout the summer the concentration values for heavy metals at South Base were higher than those at Eneide. These facts have led us to think that the sampling system is able to detect events during which some heavy metals increase their concentration in the atmosphere.

During the 1990–1991 summer for the first time the main camp was monitored by four high volume samplers. In Figures 14.2, 14.3 and 14.4 Fe, Pb and Zn concentration trends for all four samplers are shown, respectively. The results obtained are largely inconclusive, but there is some evidence of an incinerator running during the 17–20 January 1991 sampling period and the Pb response for at least two samplers (South and West Base). The different behaviours for Fe and Pb are probably due to different origins, with Fe originating from the local rocks. There are some similarities between Pb and Zn trends, although insufficient for further discussions.

The 1991–1992 campaign was characterised by heavy snowfalls, which reduced all activities at the TNB station and created specific conditions which could yield

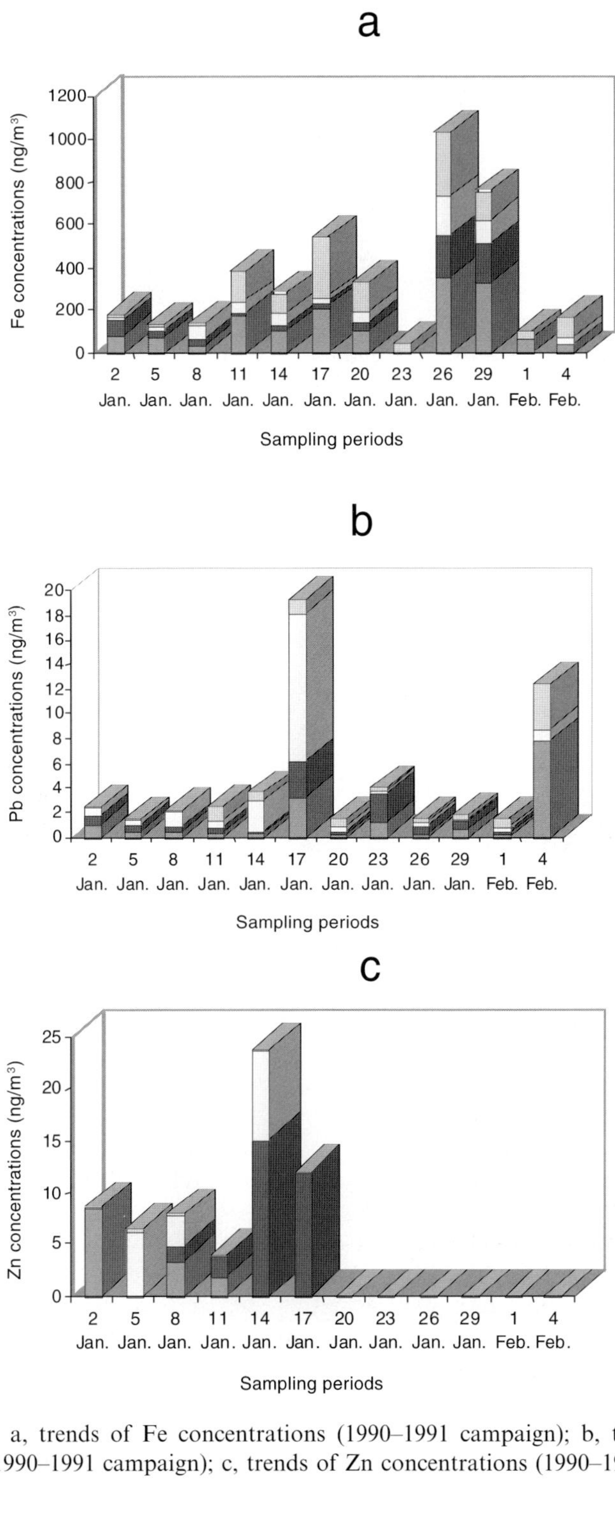

Figure 14.2. a, trends of Fe concentrations (1990–1991 campaign); b, trends of Pb concentrations (1990–1991 campaign); c, trends of Zn concentrations (1990–1991 campaign).

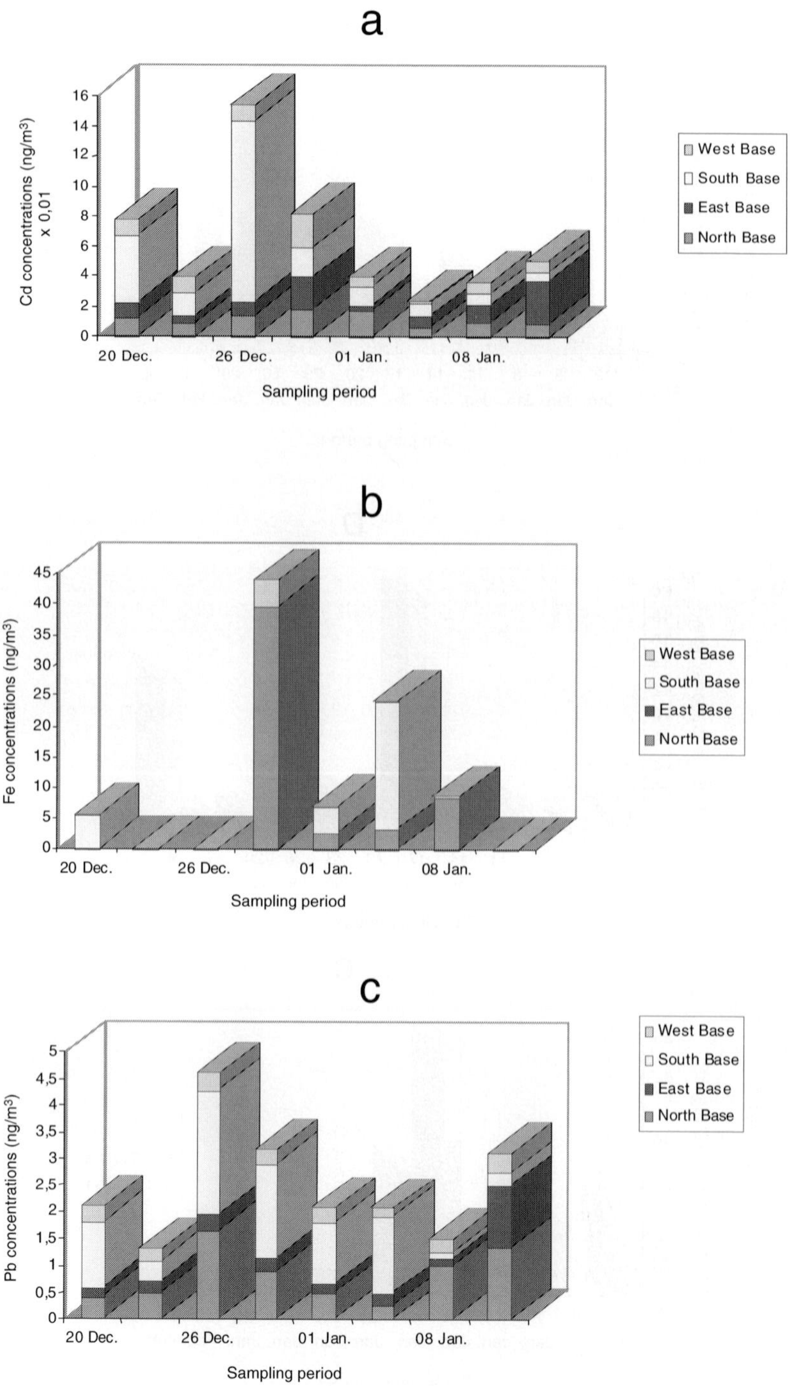

Figure 14.3. a, trends of Cd concentrations (1991–1992 campaign); b, trends of Fe concentrations (1991–1992 campaign); c, trends of Pb concentrations (1991–1992 campaign).

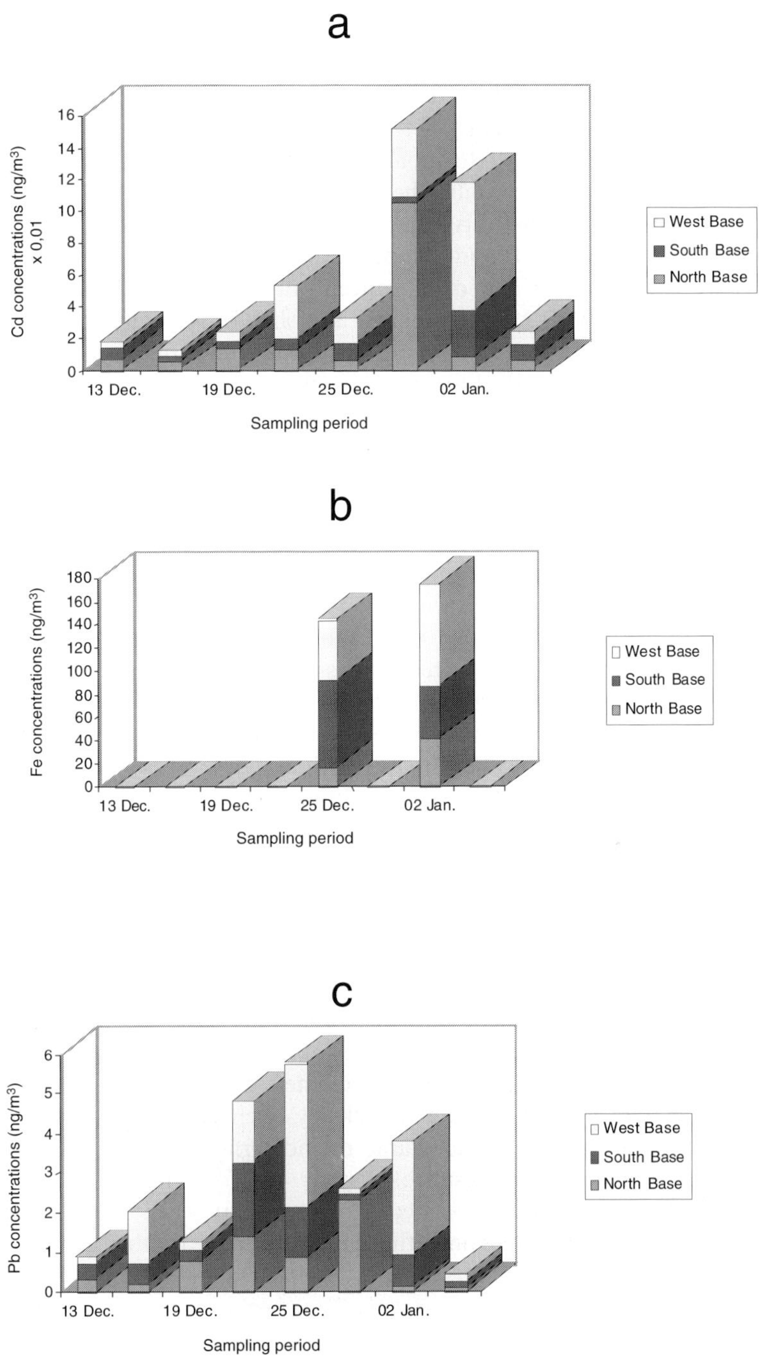

Figure 14.4. a, trends of Cd concentrations (1992–1993 campaign); b, trends of Fe concentrations (1992–1993 campaign); c, trends of Pb concentrations (1992–1993 campaign).

low or very low heavy metal concentration levels in the atmosphere. The analytical results confirmed these concentration levels, often lower than the analytical detection limits (Figure 14.5 a, b and c). The synchronization between the Cd and Pb is evident. It is thought that the origin for Cd and Pb was the same and different from that of Fe. The 1992–1993 campaign had a very reduced human activity at the TNB station (maintenance only, the campaign lasted from 13 December 1992 to 10 January 1993). During this campaign the East Base sampler was out of operation. In Figure 14.6 a, b and c concentration trends for Cd, Fe, and Pb are shown. In Figure 14.7 a, b and c Pb concentration trends relative to the 1993–1994, 1994–1995 and 1995–1996 campaigns are shown, respectively, as well as to the four samplers around the main camp. The only thing one can say is that the Pb concentration levels, for all three above-mentioned campaigns, lay in the same intervals as before (referring to the Italian station).

10. Data management

There are a number of basic requirements for any Antarctic data management system. One is an international agreement on the protocols for data collection and sample analysis. Another should be the establishment of international standards and of intercalibration activities. Quality Assurance (QA) and Quality Control (QC) play an extremely important role in the management of data systems. The road to establishing such a data management system covering all Antarctic data is long; however, a good start has been made, with a high level of international cooperation. A number of countries are developing Antarctic environmental data systems for their own data (among them Argentina, Chile, Italy, New Zealand, UK and USA). These data will in due course be listed in the Antarctic Metadata Directory.

In the Italian programme QA-QC considerations have been kept at the forefront of monitoring activities, especially with respect to the handling, shipping, storage of samples, the verification of laboratory activity, the calibration and control of measuring and testing equipment and, last, but not least, the qualification and training of the personnel involved in all stages of monitoring.

A non-conformity control is being implemented gradually. By non-conformity control is meant the verification of any deficiency in characteristics, documentation and procedures which makes unacceptable or indeterminate the quality of equipment, materials, samples or the result of a study or of an analysis. This may then entail corrective actions. This is particularly important when using facilities not uniquely dedicated to the monitoring effort, but shared with other analytical work, as is the case with the PNRA.

11. Conclusions

The Italian Antarctic research programme has been conducted by the PNRA following sound environmental principles. In designing the programme the PNRA

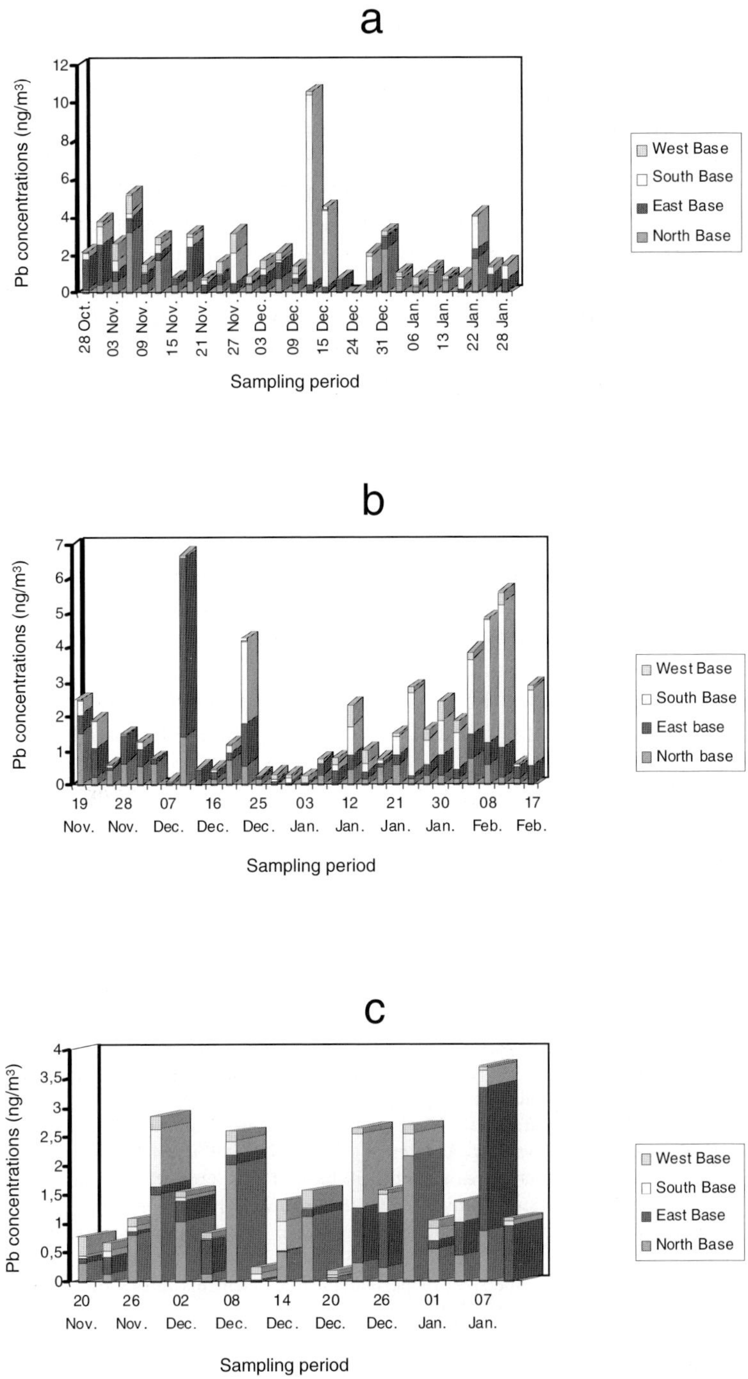

Figure 14.5. a, trends of Pb concentrations (1993–1994 campaign); b, trends of Pb concentrations (1994–1995 campaign); c, trends of Pb concentrations (1995–1996 campaign).

built on the experience of other Antarctic nations in many fields. ENEA-ANTAR, the government agency responsible for the PNRA implementation, was staffed mainly from the ENEA environmental department, which was why waste management and environmental monitoring were undertaken from the beginning of Antarctic operations, well before the consultation on the Madrid Protocol.

Although sound environmental management has absorbed human and budgetary resources, this has been advantageous not only in ensuring protection for a unique environment, but also in station management terms. A waste management plan means less waste, a cleaner site, better organized actions, less material to transport and a better exploitation of resources. The sewage treatment plant contributes to the cleanliness of sea water in the station area and makes possible marine biological research activity. Monitoring has provided data for feedback action on a number of activities, such as the adjustment of the diesel generators exhaust systems, the tuning of the engines and the continuous adjustment of the sewage treatment plant.

The data obtained on heavy metal concentrations in the air at TNB provided information on natural and/or anthropogenic events. The combination of three analytical techniques (INAA, ETA-AAS and HPLC) has proved important in the evaluation of the environmental state of the atmosphere where concentrations are so low. Finally, the elaboration and the analysis of the monitoring data continues; therefore, more definitive conclusions may be available at a later stage. The Madrid Protocol also requires the preparation of emergency plans and this contributes to the safety of stations. From the Italian experience it can be said that the Madrid Protocol is a major contribution to the protection of Antarctic environment, to the efficient management of research stations and to the safety of Antarctic personnel.

References

1. F. Francioni, T. Scovazzi, *International Law for Antarctica*, 1996, Kluwer Law International.
2. *Handbook of the Antarctic Treaty System*. 8th Edition, 1994, U.S. Department of State, Washington, USA.
3. L. Pineschi, *La protezione dell'ambiente in Antartide*, 1993, CEDAM, Padova, Italy.
4. R. A. Duce, G. L. Hoffman, W. H. Zoller, Atmospheric trace metals at remote northern and southern hemisphere sites: pollution or natural?, *Science*, 187 (1975), 59–61.
5. S. Guerzoni, R. Lenaz, G. Quarantotto, M. Taviani, Geochemistry of airborne particulate from the lower troposphere of Terra Nova Bay, Antarctica, *Tellus*, 44B (1992), 304–310.
6. P. R. Harrison, J. W. Winchester, Area-wide distribution of lead, copper and cadmium in air particulate from Chicago and Northwest Indiana, *Atm. Environm.*, 5 (1971), 863–880.
7. R. J. Lontzy, F. T. McKenzie, Atmospheric trace metals: global cycles and assessment of man's impact, *Geochim Cosmochim. Acta*, 43 (1979), 511–525.
8. J. B. Milford, C. I. Davidson, The sizes of particulate trace elements in the atmosphere: a review, *J. Air Poll. Cont. Ass.*, 35, 12 (1985), 1249–1260.
9. P. Mittner, D. Ceccato, S. Del Maschio, A preliminary characterization of the elemental composition of the aerosol coarse fraction at Terra Nova Bay (Antarctica) during the 1990-1991 austral summer, *Int. J. Environm. Anal. Chem.*, 55 (1994), 319–329.
10. N. Radlein, K. G. Heumann, Trace analysis of heavy metals in aerosols over the Atlantic Ocean from Antarctica to Europe, *Int. J. Anal. Chem.*, 48 (1992), 127–150.

11. J. E. Fergusson, *The heavy elements: chemistry, environmental impact and health effects*. 1991, Pergamon Press, Oxford, 614.
12. Z. Javorowski, M. Bysiek, L. Kownacka, Flow of metals into the global atmosphere, *Geochim. Cosmochim. Acta*, 45 (1981), 2185–2199.
13. T. A. McMahon, P. J. Devison, Empirical atmospheric deposition parameters: a survey, *Atm. Environm.*, 13 (1979), 571–585.
14. M. Murozumi, T. J. Chow, C. Patterson, Chemical concentration of pollutant lead aerosols, terrestrial dust and sea salts in Greenland and Antarctic snow strata. *Geochim. Cosmochim. Acta*, 33 (1969), 1247–1294.
15. G. Tuncel, W. H. Zoller, Sources of the Antarctic aerosols, *Boll. Geofis.*, Università degli Studi di Roma "La Sapienza", (1993), 447–471.
16. C. F. Boutron, C. Lorius, Trace metals in Antarctic snow since 1914, *Nature*, 277 (1979), 551-554.
17. C. F. Boutron, C. C. Patterson, The occurrence of lead in Antarctic recent snow, firn deposited over the last two centuries and prehistoric ice, *Geochim. Cosmochim. Acta*, 47 (1983), 1355–1368.
18. J. M. Palais, B. W. Macher, Elemental tracers in volcanic emissions in Antarctic aerosols and snow samples, *Ant. J. Rev.*, (1989), 217–218.
19. C. F. Boutron, E. W. Wolff, Heavy metals and sulphur emissions to the atmosphere from human activities in Antarctica, *Atm. Environm.*, 23, 8 (1989), 1669–1675.
20. E. D. Suttie, E. W. Wolff, Seasonal input of heavy metals to Antarctic snow, *Tellus*, 44B (1992), 351–357.
21. E. D. Suttie, E. W. Wolff, The local deposition of heavy metal emission from point sources in Antarctica, *Atm. Environm.*, 27A 12, (1993), 1833–1841.

Environmental Contamination in Antarctica
A Challenge to Analytical Chemistry
S. Caroli, P. Cescon and D.W.H. Walton, editors

Chapter 15

The duty of prior environmental impact assessment of Antarctic activities under the Madrid Protocol and its implementation in the Italian legal system

Laura Pineschi

1. Introduction

Environmental Impact Assessment (EIA) procedures are generally recognized as a useful and, in certain cases, indispensable instrument for a sound protection of the environment. As the United Nations Economic Commission for Europe (ECE)[1] recently observed:

> *Environmental impact assessment has already shown its value for implementing and strengthening sustainable development, as it combines the precautionary principle of preventing environmental damage and also arranges for public participation. EIA has also become the major tool for an integrated approach to the protection of the environment since it requires a comprehensive assessment of the impacts of an activity on the environment, contrary to the traditional sectoral approach. Moreover, EIA requires the formulation of alternatives to the proposed activity and brings facts and information on environmental impacts to the attention of the decision-makers and the public.*

Today EIA is no longer the domestic affair it used to be in the origin. EIA procedures are also regulated by a few important instruments adopted at the international level. Directive 85/337, concerning the assessment of the effects of certain public and private projects on the environment, was enacted by the European Community (EC) Council on 27 June 1985 (2). Directive 85/337 has recently been amended by Directive 97/11 of 3 March 1997 (3). Member States were requested to take the measures necessary to comply with the latter Directive by 14 March 1999 (4–8). On 25 February 1991, the United Nations ECE Convention on Environmental Impact Assessment in a Transboundary Context was concluded in Espoo (Finland) (7, 9). It entered into force on 10 September 1997. As of September 1998, Albania, Armenia, Austria, Bulgaria, Canada, Croatia, Denmark, Finland, Greece, Hungary, Italy, Latvia, Liechtenstein, Luxembourg, Netherlands, Norway, Poland,

[1]The ECE was established by the United Nations Economic and Social Council in 1947 and consists of the Eastern and Western European Members of the United Nations, the United States and Canada (1).

Republic of Moldova, Slovenia, Spain, Sweden, Switzerland, the United Kingdom and the European Community are parties to the Convention.

Within the Antarctic Treaty System (ATS) – comprehensive of the Antarctic Treaty (Washington, 1 December 1959), its associated separate international conventions and the recommendations adopted during the meetings of the Antarctic Treaty Consultative Parties (ATCPs) – several provisions on EIA have been enacted since 1975. The ATCPs are the States which can participate in the periodic meetings provided for under Art. XI of the 1959 Treaty, either as original parties or as parties carrying out substantial scientific activities in Antarctica. The ATCPs' recommendations become effective when approved by all the contracting parties whose representatives are entitled to participate in the meetings (Art. IX, para. 4) (10–14).

Some rules are rather vague and soft. For instance, the Code of Conduct for Antarctic Expeditions and Station Activities, annexed to Recommendation VIII-11 of 1975 (effective since 16 December 1978) and Recommendation XII-3 of 1983 (not effective yet), concerning man's impact on the Antarctic environment, establish a general obligation of prior EIA for a large category of Antarctic activities[2] alongside very few and vague procedural duties[3].

Other rules are more precise in their scope and content. Recommendation XIV-2 of 1987 (not yet effective at the moment of writing) specifically regulates the environmental impact assessment of scientific and logistic activities (15). Recommendation XIV-3 of 1987 (not yet effective at the moment of writing), providing safeguards for scientific drilling, and Recommendation XV-17 of 1989 (not yet effective at the moment of writing), concerning the establishment of new stations, impose additional obligations on the ATCPs and substantially reduce the discretion of states over the implementation of the ATCPs recommendations (16).

It is however with the conclusion of the Protocol on Environmental Protection to the Antarctic Treaty (Madrid, 4 October 1991, hereinafter called PEPAT), entered into force on 14 January 1998, that the ATS recorded noticeable progress in the regulation of EIA procedures. After the improvements introduced by PEPAT and its Annex I, specifically devoted to this matter, the ATS can be considered the most developed and detailed system of legally binding rules on EIA elaborated at the international level so far[4].

As of January 1998, all the ATCPs (Argentina, Australia, Belgium, Brazil, Chile, China, Ecuador, Finland, France, Germany, India, Italy, Japan, Republic of Korea, The Netherlands, New Zealand, Norway, Peru, Poland, Russian Federation, South Africa, Spain, Sweden, United Kingdom, United States, Uruguay) and Greece have ratified or approved the Protocol, but not all the contracting parties have adopted the same approach for giving effect to the provisions of PEPAT at the domestic level. Some states have enacted specific national provisions, on the

---

[2] Rec. VIII-11 applies to *major operations in the Antarctic Treaty Area*; Rec. XII-3 refers to *any scientific activity (...), including the planned provision of logistic facilities to support such activity.*

[3] See, e.g., Recommendation VIII-11, para. 4 (*a*) and Recommendation XII-3, para. 2.

[4] PEPAT is a legally binding instrument. According to Art. 26 of the 1969 Vienna Convention on the Law of Treaties: *Every Treaty is binding upon the parties to it and must be performed by them in good faith* (17).

assumption that their existing domestic laws and regulations cannot give adequate implementation to the obligations undertaken at the ATS level. Other parties have not enacted any specific national rule. This is the case of Italy, which implemented PEPAT by Law No. 54 of 15 February 1995, according to the so called *order of execution* (*ordine di esecuzione*) procedure (18). This procedure consists in the adoption of a law which provides for *the full and complete* execution of the treaty in question, whose text (in one of the official languages and in a non-official Italian translation) is reproduced in an annex to the law. This procedure is followed when the treaty is considered as *self-executing*, in the sense that it already contains all the elements which allow for its implementation in Italy and nothing more needs to be added by the legislator.

The assumption that PEPAT is a self-executing instrument is questionable. While many provisions of PEPAT are rather detailed, not all the obligations of PEPAT and its Annex I can be considered unambiguous and self-sufficient (19). Several aspects relating to the implementation of their provisions are unclear and problems of interpretation have already been highlighted by some ATCPs, Non-Governmental Organizations (NGOs) and distinguished writers.

In the following analysis, the Italian provisions on EIA will be scrutinized to determine whether the existing legislation is sufficient and adequate to give proper effect to PEPAT obligations. Before embarking on this survey, however, it is useful to illustrate briefly the scope and content of the obligations on EIA arising from PEPAT, as well as some limits, ambiguities and omissions.

2. The PEPAT regulation of the EIA procedure

In the text of the Madrid Protocol, the duty of prior EIA is included among the *environmental principles* referred to in Art. 3, para. 2 *c*). According to this provision:

> *Activities in the Antarctic Treaty area shall be planned on the basis of information sufficient to allow prior assessment of, and informed judgments about, their possible impacts on the Antarctic environment and dependent and associated ecosystems and on the value of Antarctica for the conduct of scientific research.*

It is important to note that Art. 3 of PEPAT directly provides a quite detailed list of the factors that the *informed judgments* on possible impact on the Antarctic environment shall take into full account. From this it can be inferred that adequate informed judgments are needed in every case and for every activity (therefore, not only for scientific research activities, but also for tourist and any other governmental and non-governmental Antarctic activities), as a compulsory prerequisite in order to determine the degree of the impact on the environment of a given activity (13, 20, 21). From the duty of prior EIA are excluded activities of exploitation of Antarctic seals and marine living resources, which are regulated by the Convention on the Conservation of Antarctic Seals (London, 1 June 1972) and the Convention on the Conservation of Antarctic Marine Living Resources (Canberra, 20 May 1980).

The unhappy wording of Art. 1, para. 1, of Annex I (*The environmental impacts of proposed activities referred to in Article 8 of the Protocol shall, before their commencement, be considered in accordance with appropriate national procedures)* does not consequently mean that national procedures may disregard the substance of informed judgments in question, but only that the procedural machinery to implement the evaluation of the impacts is left to the domestic legislation of the parties.

The activities fall into three categories, in the light of the informed judgments mentioned in Art. 3, according to whether they have: *i*) less than a minor or transitory impact; *ii*) a minor or transitory impact; *iii*) more than a minor or transitory impact. The consequences are the following: free proceeding, if the impact is less than minor or transitory; Initial Environmental Evaluation (IEE), if the impact is minor or transitory; Comprehensive Environmental Evaluation (CEE), if the impact is more than minor or transitory. It seems to be understood (but the correct interpretation of the conjunction/disjunction "or" is almost always a legal puzzle) that a less than a minor, but permanent, impact, as well as a transitory, but more than a minor, impact, lead directly to a CEE.

According to Art. 2 of Annex I to PEPAT, the IEE shall contain at least two elements, namely: *i*) a description of the proposed activity, including its purpose, location, duration and intensity; and *ii*) consideration of alternatives to the proposed activity and any impacts that the activity may have, including consideration of cumulative impacts in the light of existing and known planned activities. The basic elements of any IEE are therefore directly established by PEPAT. This is a relevant difference in comparison with Recommendation XIV-2, which did not lay down any specific procedural duty.

Under Art. 3, para. 2, of Annex I to PEPAT particular attention is paid to possible flaws in preparing the CEE by requesting *a description of the methods and data used to forecast the impacts of the proposed activity*, as well as an *identification of the gaps in knowledge and uncertainties encountered in compiling the information required.* This facilitates scrutiny and enquiries from both experts and non-experts, who must be informed on the name and address of the person or organization which prepared the CEE. The latter may also rely on a non-technical summary of the information provided.

Particularly important are the provisions regulating public participation in the EIA process. These provisions are much broader and more detailed than those contained in Rec. XIV-2. Under the latter, the obligation of information on CEE are conceived as a set of bilateral obligations between interested States, without any right of participation of the public. The draft CEE is made publicly available and is circulated to all parties, which also derestrict for comment (Annex I, Art. 2, para. 3). The final CEE addresses and includes, or summarizes, comments received on the draft CEE. It is made publicly available at least 60 days before commencement of the proposed activity, together with notice of any decisions and any evaluation of the significance of the predicted impacts in relation to the advantages of the proposed activity.

PEPAT also introduces a sort of international scrutiny on EIA within the Antarctic system. The draft CEE is forwarded to the Committee for Environmental Protection (CEP), an intergovernmental body to be established according to Art. 11

of PEPAT. The Committee has, *inter alia*, the function of providing advice on the application and implementation of EIA procedures and on means of minimizing or mitigating environmental impacts of activities (Art. 12, para. 1, *d*) and *e*), of PEPAT). No final decision is taken to proceed with the proposed activity unless there has been an opportunity for consideration of the draft CEE by the Antarctic Treaty Consultative Meeting (ATCM) on the advice of CEP (Annex I, Art. 3, paras. 4 and 5). However, no decision to proceed with a proposed activity shall be delayed for longer than 15 months from the date of circulation of the draft CEE

The final decision on whether to proceed with the activity in question, and, if so, whether in its original or in a modified form, pertains to the interested state. The idea of a prior collective authorization, proposed by some ATCPs during the negotiations of PEPAT, was not accepted (22). However, as any decision shall be based on the CEE as well as on other relevant considerations (Annex I, Art. 4), it is to be expected that the comments made by the public, the committee and the other states parties will have a major influence during the decision-making process.

International co-operation is promoted by providing for coordination in EIA procedures relating to jointly planned activities (Art. 8, para. 4), as well as by providing for appropriate assistance to other parties in the preparation of EIAs (Art. 6, par. 1, *b*). The wording of these provisions seems however rather vague.

Finally, all activities are subject to monitoring, in order to assess the impact that they effectively produce on the Antarctic environment once embarked upon. In this context, particularly relevant are Art. 3, para. 2 *d*), of PEPAT and Art. 5 of Annex I. The former refers to all activities, providing that *regular and effective monitoring shall take place to allow assessment of the impacts of ongoing activities, including the verification of predicted impacts.* According to this provision, if the IEE indicates that the proposed activity is likely to have no more than a minor or transitory impact, the activity may proceed, *provided that appropriate procedures, which may include monitoring, are put in place to assess and verify the impact of the activity.*

The verification of predicted impacts specifically addresses activities undertaken following the completion of a CEE. In particular it envisages the possibility of a suspension, cancellation or modification of the activity, on the basis of regular and verifiable records of the impacts of the activity.

3. A critical assessment

A closer look at the scope of PEPAT and the content of its obligations can be useful to better understand the matters where the intervention of the national legislator is necessary. This also makes it possible to assess the degree of discretion which is left to the contracting parties in the establishment of domestic procedures to ensure compliance with the Protocol.

3.1. Determination of the person in charge of the EIA

One of the points which are not specifically addressed by PEPAT is who is to take charge of undertaking the EIA. This important question is left to the sphere of the domestic jurisdiction of every contracting party. This could be a consequence of the very different solutions followed by different national legislations. In certain cases the EIA is to be made by a state organisation, in others by the developer (i.e., the applicant for authorization for a project), or by an independent firm of environmental consultants on behalf of the developer. Any of these solutions seems to be compatible with PEPAT (as a matter of fact, it could be difficult to find an independent consultant firm with adequate experience of the Antarctic environment). On this point, a specific enactment by contracting parties seems hardly avoidable. Legislation already adopted for projects to be carried out in the national territory cannot be easily presumed to meet the particular conditions of the Antarctic continent and environment.

3.2. Determination of the scope of EIA

As regards the objectives of EIA procedures under PEPAT, three points deserve particular attention. First, as a distinguished writer has already observed, the PEPAT is not aimed at prohibiting any detrimental impact on the Antarctic environment and its dependent and associated ecosystems, but just at limiting it (13). Only certain provisions of the Antarctic Treaty or PEPAT prevent particularly serious or irreversible impacts. For instance, the concern detrimental effects on the Antarctic environment which can be caused by nuclear explosions and the disposal of radioactive waste material in the Antarctic Treaty area can be inferred by Art. V of the Antarctic Treaty, which forbids these activities with no exceptions at all. Moreover, in Art. 3, para. 2 *b*), PEPAT lists six specific consequences which must be totally avoided: *i*) adverse effects on climate or weather patterns; *ii*) significant adverse effects on air or water quality; *iii*) significant changes in the atmospheric, terrestrial (including aquatic), glacial or marine environments; *iv*) detrimental changes in the distribution, abundance or productivity of species or populations of species of fauna and flora; *v*) further jeopardy to endangered or threatened species or populations of such species; or *vi*) degradation of, or substantial risk to, areas of biological, scientific, historic, aesthetic or wilderness significance.

However, the terms of the rule include terms, such as 'significant' and 'substantial' which makes the obligation less than absolute (13). This leaves the Parties a certain freedom of interpretation in the drawing up of their domestic legislations.

Second, only activities (and not plans or programmes) are subject to the EIA procedures set out by PEPAT and its Annex I. The application of EIA procedures to policies, plans and programmes has recently been considered at the international level as a substantial way for strengthening the correct environmental management (23, 24). However, in the assessment of the proposed activities, the consideration of their cumulative impact on the Antarctic environment in the light of existing and known planned activities is obligatory under Arts. 2 (IEE) and 3 (CEE) of Annex I to PEPAT. The obligation is clear and seems also self-executing. The importance of identification and consideration of possible cumulative impacts has been repeatedly stressed by ATCPs (25, 26). An interesting discussion on practical aspects of minimization and management of cumulative impact in Antartica can be found in the quoted documents (24, 27).

Third, the scope of the prior assessment of proposed Antarctic activities includes, besides the impact on the Antarctic environment and dependent and associated ecosystems, also the consideration of possible effects on *the value of Antarctica for the conduct of scientific research* (Art. 3, para. 2 *c*)). This objective is unusual if compared with other international and domestic instruments providing for EIA procedures. It is however perfectly understandable in the ATS context, where *the establishment of a firm foundation for the continuation and development of [scientific] co-operation* is one of the primary objectives of the system. The Preamble to the Antarctic Treaty states that *the substantial contributions to scientific knowledge resulting from international co-operation in scientific investigation in Antarctica* are acknowledged. Again, the obligation is clear and seems also self-executing. An invitation to give particular consideration to this objective can be inferred by the statement contained in the Final Report of the XX ATCM, where *the Meeting reiterated that in considering potential impacts on the environment during the EIA process, the values as mentioned in Article 3, paragraph 1 of the Protocol should be included* (para. 136). Among the intrinsic values of Antarctica, Art. 3, para. 1, of PEPAT includes *its value as an area for the conduct of scientific research, in particular research essential to understanding the global environment.*

3.3. The threshold of the adverse impact

PEPAT does not contain a definition of *adverse impacts on the Antarctic environment and its dependent and associated ecosystems* (Art. 3, para. 2 *a*). *Impact* and *transboundary impact* are defined in the Espoo Convention in Art. 1, sub-paras. *vii* and *viii* (28). The environmental principles that according to Art. 3 must be *fundamental considerations in the planning and conduct of all activities in the Antarctic Treaty area* are very wide in their scope. PEPAT does not supply a list of mandatory and non-mandatory projects to ensure that the most dangerous activities will always be subject to IEE or CEE (28, 29), nor does it provide guidance for a uniform interpretation of the terms *minor* and *transitory impact* contained in Arts. 1, 2 and 3 of Annex I.

The latter omission already raised discussions and proposals during the most recent ATCMs (30). For instance, on the assumption that *Minor[ness]* related to *magnitude* of the impact and *Transitory[ness]* to *duration* of the impact, New Zealand proposed the following scheme.

Duration

- very short duration (< 1 to 100 days);
- short duration [month(s) to a year – the Antarctic season perhaps];
- extended duration [year(s) to decades] with the full recognition that the actual scale used in a real situation will vary with circumstances;

Magnitude

- very low magnitude (undetectable change within or close to natural variability);
- low magnitude (detectable, i.e., above natural variability);
- appreciable magnitude (detectable and well above natural variability by a given

amount – say N% of a given parameter such as population, area, quality, etc.).

It was also proposed the establishment of a network of Environmental Officers of National Antarctic Programmes to enhance the mutual understanding of EIA processes (23, 31).

The aim of New Zealand's proposal is to clarify a vague aspect of PEPAT. This could be a first step towards a common understanding of the terms *minor* and *transitory*, which is highly desirable. Under PEPAT the final decision on whether to proceed with the activity is remitted to the interested state which, despite criticisms or contrary advice, is not barred from proceeding with the controversial activity. According to Antarctic Southern Ocean Coalition (ASOC), *the sole option remaining to halt an activity is binding dispute settlement; (....) an option that is not likely to be used* (32). It can be interesting to recall that under the Espoo Convention if the parties cannot agree whether a proposed activity listed in Appendix I is likely to produce a significant adverse impact, each of them may submit the question to an inquiry Commission in accordance with the provisions of Appendix IV (Art. 3, para. 7). The final opinion of the Commission (*based on accepted scientific principles*, Appendix IV, para. 14) does not produce binding effects on the parties. The conclusion of the inquiry procedure simply consists in the transmission of the final opinions of the Commission to the parties to the Secretariat (Appendix IV, para. 14). However, the Commission has rather wide powers of inquiry (*the inquiry commission may take all appropriate measures in order to carry out its functions*, Appendix IV, para. 6) and can render decisions by default (*absence of a party or failure of a party to present its case shall not constitute a bar to the continuation and completion of the work of the inquiry Commission*, Appendix IV, para. 9). Therefore, until an agreement is reached among the parties on the interpretation or the application of PEPAT or a practice is asserted in the application of this instrument *which establishes the agreement of the parties regarding its interpretation*, the parties maintain a wide range of discretion in the choice of the methodology to be applied to identify the appropriate level of EIA for different planned activities, as prescribed by the 1969 Vienna Convention on the Law of Treaties (Art. 31, para. 3b). As it has been already pointed out, PEPAT employs the terms of *minor* and *transitory* to identify activities as potentially having three different levels of harmful impact, i.e., less than a minor or transitory impact, minor or transitory impact, or more than a minor or transitory impact.

There is, however, at least one limit that national legislations cannot ignore. The wording of PEPAT and Annex I clearly suggests that the effect of the proposed activities must be assessed on a case-by-case basis. This implies that a domestic legislation providing for a pre-established list of activities which are *ipso jure* excluded from the duty of prior IEE or CEE cannot be considered consistent with the text and the spirit of the Protocol. On the contrary, in the absence of an agreement or an accepted practice of the parties, nothing could prevent the national legislator from providing guidance to public and private bodies with the indication of methodologies for measuring the duration and the magnitude of the potential impact of different planned activities.

3.4. Exemptions

Unlike other international instruments, Annex I to PEPAT does not contain general exemptions for particular projects (28). Only in *cases of emergency*, predetermined by Annex I (safety of human life or of ships, aircraft or equipment and facilities of high value, or the protection of the environment) can an activity be undertaken without completion of EIA procedures (Art. 7, para. 1). A duty of publicity is established by Art. 7, para. 2, for activities which are undertaken without a prior CEE, because of a state of emergency. Wider exemptions are expressly provided for by EC Directive 85/337 (see, e.g., Art. 1, paras. 4 and 5 and Art. 2, para. 3) (29). The latter Article has been amended by Directive 97/11 in order to exclude any prejudice to the duties of international co-operation expressed under Art. 7. As far as the public participation is concerned, the national authorities are allowed by the EC Directive to *respect the limitations imposed by national regulations and administrative provisions and accepted legal practices with regard to industrial and commercial secrecy [including intellectual property, according to EC Directive 97/11] and the safeguarding of public interest* (Art. 10).

The list of situations envisaged by PEPAT in Annex I is expressed in mandatory terms (*This Annex shall not apply*) and is clearly exhaustive, not illustrative. Any national provision excluding certain categories of projects from the duty of prior EIA and any rule envisaging emergency cases which are not mentioned in Art. 7 of Annex I must be considered inconsistent with PEPAT.

3.5. Public participation

From Art. 3, para. 3, of Annex I to PEPAT it should be inferred that the right for public comment and scrutiny extends to all individuals, whatever their nationality:

> *The draft Comprehensive Environmental Evaluation shall be made publicly available and shall be circulated to all Parties, which shall also make it publicly available for comment.*

As a matter of fact, in Antarctica, where there are no native inhabitants, any restrictive definition of *public participation* does not make sense. In the case of Antarctica, where the protection of the environment is *in the interest of mankind as a whole* (as stated in the preamble of PEPAT), the interested public is every individual and every NGO. When a general interest is in question and nobody is directly affected, everybody is affected.

The only procedural limit explicitly set forth in Annex I, is the period of time allowed for the receipt of comments, i.e., 90 days (Art. 3, para. 3). The domestic authorities are entitled to choose the most appropriate ways for providing the public with the information gathered for the preparation of draft CEEs and collecting the opinions expressed. However, it would not be possible to consider the duty established by PEPAT as properly complied with if the national legislator assumed a mere formalistic approach, without giving the public any clear, concrete and adequate opportunity for participation.

Extremely vague, under Art. 6, para. 2, of Annex I, is the duty to provide information in case of IEE (*any Initial Environmental Evaluation prepared in accordance with Article 2 shall be made available on request*). From the wording of the provision it is impossible to infer whether information on request is to be provided only to the other Parties or also to the public. The most restrictive approach, however undesirable, does not seem in conflict with Art. 6, para. 2.

No clear duty of public participation is expressed for the EIA in the preliminary stage. At this level, Annex I leaves the Parties completely free to assess the environmental impacts "in accordance with appropriate national procedures" (Art. 1, para. 1).

3.6. Monitoring

As it was stressed before, PEPAT and Annex I to PEPAT provide for a duty of monitoring the impact that ongoing activities effectively produce on the Antarctic environment. This is rather an important provision, since obligations on post-project analysis can hardly be found in other international instruments, including EC directives. According to Art. 7 of the Espoo Convention, a post-project analysis shall be carried out by the concerned Parties at the request of any Party. The discretion left to the Parties in the undertaking of monitoring activities is however considerable: *The concerned Parties, at the request of any such Party, shall determine whether, and if so to what extent.* There are no provisions on monitoring either in the EC Council Directive 85/337 of 27 June 1985 or in Directive 97/11 of 3 March 1997. The inclusion of an obligation to carry out a post-project analysis in the latter Directive was opposed by some EC Member States (7).

Unfortunately, however, PEPAT does not regulate the content of this obligation: according to Art. 5 of Annex I, procedures for monitoring *shall be put in place*. This probably means that the adoption of appropriate rules is deferred to a later, indefinite moment. So far, procedures for a post-project analysis have been agreed by the ATCPs (in the form, however, of a hortatory resolution) for activities for which a CEE has been prepared (33). The non-legally binding effect of the resolutions of the ATCPs can be inferred by Decision 1 (1995), adopted at the end of the XIX ATCM.

As far as the consequences at the national level are concerned, it is evident that there is a duty of monitoring possible unforeseen effects on the Antarctic environment and dependent and associated ecosystems. This obligation must be complied with for activities carried on both *within* and *outside* the Antarctic Treaty area. This is the wording of Art. 3, para 2 e) of PEPAT. However, as long as binding procedures on post-project analysis are not agreed upon at the international level, the contracting parties to PEPAT keep the right to assess the impact of ongoing activities according to their own national criteria, taking into consideration the peculiar character of the Antarctic environment and the spirit of obligations arising from PEPAT.

3.7. International co-operation

A number of recent international instruments provide for the requirement of international co-operation (information, consultation and negotiation) with other States during the EIA process. EC Directive 97/11 replaced the text of Art. 7 of the EC Directive 85/337 Article with a much more expanded regulation. Information that must be provided to another Member State likely to be affected by a proposed project is specified and provisions are set forth for specific arrangements to be made to consult other affected States and their public. A Member State can also express the desire to take part in the EIA process undertaken in another Member State. Detailed provisions on notification to and consultations with other parties during the EIA process are contained in the 1991 Espoo Convention (Arts. 3 and 5) (28). An inquiry commission can also be set up if the parties concerned cannot agree on the likelihood of a significant adverse transboundary impact (Art. 3, para. 7). In addition to this, provisions on exchange of information and consultations among the parties concerned may be found, e.g., in two other conventions (34, 35).

The practice of States in the implementation of duties of international co-operation in the case of projects which are likely to have significant effects on the environment of other States is rather disappointing. For instance, the first EC Commission's five year review on the implementation of EC Directive 85/337 within Member States clearly shows that inter-governmental co-operation is one of the fields where the implementation of the Directive has been most defective. In 1991, only a few Member States (Denmark, Germany, Greece, Ireland and Spain) had enacted rules on consultation of other Member States in case of significant transboundary impacts. According to information made available to the Commission, consultations on transboundary impacts had been undertaken for projects between Ireland and United Kingdom, Denmark and Germany, Spain and Portugal, Netherlands and its bordering States (36).

In Antarctica, where a long experience of fruitful co-operation among ATCPs is a matter of fact, the willingness of the Parties to make effective the PEPAT provisions on exchange of information on EIA is evident. In 1995, a Resolution on the EIA circulation of information among ATCPs was adopted at the end of the XIX ATCM. Under Resolution 6 (1995), the ATCPs should provide a list of IEEs and CEEs submitted to them during the preceding calendar year to the host Country of the next ATCM, not later than 1 March (37–39). The appeal to a prompt transmission of information on IEEs and CEEs prepared by contracting parties has been constantly reiterated during recent ATCMs.

Beyond the duty of transparency, however, there are also other important obligations of international co-operation established by PEPAT. For instance, Art. 6 provides for duties of assistance to other parties in the preparation of EIA (para. 1 *b*); sharing of information that may be helpful to other parties in planning and conducting their activity in the Antarctic Treaty area *with a view to the protection of the Antarctic environment and dependent and associated ecosystems* (para. 2); co-operation with parties which exercise jurisdiction in areas adjacent to the Antarctic Treaty area *with a view to ensuring that activities in the Antarctic Treaty area do not have adverse environmental impacts on those areas* (para. 3).

The effective implementation of these kinds of duties largely depends on the capacity and willingness of the institutions in charge of foreign relations of each State party to co-operate with other contracting parties. It cannot be ignored, however, that the dialogue at the international level is easier if national legislations are expressly drafted in order to facilitate international co-operation and specific provisions are introduced in order to better achieve this aim. Interesting provisions are, e.g., contained in the German implementing legislation of PEPAT of 22 September 1994. According to Art. 8 (permit procedure accompanied by CEE), the applicant must submit the results of an assessment of the activity and its environmental impact in the English and German languages. The assessment must include, *inter alia*, a description of the proposed activity, including the likely area of impact. The latter requirement, which is not mentioned in Art. 3 of Annex I to PEPAT (literally transposed in Art. 8 of the German law) should clearly facilitate the detection of adverse environmental impacts upon areas adjacent to the Antarctic Treaty area and subject to the jurisdiction of other State parties.

3.8. Responsibility and liability

The delicate issues of responsibility and liability arising from EIA procedures in Antarctica can be considered under two different aspects.

On the one hand, there are questions which fall under public international law and relate to international responsibility, i.e., the responsibility for wrongful acts arising when a State has infringed an international obligation (problems of international liability for injurious consequences arising out of acts not prohibited by international law could also occur, but they cannot be addressed in this chapter). It is well known that the matter of international responsibility is presently being codified by the United Nations International Law Commission on the basis of a draft approved in 1996 (40). The State held to have committed an internationally wrongful act is responsible *vis-à-vis* the injured State. In principle, the customary rules on international responsibility cover in general any breach to an international provision, including the provisions of PEPAT and Annex I to PEPAT. It follows that, if any provisions of PEPAT are breached or inadequately implemented within the domestic legislation, the State which has committed the internationally wrongful act is subject to all the consequences provided for by customary international law (such as cessation of wrongful conduct, reparation, restitution in kind, compensation, satisfaction, assurances and guarantees of non-repetition of the wrongful act).

It is also well known that, in the special case of Antarctica, the ATCPs are presently negotiating a draft liability annex to PEPAT. It is not possible to analyze its draft provisions hereunder (41–46). Suffice here to say that the draft annex is intended to include rules and procedures relating to liability for damage arising from activities taking place in the Antarctic Treaty Area and covered by PEPAT. When adopted and entered into force, the draft annex would prevail, as *lex specialis*, over the customary law provisions (*lex generalis*). The draft annex could also include, if agreed upon, a specific regime on responsibility and liability linked to EIA procedures. The notion of damage and its relation to the EIA procedure is dealt with in detail elsewhere (47, 48).

On the other hand, there are questions of liability for damages which usually fall under the tort law provisions of a specific domestic legislation. These questions arise among the persons involved in the EIA process (e.g., the developer, the private consultant firm, the State entity in charge of EIA, the injured person) and relate to the content, forms and degrees of liability, including the delicate questions of the determination of the amount of compensation and the possibility of compensation of the so-called environmental damage. Nothing prevents the draft liability annex to PEPAT from promoting the harmonization or unification of domestic legislations in this field and establishing a single set of rules which would cover the liability arising from EIA procedures in Antarctica. However, if this were not done by the ATS on an international basis, the domestic legislations adopted for the implementation of PEPAT should also cover the tort law aspects of EIA process in Antarctica. In this case, the need for *ad hoc* provisions seems hardly questionable.

4. Italian national legislation

In Italy, the provisions of PEPAT have been incorporated into the national system by means of a legislative act, Law No. 54 of 15 February 1995 (18). No further implementing measures have been enacted so far.

Italian general legislation on EIA is modelled on EC Directive 85/337. The application of PEPAT as *lex specialis*, instead of any other general and incompatible rule, is beyond any discussion. PEPAT aims at enhancing the protection of the very particular Antarctic ecosystem. Its provisions on EIA apply to human activities which prevalently take place in the Antarctic Treaty area and can produce adverse impacts on the Antarctic environment and its dependent and associated ecosystems.

It could be added that not only the aim, but also the spirit of the two instruments, are different. Suffice here to say that the EC Directive makes a distinction between EIA mandatory and EIA non-mandatory projects, while under PEPAT the effect of the proposed activities must be assessed on a case-by-case basis. The EC Directive provides for exemptions which do not even appear in the text of PEPAT or its Annex I.

Last but not least, the system of rules which implements the EC Directive in Italy is rather complicated, incomplete and transitory. The system is complicated, because several provisions have been subsequently adopted since the enactment by the EC Council Directive 85/337 (49–57). This causes evident problems of co-ordination among different sources of law. The system is also incomplete, because the competence to regulate EIA procedures has been recently delegated to local autonomous institutions. Presidential Decree of 12 April 1996, containing conditions, criteria and technical rules for the application of the EIA procedures to the projects listed in Annex II to the EC Directive 85/337, requires the Regions to adopt rules on EIA procedures for projects listed in its Annexes A and B within nine months of the date of its publication in the Italian Official Journal (58). The Decree gives implementation to Art. 40 of Law No. 146 of 22 February 1994, containing provisions for the enforcement of duties deriving from the participation of Italy in the European Communities (1993 Community Law). According to that Article, the Italian Government should have enacted rules to implement EIA duties with regard to projects listed in Annex II of the EC Directive within 60 days (*sic*!) from the entry into force of that Law. The preamble of the 1996 Decree recalls, *inter alia*, the need to implement *urgently and completely* the EC Directive, in consideration of a reasoned opinion given by the European Commission on 7 July 1993. In its opinion the Commission invited the Italian Government to adopt the necessary measures to subject projects listed in Annex II to a prior EIA when these projects can have a major environmental impact. The prospect of 21 different sets of regional legislation raises some doubts about the general consistency of the resulting picture.

Finally, the present regulation is transitory, since a bill containing a general and organic regime of EIA is under discussion in the Italian Parliament (59). The objective of the bill is to guarantee the certitude of law through a clarification and simplification of the national decision-making process. Furthermore, the bill is not limited to remedying the deficiencies of the past, as it contains some rules aimed at complying with EC Council Directive 97/11, which amends EC Directive 85/337. At the time of writing, the deadline (14 March 1999) for compliance with EC Directive 97/11 has not yet expired. The bill also incorporates in the EIA process some of the most innovative instruments of EC environmental law (e.g., the system of voluntary agreements with private enterprises) or provisions (e.g., on scoping and monitoring) which could not be included in the amending Directive because of lack of agreement among EC Member States (60, 61). Some provisions of the bill specify the national procedures which are to be followed to comply with the Espoo Convention, another international treaty which has been incorporated into the Italian domestic legislation by means of a simple order of execution (62).

All these remarks lead to some obvious conclusions. Under the pressure of the EC Commission control and the evolution of international treaty law or EC instruments on EIA, Italian legislation on EIA is presently under a process of re-organization and adaptation to more advanced standards of environmental protection. The results of this process may still be uncertain, as regards their consistency and effectiveness. What is however sure is that, unfortunately, there are no clear signals of parallel efforts aimed at incorporating into the domestic legislation all the specific rules required to give not only formal, but also substantive and effective implementation to PEPAT and its Annex I. At present there is only one statement from which it can be inferred that the Italian Government is aware of the need to adopt specific national regulations for giving effect to PEPAT properly. In the report annexed to the bill containing the ratification and the order of execution of PEPAT it is clearly said that the Madrid Protocol is only a framework agreement whose further specification and evolution is remitted to national measures of compliance (63).

Both the laws in force which regulate the EIA procedures for certain public and private activities in Italy and the bill which is, at the time of writing, under discussion at the Italian Parliament can hardly be useful to clarify the ambiguities of PEPAT or compensate the lack of precision of some of its provisions. From PEPAT, combined with the domestic provisions at present in force in Italy, it is difficult to detect, e.g., *who* is responsible for certifying and monitoring that a certain planned activity can have less than a minor or transitory impact, a minor or transitory impact, or more than a minor or transitory impact on the Antarctic environment and its dependent and associated ecosystems; *how* and *where* effective public participation in the EIA process is ensured and what happens if private or public bodies fail to comply with provisions on EIA for a planned Antarctic activity (64–68). In this context, e.g., the German implementing legislation of 22 September 1994 is relevant and, in particular, Art. 7, para. 4, regulating information to the public in case of IEE, Art. 9, containing measures on public inspection and objections in case of CEE and Art. 16, concerning EIA of other Parties. Under this Article, documents on EIA circulated by other Parties to PEPAT *should be laid out publicly at the headquarters of the Federal Environmental Agency for a period of three weeks* (para. 2) and *comments delivered within the prescribed period should be transmitted to the Parties concerned* (para. 3).

But even supposing that certain national provisions in force could apply through analogy to Antarctic activities, evident requirements of time- and cost-effectiveness and, above all, certainty of law impose the adoption of provisions specifically tailored for Antarctic activities. A few examples may be sufficient to illustrate this assumption.

Only one central, well identified authority should be responsible for application for, granting, revocation or suspension of a permit and the monitoring of permitted activities. A number of national legislations follow this approach, such as the Finland Act on the Environmental Protection of Antarctica of 18 October 1996 and the 1997 Japanese Law Relating to Protection of the Environment in Antartica (64–68). The involvement of different bodies in the decision-making process can only complicate and slow down the procedure. The principle of subsidiarity has no reason to be invoked in the case of Antarctic activities.

Sufficiently dissuasive sanctions should be conceived considering the particularly serious (or even irreversible) danger the Antarctic environment is exposed to by incautious action[5]. A guarantee fund could be established to recover liabilities the Italian Government could incur at the international level (as has already be done in the case of transboundary movements of hazardous wastes) (69–71).

Finally, some specific provisions are desirable in order to tackle the issue of the exercise of jurisdiction and control on human activities in the Antarctic Treaty area. This is a well-known, delicate and intricate problem, but the effective enforcement of PEPAT and its national implementing provisions inevitably run into this thorny issue. Moreover, according to Art. IX, para. 1*e*) of the Antarctic Treaty, ATCPs can adopt recommendations regarding, *inter alia*, *questions relating to the exercise of jurisdiction in Antarctica*. No measures dealing specifically with jurisdictional questions have been adopted so far (72).

[5] For example, under the New Zealand Antarctica (Environmental Protection) Act offences are differently regulated according to different activities or situations, in particular as regards Art. 15 (mineral resource activities), Art. 24 (EIA), Art. 33 (conservation of antarctic fauna and flora and protected areas), Art. 37 (waste disposal) and Art. 47 (offence to obstruct inspectors).

5. Concluding remarks

More than rights or privileges, participation in the ATS involves shouldering important responsibilities. The solemn recognition of some of these burdens was renewed on the occasion of the conclusion of PEPAT. The ATCPs have *a special responsibility* to ensure that all activities in Antarctica are consistent with the purposes and principles of the Antarctic Treaty. The development of a comprehensive regime for the protection of the Antarctic environment and dependent and associated ecosystems is *in the interest of the mankind.*

Compliance by States with international obligations relies, chiefly, on the behaviour of their bodies and the existence of appropriate national mechanisms aimed at controlling implementation of international commitments. If ATCPs (or only some of them) are not able to control public and private undertakings in Antarctica, the ambitious assertion of a special responsibility in the Antarctic Treaty area in the interest of mankind risks becoming a classic example of claimed authority without real accountability.

Failure to give the necessary or appropriate internal effect to the obligations of an international treaty can result in a breach of the treaty, which entails State responsibility under international law. Apart from any discussion on the efficacy of the mechanism of international responsibility for wrongful acts in general, it is well known that the traditional notion of State responsibility is completely inadequate as regards meeting the specific objectives of environmental protection (73–76). Furthermore, even if a liability regime for damage arising from activities covered by PEPAT is adopted within the ATS in the near future, this regime would not be able to fulfil its purpose if it were not coupled with adequate measures of prevention at the national level.

Once again the important role that State's bodies can play in the effective control on human activities in the Antarctic is evident. The physical peculiarities of Antarctica make the exercise of control in the Antarctic Treaty area a rather difficult task. Stringent control applied at the preparatory stage of these activities on the basis of clear and adequate national provisions can make this task easier. The effective protection of the Antarctic environment begins many miles away from Antarctica, where the capitals of the parties to PEPAT are located.

Acknowledgment

This research was carried out within the project National and International Legal Rules for the Protection of the Antarctic Environment, financed by the Italian National Research Programme in Antarctica (PNRA).

References

1. ECE, *Current Policies, Strategies and Aspects of Environmental Impact Assessment in a Transboundary Context*, United Nations Publication, Geneva, 1996.
2. *OJEC L 175*, 5 July 1985.
3. *OJEC L 73*, 14 March 1997.

4. A. Curatolo, in *Lessons from Europe '93,* H. G. Sevenster, A. Gasperi, F. Lulofs (Eds.), D. Anderson, Amsterdam, 1993, 25.
5. A. Nollkaemper, in *The Implementation of the EEC Directive 85/337/EEC on Environmental Impact Assessments,* H. Steiger (Ed.), Giessen, 1992, 149.
6. R. J. Cerny, W. R. Sheate, *Env. Policy Law,* 22 (1992), 154.
7. W. R. Sheate, *Eur. Env. Law Rev.,* 4 (1995), 77.
8. W. R. Sheate, *Eur. Env. Law Rev.,* 6 (1997), 235.
9. L. Pineschi, in *World Treaties for the Protection of the Environment,* T. Scovazzi, T. Treves (Eds.), Istituto per l'Ambiente, Milan, 1992, 485.
10. F. M. Auburn, *Antarctic Law and Politics.* C. Hurst & Company, London, 1982.
11. W. M. Bush, *Antarctica and International Law.* Oceana Publications, Inc., London, 1982, Vol. I.
12. A. Colella, in *International Environmental Law for Antartica,* F. Francioni (Ed.), A. Giuffrè, Milan, 1992, 203.
13. Sir A. Watts, *International Law and the Antarctic Treaty System.* Grotius, Cambridge, 1992.
14. C. Focarelli, in *International Law for Antartica,* F. Francioni, T. Scorazzi (Eds.), Kluwer Law Int., The Hague, 1996, 505.
15. *Handbook of the Antarctic Treaty System*, 8th Edition, Washington, 1994.
16. T. Scovazzi, L. Pineschi, in *International Environmental Law for Antartica,* F. Francioni (Ed.), A. Giuffrè, Milan, 1992, 149.
17. *Am. J. Int. Law*, 1969, 884.
18. *Supplement to Gazz. Uff.*, 48, 27 February 1995.
19. D. Lyons, *Polar Record,* 29 (1993), 111.
20. *A Model IEE for an Antarctic Tourism Cruise*, Doc. *XXI ATCM/IP20,* May 1997.
21. L. Pineschi, *La protezione dell'ambiente in Antartide.* Cedam, Padua, 1993.
22. F. Orrego Vicuña, in *Governing the Antarctic. The Effectiveness and Legitimacy of the Antarctic Treaty System,* O. S. Stokke D. Vidas (Eds.), Cambridge Univ. Press, Cambridge, 1996, 174.
23. *Application of Environmental Impact Assessment Principles to Policies, Plans and Programmes*, New York, 1992.
24. XXI ATCM, *Document IP61*, Christchurch, May 1997.
25. *Final Report of the XX Antarctic Treaty Consultative Meeting*, Doc. *XX ATCM/WP 37*, May 1996.
26. EC Commission Draft Directive on the *Assessment of the Effects of Certain Plans and Programmes on the Environment* of 25 March 1997, *OJEC* C 129, 25 April 1997.
27. XXI ATCM, *Document IP93*, Christchurch, May 1997.
28. Espoo Convention, 1991.
29. EC Council Directive 85/337, 27 June 1985.
30. L. Pineschi, in *International Law for Antarctica,* F. Francioni, T. Scovazzi (Eds.), Kluwer Law Int., The Hague, 1996, 261.
31. XXI ATCM, *Document IP80*, Christchurch, May 1997.
32. XVI ATCM, *Document INF21*, October 1991.
33. *Comprehensive Environmental Evaluation (CEE): Methodology for Reviewing Activities for Which a CEE Has Been Prepared, Final Report of the XXI ATCM*, Resolution 2, 1997, 134.
34. *ECE Convention on the Protection and Use of Transboundary Watercourses and International Lakes*, Helsinki, 17 March 1992, *Int. Leg. Mat.*, 1992, 1313.
35. *United Nations Convention on the Law of the Non-Navigational Uses of International Watercourses*, New York, 21 May 1997, *Int. Leg. Mat.*, 1997, 700.
36. COM (93) 28, 2 April 1993.
37. XXI ATCM, *Document IP46*, Christchurch, May 1997.
38. XXI ATCM, *Document IP57*, Rev. 1, Christchurch, May 1997.
39. XXI ATCM, *Document IP25*, Christchurch, May 1997.
40. United Nations, *Report of the International Law Commission on the Work of Its Forty-Eighth Session*, New York, 1996, 121.
41. Sir A. Watts, in *Antarctic Challenge II,* R. Wolfrum (Ed.), Duncker & Humblot, Berlin, 1986, 147.
42. F. Francioni, in *International Environmental Law for Antartica,* F. Francioni (Ed.), A. Giuffrè, Milan, 1992, 233.
43. F. Francioni, *Review of European Community and International Environmental Law,* 3 (1994), 223.

44. S. Blay, J. Green, *Env. Pol. Law,* 25 (1995), 24.
45. F. Francioni, in *International Law for Antartica,* F. Francioni, T. Scovazzi (Eds.), Kluwer Law Int., The Hague, 1996, pp. 581.
46. F. Francioni, *Recueil des Cours de l'Académie de Droit International de l'Haye* 260, 360 (1996).
47. XX ATCM, *Document INF81*, April 1996.
48. XX ATCM, *Document INF105*, May 1996.
49. Law No. 349 of 8 July 1986, *Gazz. Uff.* No. 162 of 15 July 1986.
50. DPR No. 306 of 19 June 1987, *Gazz. Uff.* No. 175 of 29 July 1987, 3.
51. Decree of the President of the Council of Ministries No. 377 of 10 August 1988, *Gazz. Uff.* No. 204 of 31 August, 6.
52. Decree of the President of the Council of Ministries of 27 December 1988, *Gazz. Uff.* No. 4 of 5 January 198, 17.
53. Law No. 146 of 22 February 1994, *Gazz. Uff.* No. 52, *Suppl. Ord.* No. 39 of 4 March 1994.
54. DPR of 12 April 1996, *Gazz. Uff.* No. 210 of 7 September 1996, 28.
55. Ministry of the Environment, Circulars GAB/96/15208 of 7 October 1996.
56. GAB/96/15326 of 8 October 1996, *Gazz. Uff.* No. 277 of 26 November 1996, 26–27.
57. DPR No. 200 of 11 March 1988, *Gazz. Uff.* No. 139 of 15 June 1988, 10.
58. *Gazz. Uff.* No. 210 of 7 September 1996, 28.
59. G. Francescon, *Riv. Giurid. Amb.*, 10 (1995), 769.
60. COM (96) 561, 27 November 1996.
61. *OJEC L 333*, 21 December 1996.
62. Law No. 640 of 3 November 1994, *Suppl. Ord. Gazz. Uff.* No. 273 of 22 November 1994.
63. *Atti Parlamentari*, Camera dei Deputati, XII Legislatura, Act No. 1458-A, 6.
64. XXI ATCM, *Document IP101*, May 1997.
65. XXI ATCM, *Document IP112*, May 1997.
66. *New Zealand Antarctica (Environmental Protection) Act*, 1994, No. 119.
67. *Norwegian Regulations Relating to Protection of the Environment in Antarctica*, 5 May 1995.
68. *United Kingdom Antarctic Regulations*, 1995 No. 490.
69. Law No. 475 of 9 November 1988, *Gazz. Uff.* No. 264 of 10 November 1988.
70. Decree of the President of the Council of Ministries No. 457 of 22 October 1988, *Gazz. Uff.* No. 256 of 31 October 1988.
71. Decree of the Ministry of the Environment of 26 April 1989, *Gazz. Uff.* No. 128 of 3 June 1989.
72. J. Rinzema, *Env. Pol. Law*, 26 (1996), 95.
73. A. Boyle, *J. Env. Law*, 3 (1991), 229.
74. B. Conforti, in *International Responsibility for Environmental Harm,* F. Francioni, T. Scovazzi (Eds.), Graham & Trotman, London, 1991, 179.
75. A. Ch. Kiss, in *International Responsibility for Environmental Harm,* F. Francioni, T. Scovazzi (Eds.), Graham & Trotman, London, 1991, 3.
76. R. Pisillo Mazzeschi, in *International Responsibility for Environmental Harm,* F. Francioni, T. Scovazzi (Eds.), 15.

Author Index

Subject Index